AF508658

Urban Climate and Adaptation Tools

Urban Climate and Adaptation Tools

Editors

Teodoro Georgiadis
Letizia Cremonini

MDPI • Basel • Beijing • Wuhan • Barcelona • Belgrade • Manchester • Tokyo • Cluj • Tianjin

Editors
Teodoro Georgiadis
National Research Council of Italy
Italy

Letizia Cremonini
National Research Council of Italy
Italy

Editorial Office
MDPI
St. Alban-Anlage 66
4052 Basel, Switzerland

This is a reprint of articles from the Special Issue published online in the open access journal *Climate* (ISSN 2225-1154) (available at: https://www.mdpi.com/journal/climate/special_issues/Urban_Climate_Adaptation).

For citation purposes, cite each article independently as indicated on the article page online and as indicated below:

LastName, A.A.; LastName, B.B.; LastName, C.C. Article Title. *Journal Name* **Year**, *Volume Number*, Page Range.

ISBN 978-3-0365-0144-4 (Hbk)
ISBN 978-3-0365-0145-1 (PDF)

Contents

About the Editors

Teodoro Georgiadis Graduated in Physics, in Astronomy, and in Environmental and Civil Engineering, Master's Degree in Territorial Planning. Senior Scientist at the Institute for the BioEconomy (former Institute of Biometeorology) of the CNR of Bologna. He deals with surface energy balances in the urban environment and with the mitigation of the effects of interactions between the atmosphere and the built environment. Member of the Capacity Building Committee of IAMAS-IUGG. Invited Scientist in the Commission of Agrometeorology of WMO.

Letizia Cremonini Postdoc at the IBE-CNR Institute of BioEconomy and freelance architect with Ph.D. in Town and Landscape Planning. Her Ph.D. thesis was entitled "The Role of Landscape in New Urban Neighborhoods in Europe". She develops projects related to the planning and ecological planning at the urban scale, maintenance of urban plant biodiversity, and studies of the relationship between vegetation and built environments and their influence on climate parameters.

Preface to "Urban Climate and Adaptation Tools"

In 1925, Le Corbusier, in his book Urbanisme, wondered if cities are still capable of preserving the dignity of human beings. That question today appears even more motivated by two phenomena that have become increasingly evident over time: urbanization and climate change.

Urbanization has led the human community to find itself, today, within the geographical perimeter of cities with growth rates that in a few decades into the future, will ensure that all men can be defined as citizens; but citizenship has a much deeper meaning than living in one place. Citizenship means being able to provide the human being with all those elements of life and well-being that reconcile all aspects of one's nature: opportunities for one's expectations, physiological well-being, social well-being, protection of one's own safety. That ability to live fully, being able to bring out all their physical and intellectual abilities. Well-being understood as the overall state of being human. Almost a century after that question, the answer cannot be positive. Although it is not possible for cities to define a precise development model, what we have seen was an urbanization that has profoundly undermined the sustainable livelihoods of places and, therefore, of the human being.

The climate–environmental problem has therefore caused all the contradictions of the modern era to explode, placing humankind in danger. Pollution, rising temperatures, intense precipitation phenomena endanger life but, above all, the well-being of citizens, with health understood as a primary good. It is the task of politics to find solutions, and the task of science and technology to find the tools to support politics in the search for solutions.

The time is actually ripe for a transition—for a transition to a more sustainable world where the human being is at the center of the system, the replacement of gross domestic product with gross domestic well-being. We already own our toolbox, which can also be improved and implemented, and this collection of works allows us to understand that the passage to the answer to that fundamental question is within reach. In reading, it will be clear that, today, we have tools to respond, we have risk reduction methodologies, we have policies based on scientific assumptions, we have the possibility of transforming a danger into new growth opportunities. The overview offered is vast, and this can only lead us to be optimistic. The real knot is not given by science or the will of politics: the real knot is given by the citizen's ability to make this hope their own, to transform it into daily awareness of what can be done in an exercise of democracy in understanding that there exists diversity that must be protected and that special actions are necessary for these; protections that are then transformed into greater resources to be redistributed for the benefit of all. The environment and climate today can represent an opportunity; by building more resilient cities, we not only protect the weakest groups but, in fact, we free up resources represented by health costs which in healthier environments are reduced, allowing access to a full life to all citizens.

The recent pandemic, with all the grief and suffering, has opened our eyes to the future and to the meaning of being a community. It has also shown us how the contradictions of modern living can explode in a very short time for us, used to talking about what the world will be like in 2050, and we understand that we don't know what tomorrow itself will be like. But we have the tools to change, and you have them in your hands right now, and now is the time to apply them.

Teodoro Georgiadis, Letizia Cremonini
Editors

Article

A New Method to Assess Fine-Scale Outdoor Thermal Comfort for Urban Agglomerations

Dirk Lauwaet [1,*], Bino Maiheu [1], Koen De Ridder [1], Wesley Boënne [1], Hans Hooyberghs [1], Matthias Demuzere [2,3] and Marie-Leen Verdonck [4]

[1] Vlaamse Instelling voor Technologisch Onderzoek (VITO), 2400 Mol, Belgium; bino.maiheu@vito.be (B.M.); koen.deridder@vito.be (K.D.R.); wesley.boenne@vito.be (W.B.); hans.hooyberghs@vito.be (H.H.)
[2] Department of Geography, Ruhr-University Bochum, 44801 Bochum, Germany; matthias.demuzere@rub.de
[3] Department of Environment, Laboratory of Hydrology and Water Management, Ghent University, 9000 Ghent, Belgium
[4] Antea Group, 2000 Antwerpen, Belgium; marie-leen.verdonck@anteagroup.com
* Correspondence: dirk.lauwaet@vito.be

Received: 6 November 2019; Accepted: 4 January 2020; Published: 6 January 2020

Abstract: In urban areas, high air temperatures and heat stress levels greatly affect human thermal comfort and public health, with climate change further increasing the mortality risks. This study presents a high resolution (100 m) modelling method, including detailed offline radiation calculations, that is able to efficiently calculate outdoor heat stress for entire urban agglomerations for a time period spanning several months. A dedicated measurement campaign was set up to evaluate model performance, yielding satisfactory results. As an example, the modelling tool was used to assess the effectiveness of green areas and water surfaces to cool air temperatures and wet bulb globe temperatures during a typical hot day in the city of Ghent (Belgium), since the use of vegetation and water bodies are shown to be promising in mitigating the adverse effects of urban heat islands and improving thermal comfort. The results show that air temperature reduction is most profound over water surfaces during the afternoon, while open rural areas are coolest during the night. Radiation shading from trees, and to a lesser extent, from buildings, is found to be most effective in reducing wet bulb globe temperatures and improving thermal comfort during the warmest moments of the day.

Keywords: thermal comfort; urban greening; urban heat island; UrbClim model; water bodies

---

1. Introduction

Rapid urban growth coupled with high population density increases the vulnerability of cities to extreme weather [1]. In cities, heat extremes are among the most important weather-related health hazards. Moreover, the effects of extreme heat are exacerbated by the presence of the urban heat island (UHI) [2]. This UHI is caused by a combination of the increased heat capacity of cities, anthropogenic heat sources, and the imperviousness of urban surfaces, which inhibit evaporative cooling [3–5]. Due to the UHI increment, cities are particularly vulnerable to heat waves, causing higher heat-related excess mortalities [6–8]. The risks of morbidity and mortality in urban areas are further increased by climate change, due to the increasing frequency of weather extremes [9,10].

In this context, integrating mitigation and adaptation measures can help avoid locking a city into counterproductive infrastructure and policies. Urban greenery has been proposed as an effective measure to mitigate the UHI and improve the urban microclimate [11–13]. Green areas are generally cooler than their surrounding built up areas, which is demonstrated by observational studies reporting instantaneous air temperature differences of 1 °C up to 7 °C [14,15]. Vegetation cools cities via shading, evapotranspiration, and alteration of the wind pattern [16]. The cooling intensity of, for example, city parks, is often largest in the evenings and during the night (like the UHI) and tends to increase with park size [17]. During the afternoon, parks with extensive tree coverage tend to be cooler due to shading effects, while at night more open parks are cooler due to greater long-wave radiative cooling [18].

Similar to parks, water features have the potential to alleviate high urban temperatures through enhanced evaporation and reduced sensible heat fluxes [19]. A number of observational studies have shown that temperatures adjacent or downwind of water bodies are reduced by around 1–2 °C compared to surrounding areas [20,21]. On specific days during the afternoon, Murakawa et al. [22] observed 3–5 °C cooler temperatures near a wide river in Hiroshima, Japan. In contrast to urban parks where the cooling effect is most pronounced during the evening and the night, these studies suggest that open water effects are most pronounced during the day, because water bodies can maintain warmer temperatures at night due to the high heat capacity and thermal inertia of water [23].

However, air temperatures are only part of the story when assessing outdoor human comfort, as humidity, wind, and radiation also play a role [24,25]. During the day in summertime, radiation is even the single most important meteorological factor influencing the human energy balance [26]. Hence, by blocking solar radiation, trees can substantially improve thermal comfort, even if air temperature reductions are small. The relationship between urban greening and outdoor thermal comfort has been the subject of many modelling studies (e.g., [12,27,28]). At the same time, new urban modelling tools have been developed that focus on the introduction of green in urban design [29,30]. Most of these studies have been performed with high resolution models (e.g., ENVI-met) that can only perform simulations for limited areas in a city and for limited time periods.

In this study, we present an offline GIS-based post processing method, coupled to the urban boundary layer climate model UrbClim [31], that is able to calculate outdoor thermal comfort for an entire urban agglomeration and for time periods of several months to years. As a heat stress indicator, we applied the wet bulb globe temperature (WBGT), the ISO standard to quantify human thermal comfort [32]. The modelling method can easily be validated, which is done by performing an extensive measurement campaign in the city of Ghent, Belgium. As an example, the model is used to assess the effectiveness of urban vegetation and open water areas in reducing air temperatures and heat stress for the city of Ghent during a particularly warm summer day.

2. Materials and Methods

2.1. The UrbClim Model

The urban boundary layer climate model UrbClim [31] is designed to cover agglomeration-scale domains at a high spatial resolution, taken as 100 m for this study. The model consists of a land surface scheme, including simplified urban physics, which is coupled to a 3-D atmospheric boundary layer model. To ensure that the synoptic forcing is properly taken into account, the boundary layer model is tied to synoptic-scale meteorological fields through the lateral and top boundary conditions. The land surface scheme used in UrbClim is based on the soil–vegetation–atmosphere transfer scheme of De Ridder and Schayes [33], extended to account for urban surface physics. This urbanization is implemented by representing the urban surface as a rough impermeable slab, with appropriate values for the albedo, emissivity, thermal conductivity, and volumetric heat capacity. The main feature of the urbanization scheme is the inclusion of a parameterization of the inverse Stanton number, which is known to be much higher in urban areas [34,35]. A full description of the UrbClim model can be found in De Ridder et al. [31].

The spatial distribution of land cover types, needed for the specification of required land surface parameters, is taken from the reference land use map for Flanders, described in White et al. [36]. The percentage of urban land cover is attributed by applying the urban soil sealing raster data files that are distributed by the European environment agency. From the normalized difference vegetation index (NDVI) acquired by the MODIS instrument on-board the TERRA satellite platform, maps of vegetation cover fraction were obtained. This fraction is specified as a function of the NDVI using a linear relationship proposed by Gutman and Ignatov [37], and then interpolated to the model grid. Model grid cells are divided into vegetation and bare soil (the complementary fraction) if they do not feature urban land use types. Where grid cells contain urban land use, the urban fraction, as derived from the soil sealing data, takes precedence over the fractional vegetation cover data, in case both sum to over 100%. In case they sum to less than that, the remaining fraction is assigned to bare soil. Terrain elevation data are taken from the GMTED2010 Dataset [38].

The UrbClim model has previously been validated regarding its energy fluxes, 2 m wind speeds, air temperatures, and urban–rural temperature differences for the cities of Antwerp, Brussels, and Ghent in Belgium, Toulouse in France, and Barcelona in Spain [31,39–41]. Also, the land surface temperatures in the UrbClim land surface scheme have already been validated with satellite data for the city of London [42]. In De Ridder [43], the urban parameterization was tested for the city of Paris, and the simulated land surface temperature compared favorably to observed values obtained from thermal infrared satellite imagery.

2.2. Outdoor WBGT Calculation

The UrbClim model has already been coupled offline to a building energy simulation model to calculate indoor heat stress, based on the WBGT [44]. Here, we present a method to calculate outdoor WBGT values from the UrbClim model results. Therefore, we follow the method of Liljegren et al. [45] to calculate the WBGT from standard meteorological variables, which is recommended for outdoor WBGT calculations in a review paper by Lemke and Kjellstrom [46]. The (outdoor) WBGT is the weighted sum of the natural wet bulb temperature T_w, the globe temperature T_g, and the dry bulb (ambient) temperature T_a:

$$WBGT = 0.7T_w + 0.2T_g + 0.1T_a \tag{1}$$

Separate calculation models for the natural wet bulb temperature and the globe temperature are used by Liljegren et al. [45], where all details on the calculations and required input parameters can be found.

In our modelling approach, hourly 2 m air temperatures, specific humidity, and wind speeds were taken from the UrbClim output data as input for the WBGT calculations. Downward solar radiation and surface pressure are also needed, and were taken from the ERA-interim re-analysis of the European centre for medium-range weather forecasting (ECMWF). Both these variables also serve as input data for the UrbClim model. To calculate the detailed amount of shade (either from buildings or trees) in a grid cell, shape files with building height and tree height were obtained from the city administration of Ghent. These files were converted to 1 m resolution raster files, and subsequently, for every hour of the period under study (see Section 2.3), the incoming solar radiation for the different solar zenith angles was calculated with the potential incoming solar radiation module of the system for automated geoscientific analyses (SAGA), an open-source geographic information system [47]. The results of these calculations were resampled to the UrbClim model grid, yielding fractional coverages of building shadow and tree shadow in every grid cell for every hour of the day. Finally, the fraction of the grid cells occupied by buildings was defined by resampling the building footprint raster to the model grid. We excluded this fraction from the calculations since we assume people do not stand regularly on their roof.

In the end, there were three types of locations in every grid cell: open, in the shadow of buildings, or in the shadow of trees. The open locations received the full amount of incoming solar radiation (both direct and diffuse, which are separately needed for the WBGT calculation), while for locations in the shadow of buildings, the direct fraction is set to 0 and only the diffuse fraction remains. In the shade of trees, the amount of direct and diffuse solar radiation was calculated according to the radiation transfer through the tree canopy scheme of De Ridder [48]. The overall WBGT value in a grid cell was taken as the weighted average of the respective WBGT values from these fractions.

Based on the approach outlined above, it is possible to obtain, in a fairly simple manner, outdoor heat stress values from general climate models, taking into account the detailed radiative shading effects of buildings and trees.

2.3. Experiment Setup

In order to evaluate the performance of the UrbClim model and the WBGT calculation, while also assessing the effectiveness of green and blue infrastructure in reducing outdoor heat stress, a dedicated measurement campaign was set up during the summer of 2015 in the city of Ghent. A measurement station was installed in the middle of a square of a public nature museum (Wereld van Kina, http://www.dewereldvankina.be/) in the city center during the months of July and August 2015. The station featured a WBGT sensor measuring 2 m (dry bulb) air temperatures, wet bulb and black globe temperatures, and relative humidity. The dry bulb temperature was measured using a highly accurate 43347 RTD temperature probe housed inside a fan-aspirated radiation shield, yielding a reported measurement uncertainty of only 0.1 °K. The relative humidity sensor has a reported measurement uncertainty of up to 5%.

During a particularly warm and sunny day (the 3rd of July 2015), two bikes were equipped with autonomous Onset HOBO U23-002 (http://www.onsetcomp.com/products/data-loggers/u23-002) temperature/humidity loggers at 2 m height and driven around the city center, through urban parks and to nearby rural areas, during the afternoon (12–16 h Local Time). The sensor tips were mounted in the same fan-aspirated radiation shields as the reference measurement station, using battery packs to power the fans during the trips. These measurements have a temporal resolution of 1 s. Furthermore, a floating fixed platform [49] was equipped with an automatic ventilated air temperature sensor (HOBO U23-002) at 2 m height to measure the cooling effect of a small lake near the city center, during the same time period as the bike measurements. Figure 1 shows the location and trajectories of all measurement devices.

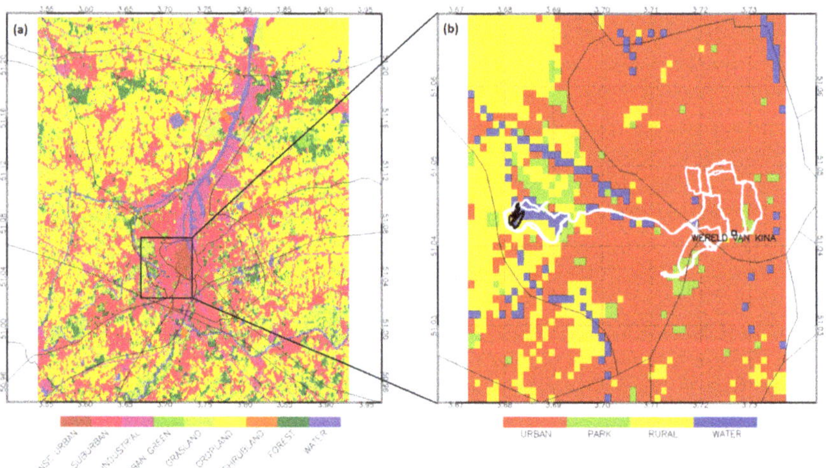

Figure 1. Extent and land use classification of the UrbClim model domain (**a**) and the focus area of this study around the measurement locations (**b**). The location of the observational station and the trajectories of the cycling (white) and boating (black) measurements are shown in the panel on the right.

Afterwards, the UrbClim model and the offline WBGT calculation were used to simulate the summer of 2015 for the wider agglomeration of Ghent, directly driven with meteorological data from the ERA-Interim re-analysis of the ECMWF, as was the setup in previous validation experiments [31,39–41]. The model domain is configured with 301 × 301 grid cells in the horizontal direction, using a spatial resolution of 100 m. Figure 1 shows the extent of the UrbClim model domain and the land use in the domain. In the vertical direction, 20 levels are specified, with the first level 10 m above the displacement height, the resolution smoothly decreasing upward to 250 m at the model top located at 3 km height. This vertical discretization closely matches that of the ECMWF host model. The simulation is initialized on 1 May at 0000 LT, resulting in a two-month spin-up before the start of the analysis on 1 July, in order to ensure model equilibrium between external forcing and internal dynamics, especially in terms of soil variables. Initial soil temperature and soil moisture data are taken from the ERA-Interim re-analysis.

3. Results

3.1. Model Evaluation

Figure 2 shows the time series and error statistics of the measured and modelled 2 m air temperatures, dew point temperatures, and wet bulb globe temperatures at the observational station for the months of July and August 2015. There is a good correspondence between the measured and simulated air temperatures, with almost no bias, root mean square errors well below 2 °C, and a correlation coefficient over 0.9. These statistics are in line with previous validation results of the UrbClim model [31,39,41]. There is more variability in the comparison of the dew point temperatures, which is to some extend due to the larger uncertainty in the measurements. The measured dew point temperature is estimated from the air temperature and the relative humidity, of which the combined measurement errors can lead to an uncertainty well of over 0.5 °C. With this in mind, the error statistics are certainly reasonable. For the wet bulb globe temperatures, there is again a very good correspondence between measured and simulated temperatures, with a small positive bias, a root mean square error just above 1 °C, and a correlation coefficient of 0.95. There are a few days (e.g., 2 July) where the simulations deviate significantly from the observations, due to some observed cloudiness that is not picked up by the ERA-Interim re-analysis.

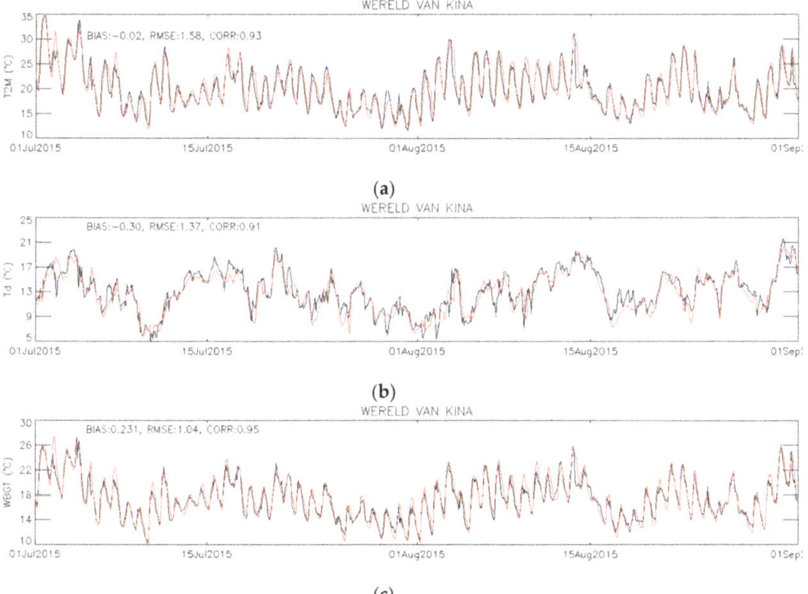

Figure 2. Time series of 2 m air temperatures (**a**), 2 m dew point temperatures (**b**), and wet bulb globe temperatures (**c**) for July and August 2015. Observations are in black, model results in red. The quantities given are the bias, root mean square error (RMSE), and correlation coefficient (CORR).

It is not straightforward to compare the simulated air temperatures to the mobile measurements. First of all, there is some local variability in land use that cannot be captured by the 100 m model resolution. Moreover, the model results have an hourly temporal resolution, whereas the observations have a temporal resolution of 1 s. To overcome these issues, we have grouped the measurements in four relevant land use classes (urban, park, rural, and water) and calculated the average value and standard deviation during a specific two hour period (14–16 h LT) of the afternoon of 3 July 2015. These values are compared to the modelled air temperatures for the same hours, taken from all grid cells of the focus area (Figure 1) for the respective land use classes. Table 1 shows the results of this comparison.

Table 1. Overview per land use class of the mean modelled and measured 2 m air temperatures, and their standard deviations (SD), for the afternoon (14–16 h Local Time) of 3 July 2015.

Land Use	Model		Measurements	
	Mean [°C]	SD [°C]	Mean [°C]	SD [°C]
Urban	30.2	0.26	30.0	0.45
Park	30.0	0.28	29.9	0.85
Rural	30.1	0.29	29.0	0.36
Water	28.4	0.43	28.4	0.32

The average model results are very close to the observed values for all land use classes, except for the rural class. The simulated rural air temperatures in the afternoon are almost as high as the urban ones, but the measured rural air temperatures are 1 °C lower than the urban air temperatures, so the model seems to underestimate, to some extent, the cooling effect of the rural locations during the afternoon. The differences between the other land use classes are well captured by the model, with the air temperatures over water being almost 2 °C lower than the urban and park air temperatures. The standard deviations of the measurements are significantly larger than these of the model results, demonstrating the large local variability within each land use class, which is difficult to capture for a 100 m resolution model.

Overall, the model performs satisfactorily compared to the measurements.

3.2. Effect of Green and Blue Areas on Air Temperatures

On the 3rd of July 2015, the air temperatures in the city of Ghent rose quickly during the day due to sunny and calm conditions, reaching over 30 °C during the day and only cooling down slightly during the evening and the night. In Figure 3 the daily cycle of modelled 2 m air temperatures for the different land use classes is presented. The results show that during these types of days, air temperature is highest in the urban areas, while urban parks show limited or no cooling effect on air temperatures during midday hours. The measurements presented in Table 1 suggested that the rural areas just outside the city are around 1 °C cooler than the city center during these warm hours, while the model estimates this effect to be a lot smaller. Both the measurements and the model show that the lowest temperatures during the afternoon are clearly found over water surfaces, being around 2 °C cooler during the warmest moment of the day.

It is important to also investigate the differences in night time temperatures, since this is the moment that the UHI is the strongest and has the biggest potential impact on human health, because the warmer urban night-time temperatures limit the recuperation of city inhabitants from heat stress during daytime [50,51]. During the night, the urban air temperatures are warmest, with the air temperatures in parks and over water being around 0.5 °C cooler. The lowest air temperatures during the night are found in the rural areas, which are around 2 °C cooler due to the greater long-wave radiative cooling in the open fields and grasslands.

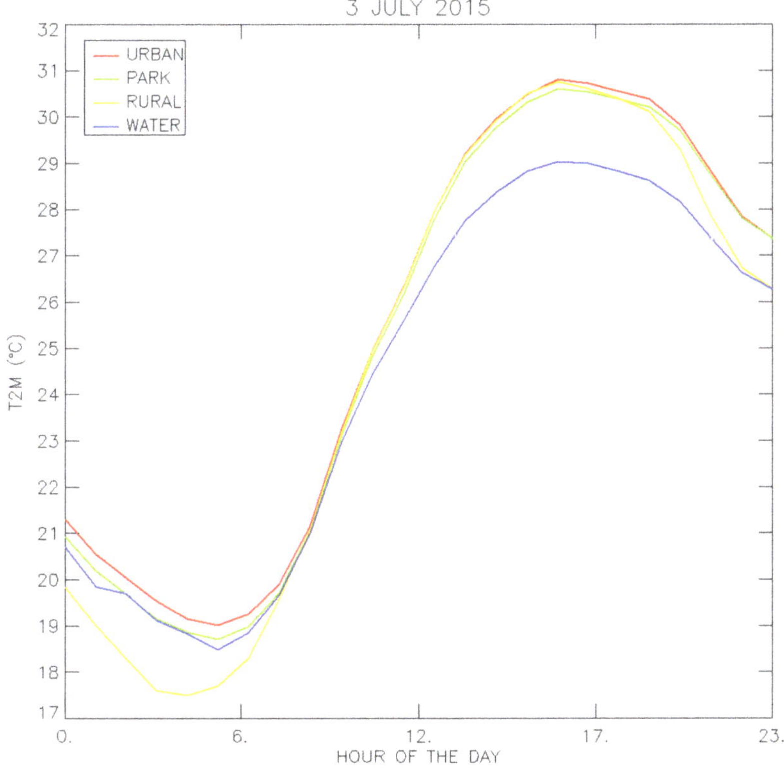

Figure 3. Daily cycle of modelled 2 m air temperatures with an hourly time resolution, averaged per land use class for the focus area.

3.3. Effect of Green and Blue Areas on Thermal Comfort

As explained in the introduction, it is important to also include humidity, wind speed, and radiation when assessing outdoor thermal comfort. Here, we apply the WBGT as a heat stress indicator, which is the ISO standard for human thermal comfort and takes these variables into account. In Figure 4, the daily cycle of the WBGT for the different land use classes is presented, differentiating between open and shaded locations.

Shading clearly plays a very important role regarding thermal comfort during the afternoon, when the differences in WBGT between all open locations are small (<1 °C), regardless of the land use. Even the water areas, where the air temperatures are several degrees cooler, have a WBGT value comparable to urban areas in the afternoon, due to the higher humidity, which offsets the cooler air temperatures. In urban areas, building shade provides a substantial cooling effect of 2 °C on the WBGT. The shade of trees in urban parks and rural areas has an even bigger cooling effect of around 3 °C. These are certainly the best places to avoid outdoor heat stress during a typical hot day in Ghent.

During the night, when there is no solar radiation, the WBGT values are in line with the air temperature results. Urban areas are warmest, with WBGT values in parks and over water being around 0.5 °C cooler. The coolest locations are the open rural areas, where minimal WBGT values are around 2 °C cooler than in urban areas due to the lower air temperatures there.

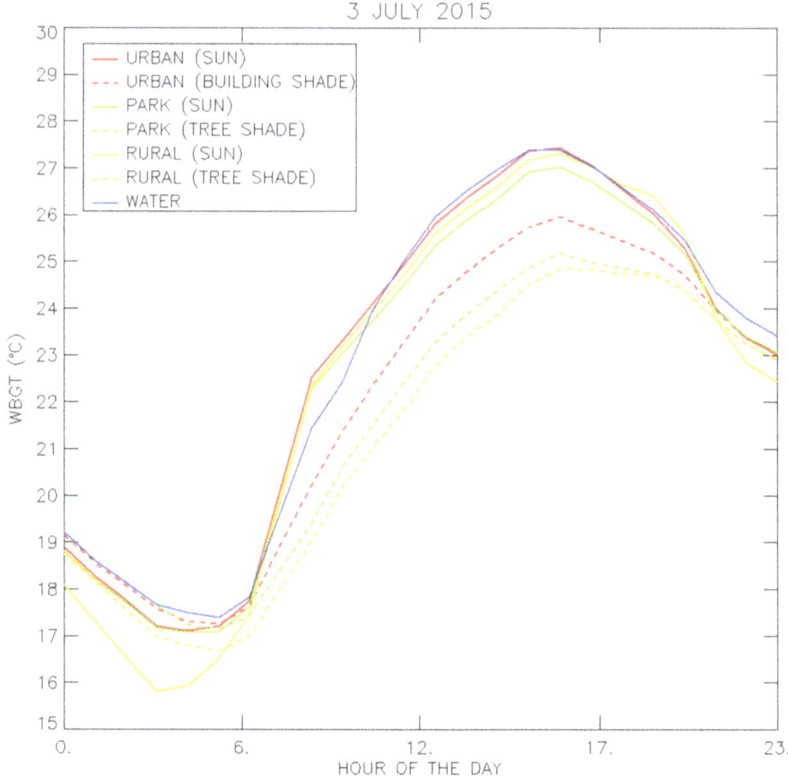

Figure 4. Daily cycle of modelled wet bulb globe temperatures with an hourly time resolution, averaged per land use class for the focus area. A distinction is made between unshaded (full lines) and fully shaded (dashed lines) areas.

4. Discussion and Conclusions

In this paper, we have presented an offline GIS-based post processing method, coupled to the urban climate model UrbClim, that can be used to calculate outdoor thermal comfort. The main advantage of our methodology is that we are able to calculate outdoor thermal comfort with a relatively high horizontal resolution for entire urban agglomerations and for long time periods (months to years), which is not possible with state of the art microclimate models. Moreover, the model can be easily validated with measurements, as we have demonstrated for the city of Ghent in Belgium, yielding satisfactorily results. As an example of the potential use of the model, the effectiveness of urban vegetation and open water areas in reducing air temperatures and heat stress for the city of Ghent during a particular warm summer day was assessed.

The modelling results regarding the effect of vegetation and water surfaces on air temperatures are in line with previous observational and modelling studies [20,21], with cooling effects of water surfaces up to 2 °C during daytime, and much smaller effects for night time temperatures when the water bodies maintain high water temperatures due to their high heat capacity. The cooling effect of park areas that is found in this study is small in comparison to reported observed values [12,13], which is probably due to the very small size of the city parks in Ghent, which, on top of that, feature a considerable amount of sealed surfaces. Related research studies on the effect of trees and water on outdoor heat stress have found similar results to the impact numbers reported here [27,28], with trees providing a substantial improvement to human thermal comfort by blocking solar radiation, even if air temperature reductions are negligible. These types of scenario studies for urban areas (e.g., [52]) are performed with limited-area microclimate models such as ENVI-met that can only model a small part of a city for a few selected days.

Our methodology allows us to map outdoor thermal comfort for an entire city. As an example, Figures 5 and 6 show the daily maximum and minimum WBGT values on the 3rd of July for the city of Ghent, respectively. The results are limited to the city administrative area, since we only have the detailed building and tree height data for this area. The figures provide further evidence for the results discussed above, as the highest maximum WBGT values are found over open water and open urban areas, whereas the lowest maximum WBGT values are found in forested areas. The minimum WBGT map reflects the UHI situation for the city of Ghent with the warmest locations found in the city center and the urbanized areas around the city, and the coolest locations being the open rural areas outside the city.

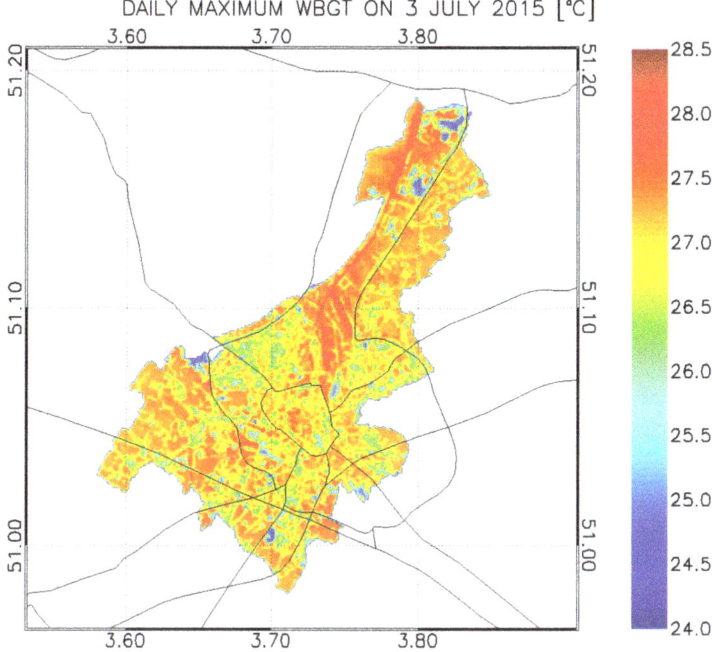

Figure 5. Map of the daily maximal wet bulb globe temperature for the city of Ghent.

Note that our analysis here focuses on a single hot day with calm and clear sky conditions. The reported cooling effects for the different land use types will differ and are lower when winds are stronger or more clouds are present. Also, in reality, the shading effect of trees will be dependent on the specific tree species and the health of the trees, which can be problematic in urban areas. Given these limitations and uncertainties, the results presented here are not suited for local risk assessment but should rather provide a coherent picture of potential outdoor heat stress benefits from water surfaces and green areas.

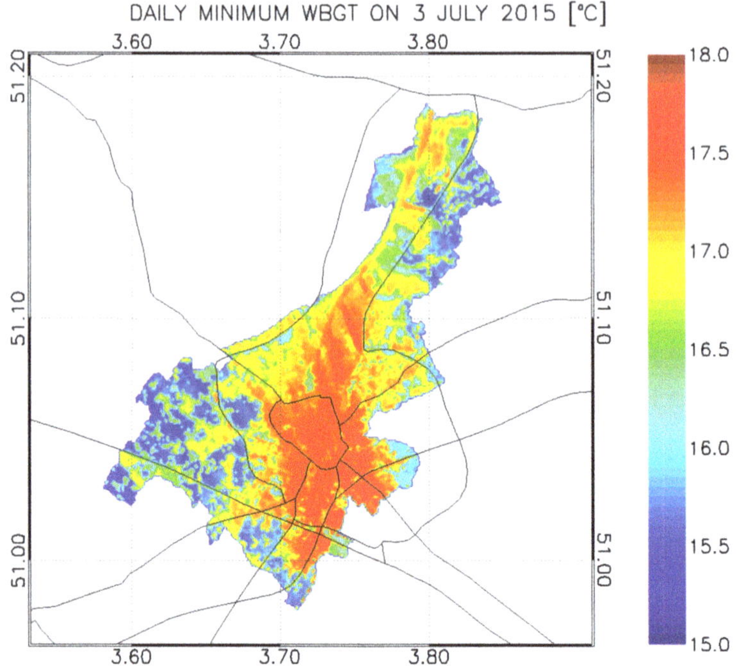

Figure 6. Map of the daily minimal wet bulb globe temperature for the city of Ghent.

Author Contributions: All authors contributed extensively to the work presented in this paper. K.D.R. supervised the project and is the creator of the UrbClim model. B.M. supervised the measurement campaign to which also K.D.R., W.B., M.D., and M.-L.V. contributed. H.H. was involved in the model creation and scripting. D.L. performed the model simulations, data analysis, wrote the manuscript, and drafted the figures. All authors discussed the results and implications and commented on the manuscript at all stages. All authors have read and agreed to the published version of the manuscript.

Funding: This research has been performed in the framework of the CORDEX.be project, funded by the Belgian Science Policy Office (BELSPO) under contract number BR/143/A2. Furthermore, the work described in this paper has received funding from the European Community's 7th Framework Programme under Grant Agreements Nos. 308497 (RAMSES) and 308299 (NACLIM) and from the European Union's H2020 Research and Innovation Programme under Grant Agreement No. 73004 (PUCS/Climate-fit.city). Marie-Leen Verdonck was funded by the Belgian Federal Science Policy Office, as part of the UrbanEARS project. Matthias Demuzere was funded by the FWO (Fund for Scientific Research) of the Flemish regional government.

Conflicts of Interest: The authors declare no conflict of interest.

References

1. Rosenzweig, C.C.; Solecki, W.W.; Romero-Lankao, P.P.; Mehrotra, S.; Dhakal, S.S.; Bowman, T.T.; Ali Ibrahim, S. *ARC3.2 Summary for City Leaders. Urban Climate Change Research Network*; Columbia University: New York, NY, USA, 2015.
2. Lemonsu, A.A.; Viguié, V.V.; Daniel, M.M.; Masson, V.V. Vulnerability to heat waves: Impact of urban expansion scenarios on urban heat island and heat stress in Paris (France). *Urban Clim.* **2015**, *14*, 586–605. [CrossRef]
3. Oke, T.R.R.; Johnson, G.T.T.; Steyn, D.G.G.; Watson, I.D. Simulation of surface urban heat islands under "ideal" conditions at night. Part 2: Diagnosis of causation. *Bound. Layer Meteorol.* **1991**, *56*, 339–358. [CrossRef]
4. Masson, V. Urban surface modeling and the meso-scale impact of cities. *Theor. Appl. Climatol.* **2006**, *84*, 35–45. [CrossRef]

5.   Lynn, B.H.H.; Carlson, T.N.N.; Rosenzweig, C.; Goldberg, R.; Druyan, L.; Cox, J.; Gaffin, S.; Parshall, L.; Civerolo, K. A modification to the NOAH LSM to simulate heat mitigation strategies in the New York City Metropolitan Area. *J. Appl. Meteorol. Climatol.* **2009**, *48*, 199–216. [CrossRef]

6.   Gabriel, K.M.A.A.; Endlicher, W.R. Urban and rural mortality rates during heat waves in Berlin and Brandenburg, Germany. *Environ. Pollut.* **2011**, *159*, 2044–2050. [CrossRef]

7.   Dousset, B.B.; Gourmelon, F.F.; Laaidi, K.K.; Zeghnoun, A.A.; Giraudet, E.E.; Bretin, P.P.; Maurid, E.E.; Vandentorren, S. Satellite monitoring of summer heat waves in the Paris metropolitan area. *Int. J. Climatol.* **2010**, *31*, 313–323. [CrossRef]

8.   Sherwood, S.C.C.; Huber, M. An adaptability limit to climate change due to heat stress. *Proc. Natl. Acad. Sci. USA* **2010**, *107*. [CrossRef]

9.   Martinez, G.S.S.; Baccini, M.M.; De Ridder, K.K.; Hooyberghs, H.H.; Lefebvre, W.W.; Kendrovski, V.V.; Scott, K.K.; Spasenovska, M. Projected heat-related mortality under climate change in the metropolitan area of Skopje. *BMC Public Health* **2016**, *16*, 407. [CrossRef]

10. Wouters, H.; De Ridder, K.K.; Willems, P.P.; Poelmans, L.; Hosseinzadehtalaei, P.P.; Tabari, H.H.; Brouwers, J.J.; Vanden Broucke, S.S.; van Lipzig, N.P.M.M.; Demuzere, M. Heat stress increase towards the mid-21st century is twice as large for cities compared to rural areas. *Geophys. Res. Lett.* **2017**, *44*, 8997–9007. [CrossRef]

11. Demuzere, M.M.; Orru, K.K.; Heidrich, O.O.; Olazabal, E.E.; Geneletti, D.D.; Orru, H.H.; Faehnle, M. Mitigating and adapting to climate change: Multi-functional and multi-scale assessment of green urban infrastructure. *J. Environ. Manag.* **2014**, *146*, 107–115. [CrossRef]

12. Tan, Z.Z.; Ka-Lun Lau, K.K.; Ng, E. Urban tree design approaches for mitigating daytime urban heat island effects in a high-density urban environment. *Energy Build.* **2016**, *114*, 265–274. [CrossRef]

13. Kong, L.L.; Lau, K.K.K.; Yuan, C.C.; Chen, Y.Y.; Xu, Y.Y.; Ren, C. Regulation of outdoor thermal comfort by trees in Hong Kong. *Sustain. Cities Soc.* **2017**, *31*, 12–25. [CrossRef]

14. Bowler, D.E.E.; Buyung-Ali, L.L.; Knight, T.M.M.; Pullin, A.S. Urban greening to cool towns and cities: A systematic review of the empirical evidence. *Landsc. Urban Plan.* **2010**, *97*, 147–155. [CrossRef]

15. Armson, D.D.; Stringer, P.P.; Ennos, A.R. The effect of tree shade and grass on surface and globe temperatures in an urban area. *Urban For. Urban Green.* **2012**, *11*, 245–255. [CrossRef]

16. Jamei, E.E.; Rajagopalan, P.P.; Seyedmahmoudian, M.M.; Jamei, Y. Review on the impact of urban geometry and pedestrian level greening on outdoor thermal comfort. *Renew. Sustain. Energy Rev.* **2016**, *54*, 1002–1017. [CrossRef]

17. Upmanis, H.H.; Chen, D. Influence of geographical factors and meteorological variables on nocturnal urban-park temperature differences—A case study of summer 1995 in Göteborg, Sweden. *Clim. Res.* **1999**, *13*, 125–139. [CrossRef]

18. Chang, C.R.R.; Li, M.H.H.; Chang, S.D. A preliminary study on the local cool-island intensity of Taipei city parks. *Landsc. Urban Plan.* **2007**, *80*, 386–395. [CrossRef]

19. Coutts, A.M.M.; Tapper, N.J.J.; Beringer, J.J.; Loughnan, M.M.; Demuzere, M. Watering our cities: The capacity for Water Sensitive Urban Design to support urban cooling and improve human thermal comfort in the Australian context. *Prog. Phys. Geogr.* **2012**, *37*, 2–28. [CrossRef]

20. Chen, Z.Z.; Zhao, L.L.; Meng, Q. Field measurements on microclimate in residential community in Guangzhou, China. *Front. Archit. Civ. Eng. China* **2009**, *3*, 462–468. [CrossRef]

21. Saaroni, H.H.; Ziv, B. The impact of a small lake on heat stress in a Mediterranean urban park: The case of Tel Aviv, Israel. *Int. J. Biometeorol.* **2003**, *47*, 156–165. [CrossRef]

22. Murakawa, S.S.; Sekine, T.T.; Narita, K.I.I.; Nishina, D. Study of the effects of a river on the thermal environment in an urban area. *Energy Build.* **1991**, *16*, 993–1001. [CrossRef]

23. Steeneveld, G.J.J.; Koopmans, S.S.; Heusinkveld, B.G.G.; Theeuwes, N.E. Refreshing the role of open water surfaces on mitigating the maximum urban heat island effect. *Landsc. Urban Plan.* **2014**, *121*, 92–96. [CrossRef]

24. Höppe, P. The physiological equivalent temperature—A universal index for the biometeorological assessment of the thermal environment. *Int. J. Biometeorol.* **1999**, *43*, 71–75. [CrossRef] [PubMed]

25. Buzan, J.R.R.; Oleson, K.K.; Huber, M. Implementation and comparison of a suite of heat stress metrics within the Community Land Model version 4.5. *Geosci. Model Dev.* **2015**, *8*, 151–170. [CrossRef]

26. Matzarakis, A.A.; Rutz, F.F.; Mayer, H. Modelling radiation fluxes in simple and complex environments—Application of the RayMan model. *Int. J. Biometeorol.* **2007**, *51*, 323–334. [CrossRef] [PubMed]

27. Ng, E.E.; Chen, L.L.; Wang, Y.Y.; Yuan, C. A study on the cooling effects of greening in a high-density city: An experience from Hong Kong. *Build. Environ.* **2012**, *47*, 256–271. [CrossRef]

28. Shashua-Bar, L.L.; Pearlmutter, D.D.; Erell, E. The influence of trees and grass on outdoor thermal comfort in a hot-arid environment. *Int. J. Climatol.* **2010**, *31*, 1498–1506. [CrossRef]

29. Coccolo, S.S.; Pearlmutter, D.D.; Kämpf, J.J.; Scartezzini, J.-L. Thermal Comfort Maps to estimate the impact of urban greening on the outdoor human comfort. *Urban For. Urban Green.* **2018**, 35. [CrossRef]

30. Reinhart, C.C.; Dogan, T.T.; Jakubiec, J.J.; Rakha, T.T.; Sang, A. UMI—An urban simulation environment for building energy use, daylighting and walkability. In Proceedings of the BS 2013: 13th Conference of the International Building Performance Simulation Association, Chambery, France, 26–28 August 2013; Available online: http://www.ibpsa.org/proceedings/BS2013/p_1404.pdf (accessed on 5 January 2020).

31. De Ridder, K.K.; Lauwaet, D.D.; Maiheu, B. UrbClim—A fast urban boundary layer climate model. *Urban Clim.* **2015**, *12*, 21–48. [CrossRef]

32. ISO. *Hot Environments—Estimation of the Heat Stress on Working Man, Based on the WBGT-Index (Wet Bulb Globe Temperature)*; ISO Standard 7243; International Standards Organization: Geneva, Switzerland, 1989.

33. De Ridder, K.K.; Schayes, G. The IAGL Land Surface Model. *J. Appl. Meteorol.* **1997**, *36*, 167–182. [CrossRef]

34. Kanda, M.M.; Kanega, M.M.; Kawai, T.T.; Moriwaki, R.R.; Sugawara, H. Roughness lengths for momentum and heat derived from outdoor urban-scale models. *J. Appl. Meteorol. Climatol.* **2007**, *46*, 1067–1079. [CrossRef]

35. De Ridder, K.K.; Bertrand, C.C.; Casanova, G.G.; Lefebvre, W. Exploring a new method for the retrieval of urban thermophysical properties using thermal infrared remote sensing and deterministic modelling. *J. Geophys. Res.* **2012**, 117. [CrossRef]

36. White, R.R.; Engelen, G.G.; Uljee, I. *Modeling Cities and Regions as Complex Systems*; The MIT Press: Cambridge, MA, USA, 2015; p. 330.

37. Gutman, G.G.; Ignatov, A. Derivation of green vegetation fraction from NOAA/AVHRR for use in weather prediction models. *Int. J. Remote Sens.* **1998**, *19*, 1533–1543. [CrossRef]

38. Danielson, J.J.J.; Gesch, D.B. *Global Multi-Resolution Terrain Elevation Data 2010 (GMTED2010)*; 2011-1073; U.S. Geological Survey: Reston, WV, USA, 2011; p. 26.

39. García-Díez, M.M.; Lauwaet, D.D.; Hooyberghs, H.H.; Ballester, J.J.; De Ridder, K.K.; Rodó, X. Advantages of using a fast urban boundary layer model as compared to a full mesoscale model to simulate the urban heat island of Barcelona. *Geosci. Model Dev.* **2016**, *9*, 4439–4450. [CrossRef]

40. Lauwaet, D.D.; Hooyberghs, H.H.; Maiheu, B.B.; Lefebvre, W.W.; Driesen, G.G.; Van Looy, S.S.; De Ridder, K. Detailed Urban Heat Island projections for cities worldwide: Dynamical downscaling CMIP5 global climate models. *Climate* **2015**, *3*, 391–415. [CrossRef]

41. Lauwaet, D.D.; De Ridder, K.; Saeed, S.; Brisson, E.; Chatterjee, F.; van Lipzig, N.P.M.; Maiheu, B.; Hooyberghs, H. Assessing the current and future urban heat island of Brussels. *Urban Clim.* **2016**, *15*, 1–15. [CrossRef]

42. Zhou, B.; Lauwaet, D.; Hooyberghs, H.; De Ridder, K.; Kropp, J.P.; Rybski, D. Assessing seasonality in the surface urban heat island of London. *J. Appl. Meteorol. Climatol.* **2016**, *55*, 493–505. [CrossRef]

43. De Ridder, K. Testing Brutsaert's temperature roughness parameterization for representing urban surfaces in atmospheric models. *Geophys. Res. Lett.* **2006**, 33. [CrossRef]

44. Hooyberghs, H.; Verbeke, S.; Lauwaet, D.; Costa, H.; Floater, G.; De Ridder, K. Influence of climate change on summer cooling costs and heat stress in urban office buildings. *Clim. Change* **2017**, *144*, 721–735. [CrossRef]

45. Liljegren, J.C.; Carhart, R.A.; Lawday, P.; Tschopp, S.; Sharp, R. Modeling the Wet Bulb Globe Temperature Using Standard Meteorological Measurements. *J. Occup. Environ. Hyg.* **2008**, *5*, 645–655. [CrossRef]

46. Lemke, B.; Kjellstrom, T. Calculating Workplace WBGT from Meteorological Data: A Tool for Climate Change Assessment. *Ind. Health* **2012**, *50*, 267–278. [CrossRef] [PubMed]

47. Conrad, O.; Bechtel, B.; Bock, M.; Dietrich, H.; Fischer, E.; Gerlitz, L.; Wehberg, J.; Wichmann, V.; Boehner, J. System for Automated Geoscientific Analyses (SAGA) v. 2.1.4. *Geosci. Model Dev.* **2015**, *8*, 1991–2007. [CrossRef]

48. De Ridder, K. Radiative transfer in the IAGL land surface model. *J. Appl. Meteorol.* **1997**, *36*, 12–21. [CrossRef]

49. Boënne, W.; Desmet, N.; Van Looy, S.; Seuntjens, P. Use of online water quality monitoring for assessing the effects of WWTP overflows in rivers. *Environ. Sci. Process. Impacts* **2014**, *16*, 1510. [CrossRef] [PubMed]

50. Changnon, S.A.; Kunkel, K.E.; Reinke, B.C. Impacts and responses to the 1995 heat wave: A call to action. *Bull. Am. Meteorol. Soc.* **1996**, *77*, 1497–1506. [CrossRef]
51. Loughnan, M.; Nicholls, N.; Tapper, N.J. Mapping Heat Health Risks in Urban Areas. *Int. J. Popul. Res.* **2012**, e518687. [CrossRef]
52. Yang, F.; Lau, S.S.; Qian, F. Thermal comfort effects of urban design strategies in high-rise urban environments in a sub-tropical climate. *Archit. Sci. Rev.* **2011**, *54*, 285–304. [CrossRef]

Article

The RainBO Platform for Enhancing Urban Resilience to Floods: An Efficient Tool for Planning and Emergency Phases

Giulia Villani [1],*, Stefania Nanni [2], Fausto Tomei [1], Stefania Pasetti [3], Rita Mangiaracina [4], Alberto Agnetti [1], Paolo Leoni [1], Marco Folegani [3], Gianluca Mazzini [2], Lucio Botarelli [1] and Sergio Castellari [5,6]

[1] Arpae SIMC—Regional Agency for Prevention, Environment and Energy of Emilia-Romagna, HydroMeteoClimate Service, 40122 Bologna, Italy; ftomei@arpae.it (F.T.); aagnetti@arpae.it (A.A.); pleoni@arpae.it (P.L.); lbotarelli@arpae.it (L.B.)
[2] LepidaScpA, Research and Prototype Area, 40128 Bologna, Italy; stefania.nanni@lepida.it (S.N.); g.mazzini@ieee.org (G.M.)
[3] MEEO srl, 44123 Ferrara, Italy; pasetti@meeo.it (S.P.); folegani@meeo.it (M.F.)
[4] NIER Ingegneria spa, 40013 Castel Maggiore (Bologna), Italy; r.mangiaracina@niering.it
[5] EEA—European Environment Agency, 1050 Copenhagen, Denmark; sergio.castellari@eea.europa.eu
[6] INGV—Istituto Nazionale di Geofisica e Vulcanologia, Sezione di Bologna, 40128 Bologna, Italy
* Correspondence: gvillani@arpae.it; Tel.: +39-051-649-7562

Received: 31 October 2019; Accepted: 11 December 2019; Published: 17 December 2019

Abstract: Many urban areas face an increasing flood risk, which includes the risk of flash floods. Increasing extreme precipitation events will likely lead to greater human and economic losses unless reliable and efficient early warning systems (EWS) along with other adaptation actions are put in place in urban areas. The challenge is in the integration and analysis in time and space of the environmental, meteorological, and territorial data from multiple sources needed to build up EWS able to provide efficient contribution to increase the resilience of vulnerable and exposed urban communities to flooding. Efficient EWS contribute to the preparedness phase of the disaster cycle but could also be relevant in the planning of the emergency phase. The RainBO Life project addressed this matter, focusing on the improvement of knowledge, methods, and tools for the monitoring and forecast of extreme precipitation events and the assessment of the associated flood risk for small and medium watercourses in urban areas. To put this into practice, RainBO developed a webGIS platform, which contributes to the "planning" of the management of river flood events through the use of detailed data and flood risk/vulnerability maps, and the "event management" with real-time monitoring/forecast of the events through the collection of observed data from real sensors, estimated/forecasted data from hydrologic models as well as qualitative data collected through a crowdsourcing app.

Keywords: web-based platform; early warning system; vulnerability simulations; flood risk maps; rainfall estimates; microwave links; CML; crowdsourcing; sensible targets

---

1. Introduction

Climate change affects the water cycle by intensifying it and this can change the magnitude, frequency, and timing of river floods in areas of the planet [1] such as some parts of Europe [2]. In particular, Blöschl et al. [3,4] by analyzing a pan-European database over the past five decades found clear patterns of change in flood timing in Europe due to changes in climate. However, the increasing trend of disasters due to floods in Europe is due also to non-climatic drivers, such as continued socio-economic growth, which induces population growth, economic wealth, and unplanned urbanization. Furthermore, the projected change in frequency of discharge extremes in Europe is likely to have a large impact on the flood hazard, e.g., current 100-year flood peaks are projected to double in frequency within three decades [5].

The Floods Directive 2007/60/EC [6] published by the European Commission (EC), driven from the rising of human and economic losses of natural hazards, such as floods in Europe, aims to enhance the prevention, preparedness, protection, and response to flooding events and to increase the awareness of risk prevention measures within society. The early warning systems (EWS) are mentioned in this Directive as a relevant part of this disaster risk management cycle to support effective preparedness towards floods. Hence, it is fundamental for the European Member States to better identify the risk and occurrence of river floods and to better monitor the vulnerability of the society in order to establish effective early warning systems.

The usage of EWS as a useful tool to prevent damage, enhance the resilience of a society from natural hazards has been highlighted in the recent global policy treaties for climate change and disaster risk reduction in the last decade (e.g., the Paris Agreement in Articles 7 and 8 and the Sendai Framework for Disaster Risk Reduction in Priority 3 and 4). Furthermore, several success stories have shown that major EWS developments have been carried out due to technological advances (e.g., better forecasts made possible from radar nowcasting, ensemble weather models, high-resolution satellite data, more effective communication and sharing of information) [7–11]. Early warning systems can have a significant positive cost/benefit ratio in both disaster risk reduction and climate change adaptation [8,11–13]. In addition, the 2030 Agenda for Sustainable Development [14] has endorsed early warning systems as an essential action to be financed for protecting lives and property, thus contributing to sustainable development [15].

Early warning systems are defined by the United Nation Office for Disaster Risk Reduction (UNDRR) as "integrated system of hazard monitoring, forecasting and prediction, disaster risk assessment, communication and preparedness activities systems and processes that enables individuals, communities, governments, businesses and others to take timely action to reduce disaster risks in advance of hazardous events." All these activities need to be coordinated in specific areas under interest and across multiple levels of governance to work effectively and to provide input in short times.

The challenge is to develop and implement early warning systems for any kind of scale of river catchments and especially for small catchments with a drainage area of a few hundred square kilometers that can be subjected to flash floods causing large amounts of damage in the urban context.

General review studies have discussed advances in river flood and river flash flood monitoring/forecasting and confirm the great hydrological challenge in developing and implementing an efficient early warning system despite the reliability of forecasts having increased due to the efficient integration of meteorological and hydrological modeling capabilities in the last years, especially in the developed countries [16,17].

Alfieri et al. [7] reviewed the current European operational warning systems for water-related hazards (e.g., river floods and coastal floods, flash floods, debris flows, mudflows, rainfall-induced landslides) due to severe weather conditions. This study identified those EWS including a weather prediction component to detect extreme events with considerable lead time, which can support early preparedness and effective disaster risk reduction. Further work is needed to better exploit the benefits provided by such systems [7].

Acosta-Coll et al. [18] conducted a systematic review to define the adequate structure of EWS for pluvial flash floods in urban areas. They identified the need for people-centered EWS with fail-safe systems able to guarantee the dissemination and communication of timely alerts during rainfall events. For doing so, the amount of rain and the water level need to be monitored and processed in real-time through the ultrasonic or radar sensors, which are more suitable for these applications. Finally, Acosta-Coll et al. [18] divided the EWS into four structures: "Disaster Risk Knowledge," "Forecasting," "Dissemination and Communication of Information," and "Preparedness and Response."

The presence of large complicated areas with barriers blocking, stormwater flow, insufficient drainage capacity, or the vicinity of a river, along with population growth and climate change impacts makes the urban areas highly vulnerable to flash floods. Reliable EWS for flash flood forecasting in urban watersheds are challenging and need special attention to the overall control of all components to reduce potential severe losses and as well are key for more effective and coherent implementation of existing European Union (EU) policies on disaster risk reduction and climate change adaptation [19,20].

Finally, recent European projects started to find new solutions for adapting to intense river floods in urban areas by addressing the different aspects of monitoring and forecasting. The EU Interreg project URBAN-PREX (monitoring, forecasting, and development of online public early warning systems for extreme precipitation and pluvial floods in urban areas in the Hungarian–Serbian cross-border region) [21] supported, as components of online public early warning systems for citizens and public authorities, the implementation of monitoring precipitation networks through rain gauges in urban areas in the Hungarian–Serbian cross-border region. On the other hand, the EU Interreg project RAINGAIN [22] used radar technology to provide high-resolution estimates of rainfall in cities to forecast pluvial floods in test urban sites in the UK, Netherlands, Belgium, and France. The EU FP7 project STAR-FLOOD (STrengthening And Redesigning European FLOOD risk practices: Towards appropriate and resilient flood risk governance arrangements) [23] aimed to improve the implementation of flood risk strategies in urban areas which are vulnerable to pluvial floods in different European countries (UK, Sweden, Poland, Netherlands, Belgium and France) by designing policies for an appropriate and resilient flood risk governance. Among its different outcomes, this project highlighted the need for further improving systems for forecasting, warning and emergency responses for urban floods that are proactive, risk-based and use collaborative approaches, for instance by optimizing the use of ICT (apps).

The need to improve the EWS and emergency communications related to flood risk in urban areas has been addressed from the EU Horizon 2020 project FLOOD-serv (Public FLOOD Emergency and Awareness SERVice) [24]. This project aimed to develop a collaborative platform that links citizens, public authorities, and other stakeholders and to enable the public to be warned in due time to reduce the adverse effects of floods.

Taking into consideration past studies and experience, the project RainBO focuses on the improvement of knowledge, methods and tools for the monitoring and forecast of extreme precipitation events and the assessment of the associated flood risk for small and medium watercourses in urban areas.

RainBO, funded by the EU Life Program, is a follow-up of the BLUEAP (Bologna Local Urban Environment Adaptation Plan for a Resilient City) LIFE project and T-Rain, a Climate-KIC (Knowledge and Innovation Community) project, in terms of implementing a reliable service based on big data coming from cellular networks.

The partners within the RainBO project, under the coordination of Lepida scpa (the in-house company of the Emilia-Romagna Region in charge of planning and implementing telecommunication infrastructures and IT services), comprise Arpae SIMC (the HydroMeteoClimate Service of the Regional Agency for Prevention, Environment and Energy of Emilia-Romagna is involved), the municipality of Bologna, MEEO (a small and medium enterprise with expertise in remote sensing and Commercial microwave links technology) and NIER (a consulting company with expertise on environmental analysis and risk evaluation).

The present article aims to describe the RainBO project and its main outcome: a webGIS (Geographic Information System) modular platform addressed local administrations to provide information from observed data, forecasts, and models before and during extreme events of precipitation in vulnerable river basins. On the platform, weather data are collected from traditional and innovative monitoring systems, weather forecasts are gathered from existing meteorological models whereas hydrological forecasts are provided by operational and newly developed models. The platform has been tested on two Italian urban areas, as detailed in the following.

The structure of the article is as follows: Section 2 describes an examination of existing data and useful models for the purposes of the RainBO platform, and afterward, the innovative methodologies developed within the project are explained. In Section 3, the platform and the main results achieved during the project are then presented, with a description of its main structure, databases, and modules. Finally, remarks for further developments are described.

2. Materials and Methods

2.1. Study Areas

The application of the RainBO platform is performed on two test areas, the cities of Bologna and Parma, located in the Emilia-Romagna region, Italy. The climatic features referred to 1991–2015 period for these two cities are shown in Table 1 [25].

Table 1. Annual climatic features of Bologna and Parma (reference climate: 1991–2015).

Variable	Bologna	Parma
Minimum temperatures (unit: °C)	9.88	8.76
Maximum temperatures (unit: °C)	19.27	19.14
Mean temperatures (unit: °C)	14.59	13.95
Precipitation (unit: mm)	775.7	795.3

Both these urban areas are crossed by watercourses: the Ravone creek flows through Bologna, the Parma river flows through Parma (Figure 1).

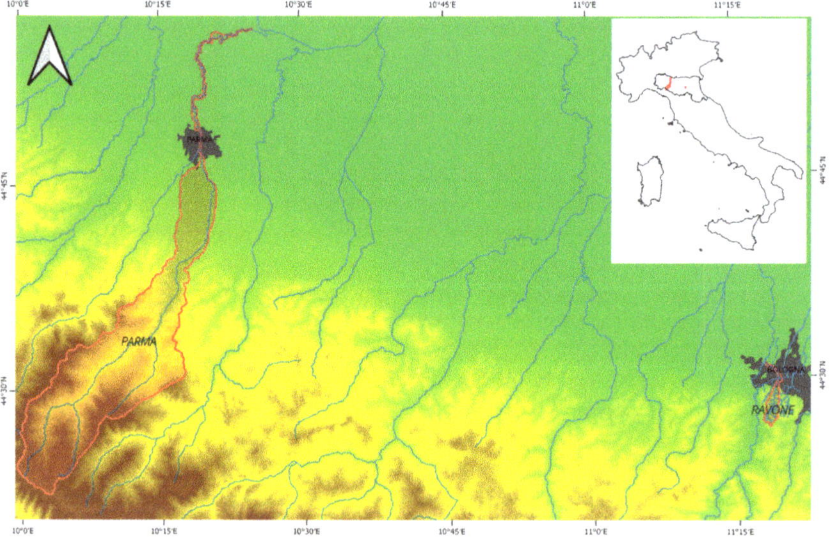

Figure 1. Study areas of the RainBO project.

As shown in the figure, the catchment of the Ravone creek is small, whereas the catchment of the Parma river is medium-large. The different sizes of the catchments are reflected in two different modeling approaches of the hydrologic forecasts, as explained below.

It is worthy to mention that each hydraulic section of the Emilia-Romagna catchments has three specific thresholds/levels of alarm: yellow threshold/warning, orange threshold/pre-alarm and red threshold/alarm. These thresholds are site-specific and defined according to Civil Protection purposes for public safety, taking into account the geometry of the hydraulic section of the watercourse and the statistical distribution of historical recordings.

2.1.1. Ravone Catchment

The Ravone catchment, together with the Aposa catchment, is the largest river basins at SW of Bologna. They flow from the hills directly across the central part of the city. The upper part of the Ravone catchment is characterized by hills and steep slopes, with a prevalent vegetation coverage (grass, shrubs, and forest) and partial urbanization in the stream valley. In contrast, the lower portion is flat and densely urbanized, the drainage network is mainly artificial, and the watercourse is connected with the main urban drainage system in a critical spillway crosspoint, flowing in a culvert underneath the city's urban area before joining the Reno river. The length of the natural reach of the Ravone is approximately 4 km covering an area of 6 km^2.

The Ravone catchment has been chosen as a study area of RainBO because, as recorded in historical data, the response of the catchment to extreme rainfall past events caused significant damages. For instance, during a flood that occurred on July 22nd, 1932, a victim and severe damages to streets and houses in the Southern part of the city were recorded [26].

The catchment is equipped by an existing monitoring network (Figure 2), including a weather station 500 m far from the Ravone catchment equipped with a rain gauge on S. Luca hill where data have been collected since the early 1930s. To integrate the available dataset and to better monitor the Ravone catchment, in 2014 a second rain gauge was installed at the catchment upstream end of mount Paderno and a water level gauge was installed at the culvert entry of the Ravone creek (Figure 3).

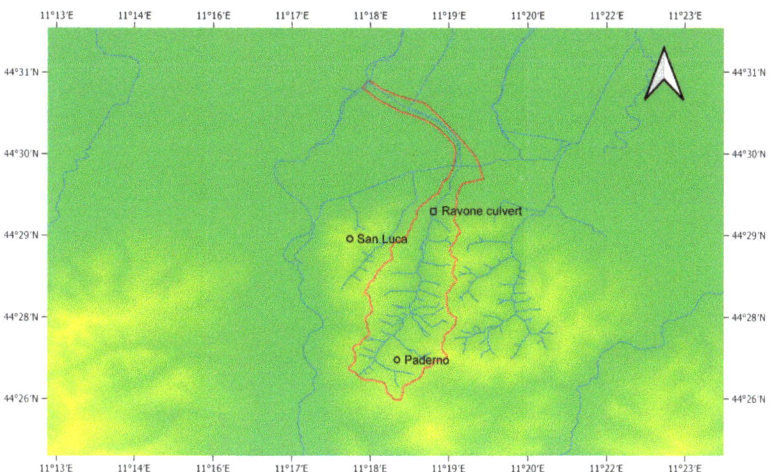

Figure 2. Map of existing monitoring network on the Ravone area. The circles are the rain gauges, the square is the water level gauge.

Figure 3. The entry of the culvert of the Ravone creek where the water level gauge is installed and the alarm thresholds are highlighted.

2.1.2. Parma Catchment

Parma river is the main river flowing through a homonymous city in Emilia-Romagna; it starts from Mount Marmagna at 1842 m.a.s.l. and flows in an N-NE direction joining the River Po near the city of Colorno as the right tributary. Usually, the hydrological responses for these basins are characterized by high discharges in the spring and autumn and low discharges in the summer.

The small catchment area and the steep part of the valley give high hydrological response during severe storms, generating, under particular conditions, flash flood events. Its major tributary is the Baganza River which has a very similar course and joins the main river on its left in the city of Parma (Figure 4).

Figure 4. Baganza river in Parma after the October 2014 event.

The available monitoring network for the Parma basin is composed of 15 rain gauges, 11 thermometers, and 6 water level gauges, these sensors are sufficient for the application of the RF Model, hence, no other installation was required (Figure 5).

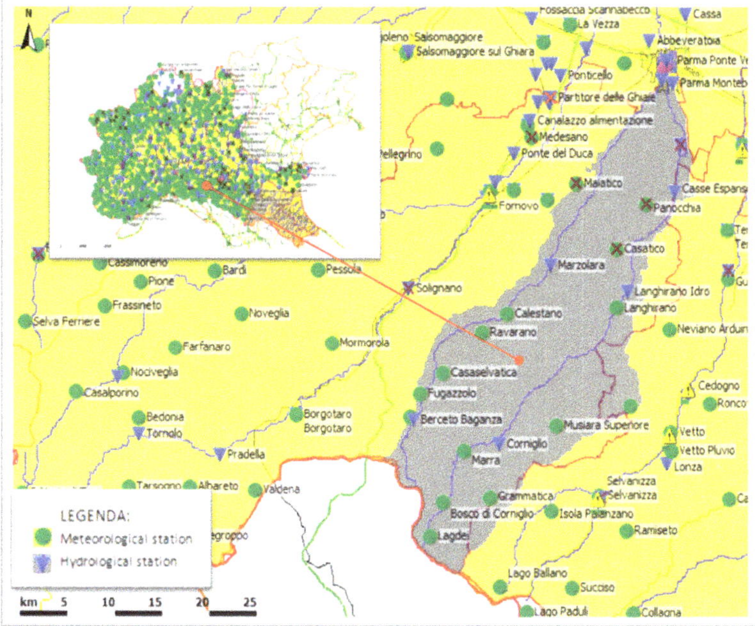

Figure 5. Parma-Baganza basin and monitoring network.

2.2. Background of the RainBO Project

The existing data and models at the beginning of the project identified as necessary for the development of the RainBO platform are described.

2.2.1. Territorial Data

According to the existing national and regional guidelines for the planning of Civil Protection, the following maps on the Emilia-Romagna region are available:

- Regional technical cartography (CTR) 1:5000 updated in 2013 with the topographic database (TIFF format)
- Ortho-photo Agea2014 (TIFF format) resolution 50 m
- Digital Surface Model Agea2008 (TIFF format) resolution 5 m × 5 m
- Digital Terrain Model Agea2008 (TIFF format) resolution 5 m × 5 m
- River catchments from numerical data 1:10.000 (shapefile format)
- Toponymy (shapefile format) 1:5000

These data are projected in the WGS84 UTM32N reference system.

2.2.2. Hazard, Vulnerability, and Risk Maps

To set up a tool for the management of flood risk in the Member States of the European Union, the main legislative reference is the Floods Directive [6]. In the directive, the hydraulic risk is the product of the hazard and potential damage at a specific event:

$$R = P \times E \times V = P \times Dp, \tag{1}$$

where:

- P (Hazard): it is the probability of occurrence, within a certain area and in a certain time interval, of a natural phenomenon of assigned intensity
- E (Exposure): it represents people and/or assets (structures, infrastructures, etc.) and/or activities (economic, social, etc.) exposed to a natural event
- V (vulnerability): the degree of capacity (or incapacity) of a system/element to resist at the natural event
- Dp (potential damage): it is considered as the degree of foreseeable loss following a natural phenomenon of a given intensity, the function of both value and vulnerability of the exposure
- R (risk): expected number of victims, injured persons, damage to property, cultural assets e environmental, destruction or interruption of economic activities, as a result of a natural phenomenon of assigned intensity

Emilia-Romagna Region has developed the Flood Risk Management Plan (FRMP), to be compliant with the Floods Directive [6] and Legislative Decree 49/2010, which requires these flood risk management plans to include measures to reduce the probability of flooding and its potential consequences and to address all phases of the flood risk management cycle but in particular the prevention, the protection, and the preparedness. As the causes and consequences of floods are different in the different member states of the Community, the Management Plans take into account the specific characteristics of the territories and propose specific objectives and measures tailored to the needs and priorities.

The FRMP of the Emilia-Romagna Region is represented by three projects, one for each hydrographic district (Po River, Northern Apennines, and Central Apennines).

Existing hazard maps represent the potential extent of flooding caused by watercourses (natural and artificial) with reference to three scenarios (rare floods, infrequent, and frequent) colored with three different shades intensity of blue, depending on the frequency of flooding as follows:

- rare floods of extreme intensity: return time up to 500 years from the event (low probability)
- infrequent floods: return time between 100 and 200 years (average probability)
- frequent floods: return time between 20 and 50 years (high probability)

The hazard maps from the Floods Directive are available on the whole Italian territory. They constitute the reference hazard maps for the computation of hydraulic risk and, for the purpose of the RainBO project, they have been selected on the study areas of Bologna and Parma.

Moreover, for the Bologna study case, in addition to the reference hazard map where the Ravone and other small streams are not included, a specific map for the Ravone creek is available [27]. This has been obtained through scenarios of flood analysis [28].

The existing vulnerability maps from the Floods Directive have been clipped on the study areas of Bologna and Parma. It should be noted that these reference maps do not consider in detail population distribution issues, as well as risk maps, as a consequence. In particular, to define the expected damages of a flood, the Directive suggests including the following main items:

- urban areas and urban expansion areas
- industrial and technological areas
- environmental heritage and cultural assets of significant interest
- presence of critical infrastructures such as transport, communication, utility networks
- presence of public and private services: sports plant, recreational facilities, accommodation facilities

The risk maps indicate the presence of potentially exposed elements (population involved, services, infrastructure, economic activities, etc.) which fall within floodable areas by means of a classification in 4 risk categories, represented by a color scale: yellow (moderate or no risk), orange (medium risk), red (high risk), purple (very high risk).

Existing risk maps from Floods Directive, created by the integration of hazard maps and vulnerability ones, identify static situations and do not take into account urban territory resilience peculiarity. In this case, risk maps have also been selected for the study areas of Bologna and Parma.

2.2.3. Historical Events

A catalog of historical events from 1981 is available for the Parma and Reno, including the Ravone catchment. A historical event is defined as the exceedance of at least one pre-alarm threshold. For each event, the involved catchment, the exceeded thresholds of water level gauge, as well as the description of the flooded area are recorded.

Moreover, where available, for each event further information (i.e., the number of people evacuated, dead or wounded people, emergency state—if requested, the possible assessment of economic damages) were collected. Moreover, the reports for all the events that occurred in Emilia-Romagna are available [29].

2.2.4. Observed Meteorological Data

A complex infrastructure for environmental monitoring and for an early-warning system is based on the integration of different data from many sources. The hydro-pluviometric network of Arpae collects data from rain gauges, thermometers, and water level gauges on the Emilia-Romagna region (Figure 6).

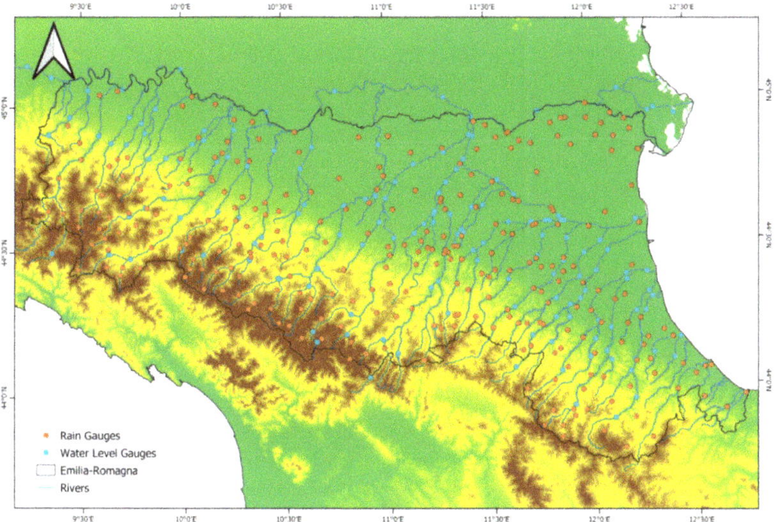

Figure 6. The hydro-pluviometric network of Emilia-Romagna.

For the RainBO purposes, the key variables are water level gauges (242 sensors) and rain gauges (282 sensors). Data coming from the hydro-pluviometric monitoring system are available on the Arpae ftp server. These data are acquired from the network every 15 min, the units are millimeters (mm) for precipitation and meters (m) for water level. In general terms, the series starts from 2003, the historical series for some sensors start from 1980.

These data are integrated into the Sensornet platform through a web service. Sensornet is the Internet of Things Platform of the Emilia-Romagna Region, that collects data and information from thousands of sensors distributed on the territory and builds a digital map over time [30].

The development of the RainBO platform implied also the empowerment of the monitoring network in critical areas not properly covered by sensors such as the Ravone catchment. For this reason, new monitoring sensors were installed in the Ravone area:

- a new real-time water level gauge has been installed in the upper part of the river basin
- a new weighing rain gauge has been located in a public property on the right side of the valley

These new monitoring sensors installed in the Ravone area are aimed at measuring environmental variables to be collected in the RainBO platform and at monitoring the Ravone creek with a high degree of accuracy.

2.2.5. Forecast Meteorological Data

The COSMO-LAMI forecasts provided by Arpae-SIMC are available on an open data platform as GRIB files over Emilia-Romagna [31].

Meteorological COSMO-LAMI forecasts are based on the operative non-hydrostatic limited-area atmospheric COSMO (Consortium for Small-scale Modelling) model, nested on the ECMWF (European Centre for Medium-Range Weather Forecasts) operational global forecast. This system is developed and maintained by the homonymous European consortium, and managed by Arpae-SIMC on the basis of the LAMI (Limited Area Model Italia) agreement. The forecast is issued twice a day (00 and 12 UTC) with a time range of 72 h on a regular 5 × 5 km grid covering the Mediterranean area.

2.2.6. Estimated Data

Commercial microwave links and radar data are estimated data, essential for the purposes of RainBO in order to monitor precipitation events.

The microwave links based on the rainfall monitoring system is an innovative but already tested technology, exploiting the microwave links used in commercial cellular communication networks (so-called commercial microwave links, CML).

Conventional rain gauges are not so effective during intense precipitation as their operating principle is based on mechanical tilting parts, which makes their measurements unreliable during this type of phenomenon. Rain gauges provide point-like measurements of the amount of rain fallen within the instrument sampling area, cumulated on time intervals, usually ranging from one minute to one day, with well known instrumental [32] and representativeness [33] limitations.

A relatively new and independent approach to the estimates of precipitation at the ground became available in the last decades with the broad diffusion of CMLs for cellular communication: integral precipitation content along a line path between two antennas can be estimated by measuring the attenuation of the microwave signal along the same path [34].

Heavy rain causes electromagnetic signal attenuation (from the transmitting antenna to the receiving one) and, subsequently, path-averaged rainfall intensity can be retrieved from the signal's attenuation between transmitter and receiver by applying, almost in real-time, a rainfall retrieval algorithm.

A distributed monitoring system can be developed by using received signal level data from the massive number of CMLs used worldwide in commercial cellular communication networks.

The first studies on this technology were concentrated on algorithms for spatial-temporal interpolation [35] from the joint analysis of multiple CMLs. The great potential of CMLs for ungauged regions was demonstrated by the Burkina Faso application [36]. In 2012, Overeem [37] demonstrated that processing algorithms are capable of providing real-time rainfall maps for an entire country, in this case, the Netherlands.

With regard to radar data, an existing dataset on Emilia-Romagna is based on hourly precipitation estimates obtained from the merger of the regional radar network managed by Arpae-SIMC. This network is composed of two C-Band systems, one located at San Pietro Capofiume (Bologna) and the other in Gattatico (Reggio Emilia). Every 5 min during precipitation events, the radars provide reflectivity data that are processed by several algorithms. The reflectivity value is correlated to the precipitation intensity. Reflectivity data are provided by Arpae-SIMC in open data [38].

For the RainBO project, an ad hoc stream was triggered for the automatic and periodic distribution of both reflectivity maps and hourly precipitation maps on the ftp server provided by Lepida, according to the RainBO project goals. Sent images are a merge of the two systems (reflectivity radar maps and accumulated hourly precipitation). Every new capture of the reflectivity (the frequency of observations is related to the weather conditions at the time) generates a merged image, which is then sent to the Lepida ftp server.

2.2.7. Crowdsourcing

A platform addressed to provide information for planning and management of extreme events and flash floods in urban areas should comprise also observations and contributions from citizens. These types of data, usually collected by means of applications, are a source of additional information during ongoing events but they are also conceived in order to engage and raise awareness of citizens. In this regard, the system Rmap [39] is a participatory monitoring and exchange system promoted by Arpae-SIMC, based on open hardware and software infrastructures, to collect and share meteorological data gathered by citizens between public and private institutions.

The Rmap application is mainly addressed to users and citizens with meteorological domain expertise and collects automatically the weather data in a WMO (World Meteorological Organization) binary data software (Binary Universal Form for the Representation of meteorological data—BUFR) through dedicated devices based on open hardware and free software. The weather information data can also be uploaded manually by expert users by using an on-line application, but the graphical user interface is not conceived as a smart tool for the general public.

It defines a set of standards for meteorological data sensing (security, reliability, elaboration) and for the transmission data system (transmission protocols, data formats, metadata formats, etc.).

The Rmap project has been promoted by Arpae-SIMC for some years, as it is an interesting project with the objective of defining methods, protocols, and formats to collect and share environmental data. The project is also promoted by Arpa Veneto, Cineca and the Computer Science Department of the University of Bologna and the RaspiBO network.

The Rmap system was taken into consideration because of this robust partnership and the relevant effort spent in its standardization.

The Rmap project adopts indeed a scientific approach based on standards defined by the WMO, in particular using their elaboration and classification process.

2.2.8. Models

The models used for the development of the RainBO services are:

- CRITERIA-1D
- CRITERIA-3D
- RANDOM FOREST

CRITERIA-1D [40,41] is a one-dimensional model developed by Arpae simulating the soil water balance, nitrogen balance, and crop development. The model is usually applied to agricultural case studies, nonetheless one of its main outputs (i.e., soil moisture) is a crucial variable for hydrological purposes. The CRITERIA-1D model simulates soil water movement by using a simplified model (tipping bucket) or a numerical model. It requires as an input at least daily data of temperature and precipitation, soil features, and crop information.

CRITERIA-3D [42] is a physically-based model developed by Arpae that works at the catchment-scale and solves equations of surface and subsurface water flow in a three-dimensional domain. The hydrologic component is a dynamic link library integrated into the other Arpae software, implemented within a comprehensive model that simulates the physical processes occurring in the catchment: surface energy, radiation budget, snow accumulation and melt, potential evapotranspiration, plant development, and plant water uptake.

The two models are under development and they are available as open-source code [43].

After the flooding of the Parma and Baganza rivers in Parma on October 2014, caused by heavy rains, the Civil Protection Agency of the Emilia-Romagna region required Arpae to develop a hydrological simulation model capable of promptly evaluating in advance the probability of overcoming the alert thresholds, especially for rapid or flash flood events. The RANDOM FOREST (RF) algorithm, applied in a hydrological context, provides the probability of overcoming the alert thresholds of some observation points for basins at small and medium scales in the oncoming next 6–8 h.

The RF model was added beside the existent hydrological-hydraulic model applied in real-time into the Flood Early Warning System Emilia-Romagna (FEWS EMR) to provide a fast and preliminary response during flash flood or extreme rainfall events in the Emilia-Romagna basins.

The model is an ensemble learning method that operates by constructing a multitude of decision trees at training time and outputting the mode of the classes (classification) or mean prediction (regression) of the individual trees. Each tree classifies the dataset using a subset of variables. The number of trees in the forest and the number of variables in the subset are hyper-parameters and, for this reason, they have to be chosen a priori.

The number of trees is in the order of hundreds, while the subset of variables is quite small, if compared to the total number of variables, in Figure 7.the final Random Forests tree generated for Parma River at Ponte Verdi is shown, all paths end with terminal node that contains the probability of exceedance for each H.T.A. (hydrometric thresholds alert).

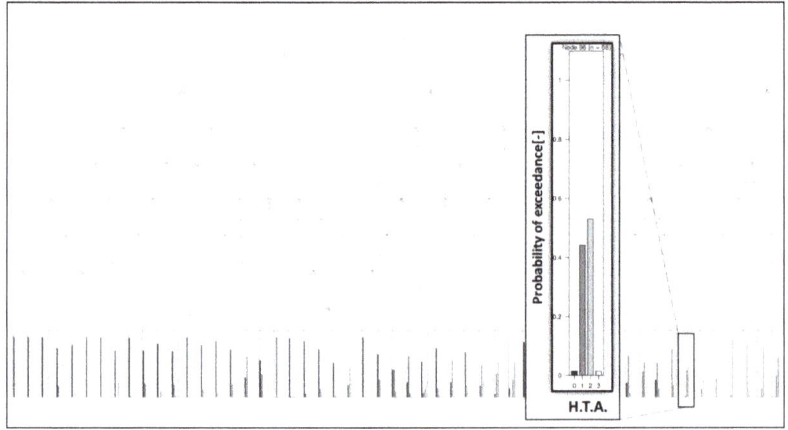

Figure 7. Random Forest tree generated for the main Parma River section (Ponte Verdi).

RF also provides a natural way to assess the importance of input variables (predictors). This is achieved by removing a variable at a time and assessing whether the out-of-bag error changes or not. If the out-of-bag error changes, the variable is important for the decision [44].

Model parameterization was performed using historical data (2003–2016), while recent data (2017–2019) are used for validation. Model parametrization was accomplished primarily by extracting historical events for each river section, hence defining the reference response time of the basin. This value was then used in the RF model, defining the maximum aggregation time for rainfall (Figure 8). For the RainBO project, the model was implemented using as input the same observed date used for the other Hydrological-Hydraulic model: observed mean hourly rainfall and discharge data.

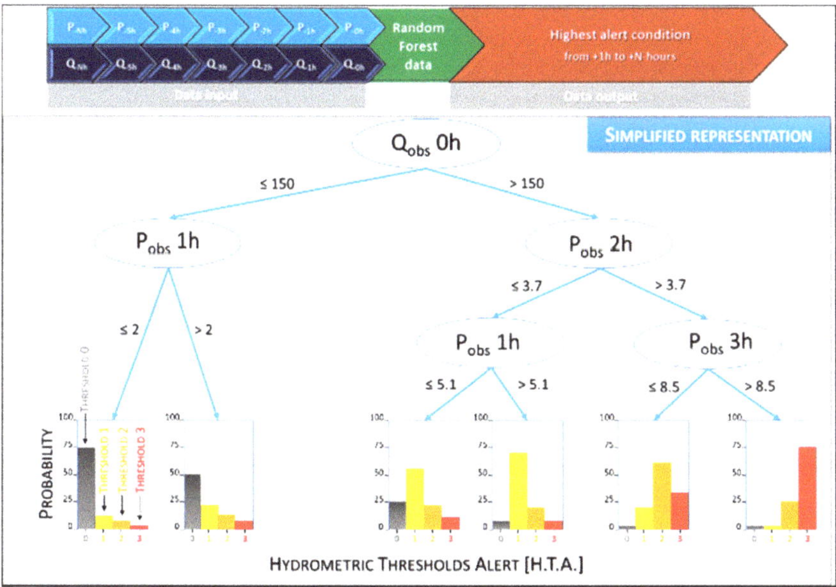

Figure 8. Random Forest model schematization. P_{obs} = observed mean hourly rainfall (unit: mm), Q = discharge (unit: m^3/s).

2.3. Foreground of RainBO Project: Innovation and Development

The development of the RainBO infrastructure started from existing data and models, then it focused on the application of new models specifically developed and new technologies, such as CML to enhance the monitoring systems during extreme events.

The project includes the development of a crowdsourcing web application for collecting and sharing information on the observed weather and its possible local effects/impact.

2.3.1. Commercial Microwave Links

The rainfall monitoring system proposed in the RainBO project exploits the Commercial Microwave links (CML or simply Microwave links—MWL) used worldwide in commercial cellular communication networks. Rain-induced attenuation and, as a consequence, path-averaged rainfall intensity can be retrieved from the signal's attenuation by applying, almost in real-time, a rainfall retrieval algorithm. The algorithm chosen for this purpose is the RAINLINK retrieval algorithm [37]. The code is an open-source R package available for free download on GitHub [45].

The implementation of the algorithm for the specific needs of the Italian test areas, as well as code debugging and improving, was carried out within the RainBO project.

To validate the algorithm, a dataset was provided by Vodafone Italia on Bologna and Parma urban areas from February 2016 to June 2016. Other (not commercial) microwave link data were supplied by Lepida to cover the Apennines area, from March 2016 up to date, providing near-real-time data.

During the project, the CML data validation on the Bologna area was performed by comparing the quantitative precipitation estimates (QPE) from CML, radar, and rain gauges.

Radar and rain gauges data were chosen for the validation because they are currently validated, used, and published by ARPAE in their operative meteorological services. It resulted that the coherence between CML and the other estimates is quite promising even if it requires a tuning activity to integrate the dataset with existing technologies (like rain gauges or radar).

Excellent results are achieved mainly in a convective event as shown in Figure 9. On 11 May 2016 precipitation occurred with a well-defined gradient in the West–East direction and some local maxima in the North-West and South-West part of the province. Microwave accumulation slightly underestimated the rainfall field while an overestimation is recorded in the non-adjusted radar. The adjustment procedure well calibrates radar data as displayed in the top right panel of Figure 9. Fine-scale structure of the daily amount is well detected in both remote sensing maps.

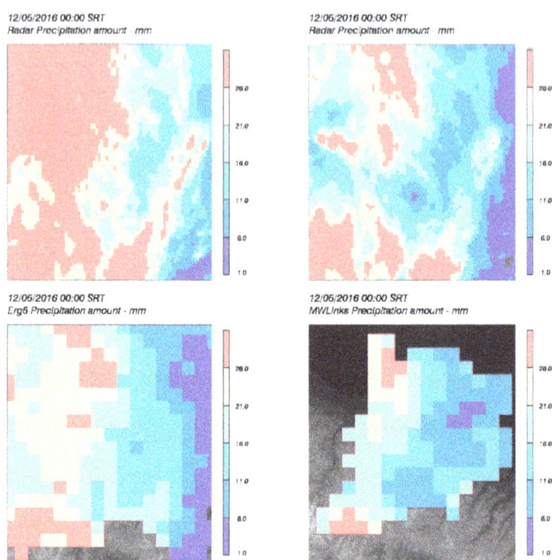

Figure 9. Quantitative precipitation estimates accumulated from 11/05/2016 00 UTC to 12/05/2016 00 UTC. (top left) Radar quantitative precipitation estimates (QPE), (top right) radar adjusted QPE; (bottom left) ERG5; (bottom right) microwave links QPE.

A qualitative analysis showed that the performance of the CLM estimate seems to increase in the second half of the analyzed period (May and June), mainly characterized by convective storms, even embedded in frontal patterns, where showers and heavy rain play an important role. As the microwave estimate is based on the attenuation that occurred in the link path, it is strongly related to rainfall intensity and the signal is, normally, stronger in the convective season. This behavior was confirmed by the quantitative analysis done by using statistical indicators that confirm a good rainfall estimation by CML data during summer and spring seasons. The validation has pointed out that CML estimation slightly underestimates precipitation occurrence both in spatial coverage and point amount, while radar adjusted has a complementary feature.

The precipitation monitoring infrastructure implemented during the RainBO project, based on the CML data, was called Rainlink4EMR (as the RAINLINK algorithm was applied to the Emilia-Romagna region) and a near-real-time module was implemented, which downloads the power attenuation raw data (CML data) from the Lepida telecommunication infrastructure and provides rainfall estimates on the midpoint of the links with a delay of a few minutes.

The validation study allowed us to achieve a first operational version of the related service, providing satisfactory reliability in monitoring convective events.

2.3.2. Crowdsourcing App

A crowdsourcing web application was implemented in RainBO to collect and display information regarding the observed current weather from expert users, as well as from people without any technical skills. This webApp was created by means of a networking activity with the Rmap project [39] and, thus, it was called Rmap4RainBO.

The Rmap application is mainly addressed to users with meteorological domain expertise and collects automatically data coming from standard hardware devices and weather observation manually uploaded. The RainBO crowdsourcing application, called Rmap4RainBO, develops and improves the Rmap functionalities for the uploading and visualization of the observed weather and it keeps the WMO standard in the weather code. The RainBO crowdsourcing component, differently from the Rmap one, aims to address people without any technical skills so, as a consequence, it was developed as an intuitive and smart application that can be accessed through the RainBO project homepage [46].

The Rmap4RainBO application can benefit from the Rmap data and vice versa as the crowdsourcing information uploaded by citizens through Rmap4RainBO feeds the Rmap database.

Differently from the Rmap, the RainBO crowdsourcing webApp records the impacts, meant as effects of weather on the territory (damaged roads, fallen trees, ice on the road, etc.). The codes used to label the impacts were defined internally at the RainBO consortium as any official code was found on WMO standards; it represents a further innovative contribution by the RainBO project.

2.3.3. RainBO Vulnerability Model

The vulnerability module calculates the degree of vulnerability of exposed items over flood events.

The vulnerability reference maps do not consider in detail population distribution issues, whereas the vulnerability model developed for RainBO includes the presence of sensible targets in the territory (e.g., schools, nurseries, hospitals) and critical targets that can worsen a scenario reducing resilience, such as the fire brigade building.

To take into account in a realistic way the distribution of people on a territory, the developed algorithm considers different time frames: for example, the distribution of the population is supposed to be different in night hours (mainly in houses) with respect to working hours (mainly in workplaces). In a similar way, during morning time, students and teachers are supposed to be in schools, while during the afternoon school users decrease, and during the night no one is supposed to occupy these target buildings.

Vulnerability maps calculated by the vulnerability module are based on territorial data collected on the platform. Moreover, the maps, in summary, are calculated as a function of:

- time frame
- resident population distribution (based on land use—Copernicus, Urban Atlas 2012)
- employees distribution of industrial, commercial and agricultural sectors (based on land use—Copernicus, Urban Atlas 2012)
- users of sensible targets
- presence of critical targets as institutional site and first aid structures that could reduce the resilience of a territory, if they are involved by emergency events
- presence of critical targets such as industrial areas and utility networks, which could produce a domino effect if they are involved by emergency events

The platform provides 32 vulnerability maps (20 × 20 m grids in raster format), corresponding to 32 reference time frames.

This information is useful not only to compute more realistic vulnerability maps but also because it can be used by the early warning module. Through these data, the priority of targets to be warned during an ongoing emergency can be identified.

2.3.4. Hydrologic Forecast for Small Catchments

One of the main activities of the project has been the development of a model of the hydrological forecast for small basins, by using Ravone as a test case.

Within the available dataset (from 2014 to 2018) of observed water level at the culvert entry there are not events that exceed the alarm threshold. For this reason, CRITERIA-3D model has been used to simulate scenarios in order to assess the effects of severe rainfall events potentially able to exceed the alarm threshold (Figure 10), starting from initial conditions of soil moisture corresponding to the most remarkable event within the dataset (recorded on March 25th, 2015). The scenarios include three possible precipitation sum (70, 85, and 100 mm per event) with two possible event lengths (9 or 14 h) and two possible precipitation hyetographs (triangular and trapezoidal). These choices correspond to a discretization of the precipitation intensities recorded during the most remarkable past events (including the event recorded in 1932) when water level gauges were not installed. Thus, 12 possible precipitation scenarios on the Ravone catchment have been simulated with the CRITERIA-3D model. This discretization is a compromise between the need to simulate as many cases as possible and to run simulations within an acceptable computational time.

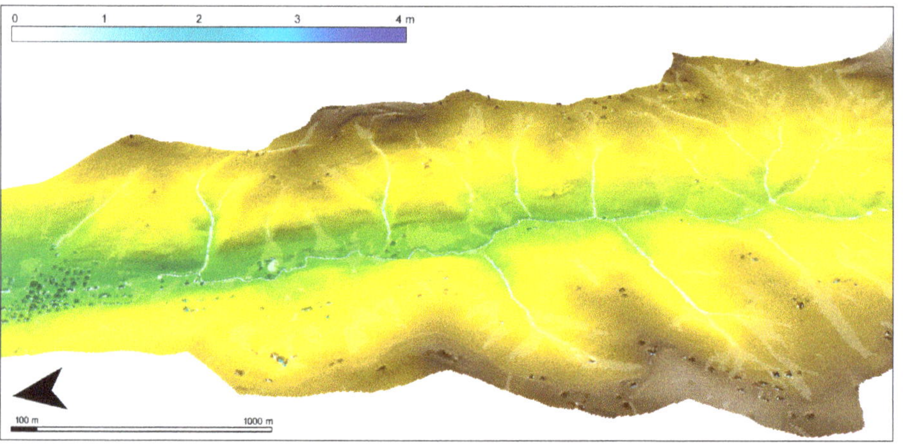

Figure 10. CRITERIA-3D simulation of the surface water flow on the Ravone catchment during a rainfall event. The color scale represents the surface water level (unit: m).

The water levels simulated by CRITERIA-3D have been integrated with observed data. On the resulting dataset, a statistical analysis has been performed, taking into account water levels, precipitation and soil moisture (defined as water holding capacity, see below for further details). As a result, a significant logistic regression between these variables has been identified.

The hydrological forecast is based on this regression using as input the forecast of precipitations of the COSMO-LAMI model, presenting the best available resolution in open data and calibrated on the study area. In order to include the spatial variability of the event, the computation is performed on the Cosmo grid cell containing the prevailing area of the Ravone catchment and on the 8 neighboring cells.

To validate the forecast algorithm, a hindcast analysis using the COSMO-LAMI forecasts for the period 2015–2016 was carried out. The analysis showed that the maximum intensity of precipitation forecast on the area is mainly underestimated, with an average underestimation of approximately a third of the observed value. To compensate for this underestimation, that could be cause missing alarms, the operational dataset used as input for the hydrologic forecast is integrated with a second forecast series where the maximum hourly intensity of precipitation is increased by 33%.

Therefore, an ensemble of forecast scenarios of precipitation is produced and an ensemble of the hydrological forecast is derived. The statistical distribution of this output is computed to provide a boxplot of the hydrological forecast.

In addition to the precipitation, the crucial variable for the hydrological forecasts is the soil moisture of the catchment. We decided to use an estimated value of this variable instead of measured one because it has the advantage that it is not affected by sensor lacking or failures and local peculiarity. To estimate the soil moisture for catchments of small dimensions as Ravone, it is possible to assess the mean soil moisture of the area by means of a mono-dimensional soil water balance model, as CRITERIA-1D. For the development of the algorithm for the forecasts of exceeding the hydrometric threshold, a new output variable named water holding capacity (WHC) has been added to the CRITERIA-1D model. WHC provides the maximum amount of water the soil can retain before the runoff starts, given the current conditions of soil moisture. For the Ravone study case, CRITERIA-1D is set with the parameters of the prevailing soil on the catchment (silty loam) and the parameters of prevailing crop coverage in the area (fallow); the WIIC index is computed on the upper soil layer (30 cm). Weather data (daily temperature and precipitation) used as input are the values of the analysis grid ERG5 on the Emilia-Romagna region.

3. Results and Discussion

The main output of RainBO has been the integration of the data and models described in the Materials and Methods encapsulated in an organic platform [47], as presented in the next paragraph.

3.1. The RainBO Platform

The RainBO platform consists of the following key elements:

- database containing monitoring, territorial, and historical data
- software modules, which are the platform intelligence
- graphic interface, which is the platform output

One of the most important features of the platform is the database containing the monitoring data, whose functionality is to integrate data collected from different monitoring infrastructures, both conventional and unconventional, as well as forecast data, hydrological, and meteorological models, and estimated ones.

In particular, the implementation of an advanced monitoring infrastructure within the RainBO Life project consists of the integration of these types of data:

- real sensors data (e.g., weather stations)
- "virtual sensors" data, not associated with observed measurements from physical sensors, but obtained indirectly through the estimation of correlated data or from simulation models
- forecast data, provided by simulation models

This structure allows us to monitor extreme precipitation events, their evolution, and to generate early warnings. It is worthy to mention that the concept of "virtual sensor" allows us to integrate information from observed and not observed data sources, georeferencing them with the same reference system and synchronizing them over time.

The integration of these new virtual sensors into the RainBO platform has been accomplished in a simple way, using the same data model defined for the physical sensors, without any extension or specialization and providing the platform with an enhanced monitoring infrastructure.

The territorial database hosts both the input data and the output data coming from the processing of the application modules, as well as, obviously, the data necessary to describe the territorial characteristics.

The RainBO platform architecture has been designed according to the following attributes:

- open: each module exposes standard interfaces (web services) to ensure system generality and replicability as well as interoperability and integration with other platforms
- centralized: each DB is centralized and enables data sharing, managed, and updated by different users

- scalable: each module is developed so as to be implemented on different machines
- modular: the platform is formed by individual modules ensuring more flexibility, maintainability over time, as well as platform evolution as each module can evolve or be replaced independently from each other
- configurable: each module is configurable, i.e., the operating parameters must be read from the table and not written in code

RainBO platform has been conceived according to two operational modes:

1. Planning support
2. Event management

Both the modes display a time bar with different time ranges according to the selected operational mode: the menu bar for the planning support mode refers to historical events whereas the menu bar for the event management mode refers to monitoring data (from −24 h to +72 h), in addition to menu bars and GIS maps specifically for each mode (Figure 11).

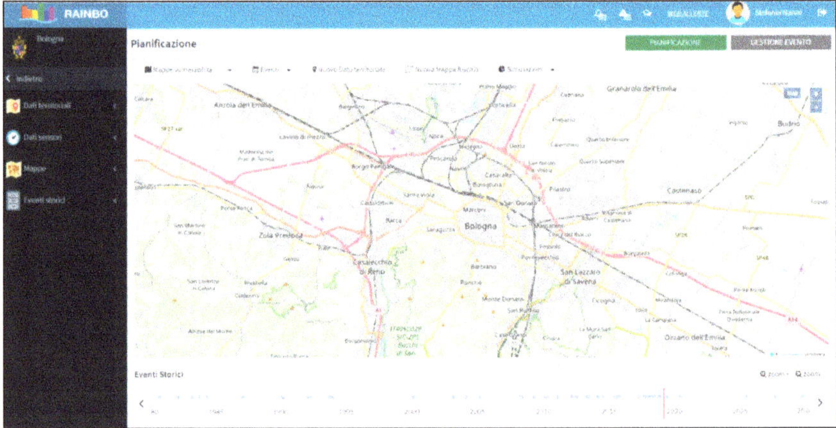

Figure 11. The opening screen of the RainBO platform interface in the planning support mode. On the left bar the thematic maps can be selected and displayed, on the bottom time bar the blue circles, corresponding to historical events, can be clicked and displayed on the map. The two buttons (green and gray) on the upper right corner allows switching from planning support mode to event management mode.

3.2. Territorial Data

To collect territorial data, the data model was defined according to the existing national and regional guidelines for civil protection emergency plans. It requires the mapping of all the critical, sensible, and strategic items, flood maps, river and territory maps (e.g., land use, network infrastructures, buildings, factories, parks).

The territorial database of the RainBO platform contains the existing data listed above: basic regional maps, hazard maps from the Floods Directive, and territorial data at municipality level (e.g., hospitals, schools, emergency areas). The territorial database also contains the maps resulting from the vulnerability module elaborations.

The definition of the spatial data model required an important standardization work to specify its structure, name, and format, to define a standard at the implementation level, both for the RainBO platform and as a reference for data coming from third-party systems.

For instance, the attributes to be provided for schools are the location, the polygon of the area, the number of students, employees and disabled persons, the number of floors of the building, the contacts of the school manager, including the phone numbers. The set of this information, as will be explained below, is necessary both to define the degree of vulnerability of critical sites, but also for the set up of the early warning system.

3.2.1. Hazard, Vulnerability, and Risk Maps

The RainBO platform includes reference hazard maps as:

- hazard maps deriving from the Floods Directive: they represent the potential extent of flooding caused by (natural and artificial) watercourses or by sea, with reference to three scenarios (rare floods (P1—L), infrequent (P2—M) and frequent (P3—H)) represented with three different shades of blue, where the decrease of frequency of flooding corresponds to the decrease in intensity of color. The Floods Directive hazard maps derive from the national hydro-geological management plan (PAI) and they are available for the main basins
- hazards maps from specific hydraulic model/studies for small basins, not included in the Floods Directive maps
- historical events maps that are maps of the flooded areas due to past events. These maps represent an important source of additional information to compare reference maps listed before and real ground effects expected in case of an event.

Concerning these maps, Figure 12a shows the hazard map related to the Parma river in the city of Parma from the Floods Directive. Figure 12b shows the same map integrated with the flooded area during the flood of October 13, 2014. This additional information extends the standard risk area to a secondary one.

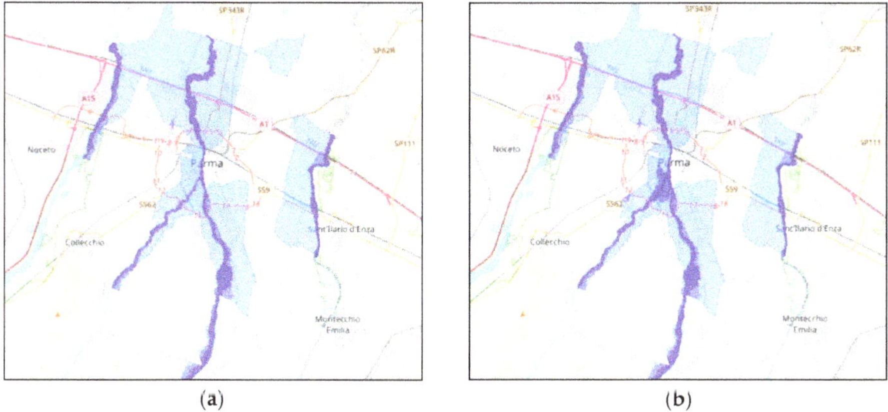

(a)   (b)

Figure 12. (a) Parma River hazard map from Floods Directive; (b) Parma River hazard map integrated with the flooded area of October 2014. The color from light blue to dark blue shows three different level of hydraulic hazard: P1—L (Low probability of floods or extreme event scenarios), P2—M (infrequent floods: return time between 100 and 200 years—medium probability), P3—H (frequent floods: return time between 20 and 50 years—high probability).

Figure 13 shows the hazard map related to the Reno River in the city of Bologna from the Floods Directive, integrated with the additional hazard map for the Ravone creek.

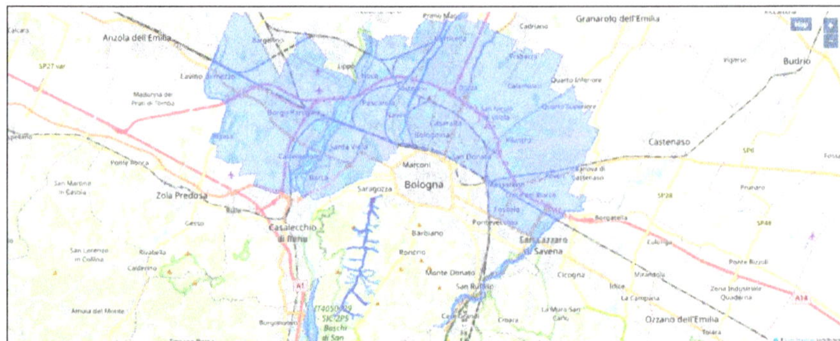

Figure 13. Reno River and Ravone creek hazard map. The color from light blue to dark blue shows three different level of hydraulic hazard: P1—L (low probability of floods or extreme event scenarios), P2—M (infrequent floods: return time between 100 and 200 years—medium probability), P3—H (frequent floods: return time between 20 and 50 years—high probability).

The vulnerability module calculates the degree of vulnerability as described in Section 2.3.3. As a result, 32 vulnerability maps, corresponding to 32 predefined time frames, are produced.

Furthermore, vulnerability maps are a support for territorial planning in prevention and preparedness stages also according to the regional law of December 21, 2017, n.24, which requires the preliminary assessment of the risk, and therefore of the vulnerability, with respect to different types of events, including the hydraulic one, for the purpose of defining the regional urban plan.

In case of an ongoing event, or forecast event, the RainBO platform can select and make available the vulnerability map of the corresponding time frame, providing support for its management.

In more detail, the main difference between the 32 vulnerability maps concerns the distribution of residents, workers, and users of sensible targets during the whole day. By way of example, during the night sensible targets, workplaces, and facilities are usually closed, therefore it is supposed that most of the population are in residential areas. As a consequence, in the vulnerability map referring to this time frame, the urban residential areas are represented in red, whereas during the morning of a working day these areas are green; during the night frame, also sensible targets as schools, gyms, or museums are green areas, whereas during opening hours these areas are red.

Figure 14 shows the vulnerability map of an area of the city of Bologna corresponding to a working day at 4 a.m. The vulnerability level shown in the following pictures is:

- Red for high vulnerability, which means the presence of many people in the grid cell.
- Orange is medium-high vulnerability
- Yellow is medium vulnerability
- Green is low vulnerability, due to the presence of few people in the grid cell

As a result, the RainBO platform provides also the corresponding 32 risk maps. In case other up-to-date or detailed hazard maps (besides the Flood Directive maps) are available, the platform allows us to specify the hazard map on which the risk can be calculated.

As described above, RainBO risk maps based on the vulnerability maps are more detailed and focused on the distribution of sensible and strategic targets and on the time depending presence of citizens in an urban area, with respect to Floods Directive data.

As a result, in case of a forecast event, the risk map, derived from the combination of the hazard map (now from the Floods Directive) and the vulnerability map (selected according to the time frame by the timing of forecast alert between the 32 maps of default) could better support decision-makers to prioritize the warning from the red areas to the green ones.

Moreover, in the planning phase, the capacity of the software to calculate new risk maps from vulnerability maps can support municipal technical and planning offices to evaluate territorial planning choices, as a preventive measure.

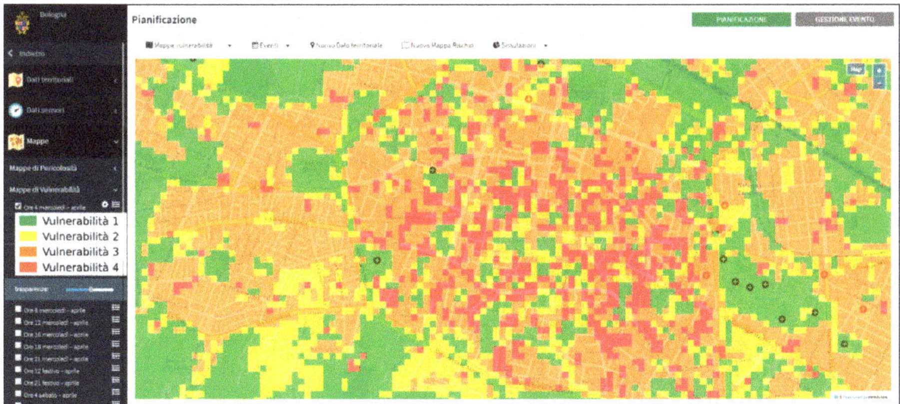

Figure 14. Vulnerability map at 04:00 of a working day. In the color legend, green ("vulnerabilità 1") is low vulnerability, yellow ("vulnerabilità 2") is medium vulnerability, orange ("vulnerabilità 3") is medium-high vulnerability, red ("vulnerabilità 3") is high vulnerability.

Figure 15 shows the risk map at 4:00 a.m. of a working day related to the Ravone creek, produced by using the hazard map of the catchment integrated within the RainBO platform and not included in the Floods Directive.

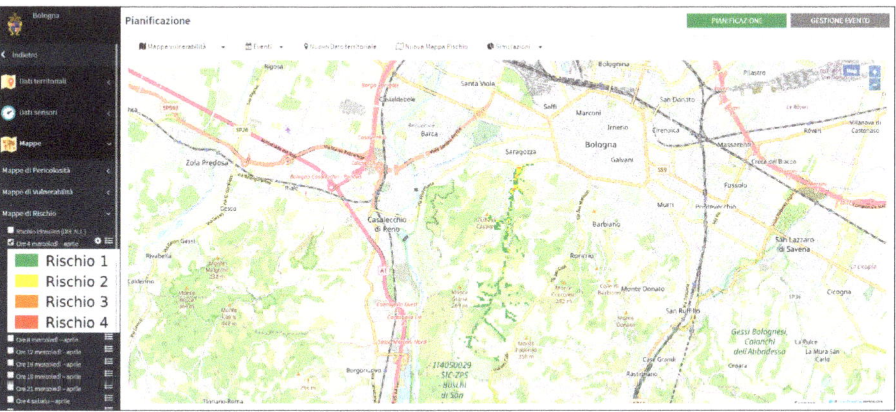

Figure 15. Risk map at 04:00 of a working day referred to the Ravone creek. In the color legend, green ("rischio 1") is low risk, yellow ("rischio2") is a medium risk, orange ("rischio3") is medium-high vulnerability, red ("rischio3") is high risk.

3.2.2. Historical Events

RainBO platform integrates a database where the most significant past flood events on test catchments provided by Arpae are stored. This database is aimed at collecting meaningful information for the management of future events. The time bar allows us to identify and explore the flood events in the planning support mode. The most important information linked to the event (e.g., data catchment of interest, the exceeded threshold of water level gauge, the description of flooded area uploaded on the platform as vectorial data) are highlighted in one view Figures 16 and 17.

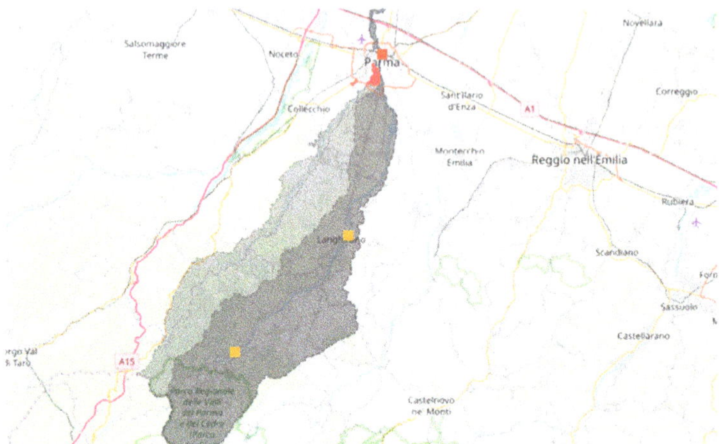

Figure 16. Past flood event on the Parma catchment. The colored squares represent past events where the red (alarm) and orange (pre-alarm) thresholds were exceeded.

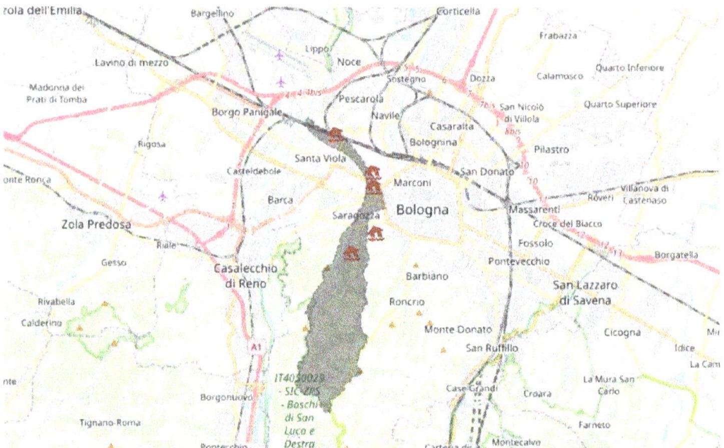

Figure 17. Map of critical points referred to an extreme event occurred on 2015, March 25th in Ravone catchment. The house icons represent the points where damages due to extreme events are recorded.

3.3. Observed, Forecast, and Estimated Data

With regard to the observed data, the RainBO platform collects data from heterogeneous sensors: inclinometers for landslide monitoring, rain gauges, and water level gauges for hydro-pluviometric monitoring of the Arpae network (including the new monitoring stations installed for the RainBO project), inductive-loop detectors for traffic monitoring. The RainBO platform integrates more than 1500 sensors of different types and technologies (Figure 18).

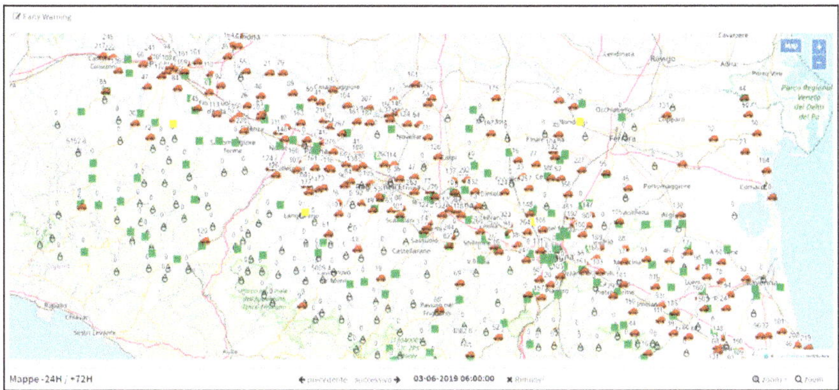

Figure 18. RainBO sensors of observed data: water level gauge (rectangles), rain gauges (drops), and traffic data (cars).

Precipitation forecast maps of the COSMO-LAMI model (Figure 19) are available in a specific section of the platform. Forecasts are summed on 3 h and can be displayed by means of the time bar until the following 72 h.

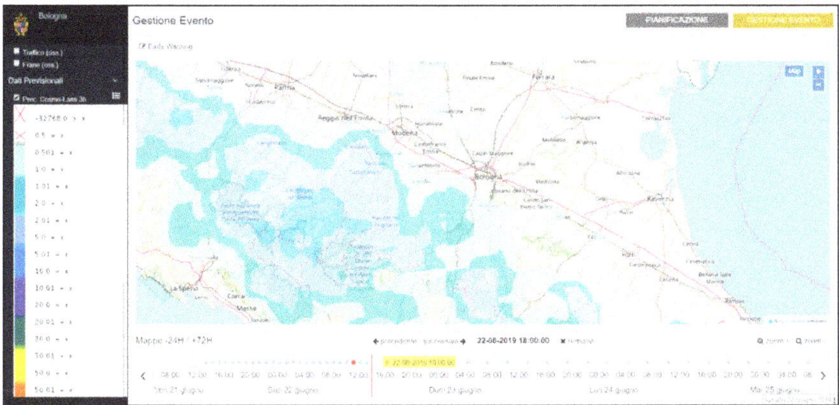

Figure 19. Map of the COSMO-LAMI precipitation forecast. The left-sided color legend represents the quantity of precipitation forecast (unit: mm).

With regard to the estimated data, the system displays maps of radar reflectivity (Figure 20) and radar-estimated precipitation. These two variables are directly related. The hourly maps of these variables can be explored using the time bar for the previous 24 h and it allows the monitoring of ongoing events.

Concerning CMLs, the RainBO platform integrates 73 microwave virtual sensors, corresponding to the midpoint of radio links on the Lepida wireless network (Figure 21).

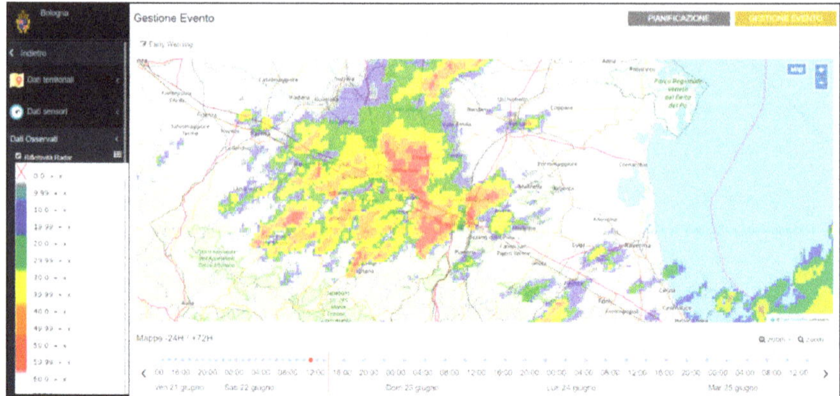

Figure 20. Map of radar reflectivity, a proxy variable to estimate precipitation. The left-sided color legend shows the scale of reflectivity (unit: dBZ).

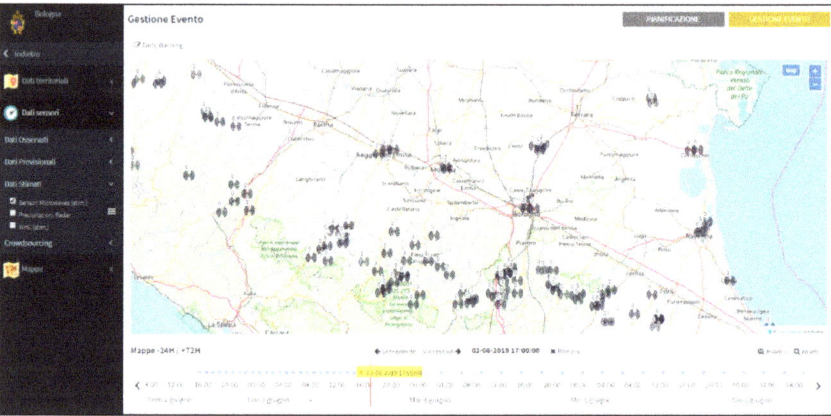

Figure 21. Virtual sensors of precipitation estimated by commercial microwave links (CMLs). The radio wave icons represent the midpoint of radio links on the Lepida wireless network.

3.4. Hydrologic Forecast

RainBO platform provides the forecast of exceeding critical thresholds on the two test catchments, the River Parma for the Parma municipality and the creek Ravone for Bologna municipality. The forecast refers to specific hydraulic sections of the catchments: the culvert entry for Ravone creek, Ponte Verdi, and Ponte Nuovo hydraulic sections for the Parma River.

The forecast covers different time ranges, 6 h for the Parma and 72 h for Ravone, as it is produced by different forecast models, as explained before. In the following, two examples of the RainBO Platform functioning in Event management mode during two extreme events are presented. With regard to the Bologna pilot case, during the afternoon of May 18th, 2019, the Ravone creek exceeded the warning threshold with two water level peaks of 0.82 m at 14 UTC and 0.69 m at 17 UTC.

The water level forecast at the culvert of Ravone issued by the RainBO platform on the morning of May 17th (Figure 22a) matched with the observed values: the exceeding of warning threshold was foreseen with a low probability to exceed the pre-alarm threshold. However, the timing of the event was not correctly forecasted. The hydrological model uses as input the COSMO-LAMI model that forecasted the precipitation peak during the early morning, whereas it occurred about 8 h later. Therefore the hydraulic forecast shows the same time shift. (Figure 22b).

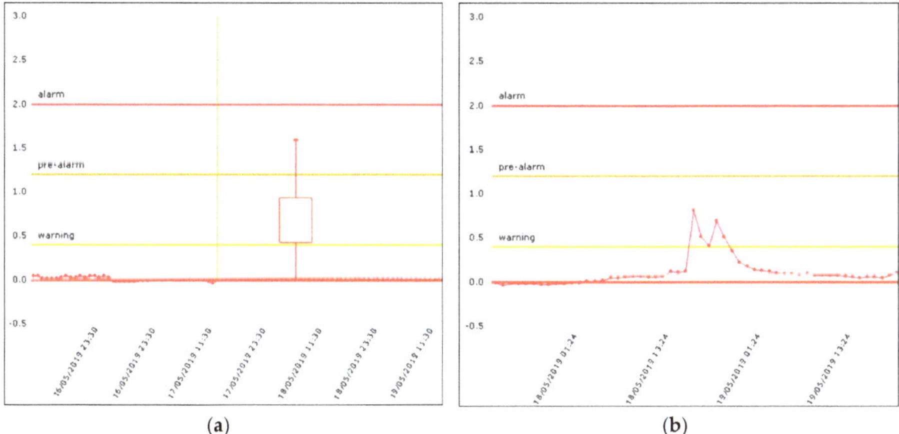

Figure 22. (**a**) Operational forecast of the maximum water level (m) at the culvert entry of Ravone delivered by the platform on the morning of May 17th, 2019. The green line is the moment of the forecast, on the left, the red dots are the observed data, on the right, the boxplot is the forecast distribution. The box of boxplot represents the interval between the 25° and 75° percentile, the tails are the 5° and 95° percentiles; (**b**) water level (m) observed at the culvert entry of Ravone, as displayed on the RainBO platform in the event management mode.

Concerning the Parma case study, in 2017, December 12th, the Parma river basin was interested in an intense and prolonged rainfall event, starting from the 14:00 UTC of the 10th of December a weak rainfall interested the basin till the day after when the rainfall started growing in intensity and space. The peak in terms of mean area rainfall was at midnight on December 11th with about 8 mm/hr (Figure 23).

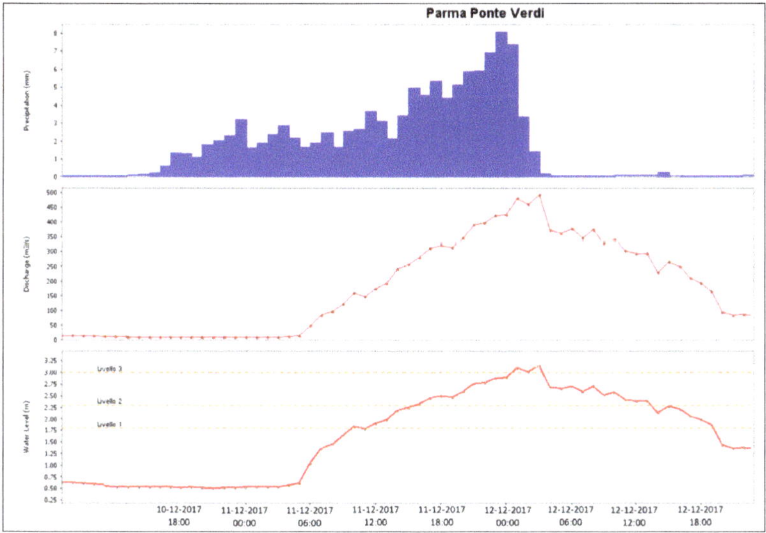

Figure 23. Mean rainfall, discharge, and water level during the December 2017 event.

Looking at the riverside we observed, at 05:00 UTC, the first raise of levels in the river Parma at Ponte Verdi with a level 1 threshold crossing event on the same day at 11:00 UTC. Levels raised continuously in the next hours due to the prolonged rainfall in the basin, reaching the threshold 2 at 16:00 UTC and a peak of 3.14 m (meter above local reference) at 03:00 UTC of the 11th of December, and at the same time, the rainfall in the basin ended with a total amount of mean area rainfall for the entire event of about 116 mm.

The forecast probability generated by the RF model during this event is shown in Figure 24. We can notice that all threshold reached a high probability, between 90–100%, this is obviously related to the high level reached by the river during the event, but we can also consider the forecast lead time provided by the model, in fact, if we look at Table 2, we can find a good consistency between probability and observed effects few hours later, especially for threshold 3 where at 18:00 UTC the probability raise from 12% to 44%, and lately at 20:00 reached a 73%, about 6 h before the effective threshold crossing event, observed at 02:00 UTC (red color).

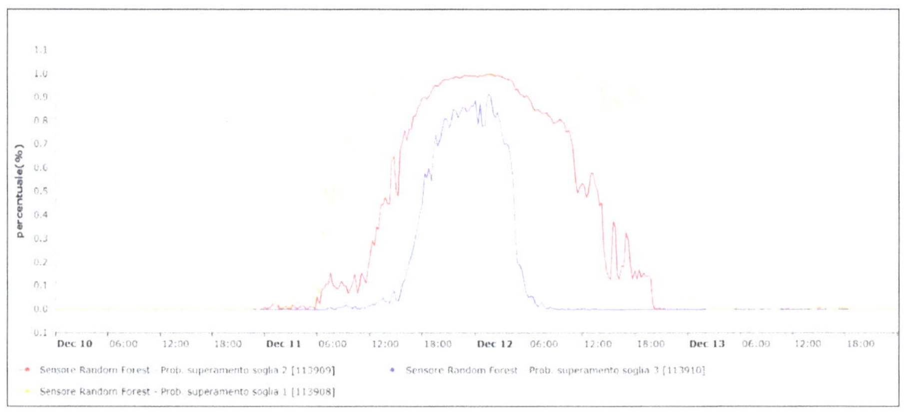

Figure 24. RainBO platform RF results for the Parma River during the event of December 2017.

Table 2. Threshold crossing probability for each level (Green < Threshold 1, Yellow < Threshold 2, Orange < Threshold 3, Red > Threshold 3).

Thr/hrs	07:00 A.M.	08:00 A.M.	09:00 A.M.	10:00 A.M.	12:00 A.M.	02:00 P.M.	04:00 P.M.	06:00 P.M.	08:00 P.M.	10:00 P.M.	00:00 A.M.	02:00 A.M.	03:00 A.M.
1	52	54	67	87	90	96	100	100	100	100	100	100	100
2	11	10	10	11	22	45	76	90	96	99	99	99	99
3	0	0	1	0	1	4	12	44	73	81	89	83	74

3.5. Early Warning Module

An early warning module was developed in the RainBO platform in order to identify the expected scenario as a function of monitoring, forecast and vulnerability maps. A filtering and signaling system allows to select and group essential information useful for the users during extreme events.

In more detail, in the event management mode, two specific icons conceived for early warning are included. The icons are highlighted if a warning occurs. The first icon refers to ongoing events (alert is triggered by observed data of threshold exceeding recorded by water level gauges), whereas the second icon refers to forecasts (alert is triggered by hydrologic models).

In the presence of an alert, a specific dashboard is opened by clicking on the reported event. The dashboard includes only the essential and useful information for the Event management, as the list of catchments where critical thresholds are exceeded (Figure 25), the link to the corresponding charts (Figure 26), the vulnerability map corresponding to the current or forecast scenario.

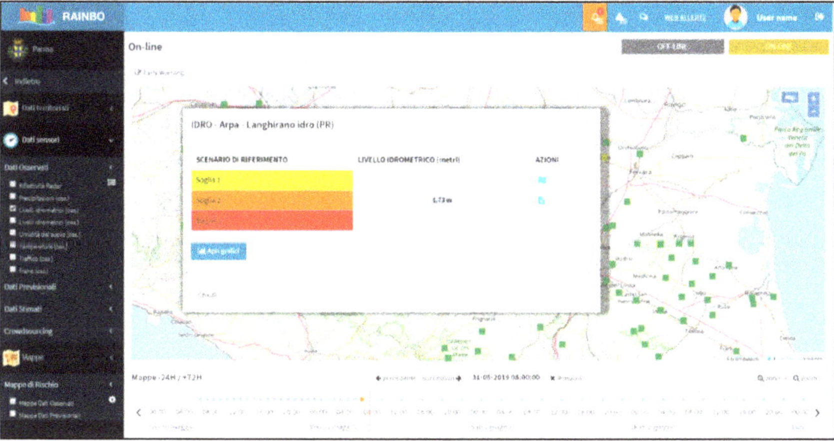

Figure 25. Alert of an ongoing event. The icon on the right side of the platform is highlighted and the dashboard displays the information connected with the event.

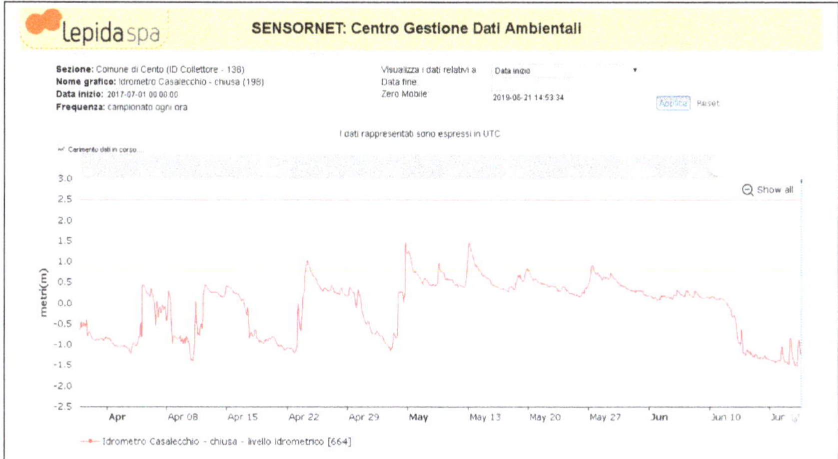

Figure 26. Example of charts of water level gauge and its alarm thresholds.

Furthermore, in the event management mode, a specific tool allows selecting an area of interest where critical targets and sensible targets (with users) are listed, so that reference persons can be identified in order to contact them (Figure 27).

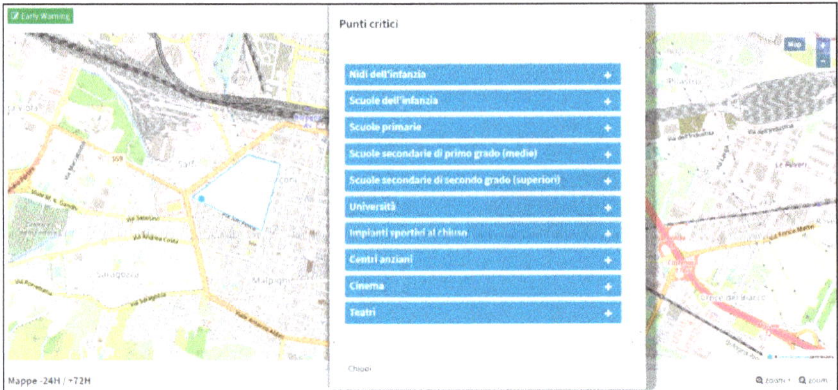

Figure 27. Critical sites list and corresponding information displayed by selecting an area of interest.

3.6. Discussion

In general terms, the flood risk in urban areas can be reduced through the implementation of planning, and monitoring and forecasting phases. These activities are related to different time scales and can be supported through a tool that uses multiple information collected and which can display them easily.

The RainBO platform integrates meteorological and territorial data in time/space and provides a holistic interface for the planning phase and disaster risk reduction, allowing to prevent impacts of flood events and reduce the residual risk during their occurrence.

As flooding in urban areas in Europe is increasing, European projects tackled this issue in urban environments with different features. The EU Interreg project URBAN-PREX aimed at monitoring/forecasting and developing online public EWS for extreme precipitations and pluvial floods in urban areas. URBAN-PREX focused on precipitation monitoring by means of rain gauges located in urban areas and precipitation forecasts.

With respect to this project, the approach of RainBO includes further modules that deal with the planning, monitoring, and forecasting phases. Maps of hazard, vulnerability, and risk are available in the planning phase; data and innovative technology such as CML have been included to monitor precipitation events in the monitoring phase; finally, there is a very clear focus on specific hydrological modeling in the forecasting phase.

Another EU Interreg project named RAINGAIN sought to obtain detailed rainfall data at an urban scale, to predict urban flooding and to implement the use of rainfall and flood data in urban water management practice. In this case, radar technology is used to provide estimates of rainfall in time and space in cities.

RainBO provides estimates of precipitation from different sources such as rain gauges, radar data and CML, and provides a forecast of precipitation from a limited area meteorological model. These data related to precipitation are integrated with other processes (e.g., hydrological modeling) and with territorial data useful before and during the event.

Furthermore, the EU FP7 project STAR-FLOOD aimed to improve the implementation of flood risk strategies in urban areas by building appropriate and resilient flood risk governance. This project stressed the need to have proactive and risk-based systems for forecasting, warning and emergency responses for urban floods that also need to use collaborative approaches. Therefore STAR-FLOOD is more focused on the planning phase, whereas RainBO comprises also a consistent part devoted to forecasting and event management.

Finally, the EU Horizon 2020 project FLOOD-serv aimed to develop a collaborative platform among citizens, public authorities and other stakeholders, which enable alerts in due time to reduce the adverse effects of the floods. In this regard, it is worthy to mention that the RainBO platform is addressed only to decision-makers; decision-makers of a municipality can support both territorial planning and early warning systems through this platform.

RainBO platform has the added value of combining in only one integrated tool territorial data, historical data, real-time monitoring, a crowdsourcing system, hydraulic models for small creeks and medium rivers and the early warning system.

The strengths of the project both in terms of the development of new products and enhancements of existing ones are:

- Vulnerability map calculation module (for hydraulic risk purposes)
- Integration of observed, estimated, and predicted data
- Hydrological simulation model for small basins

Moreover, a first operational level has been defined for a rainfall monitoring system based on CML and it consists of satisfactory reliability in monitoring convective events, mainly during the summer and spring seasons.

4. Conclusions

The RainBO platform is open, centralized, scalable, modular and configurable; it means that it can include various data and models in a different time and space scale, providing an opportunity of replicability and scalability in other geographical and administrative contexts and also for different purposes.

Furthermore, the effectiveness of the RainBO platform has been confirmed by the other municipalities in the Emilia-Romagna Region (i.e., Cento, Comuni della Val Samoggia, Anzola dell'Emilia) keen to demonstrate this platform.

Some limitations of the RainBO platform are to be mentioned:

- The platform is addressed to decision-makers but not to citizens (the citizens can only send information through crowdsourcing)
- The hydrologic models used within the platform have to be calibrated and validated if applied in different river basins, therefore they are not easily replicable in other contexts
- A very large amount of information can be challenging to manage

From a future perspective, the extension of this platform can be foreseen in other urban areas due to the diffusion of open data. As an increasing number of territorial open data is now available, one of the benefits of the platform is also the convergence on the same hub of these data organized in a functional and rational way.

The specific features of this platform allow the upload of datasets provided from international programs (e.g., Copernicus) to supply the lack of local information or infrastructures.

Another future development could be the re-engineering of the platform with a simplified set of information, conceived for the citizens of specific urban areas.

In the future, an effort should be made to enhance the robustness of this platform, i.e., improving the operational functioning of the monitoring and forecasting phases.

Author Contributions: Conceptualization, S.N. and F.T.; methodology, S.N., F.T., R.M., A.A. and P.L.; software, S.N., S.P. and R.M.; supervision, M.F., G.M., L.B. and S.C.; writing—original draft, G.V., S.N., R.M. and S.C.; writing—review and editing, G.V.

Funding: This project has received funding from the European Union's LIFE programme under project No. LIFE15/CCA/IT/00035.

Acknowledgments: This article results from the RainBO project.

Conflicts of Interest: The authors declare no conflict of interest.

References

1. IPCC. *IPCC Special Report on Managing the Risks of Extreme Events and Disasters to Advance Climate Change Adaptation (SREX)*; Cambridge University Press: Cambridge, UK, 2012.
2. EEA. *Climate Change, Impacts and Vulnerability in Europe—An Indicator-Based Report*; EEA: Copenhagen, Denmark, 2017.
3. Blöschl, G.; Hall, J.; Parajka, J.; Perdigão, R.A.; Merz, B.; Arheimer, B.; Čanjevac, I. Changing climate shifts timing of European floods. *Science* **2017**, *357*, 588–590. [CrossRef] [PubMed]
4. Blöschl, G.; Hall, J.; Zivkovic, N. Changing climate both increases and decreases European river floods. *Nature* **2019**, *573*, 108–111. [CrossRef] [PubMed]
5. Alfieri, L.; Burek, P.; Feyen, L.; Forzieri, G. Global warming increases the frequency of river floods in Europe. *Hydrol. Earth Syst. Sci.* **2015**, *19*, 2247–2260. [CrossRef]
6. EU, Directive 2007/60/EC of the European Parliament and of the Council of 23 October 2007 on the Assessment and Management of Flood Risks. 6.11. 2007, pp. 27–34. Available online: https://eur-lex.europa.eu/legal-content/EN/TXT/?uri=CELEX:32007L0060 (accessed on 26 November 2019).
7. Alfieri, L.; Salamon, P.; Pappenberger, F.; Wetterhall, F.; Thielen, J. Operational early warning systems for water-related hazards in Europe. *Environ. Sci. Policy* **2012**, *21*, 35–49. [CrossRef]
8. UNEP. *Early Warning Systems: A State of the Art Analysis and Future Directions*; UNEP: Nairobi, Kenya, 2012.
9. UNFCCC. *Adoption of the Paris Agreement*; UN Framework Convention on Climate Change: Bonn, Germany, 2015.
10. UNISDR. *Sendai Framework for Disaster Risk Reduction 2015–2030*; United Nations: Geneva, Switzerland, 2015.
11. UNDRR. *Global Assessment Report on Disaster Risk Reduction*; United Nations: Geneva, Switzerland, 2019.
12. Golnaraghi, M. *Institutional Partnerships in Multi-Hazard Early Warning Systems*; Springer: Berlin/Heidelberg, Germany, 2008; pp. 1–8.
13. Bouwer, L.; Papyrakis, E.; Poussin, J.; Pfurtscheller, C.; Thieken, A. The costing of measures for natural hazard mitigation in Europe. *Nat. Hazards Rev.* **2014**, *15*, 04014010. [CrossRef]
14. UN. *Transforming Our World: The 2030 Agenda for Sustainable Development*; United Nations: New York, NY, USA, 2015.
15. UNISDR. *Disaster Risk Reduction and Resilience in the 2030 Agenda for Sustainable Development*; United Nations: Geneva, Switzerland, 2015.
16. Hapuarachchi, H.A.P.; Wang, Q.J.; Pagano, T.C. A review of advances in flash flood forecasting. *Hydrol. Process.* **2011**, *25*, 2771–2784. [CrossRef]
17. Jain, S.K.; Mani, P.; Jain, S.K.; Prakash, P.; Singh, V.P.; Tullos, D.; Kumar, S.; Agarwal, S.P.; Dimri, A.P.A. Brief review of flood forecasting techniques and their applications. *Int. J. River Basin Manag.* **2018**, *16*, 329–344. [CrossRef]
18. Acosta-Coll, M.; Ballester-Merelo, F.; Martinez-Peiró, M.; De la Hoz-Franco, E. Real-Time Early Warning System Design for Pluvial Flash Floods-A Review. *Sensors* **2018**, *18*, 2255. [CrossRef] [PubMed]
19. EU. *Establishing the Urban Agenda for the EU 'Pact of Amsterdam'. Agreed at the Informal Meeting of EU Ministers Responsible for Urban Matters on 30 May 2016 in Amsterdam, The Netherlands*; EU: Brussels, Belgium, 2016.
20. UN. *New Urban Agenda*; United Nations: Geneva, Switzerland, 2017.
21. URBAN-PREX–Interreg Project. Available online: http://www.urban-prex.org/ (accessed on 18 November 2019).
22. RAINGAIN–Interreg Project. Available online: http://www.raingain.eu/en/interreg-ivb-north-west-europe-programme-1 (accessed on 18 November 2019).
23. STAR-FLOOD FP7 Project. Available online: https://www.starflood.eu/ (accessed on 18 November 2019).
24. FLOOD-SERV Horizon 2020 Project. Available online: http://www.floodserv-project.eu/ (accessed on 18 November 2019).
25. Climatic Tables of Emilia-Romagna. Available online: https://www.arpae.it/dettaglio_generale.asp?id=4143&idlivello=1591 (accessed on 26 November 2019).
26. Grazzini, F.; Dottori, F.; Di Lorenzo, M.; Spisni, A.; Tomei, F. *Nubifragi e Rischio Idraulico Nella Collina Bolognese: Il Caso Studio del Torrente Ravone*; Public Report of Arpae; Arpae: Bologna, Italy, 2013.
27. Bracaloni, A. Analisi del Rischio Idraulico in Ambiente Urbano: Il Caso del Torrente Ravone a Bologna. Master's Thesis, University of Bologna, Bologna, Italy, 10 March 2016.

28. Dottori, F.; Grazzini, F.; Di Lorenzo, M.; Spisni, A.; Tomei, F. Analysis of flash flood scenarios in an urbanized catchment using a two-dimensional hydraulic model. In Proceedings of the International Association of Hydrological Sciences, Bologna, Italy, 4–6 June 2014.

29. Technical Reports of Hydro-Meteo Events in Emilia-Romagna. Available online: https://www.arpae.it/documenti_find.asp?parolachiave=sim_rapportoradar&cerca=si&idlivello=64 (accessed on 18 October 2019).

30. Nanni, S.; Mazzini, G. Advanced Infrastructure for Environmental Monitoring and Early-Warning System Integration, Sensor Networks. In *International Conference on Sensor Networks*; Benavente-Peces, C., Cam-Winget, N., Fleury, E., Ahrens, A., Eds.; Springer Nature: Basel, Switzerland, 2019.

31. Numerical Weather Forecasts of Emilia-Romagna. Open Data of ARPAE. Available online: https://dati.arpae.it/fa_IR/dataset/previsioni-meteorologiche-numeriche-emilia-romagna (accessed on 18 October 2019).

32. Lanza, L.; Stagi, L. Certified accuracy of rainfall data as a standard requirement in scientific investigations. *Adv. Geosci.* **2008**, *16*, 43–48. [CrossRef]

33. Porcù, F.; Milani, L.; Petracca, M. On the uncertainties in validating satellite instantaneous rainfall estimates with raingauge operational network. *Atmos. Res.* **2014**, *144*, 73–81. [CrossRef]

34. Casicci, L.; Ioiò, C.; Pecora, S. An operational system for the Po flood forecasting in Italy. In Proceedings of the 7th International Conference on Hydroinformatics, Nice, France, 4–8 September 2006.

35. Zinevich, A.; Messer, H.; Alpert, P. Frontal Rainfall Observation by a Commercial Microwave Communication Network. *J. Appl. Meteorol. Climatol.* **2009**, *48*, 1317–1334. [CrossRef]

36. Doumounia, A.; Gosset, M. Rainfall monitoring based on microwave links from cellular telecommunication networks: First results from a West African test bed. *Geophys. Res Lett.* **2014**, *41*, 6016–6022. [CrossRef]

37. Overeem, A.; Leijnse, H.; Uijlenhoet, R. Country-wide rainfall maps from cellular communication networks. *Proc. Natl. Acad. Sci. USA* **2013**, *110*, 2741–2745. [CrossRef] [PubMed]

38. Radar Meteo. Open Data of ARPAE. Available online: https://dati.arpae.it/dataset/radar-meteo (accessed on 18 October 2019).

39. Rmap Participatory Monitoring and Exchange System. Available online: http://rmap.cc/ (accessed on 18 October 2019).

40. Marletto, V.; Ventura, F.; Fontana, G.; Tomei, F. Wheat growth simulation and yield prediction with seasonal forecasts and a numerical model. *Agric. For. Meteorol.* **2007**, *147*, 71–79. [CrossRef]

41. Tomei, F.; Antolini, G.; Bittelli, M.; Marletto, V.; Pasquali, A.; Van Soetendael, M. Validazione del modello di bilancio idrico CRITERIA. *Ital. J. Agrometeorol.* **2007**, *1*, 66–67.

42. Bittelli, M.; Tomei, F.; Pistocchi, A.; Flury, M.; Boll, J.; Brooks, E.S.; Antolini, G. Development and testing of a physically based, three-dimensional model of surface and subsurface hydrology. *Adv. Water Resour.* **2010**, *33*, 106–122. [CrossRef]

43. CRITERIA-3D. Open Source Code of ARPAE. Available online: https://github.com/ARPA-SIMC/CRITERIA3D/wiki/CRITERIA3D (accessed on 18 October 2019).

44. Breiman, L. Random Forests. *Mach. Learn.* **2001**, *45*, 5. [CrossRef]

45. RAINLINK Retrieval Algorithm Open Source Code. Available online: https://github.com/overeem11/RAINLINK (accessed on 30 October 2019).

46. Crowdsourcing App Rmap4RainBO. Available online: https://partecipa.RainBOlife.eu (accessed on 30 October 2019).

47. RainBO Platform. Available online: http://webgis.rainbolife.eu/prototipo/01-tempo-di-pace.html (accessed on 18 October 2019).

Article

Land Use Changes in a Peri-Urban Area and Consequences on the Urban Heat Island

Marianna Nardino [1,*] **and Nicola Laruccia** [2]

[1] National Research Council-Institute for Bioeconomy (CNR-IBE), Via Gobetti 101, 40129 Bologna, Italy
[2] Emilia Romagna Region-Servizio Programmazione e Sviluppo locale Integrato, Viale della Fiera 8, 40127 Bologna, Italy; nicola.laruccia@regione.emilia-romagna.it
[*] Correspondence: Marianna.Nardino@ibe.cnr.it; Tel.: +39-051-639-9001

Received: 17 October 2019; Accepted: 12 November 2019; Published: 13 November 2019

Abstract: The effect of urbanization on microclimatic conditions is known as "urban heat islands". In comparison with surrounding rural areas, urban climate is characterized by higher mean temperature, especially during heat waves and during nights. This results in a higher energy requirement for air conditioning in buildings and in a greater bioclimatic discomfort for urban populations. The reasons of this phenomena are ascribable principally to the increase of solar radiation storage and to the decrease of dissipation of water by evapotranspiration in urban environment respect to rural ones. The aim of this paper is to give a quantification of the air temperature increase due to an urbanization process. This quantification is conducted by comparing surface energy balance (incoming and outcoming radiation and turbulent fluxes) in urbanized area versus rural areas. This quantitative approach will be validated using a fluidodynamic model (Envi-Met) in a case study area representative of one among the various regional models of urban area growth. In particular, the model of expansion of small towns around big cities (2003–2008 land use changes) of a plain near-urban area in the Po Valley region (Italy) was used.

Keywords: urban heat island; urbanization; urban surface energy balance; fluidodynamic modeling; Envi-Met

1. Introduction

The urbanization effect refers to a general increase in population and in the amount of industrialization of a settlement, due principally to the increase, in number and extent, of cities and the movement of people from rural to urban areas. The "urban sprawl" is used to define the increase in spatial scale or in the peripheral area of the cities. At the present, more than half of the global population lives in cities and cities themselves are growing to unprecedented sizes [1].

The high density of population and the consequent use of primary resources by urban residents, especially in the North Hemisphere, make cities and their inhabitants key drivers of global environmental changes [2,3].

Land-use and land-cover changes are recognized as causes of local, regional and global warming: The urban areas are the major sources of anthropogenic carbon dioxide emissions from the burning of fossil fuels for heating, from industrial processes, from transportation of people, etc. [4,5]. The main effects of land use changes (from rural to urban) can be found on the surface energy and water balances changes. The partitioning of sensible and latent heat fluxes is a function of varying soil water content and vegetation cover [6,7]. Water balance is strongly influenced by the soil sealing: Urbanization makes the surface permanently covered by impermeable artificial material (e.g. asphalt and concrete), for example through buildings and roads.

An increase of settlement areas over time is defined as land take, also referred to as land consumption. This process includes the development of scattered settlements in rural areas, the expansion of urban areas around an urban nucleus (including urban sprawl), and the conversion of land within an urban area (densification) [8,9]. Depending on local circumstances, a greater or smaller part of the land take will result in actual soil sealing. Urban sprawl can be defined as the unplanned incremental urban development, characterised by a low-density mix of land uses on the urban fringe. It is important to underline that even planned urban development may result in land take and soil sealing.

Soil sealing and realization of buildings on it alter the surface energy balance with two different main mechanisms:

(1) Soil sealing reduces the vegetation cover and prevents the storage of rainwater and consequently the amount of water stored into the soil. Consequently, water available for evapotranspiration processes is much lower than in a natural surface. It follows that the latent heat dissipated from urban surfaces is close to zero and the amount of advanced energy will be available for other processes usually related to an increase of the thermal field [10].

(2) Materials used for the buildings may, according to their specific thermal and optical properties, store energy and radiation in form of heat when the radiation budget (R_n) is positive and release energy when R_n is negative.

The combination of these two different mechanisms results in a different thermal trend that is observed in the built environment compared to the surrounding rural areas.

One of the most prominent feature of the urban climate is the urban heat island (UHI) effect, which is strongly tied to the geometry and dimensions of building, land use patterns, vegetation cover and the intensity of the anthropogenic heat release [11,12]. The UHI makes the city warmer than its rural surroundings and it is stronger at night than during the day. This effect also decreases with increasing wind speed and cloud cover and it's less pronounced in summer and winter [11].

On average, urban temperatures may be 1–3 °C warmer than surrounding urban environment [6].

The aim of this work is to quantify the change in temperature range that can be expected on the basis of different energy balances resulting from the transformation of rural areas in urban residential or productive areas. The changes in the soil use (from natural surface to no permeable surface) returns an environment where the storage of heat during the day by building materials is released during the night, increasing the urban heat island. This effect is the main subject of the present work.

2. Materials and Methods

2.1. Energy Balance Method

The Earth surface radiation balance is described by the following equation:

$$R_n = (Sw_{in} - Sw_{out}) + (Lw_{in} - Lw_{out}) \tag{1}$$

where R_n is the net radiation, Sw_{in} is the incoming shortwave (visible) radiation, Sw_{out} is the outcoming shortwave radiation, Lw_{in} is the incoming longwave (infrared) radiation and Lw_{out} is the outcoming longwave radiation.

The term $(Sw_{in} - Sw_{out})$ is the net shortwave radiation and it can be also described as $(1 - \alpha)Sw_{in}$, where α is the surface albedo used to quantify the solar radiation reflected by a surface. Albedo is a characteristic of the specific surface and depends on the optical characteristics of the reflecting surface.

Lw_{in} and Lw_{out} depend on the atmosphere and surface temperature following the black body equation:

$$Lw = \varepsilon \, \sigma \, T^4 \tag{2}$$

where ε is the body infrared emissivity, σ is the Stefan–Boltzmann constant (5.67 10^{-8} W m^{-2} K^{-4}) and T is the body temperature expressed in K.

The net radiation given by (1) is then utilized to the partitioning processes at the surface, following the surface energy balance equation [12]:

$$R_n = H + LE + G \tag{3}$$

where H is the sensible heat flux, LE is the latent heat flux and G is the ground heat flux.

Usually the greatest part of the net radiation is used for sensible and evapotranspiration processes while the G term represent only the 10% of the total available energy (R_n).

To estimate the lower dissipation of latent heat flux due to the transformation of cultivated land in urbanized area, it is necessary to quantify the evapotranspiration (ET) before and after the transformation. The ET of a cultivated area depends on climatic characteristics of the site and on eventual water contributions from irrigation, on vegetable cover and on soil type. As vegetable cover, for this study, it was assumed the cultivation of winter wheat, not irrigated and growing in optimal pedological and climatic conditions to ensure a good water supply during vegetation season (November to June). Wheat is one of the more widespread dry crops in the Emilia-Romagna Region plain.

From a sealed surface, the evaporation of rain water can be considered negligible. On the other hand, from a winter wheat crop, in the case of optimal soil and climatic conditions, about 700 lm^{-2} of water per year changes from liquid into vapor phase.

A very simplified, but amply adopted method to calculate potential evapotranspiration (ETc) of a crop consists on multiplying reference evapotranspiration (ET$_0$) per crop cultural coefficient (Kc):

$$ETc = Kc \, ET_0 \tag{4}$$

In this study ET$_0$ was calculated according to Hargreaves method [13] using the data of a meteorological station relatively close to the case study area, San Pietro Capofiume (44.648993 °N, 11.650055 °E, 11 m a.s.l.). The values of the cultural coefficient Kc for wheat were adopted following the [14] suggestions and reasonable Kc's for spontaneous grasses, growing after crop harvesting, were chosen.

If a land use change occurs, such as a significant urbanization, the terms of this budget equation change significantly and other terms, depending on different processes, must be taken into account. The Equation (3) of the surface energy balance can therefore be rewritten in this way:

$$R_n = H + LE + G + Qs \tag{5}$$

where Qs is the storage heat flux, which is strongly dependent on the ratio between green and sealed areas and on the building geometrical, optical and thermal characteristics.

Oke [6] proposed these equations for Qs:

$$\text{day } Qs = (0.20\lambda v + 0.33\lambda p) \, R_n + 3\lambda v + 24\lambda p \tag{6}$$

$$\text{night } Qs = (0.54\lambda v + 0.90\lambda p) \, R_n \tag{7}$$

where λv is the green area fraction and λp is the building area fraction.

The reconstruction of the thermal change given by the lower amount of evapotranspiration of a sealed surface was performed following the Fick law [12]:

$$\Delta T = Q \, Dz \, k^{-1} \tag{8}$$

where

Q = Latent Heat Flux + Storage Heat Flux (W m^{-2})
Dz = reference height (2 m)
k = air thermal conductivity (0.026 W m^{-1} K^{-1})

2.2. Fluidodynamic Simulation (Envi-Met): The Case Study

ENVI-met [15] is a three-dimensional non-hydrostatic microclimate model designed to simulate the surface–plant–air interactions within daily cycle in urban environment with a typical resolution of 0.5 to 10 m in space and 10 s in time.

The model has been widely used in many previous studies to simulated flow around and between buildings, exchange processes of heat and vapor at the ground surface and at the walls, turbulence exchange of vegetation and vegetation parameters, bioclimatology, and particle dispersion [16]. The model was validated as reported in [17].

In order to run the model, the detailed data on soil characteristics, buildings, vegetation, and initial atmospheric conditions for the area of interest were inserted.

After this input phase, the model is ready to run and the desired variables have been selected and saved into the output files [16].

ENVI-met can be used for several studies to test various urban canyon aspects as well as ratios and orientation effects on outdoor thermal comfort, the role of vegetation in the mitigation of the urban heat island effect, and other factors.

3. Results and Discussion

Hourly measurements of the surface radiation balance components (Sw_{in}, Sw_{out}, Lw_{in}, Lw_{out}) are available in stations relatively close to the case study area. San Pietro Capofiume was used as the representative station for rural open space. In this site CNR-IBE performed radiation components measurements for the period 1 January 2002–31 December 2003.

Bologna urban ARPAe (HydroMeteorological Service of the Emilia-Romagna Regional Agency for Environmental Protection) station (44.500754 °N, 11.328789 °E, 78 m a.s.l.) has been considered as the representative station for sealed conditions (measurements period 1 January 2006–31 December 2009). Although this data represents only few years, furthermore non-coincident, from Table 1 it can be inferred that the values of the two stations are substantially super imposable with regard to the Sw_{in} and the Sw_{out} values, while the differences between the values of Lw_{in} and Lw_{out} are attributable to the different surface types (rural and urban), principal aim of this paper.

Table 1. Annual mean values of the surface radiation balance components for rural (San Pietro Capofiume) and urban (Bologna urban) sites.

Station	Sw_{in} (W m^{-2})	Sw_{out} (W m^{-2})	Lw_{in} (W m^{-2})	Lw_{out} (W m^{-2})
Bologna urban	148	26	325	396
San Pietro Capofiume	156	27	305	361

Throughout the hourly measurements, in Table 2 are reported the diurnal and nocturnal R_n values obtained for the two stations and for each month of the year.

Table 2. Annual mean of net radiation values in nocturnal and diurnal hours for rural (San Pietro Capofiume) and urban (Bologna urban) sites.

Month	Nocturnal R_n (W m^{-2})		Diurnal R_n (W m^{-2})	
	San P. Capof.	Bologna Urban	San P. Capof.	Bologna Urban
January	−17	−39	78	52
February	−31	−57	138	124
March	−34	−63	253	201
April	−28	−62	246	278
May	−24	−52	288	280
June	−36	−60	287	285
July	−51	−71	279	293
August	−39	−67	255	239
September	−28	−62	198	189
October	−26	−55	157	141
November	−18	−44	77	71
December	−19	−47	68	48

Table 3 shows the values of mean monthly latent heat flux which would be missed in case of transformation of a wheat field into a sealed surface.

Table 3. Monthly mean values of reference evapotranspiration (ET$_0$), crop coefficient (Kc), crop potential evapotranspiration (ETc) and corresponding latent heat flux.

Month	ET$_0$ (mm)	Kc (Wheat and Spon. Grass)	ETc (mm)	Latent Heat Flux (W m^{-2})
January	19.8	0.4	7.9	6.9
February	36.3	0.5	18.2	17.4
March	66.9	0.7	46.8	40.8
April	109.1	0.9	98.2	88.4
May	153.1	1.1	168.4	146.7
June	178.4	1.0	178.4	160.7
July	204.0	0.2	40.8	35.6
August	166.1	0.3	49.8	43.4
September	110.8	0.5	55.4	49.9
October	64.3	0.4	25.7	22.4
November	31.0	0.4	12.4	11.2
December	19.1	0.4	7.7	6.7

The area chosen as a case study is located in San Giovanni Persiceto (44.6283 °N, 11.1992 °E, 21 m a.s.l.), a small town close to Bologna city. The urbanization occurred in this area from 2003 to 2008 was estimated comparing aerial images (orthophoto by AGEA for 2008 and QuickBird satellite orthoimages by Digital Globe™–Telespazio for 2003) (Figure 1). The built-up surface varied from 5.5% in the 2003 to 30% in 2008 and the natural cover varied from 94.5% in 2003 to 70% in 2008. Inserting these values as λv and λp in (6) and (7), the difference of the storage heat flux Qs between 2003 and 2008 was obtained, separately for night and daytime hours. The annual difference in the cumulated storage heat flux is quite similar between night and day (+80 W m^{-2} for the diurnal hours and −84 W m^{-2} for nocturnal ones) which means that all the heat stored during the day by urban materials is re-emitted into the atmosphere during the night. On a monthly basis, this effect is visible in the winter months and during months with high energy availability (May, June, and July) the day energy accumulation is greater than the night emission (Table 4).

Figure 1. QuickBird satellite orthoimages (®Digital Globe™–Telespazio, 2003) (left) and aerial orthophotographs (AGEA) for 2008 (right).

Table 4. Monthly mean values of storage heat flux during diurnal and nocturnal hours for 2003 and 2008 and relative differences for each year.

	Diurnal Qs (W m^{-2})			Nocturnal Qs (W m^{-2})		
Month	2003	2008	Difference	2003	2008	Difference
January	23.4	23.1	−0.3	−10.1	−15.2	−5.1
February	37.9	42.4	4.4	−18.1	−25.1	−7.0
March	66.0	62.9	−3.2	−20.0	−27.8	−7.8
April	64.5	83.6	19.2	−16.6	−24.6	−8.0
May	74.7	84.0	9.3	−14.5	−21.1	−6.6
June	74.3	85.4	11.1	−20.6	−27.8	−7.1
July	72.6	87.5	14.9	−29.4	−37.1	−7.7
August	66.7	73.0	6.3	−22.5	−30.6	−8.1
September	52.7	59.9	7.2	−16.6	−24.6	−8.0
October	42.5	46.9	4.4	−15.3	−22.3	−7.0
November	22.9	28.3	5.4	−10.9	−16.6	−5.8
December	20.8	22.1	1.3	−11.7	−17.8	−6.2

Since the evapotranspiration processes occur only during diurnal hours, the difference between the two reference years in terms of latent heat flux is assigned entirely to daylight hours. The greater energy availability for 2008 for heat transfer processes was obtained summing the diurnal latent heat flux differences (proportioned according to the fractions of cultivated and urban areas in 2003 and 2008) with the differences in diurnal storage heat flux. Nocturnal thermal variation depends only on variations in storage heat flux reported in Table 4.

Table 5 illustrates the total monthly differences (2008 vs 2003) in energy budget for diurnal hours and the corresponding diurnal, nocturnal and mean temperature variations calculated according to (8).

The largest increases in temperature due to an urbanization process occur during the summer months, when the energy available for exchange processes is greater.

This approach is an estimate of changes in energy balance and this leads at uncertainties, but in general the results show an annual mean increment in air temperature of 0.36 °C due to this urbanization process. Again, the single months show the differences due to energy availability, giving greater values of air temperature increment during summer months.

Table 5. Total difference (2008 vs 2003) in energy budget for diurnal hours and corresponding diurnal thermal variation (ΔT day), nocturnal thermal variation (ΔT night) and mean thermal variation (ΔT mean).

Month	Δ LE + Diurnal Δ Qs (W m^{-2})	ΔT day (°C)	ΔT night (°C)	ΔT mean (°C)
January	2.1	0.11	0.27	0.19
Febrary	0.1	0.00	0.37	0.19
March	13.7	0.71	0.41	0.56
April	3.8	0.20	0.42	0.31
May	28.7	1.49	0.34	0.92
June	30.6	1.59	0.37	0.98
July	−5.7	−0.29	0.40	0.05
August	5.0	0.26	0.42	0.34
September	5.7	0.30	0.42	0.36
October	1.4	0.07	0.36	0.22
November	−2.5	−0.13	0.30	0.09
December	0.4	0.02	0.32	0.17

The same area was studied with the Envi-Met fluidodynamic model in order to have a modeling feedback of estimation method based on the surface energy balance.

The two input areas (2003 and 2008), shown in Figure 2, were inserted through the program ENVI-met Eddie, taking into account the geographical location, building dimensions, land use patterns. The vegetation was assumed to be winter wheat. The domain model consisted of a 140 × 140 × 20 grid with a spatial resolution of 5 m × 5 m × 2 m, resulting in a horizontal area of 700 × 700 m^2 with a 40 m vertical extent.

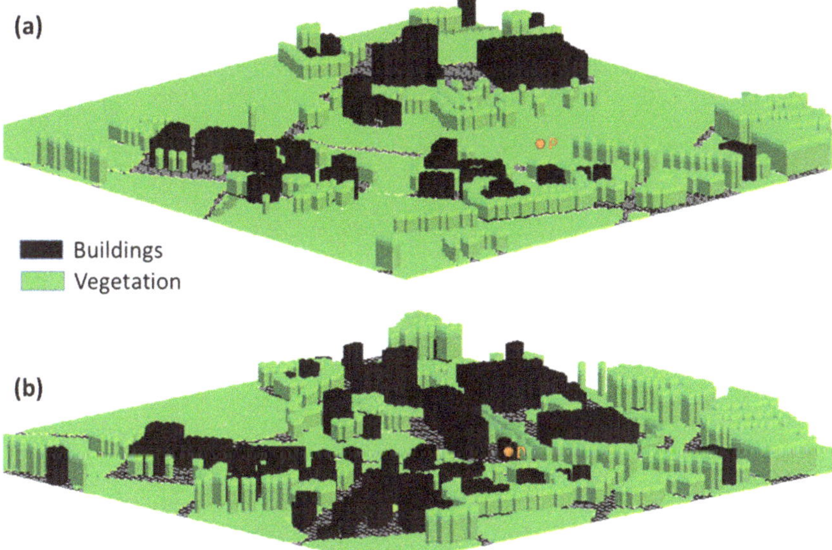

Figure 2. Input maps inserted into Envi-met model for 2003 (**a**) and 2008 (**b**).

The case study reported here focused on the simulation of a typical hot day whose values of meteorological variables were obtained considering the data of the nearest ARPAe weather station (San Pietro Capofiume).

The simulation started at 9:00 a.m. and lasted for 24 hours. The configuration file contained the atmospheric values:

- Speed and wind direction at 10 m: 1.4 m s^{-1}, 90 °;
- Surface roughness length (z0): 0.1 m;
- Air temperature: 301.7 K
- Specific humidity at 2500 m: 7 g water/kg air
- Relative humidity at 2 m: 50%

As first results the potential temperature (the temperature that a sample of air attains if reduced to a pressure of 1000 millibars without receiving or losing heat) to the environment difference between the two simulation years (2003 and 2008) were plotted (Figure 3) at 14.00 p.m. The urbanization process carried out from 2003 to 2008 led to a warming of some areas especially those in which it has had a greater overbuilding. In these areas the 2008 potential temperature at 1.6 m height is about 0.4–0.6 °C higher than 2003.

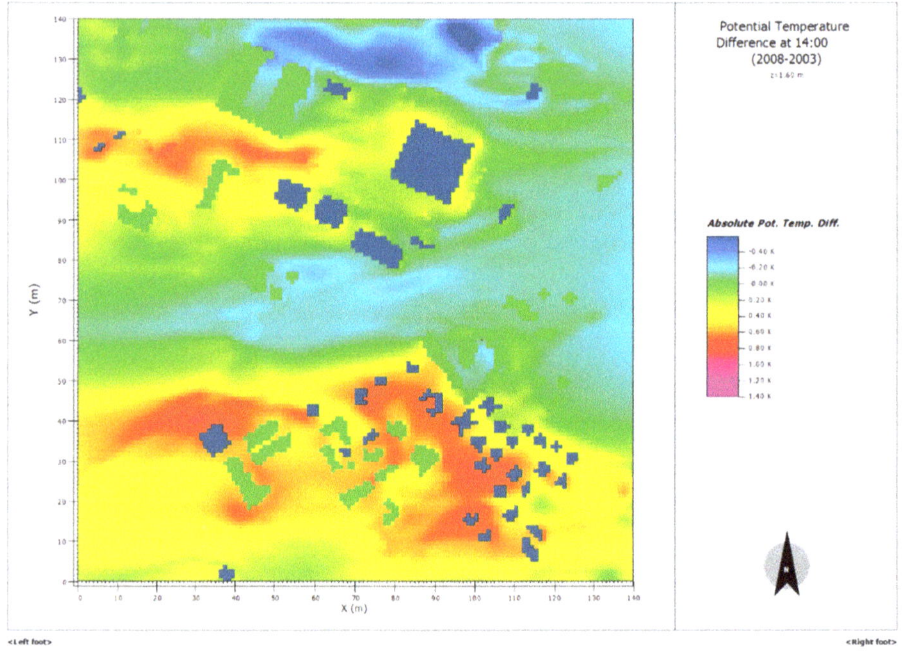

Figure 3. Potential temperature difference (2008–2003) at 14:00 pm at 1.60 m height for the considered case study.

During the night (Figure 4 at 2:00 a.m.) the potential temperature difference between the two years reaches 1 °C. This is typical of the urban heat islands, where the greatest effect is recorded during the night, so substantial increase occurs in daily minimum temperatures [18]. In addition, the total area tends to be warmer in 2008 than in 2003, and this effect is more evident during the night hours than daytime.

As a matter of fact, if the air temperature trends is plotted for a strategic point of the considered area (a point which was completely rural in 2003 and that became urban in 2008, marked as P in Figure 2) it can be seen that during the day air temperature tends to be the same for the two years and during the night the change turns out to be even 1 °C (Figure 5). The daily mean of air temperature differences is 0.35 °C that, looking at Table 5 for June month, is comparable with the night values, but not with the daytime ones. Probably the high available energy during this month means that further exchange processes come into play, that the balance method, used in this work, does not account for. A better diurnal values estimate is surely necessary, even if during August and September (months similar to June as far as concerns air temperature values) the obtained delta temperature (0.34 °C and 0.36 °C) is very close to the model result.

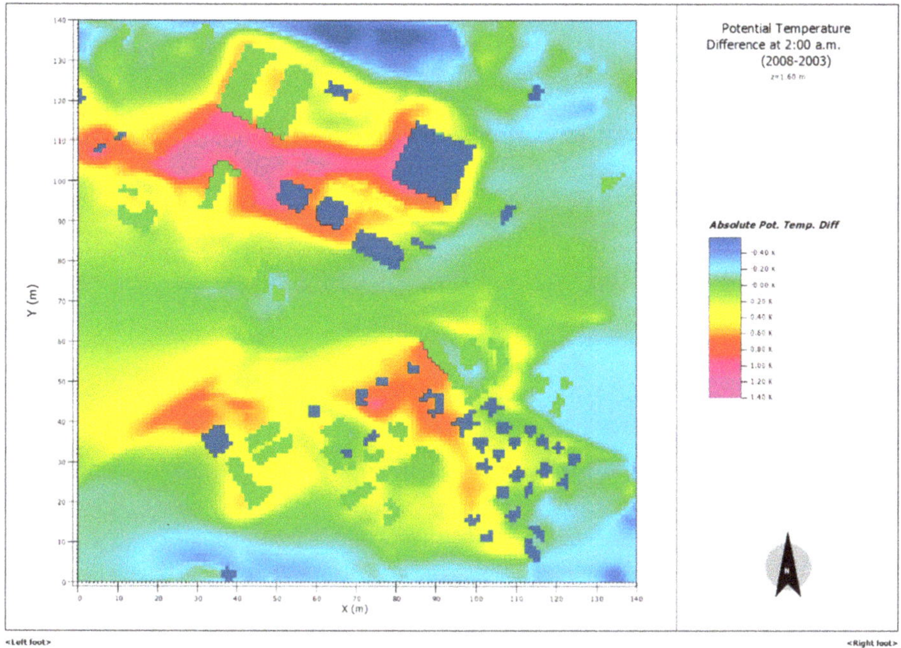

Figure 4. Potential temperature difference (2008–2003) at 2:00 a.m. at 1.60 m height for the considered case study.

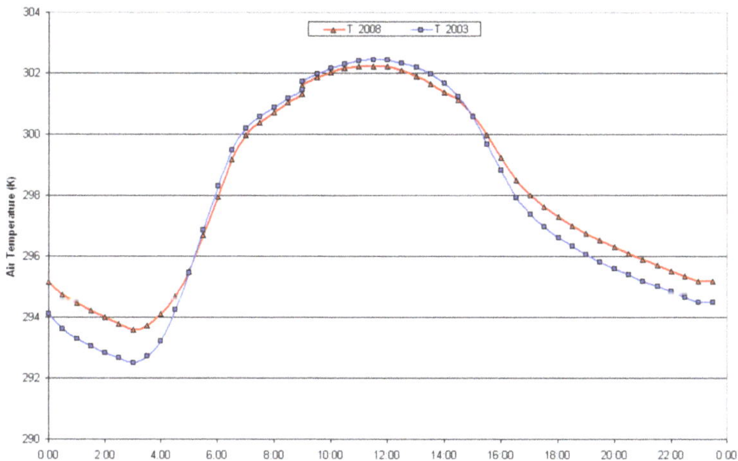

Figure 5. Air temperature for 2003 and 2008 in the P point marked in Figure 2.

June is the month in which the budget method assigned greater evapotranspiration values, but the fluidodynamic model does not consider specifically crop type and evapotranspiration needs. For this reason, the strongest disagreement results especially in this month.

On the other hand, the model results are in strong agreement with the annual average increase obtained with the method of the energy budget (0.36 °C).

This suggests a profitable use of this methodology to estimate the increase in air temperature due to urbanization processes on a regional scale.

4. Conclusions

The effect of an urbanization in a small town shows the consequences of the well-defined and studied "urban heat island": The impact of such kind of urbanization leads an increment in the air temperature, and therefore a strengthening of the urban heat island, close to 1 °C.

The energy balance method is verified and supported by a fluidodynamic model to have the right data interpretation. The idea is to develop a simple methodology to consider the urbanization effects in terms of temperature increase, economic cost that goes with it and biometeorological uncomfortable for the people.

This methodology could be applied at regional scale to improve and develop spatial planning but taking into considerations the mitigating effects strategies (i.e. urban green, special building materials, study of the orientation of buildings, and shadows).

The radiation and energy budget method utilized to calculate the air temperature differences due to an urbanization process showed some limits and some uncertainties.

Surely, the different approaches to compute evapotranspiration used in the budget method and in the fluidodynamic model and some hypothesis adopted (absence of water deficit in budget method), strongly influenced the results.

Anyway, the fluidodynamic model confirms an air temperature increment of the same magnitude order of the energy budget method, suggesting this last methodology can be considered a sufficiently reliable method to estimate and predict the variation in the thermal field due to changes in land-use. Its application to a regional scale should take into account the different climatic zones and the different models of urban growth (existing residential or production areas densification, expansion of urban area with new buildings, etc.). The storage heat flux computation can be improved considering further variables: volumetric ratio between sealed and unsealed surfaces and optical and thermal properties of building materials.

Land use change, and in particular urbanization process, results in air temperature increase. The methodology developed in this work can be improved to furnish grater information during a regional territory planning. Air temperature increment give a surplus of energy needs of air conditioning systems: throughout thermotechnical computations it is possible to obtain the economic cost to reinstate indoor air temperature after urbanization process.

Moreover, the bioclimatic discomfort for the population caused by urban land use increment can be computed and successively taken into account during the initial design of an urban area.

Author Contributions: The authors contribute at all in this study thanks to different thanks to the different skills: fluid dynamics and agronomics. The manuscript was written together.

Funding: This research received no external funding.

Acknowledgments: The authors would like to thank IdroMeteorological Service (ARPAe) of Emilia-Romagna Region for furnishing meteorological data through DEXTER service, AGEA for orthophotographs and Digital Globe™–Telespazio for QuickBird orthoimages, acquired through internal Map Service of Emilia-Romagna Region.

References

1. World Urbanization Prospects: The 2018 Revision. Available online: https://population.un.org/wup/Publications/Files/WUP2018-KeyFacts.pdf (accessed on 13 November 2019).
2. Grimmond, S. Urbanization and global environmental change: Local effects of urban warming. *Geogr. J.* **2007**, *173*, 83–88. [CrossRef]

3. Chapman, A.; Watson, J.E.M.; Salazar, A.; Thatcher, M.; McAlpine, C.A. The impact of urbanization and climate change on urban temperature: A systematic review. *Landsc. Ecol.* **2017**, *32*, 1921–1935. [CrossRef]

4. Lim, Y.K.; Cai, M.; Kalnay, E.; Zhou, L. Observational evidence of sensitivity of surface climate changes to land types and urbanization. *Geophys. Res. Lett.* **2006**, *32*, L22712. [CrossRef]

5. Li, Y.; Zhao, M.; Motesharrei, S.; Mu, Q.; Kalnay, E.; Li, S. Local cooling and warming effects of forest based on satellite data. *Nat. Commun.* **2015**, *6*, 6603. [CrossRef] [PubMed]

6. Oke, T.R. The surface energy budgets of urban area. In *Modelling the Urban Boundary Layer*; AMS: Boston, MA, USA, 1981.

7. Fischer, E.M.; Seneviratne, S.I.; Vidale, P.L.; Lüthi, D.; Schär, C. Soil moisture-atmosphere interactions during the 2003 European summer heat wave. *J. Clim.* **2007**, *20*, 5081–5099. [CrossRef]

8. Liu, Y. Introduction to land use and rural sustainability in China. *Land Use Policy* **2018**, *74*, 1–4. [CrossRef]

9. Liu, Z.; Liu, Y.; Baig, M.H.A. Biophysical effect of conversion from croplands to grasslands in water-limited temperate regions of China. *Sci. Total Environ.* **2019**, *648*, 315–324. [CrossRef] [PubMed]

10. Moulai, M.; Khavari, F.; Shahhosseini, G.; Zanjani, N.E. A study of the urban heat island mitigation strategies: The case of two cities. *Int. J. Urban Manag. Energy Sustain.* **2017**, *1*, 1–7.

11. Memon, R.A.; Leung, D.Y.C.; Chunho, L. A review on the generation, determination and mitigation of Urban Heat Island. *J. Environ. Sci.* **2008**, *20*, 120–127.

12. Stull, R.B. *An Introduction to Boundary Layer Meteorology*; Kluwer Academic Publishers: Dordrecht, The Netherlands, 1988; p. 666.

13. Hargreaves, G.H.; Samani, Z.A. Estimating potential evapotranspiration. *Tech. Note J. Irrig. Drain. Eng.* **1982**, *18*, 980–984.

14. Allen, R.G.; Pereira, L.S.; Raes, D.; Smith, M. Crop Evapotranspiration (Guidelines for Computing Crop Water Requirements). In *FAO Irrigation and Drainage Paper No. 56*; FAO: Rome, Italy, 1998.

15. Bruse, M.; Fleer, H. Simulating surface-plant-air interactions inside urban environments with a three dimensional numerical model. *Environ. Model. Softw.* **1998**, *13*, 372–384. [CrossRef]

16. ENVI-Met Software Home Page. Available online: http://www.envi-met.com/ (accessed on 13 November 2019).

17. Yang, X.; Zhao, L.; Bruse, M.; Meng, Q. Evaluation of a microclimate model for predicting the thermal behaviour of different ground surfaces. *Build. Environ.* **2013**, *60*, 93–104. [CrossRef]

18. Giridharan, R.; Lau, S.S.Y.; Ganesan, S. Nocturnal heat island effect in urban residential developments of Hong Kong. *Energy Build.* **2005**, *37*, 964–971. [CrossRef]

 *climate*

Article

Air Pollution Flow Patterns in the Mexico City Region

Alejandro Salcido *, Susana Carreón-Sierra and Ana-Teresa Celada-Murillo

Instituto Nacional de Electricidad y Energías Limpias, Reforma 113, Palmira, Cuernavaca 62490, Morelos, Mexico; susana.carreon@ineel.mx (S.C.-S.); atcelada@ineel.mx (A.-T.C.-M.)
* Correspondence: salcido@ineel.mx; Tel.: +52-777-362-3811 (ext. 7087)

Received: 18 September 2019; Accepted: 31 October 2019; Published: 5 November 2019

Abstract: According to the Mexico City Emissions Inventory, mobile sources are responsible for approximately 86% of nitrogen oxide emissions in this region, and correspond to a NOx emission of 51 and 58 kilotons per year in Mexico City and the State of Mexico, respectively. Ozone levels in this region are often high and persist as one of the main problems of air pollution. Identifying the main scenarios for the transport and dispersion of air pollutants requires the knowledge of their flow patterns. This work examines the surface flow patterns of air pollutants (NO_2, O_3, SO_2, and PM_{10}) in the area of Mexico City (a region with strong orographic influences) over the period 2001–2010. The flow condition of a pollutant depends on the spatial distribution of its concentration and the mode of wind circulation in the region. We achieved the identification and characterization of the pollutant flow patterns through the exploitation of the 1-hour average values of the pollutant concentrations and wind data provided by the atmospheric monitoring network of Mexico City and the application of the k-means method of cluster analysis. The data objects for the cluster analysis were obtained by modeling Mexico City as a 4-cell spatial domain and describing, for each pollutant, the flow state in a cell by the spatial averages of the horizontal pollutant flow vector and its gradients (the divergence and curl of the flow vector). We identified seven patterns for wind circulation and nine patterns for each of NO_2, O_3, PM_{10}, and SO_2 pollutant flows. Their seasonal and annual average intensities and probabilities of occurrence were estimated.

Keywords: pollution flow patterns; wind circulation patterns; emission inventory; criteria pollutants; Mexico City

---

1. Introduction

In cities and large urban settlements, tropospheric ozone and particulate matter (PM_{10} and $PM_{2.5}$) are the most dangerous air pollutants for human health [1]. Air pollution is a major environmental issue in urban areas, which can affect the well-being and quality of life of citizens. More and more frequently, there are detected chronic diseases of great importance that are associated with continuous exposures to high concentrations of air pollutants. Epidemiological studies have reported that exposure to air pollutants such as particulate matter, nitrogen oxides (NOx), sulfur dioxide (SO_2), and surface ozone (O_3) associates with an increase in mortality and hospital admissions predominantly related to respiratory and cardiovascular diseases [1,2]. This critical issue concerns the 20 million people (including 9 million children) living in the Mexico City Metropolitan Area (MCMA). Despite the reductions in the emissions of common air pollutants in MCMA since the early 1990s, millions of people remain exposed to concentrations above the critical levels associated with increased risks for cardiovascular and respiratory diseases. The anthropogenic sources still produce and release to the atmosphere every year large amounts of carbon monoxide (CO), nitrogen oxides (NOx), sulfur dioxide (SO_2), particulate matter (PM_{10} and $PM_{2.5}$), and volatile organic compounds (VOC) [3]. However, because of the frequency of occurrence of high levels, persistence, and spatial distribution, the most critical air pollutants in MCMA are by far ozone and PM_{10} [4–6].

At the MCMA, the complexity of the air pollution problem is also strongly related to other essential factors such as the geographical setting, meteorology, and topography. MCMA lies inside a subtropical basin with latitudes between 19.2 and 19.6 °N and longitudes between 98.9 and 99.4 °W, and an average altitude of 2240 m, surrounded by high mountains (Figure 1). In the north direction, the basin extends into the Mexican plateau and the arid interior of the country, with the Sierra de Guadalupe creating a small 800 m barrier above the basin floor. Its climate classification comprises two seasons: the rainy season from May to October and the dry season from November to April. This classification stems from the two main meteorological patterns on the synoptic-scale: dry westerly winds with anticyclone conditions from November until April, and moist flows from the east due to the weaker trade winds along the other six months [7]. The MCMA meteorology, however, is by far more complicated than this simple classification expresses it. The basin interacts with the Mexican plateau and the lower coastal areas. Moreover, due to the MCMA location, large-scale pressure gradients are generally weak, and intense solar radiation is registered here throughout the year [7–11]. These conditions and the presence of high mountains in the surroundings are ideal for the development of thermally driven winds.

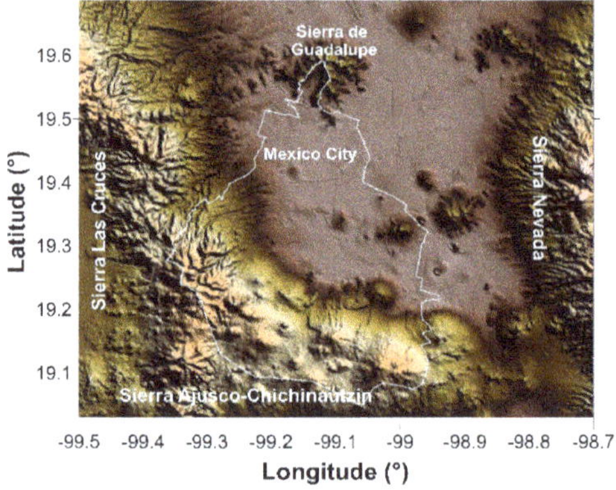

Figure 1. Complex topography in the Mexico City region.

The knowledge of wind circulation and air pollutant spatial distributions constitutes a relevant element to understand the flow dynamics of urban air pollution. It allows knowing how the air pollutant emissions produced in an urban settlement may disperse in the atmosphere, how these air pollutants may be distributed spatially in the city, and how the winds may transport them towards neighboring sites [12,13]. From the standpoint of fluid dynamics, the pollutant-flow vector, defined as the product of the fields of the pollutant concentration and wind velocity,

$$\mathbf{J}(\mathbf{x}, t) = \mu(\mathbf{x}, t)\mathbf{v}(\mathbf{x}, t), \tag{1}$$

describes the transport of a given pollutant in the atmosphere. Then, the surface flow condition of a pollutant depends on both factors: the surface distribution of its concentration and the local wind circulation.

Local wind circulation in Mexico City and its relation to driving forces and air pollution have been studied extensively for almost three decades. However, most of the studies have been performed on an episode-by-episode basis and using small data sets obtained from short-term experimental campaigns with different approaches [14–21]. Some of the main exceptions are the first long-term micrometeorological campaign carried out by Salcido et al. [11] in this region during 2001, and the studies of Klaus et al. [22], de Foy et al. [23], Salcido et al. [24], and Carreón-Sierra et al. [25], which were carried out using data provided by the atmospheric monitoring network of Mexico City (SIMAT:

Sistema de Monitoreo Atmosférico de la Ciudad de México [26]). Klaus et al. [22] reported a principal component analysis of air quality and meteorological data for the period February–December 1995. The study of de Foy et al. [23] reported an examination of wind transport during the experimental campaign of the Megacity Initiative: Local and Global Research Observations (MILAGRO) project; they used cluster analysis for making a comparison to climatology between the period of March 2006 and the period 1998–2006 using hourly wind data of the warm, dry season. In their work, Salcido et al. [24] used a lattice wind model at a meso-β scale to carry out a description and characterization of Mexico City local wind events of the period 2001–2006. Furthermore, Carreón-Sierra et al. [25] proposed an extension of the local wind state using a non-local description based on the wind velocity and its gradients and applied hierarchical cluster analysis to recognize and characterize the Mexico City wind circulation patterns in the period 2001–2006. As far as we know, the last paper reported the first work where a non-local approach is used to characterize the wind condition for identifying patterns with more detail than the simple results produced by the wind rose method.

In this paper, extending our previous work [25], we examine the main surface patterns of wind circulation and pollutant-flow of NO_2, O_3, SO_2, and PM_{10} in the Mexico City region over the period 2001–2010. We achieved the identification and characterization of the pollutant-flow patterns through the exploitation of the 1-hour average values of the pollutant concentrations and wind data provided by the atmospheric monitoring network of Mexico City (SIMAT [26]) and the application of cluster analysis as a data mining procedure. First, we considered Mexico City as a 2D lattice domain and defined the flow condition (or flow state) of a given pollutant in each lattice cell in terms of a flow vector field and its gradients (precisely, the divergence and curl of this flow vector). Second, we used the 1-h average values of the pollutant concentrations and wind data provided by SIMAT to estimate, using Kriging interpolation, discrete representations of the pollutant-flow vector field and its gradients over the lattice domain. Finally, we used the k-means method of cluster analysis [27–29] to identify and characterize the pollution-flow patterns from the clustering modes of the flow states. We identified seven patterns for wind circulation and nine patterns for each of NO_2, O_3, PM_{10}, and SO_2 pollutant-flows. Their seasonal and annual average intensities and probabilities of occurrence were also estimated.

2. Materials and Methods

In this section, we detail the study domain, data, and procedure that we used to identify and characterize the wind and pollution-flow patterns.

2.1. Study Domain

We considered the part of Mexico City located at 19.3°–19.6°N and 99.0°–99.3°W as the study domain. In Figure 2, we showed this region as enclosed by a solid line square. This region contains 75% of the stations of the wind-monitoring network (REDMET) of SIMAT, 95% of the NO_2 stations, 83% of the O_3 stations, 93% of the PM_{10} stations, and 86% of the SO_2 stations of the air quality monitoring network (RAMA) of SIMAT (Figure 3). Because of the topographic complexity of the mountains surrounding Mexico City, we examined the pollutant flows in this region, assuming it divided into quadrants (NE, NW, SW, and SE). We defined these quadrants using a reference frame with origin at (19.43°N, 99.13°W) and the axes along the west to east (W–E) and south to north (S–N) directions (dotted lines in Figure 2). The origin of the reference frame is the geometric centroid of the REDMET stations. The positions of the geometric centers of the city quadrants are: (19.5°N, 99.1°W), (19.5°N, 99.2°W), (19.4°N, 99.2°W), and (19.4°N, 99.1°W) for the NE, NW, SW, and SE quadrants. The dimensions of each quadrant were 15.66 km length in the west–east direction and 16.56 km length in the south–north direction (each quadrant is a square of 540 × 540 arcsec). In principle, we expect to detect the ventilation effects due to the openings located at the west and east sides of the Sierra de Guadalupe (north of the city) at NW and NE quadrants, respectively, and to observe in the SE quadrant the effect of the gap situated in the southeast of the México basin. Also, in the SE and NE quadrants, we expect to recognize the effects of the winds blowing along the ventilation channel determined by the volcanos (Sierra Nevada) on the east side of the city. Moreover, we expect to recognize the wind effects due to the Mexico City mountain-valley system in all quadrants, but mainly in the NW and SW quadrants.

2.2. Data

Figures 2 and 3 illustrate the study domain and the spatial distributions of the stations of the monitoring networks of SIMAT for wind, nitrogen dioxide, ozone, PM_{10}, and sulfur dioxide. These monitoring networks provided the 1-hour average values of the meteorological and air quality variables systematically and made them publicly available through its web site [26]. For the present work, we collected the wind data (wind speed and wind direction) and the air quality data (NO_2, O_3, PM_{10}, and SO_2 surface concentrations) for the period 2001–2010, which constitute a database with 87,600 records, approximately, for each hourly averaged variable.

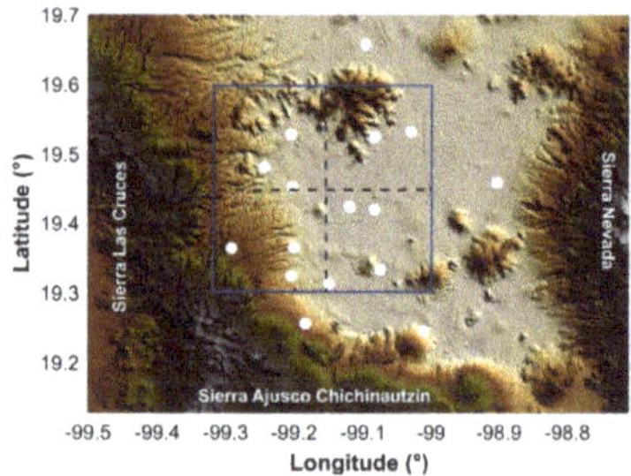

Figure 2. Study domain for Mexico City and spatial distribution of the stations (white dots) of the wind-monitoring network (REDMET) of the atmospheric monitoring network of Mexico City (SIMAT). The REDMET provides 1-hour average values of wind speed and wind direction, and other properties such as temperature and relative humidity.

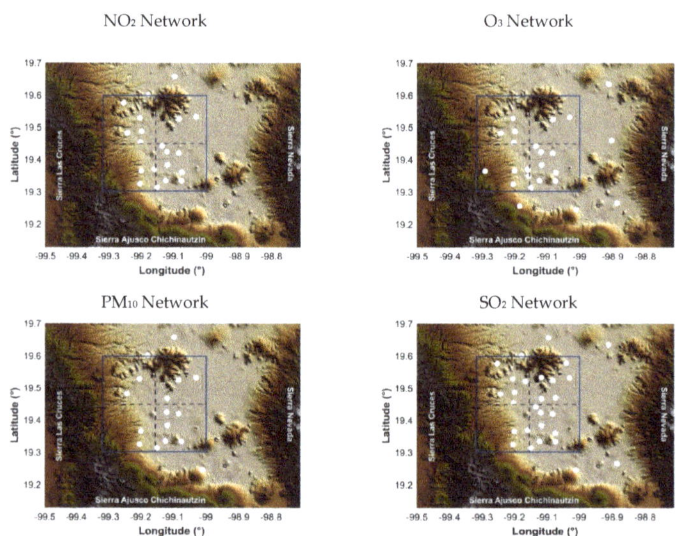

Figure 3. Spatial distribution of the stations (white dots) of the NO_2, O_3, PM_{10}, and SO_2 monitoring networks (RAMA) of SIMAT. The RAMA provides 1-hour average values of the surface concentrations of NO_2, O_3, PM_{10}, and SO_2, among other chemical compounds.

2.3. Study Database

In a first step, we represented the study domain (D) as a square grid G with 289 nodes G_{rs}, where the subscripts $r = 1, 2, \ldots , 17$ and $s = 1, 2, \ldots , 17$ vary along with the West–East (WE) and South–North (SN) directions, respectively. Using the wind speed and wind direction data supplied by the REDMET stations, and the NO_2, O_3, PM_{10}, and SO_2 concentration data supplied by the RAMA stations, we estimated (using Kriging interpolation) the horizontal components of the flow vector field at each grid node G_{rs} and time t,

$$J_{WE}(r,s,t) = \mu(r,s,t)v_{WE}(r,s,t), \qquad J_{SN}(r,s,t) = \mu(r,s,t)v_{SN}(r,s,t) \tag{2}$$

For a given pollutant, at the node G_{rs} and time t, $\mu(r,s,t)$ is the pollutant concentration, and $v_{WE}(r,s,t)$ and $v_{SN}(r,s,t)$ are the wind velocity components. Here, time t is an integer running from 1 to the number H of hours (87,600, approximately) in the period 2001–2010.

Then, we defined a 2D lattice L composed by 64 regular non-overlapping cells C_{ij} covering D, with $i = 1, 2, 3, \ldots , 8$ and $j = 1, 2, 3, \ldots , 8$, such that each cell C_{ij} is surrounded by nine grid nodes of G, as Figure 4 illustrates it.

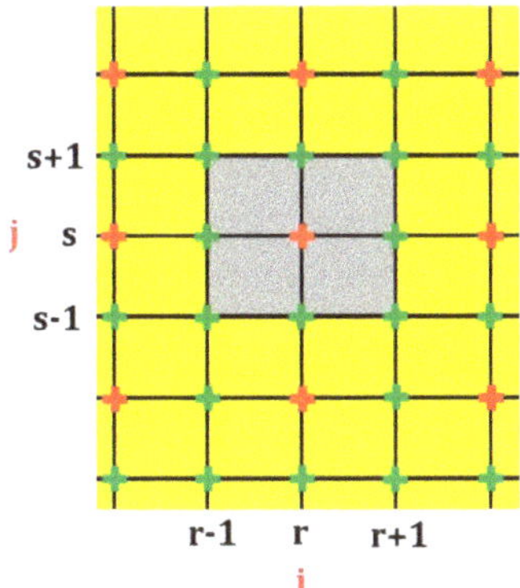

Figure 4. Arrangement of the calculation grid nodes G_{rs} and the lattice cells C_{ij}. The green crosses represent the grid nodes, and the red crosses indicate the centers of the lattice cells. In gray, we show the cell C_{ij}.

The estimations, for each pollutant, of the flow vector components at the grid nodes G_{rs}, allow calculating (numerically) the spatial averages of the components of the pollutant-flow vector and its gradients at the cells C_{ij} of the lattice L using their 2D discrete definitions in terms of centered finite differences [24,30]. For a given pollutant, we will denote the flow vector components at C_{ij} by $j_x(i, j, t)$ and $j_y(i, j, t)$, and the associated divergence and curl of the flow vector will be denoted by $\Gamma(i, j, t)$ and $\Omega(i, j, t)$. Here, x and y denote the WE and SN directions, respectively. At this level, we will be referring to this non-local discrete representation of the flow condition of the study domain as the 64-cell model. It involves 256 (= 4 × 64) parameters to define an *event* of the system (i.e., to describe the flow condition over the complete spatial domain). In this work, for simplicity in the handling and interpretation of the results, we considered the reduction of the 64-cell model to the cases of the 4-cell and 1-cell models, which we obtained using a spatial averaging procedure. In any case, the four parameters $(j_x, j_y, \Gamma, \Omega)$ define the flow state at a lattice cell; therefore, the 4-cell and 1-cell models respectively entail 16 and 4 parameters to describe an *event* of the system. It is convenient to note that the introduction of the additional variables (Γ, Ω) in the flow state definition, allows recovering some of the information lost during the spatial averaging of the flow vector components. This assumption gives a slight non-local character to this description [31].

For each cell C_{ij}, we calculated, hour by hour, the average of the four state-variables over the 2001–2010 period, defining a one generic year database for each pollutant and cell model considered. The results were normalized and then organized in seasonal groups: January–March (winter), April–June (spring), July–September (summer), and October–December (autumn).

We performed data normalization as follows. We determined the maximum absolute values of the magnitude of the pollutant-flow vector $|j|_{max}$, and the divergence $|\Gamma|_{max}$, and curl $|\Omega|_{max}$ of the flow vector, in the generic year database. Then the normalized and dimensionless state variables (all of them now ranging in the interval $[-1, +1]$) were calculated using the following expressions:

$$\hat{j}_x(i, j, t) = \frac{j_x(i, j, t)}{|j|_{max}}, \tag{3}$$

$$\hat{j}_y(i, j, t) = \frac{j_y(i, j, t)}{|j|_{max}}, \tag{4}$$

$$\hat{\Gamma}(i, j, t) = \frac{\Gamma(i, j, t)}{|\Gamma|_{max}}, \tag{5}$$

$$\hat{\Omega}(i, j, t) = \frac{\Omega(i, j, t)}{|\Omega|_{max}}. \tag{6}$$

2.4. Clustering Method

We used the k-means method of cluster analysis as a procedure of data mining to identify and characterize the main patterns of wind and flow of pollutants in the Mexico City region.

Cluster analysis comprises a broad set of techniques for finding non-overlapping subgroups of data objects in a dataset. When clustering the data objects of a dataset, we seek to organize them into a given number of distinct subsets (groups, or clusters), so that they constitute a partition of the initial dataset; this way, the data objects within each subset are quite similar to each other, while data objects in different groups are quite different from each other [29]. Clustering is an unsupervised problem because we are trying to discover structure (distinct clusters) based on a dataset, without being trained by a response variable.

The k-means clustering method is the simplest and the most frequently used technique of cluster analysis for partitioning a dataset into a set of k groups. In k-means clustering, we seek to partition the set of data objects into a pre-specified number of clusters, so that the total within-cluster variation (TWCV) is as small as possible.

Suppose we have a dataset $X = \{x_1, x_2, \ldots, x_N\}$, $x_i \in \mathbf{R}^d$, and we try to partition X into M disjoint clusters C_1, C_2, C_M. Then, the k-means algorithm detects local optimal solutions concerning the TWCV defined as the sum of squared Euclidean distances between each data object x_i and the centroid m_k of the cluster C_k that contains x_i. Analytically, the TWCV is given by

$$\varepsilon(m_1, m_2, \ldots, m_M) = \sum_{i=1}^{N} \sum_{k=1}^{M} \lambda_{ik} \|x_i - m_k\|^2 \tag{7}$$

where $\lambda_{ik} = 1$ if $x_i \in C_k$, and 0 otherwise. The TWCV measures the clustering goodness, and the optimization problem that defines the k-means clustering resides in minimizing TWCV. In practice, we used the software **DataLab** developed by Hans Lohninger [32] to perform the cluster analysis with the k-means algorithm.

For each air pollutant considered, we define its set of data objects as

$$H = \{Q(t)|t = 1, 2, \ldots, T\}. \tag{8}$$

where T = 8760 is the number of hours of the generic year, and

$$Q(t) = \left\{ Q_{ijt} \equiv (\hat{j}_x, \hat{j}_y, \hat{\Gamma}, \hat{\Omega})_{ijt} \middle| i = 1, 2, \ldots, Nx; j = 1, 2, \ldots, Ny \right\}, \tag{9}$$

with $N_x = 2$ ($N_x = 1$) and $N_y = 2$ ($N_y = 1$) in the case of the 4-cell (1-cell) model. We organized the dataset H in the seasonal subsets H_{winter}, H_{spring}, H_{summer}, and H_{autumn} comprising, respectively, the data objects of the periods January–March (winter), April–June (spring), July–September (summer), and October–December (autumn). We identified each data object of these subsets with a label of the form MMDDHH that specifies the date (relative to the generic year) and hour of occurrence.

To carry out the cluster analysis for each seasonal period, we build a data matrix structured as described in Table 1.

Table 1. Structure of the data matrix for the cluster analysis.

Columns	Variables
1	DateTime (MMDDHH)
2–5	The four values of the 1-cell model parameters: $(\hat{j}_x, \hat{j}_y, \hat{\Gamma}, \hat{\Omega})_{1C}$
6–21	The 16 values of the 4-cell model parameters: $(\hat{j}_x, \hat{j}_y, \hat{\Gamma}, \hat{\Omega})_{C0}$, $(\hat{j}_x, \hat{j}_y, \hat{\Gamma}, \hat{\Omega})_{C1}$, $(\hat{j}_x, \hat{j}_y, \hat{\Gamma}, \hat{\Omega})_{C2}$, $(\hat{j}_x, \hat{j}_y, \hat{\Gamma}, \hat{\Omega})_{C3}$

For each pollutant, using the respective data matrixes of the seasonal subsets, we can apply the k-means algorithm for clustering the associated data objects considering the parameters of the 1-cell or 4-cell model as the clustering variables. Although we can expect that the 4-cell model variables will produce a physically more detailed clustering of the wind and pollutant-flow events, in this work, for easiness in handling the results, we used only the 1-cell model parameters as clustering variables. The application of the clustering procedure to the 1-cell model data objects will organize them, assigning the cluster number to the date-time labels that identify the events. Consequently, the 4-cell model data objects also receive cluster numbers correspondingly. It means that the data objects of the 4-cell model were also organized into clusters by the same clustering procedure. Therefore, for each pollutant and seasonal period, we have clusters of events characterized by the sixteen parameters of the 4-cell model:

$$(\hat{j}_x, \hat{j}_y, \hat{\Gamma}, \hat{\Omega})_{C0}, \ (\hat{j}_x, \hat{j}_y, \hat{\Gamma}, \hat{\Omega})_{C1}, \ (\hat{j}_x, \hat{j}_y, \hat{\Gamma}, \hat{\Omega})_{C2}, \ (\hat{j}_x, \hat{j}_y, \hat{\Gamma}, \hat{\Omega})_{C3} \tag{10}$$

where C0, C1, C2, and C3 are the cells of the model.

2.5. The Number of Clusters

The k-means clustering is a simple and fast algorithm that can efficiently deal with large datasets. However, the k-means approach requires pre-specifying the number of clusters; and preferably, we would like to use an optimal number of clusters that we defined according to some given criterion. In this work, because of the motivation expressed in the following two paragraphs, we applied the k-means clustering method to organize each seasonal period in six clusters of data objects (events).

2.5.1. Physical Motivation

In Figure 5, we presented the mean diurnal behavior of solar radiation, temperature, and wind speed observed during 2001. We obtained these results from data we measured at an urban site southeast Mexico City (Xochimilco) [11,25]. The plots evidence that the meteorological events comprised in the six time-periods 0–4, 4–8, 8–12, 12–16, 16–20, and 20–24 h, are quite different from each other. The periods 0–4 and 20–24 h show the cooling phase of the atmosphere during the night. In the period 4–8 h, we observe the sunrise occurrence, and that temperature and wind speed reach their minimum values. The period 8–12 h shows the growing phases of solar radiation, temperature, and wind speed, with solar radiation reaching its maximum. The period 12–16 h depicts the temperature reaching its maximum, while wind speed keeps growing, and solar radiation starts to decrease. Finally, the period 16–20 h shows the sunset occurrence, the wind speed reaching its maximum, and temperature decreasing. Appreciation of these different behaviors suggests searching for six groups during the application of the k-means algorithm for clustering the data objects defined by the events of wind circulation and pollutant-flow in Mexico City.

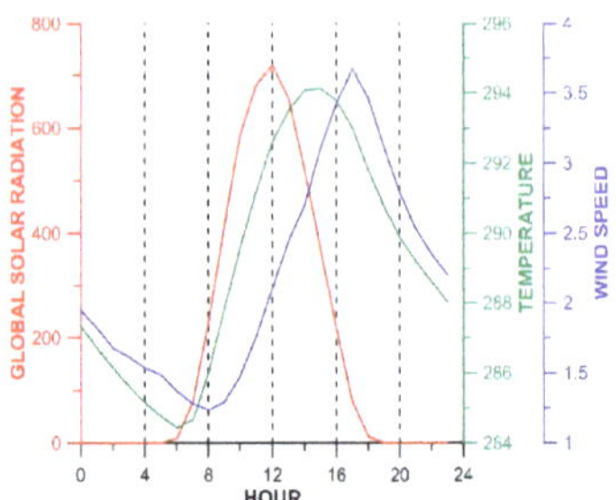

Figure 5. Average diurnal behavior of solar radiation (red line), temperature (green line), and wind speed (blue line) in an urban site (Xochimilco) of Mexico City during 2001 [11,25]. We observe that the meteorological events comprised in the six intervals of 0–4, 4–8, 8–12, 12–16, 16–20, and 20–24 h are different from one period to another. This appraisal suggests searching for six clusters during the cluster analysis of the wind and pollutant-flow events.

2.5.2. The Elbow Method

From the standpoint of cluster analysis, one of the most popular methods for determining the optimal clusters is the *elbow method*. It provides a simple solution. The basic idea is to compute the k-means clustering algorithm using different numbers, k, of clusters. The next step is drawing the total within-cluster variation (or total within-cluster sum of squares) versus the number of clusters. Commonly, the analysts consider the location of a bend in the plot as an indicator of the appropriate number of clusters. However, it is not always clear where the bend point position is. We preferred to calculate the percentage reduction (δ) of the total within-cluster sum of squares of a given k, relative to its value for the previous number of clusters $k - 1$, and then to plot δ as a function of k. As our optimal number of clusters, we selected the value of k from which on the reduction δ is less than 10% each unit step in increasing k. For example, according to this procedure, Figure 6 shows that we must select $k = 6$ when clustering the winter wind data objects of the present work.

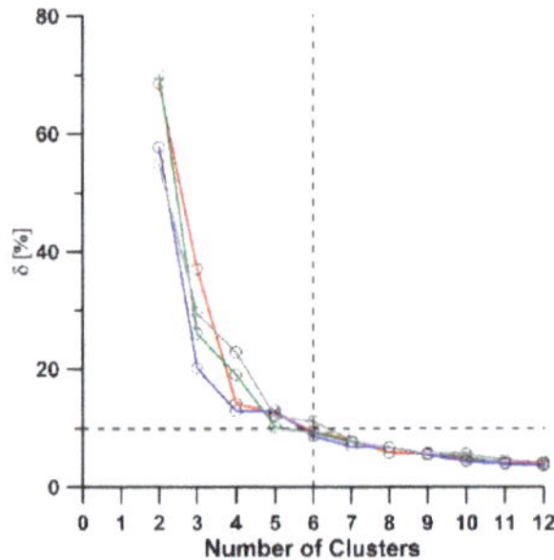

Figure 6. Percentage reduction (δ) of the total within-cluster sum of squares, relative to its previous value, as a function of the number of clusters. Here, for the clustering of the winter wind data objects, we observe that from 6 clusters on, the reduction δ is less than 10% each unit step in the increase of the number of clusters: $0\% \leq \delta \leq 10\%$.

3. Results and Discussion

Figure 7 summarizes the application of the k-means clustering method (with k = 6) to the seasonal subsets, H_{winter}, H_{spring}, H_{summer}, and H_{autumn}, of the data objects (events) of wind circulation and pollutant-flows. This figure shows an arrangement with four columns (one per seasonal period) and five rows (one for wind circulation, and four for the pollutant-flows of NO_2, O_3, PM_{10}, and SO_2). Each graph shows six plots (one per cluster) differentiated by the line color. The algorithm enumerates the clusters according to their sizes, from the larger to the smaller, so colors mean no more. Each plot presents the *hourly population* of one cluster, that is to say, the number of data objects belonging to the cluster as a function of the hour of the day (Mexico City local time, UTC-6 h).

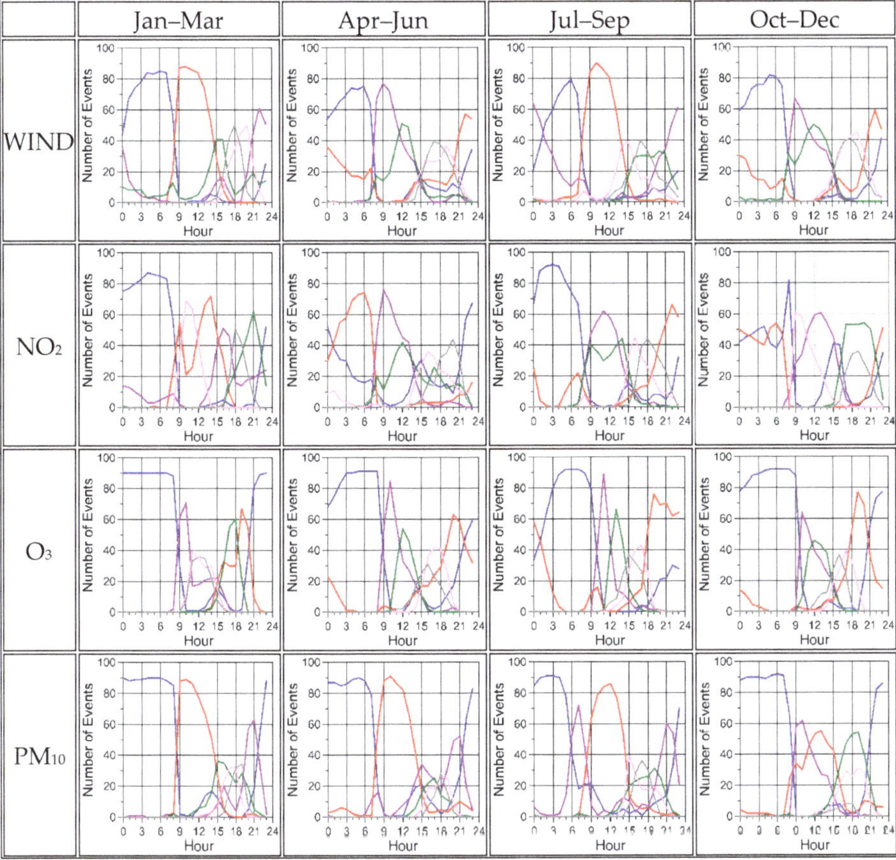

Figure 7. *Cont.*

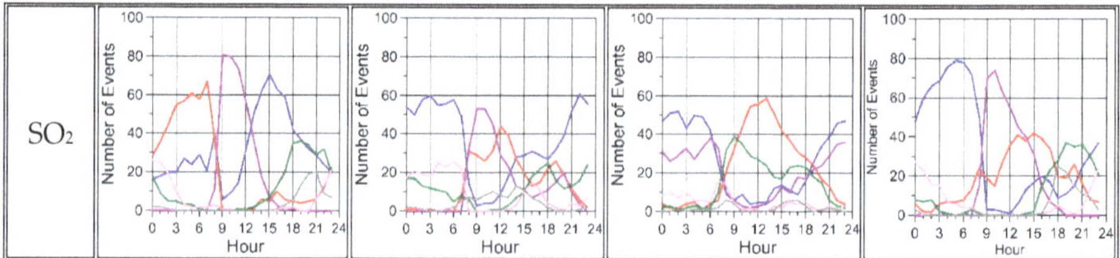

Figure 7. Application of the k-means algorithm (with k = 6) to the seasonal subsets of the data objects of the wind circulation (first row) and pollutant-flows (next four rows) events. Each graph shows six plots, one per cluster, individualized by the color. Each plot expresses the *hourly population of the cluster*, i.e., the number of data objects (or events) of the cluster as a function of the hour of the day.

For wind circulation or any given pollutant-flow, a qualitative comparison of graphs in the respective row allowed recognizing some seasonal similarities and differences between the clusters. For example, for the wind circulation (first row), if we pay attention to the clusters represented by a blue-line plot in the four seasons, we can observe an evident regularity: the hourly populations have similar trends and indicate that the majority of their events occurred during the night, mainly from midnight until dawn. We interpreted this regularity as the possible existence of a wind pattern. The plots of the pollutant-flows also show a similar regularity.

The trends of the hourly populations of the clusters provide useful but insufficient information for discovering patterns in sets of events of wind circulation or pollutant-flows. In [25], the authors detected two wind clusters with similar trends in the hourly populations (both in the number of events and in the times of occurrence), but in one case the winds were blowing from the north and in another one from the south, indicating different patterns of wind circulation. However, we can use the information provided by the values of the 16 parameters of the 4-cell model to complement the pattern identification process.

Eight of the sixteen parameters of the 4-cell model are the components of the flow vectors in the quadrants (NW, NE, SW, and SE) of the MCMA. This complementary information and the hourly population plots allowed recognizing the wind circulation patterns shown in Figure 8, and the pollution-flow patterns shown in Figures 9–12. We found seven patterns for wind circulation and nine patterns for each of the pollutant-flows (NO_2, O_3, PM_{10}, and SO_2). In Figure 8, we show the hourly population of each wind pattern and a graphic representation of the wind velocity at the quadrants. Figures 9–12 show the hourly population of each pattern, a graphic representation of the flow vector at the quadrants, the surface distribution of the pollutant concentration, and the mean spatial pollutant concentration (expressed in ppb for NO_2, O_3, and SO_2, and in μgm^{-3} for PM_{10}). In these figures, the hourly population of a pattern is an average over the seasons the pattern occurred; the vectors (the wind velocity and the pollutant-flow vectors) are averages over the elements of the pattern and the seasons; the spatial distribution of the pollutant concentrations is the average over the pattern elements and seasons; and the mean spatial concentration of a pollutant is the average over the positions of the average pollutant distribution. The lengths of the arrows that represent the wind and flow vectors were scaled to get the larger one fitted in the square that represents the quadrant, allowing a qualitative comparison of the vector magnitudes among the patterns of wind circulation or the patterns of any of the pollutant flows.

3.1. Wind Circulation Patterns

The proposed methodology for examining the database of the wind events that occurred in Mexico City from 2001 to 2010 allowed discovering there the presence of seven wind circulation patterns and the estimation of their seasonal and annual occurrence frequencies, which we summarized in Table 2. We denoted these patterns as WIND-Pn, with n = 1, 2, … , 7, and enumerated them trying to follow the order of their appearance during the day. In Figure 8, we presented the hourly population of each wind pattern, including a sketch out of the corresponding wind velocity in the city quadrants.

Table 2. Seasonal and annual frequencies of the wind circulation patterns (%).

Pattern	Jan–Mar (Winter)	Apr–Jun (Spring)	Jul–Sep (Summer)	Oct–Dec (Autumn)	Annual
WIND-P1	33	31	25	32	30
WIND-P2	25	18	24	16	21
WIND-P3	0	12	9	14	9
WIND-P4	20	0	0	0	5
WIND-P5	9	9	20	11	12
WIND-P6	14	21	22	17	18
WIND-P7	0	9	0	10	5

In the following paragraphs, we provide brief descriptions of the wind circulation patterns. We used the Mexico City local time (UTC-6 h) in all figures and descriptions.

WIND-P1: early morning downslope winds. WIND-P1 was the wind pattern with the highest annual frequency (30%). It was observed systematically throughout the year during the night, showing a large peak around hour 6 (Figure 8) and a small peak around hour 15. The mean wind velocities at the quadrants suggest downslope winds from the surrounding mountains, converging towards Mexico City basin. The converging winds represented by the small peak reflect the presence of convective upwards winds due to the urban heat island (UHI) in the Mexico City region. Jauregui [7] and Klaus et al. [33] pointed out that the Mexico City downslope winds are reinforced by UHI.

WIND-P2: northeasterly and easterly winds. The annual average frequency of this wind pattern was 21%. It is observed systematically throughout the year, especially during winter (25%) and summer (24%). The mean velocities correspond to winds blowing from NE in almost all the quadrants of the city (trade winds). The wind events of this pattern occurred from the sunrise (hour 7) up to the hour 17. Its hourly population presents a high peak between hours 10 and 11.

WIND-P3: midday northerly winds. This pattern had an annual frequency of 9%. It was observed during spring (12%), summer (9%), and autumn (14%) from hour 7 to hour 21, with a peak around hour 13. The mean wind velocities in the quadrants indicate winds blowing from the north and northeast sectors.

WIND-P4: westerly and southerly winds. This wind pattern was observed only winter season with a seasonal frequency of 20%; its events occurred throughout the day, but mainly from noon to midnight, reaching its maximum population around hour 17. The mean wind velocities represent winds with a westerly component in the west quadrants and southerly winds in the east quadrants. These velocities correspond to winds blowing from the south and southeast city sectors through a gap between the mountains of the Sierra Ajusco-Chichinautzin and the Sierra Nevada, following the ventilation channel S–N located at the east side of the city (Figure 1). The local winds of this pattern could be related to the subtropical jet stream of winter [34] or with the westerlies that are permanently occurring in subtropical and middle latitudes, but coupled with the winds driven by the gap between the mountains located at the southeast. Doran and Zhong [35] described the main characteristics of a gap wind system in the southeastern corner of Mexico City that produces low-level jets occurring regularly during the late winter.

WIND-P5: afternoon northerly winds. This pattern was observed throughout the year with an annual frequency of 12%. It occurred from noon to midnight, with its maximum around hour 19. The mean wind velocities at the quadrants indicate winds blowing from the north sectors.

WIND-P6: midnight downslope winds. This pattern was observed systematically throughout the year, mainly during the nighttime hours, from the sunset (hour 18) to the sunrise (hour 6 of the next day), with its maximum around midnight. It got an annual average frequency of 18%. The population plot of this pattern has a small peak close to hour 15, which could be related to the UHI effect of the city. The wind velocities in the western quadrants of the city represent downslope flows from Sierra de las Cruces (at west) and Sierra del Ajusco-Chichinautzin (located at SW and S) toward the town. In the eastern quadrants, otherwise, late afternoon northerly gap winds are observed guided by the S–N ventilation channel of the city.

WIND-P7: UHI-driven winds. This wind pattern was observed only during the spring (9%) and autumn (10%) seasons, with a low annual occurrence frequency of 5%. The wind events of this pattern occurred from the hour 14 to the hour 22, and its population reached a maximum value close to the hour 17. The mean wind velocities of the city quadrants indicate cyclonic winds converging towards Mexico City downtown, which seems to be related to an upwards flow driven by the UHI.

Summarizing, the wind patterns with the most significant frequencies were WIND-P1 and WIND-P2, with annual values of 30% and 21%, respectively. These two patterns, however, correspond to small velocities in comparison with the other five wind patterns. An outstanding feature of WIND-P1 is that, although its events occurred mainly during the night with converging winds, it also comprises daylight events of winds with the same characteristic of convergence towards the city downtown. We cannot interpret these last events as katabatic winds produced by the mountain-valley system; instead, we must understand them as winds associated with convective upwards flows driven by the urban heat island (UHI) effect. The pattern WIND-P6 also occurred during the night, but it peaked at midnight. This wind pattern comprises the downslope winds of the first hours of the night and also reflects the effect of the late afternoon northerly gap winds. The patterns WIND-P3 and WIND-P5 reflect the well-known prevailing winds from N and NE of the Mexico City region, and WIND-P2 evidence the occurrence of the trade winds.

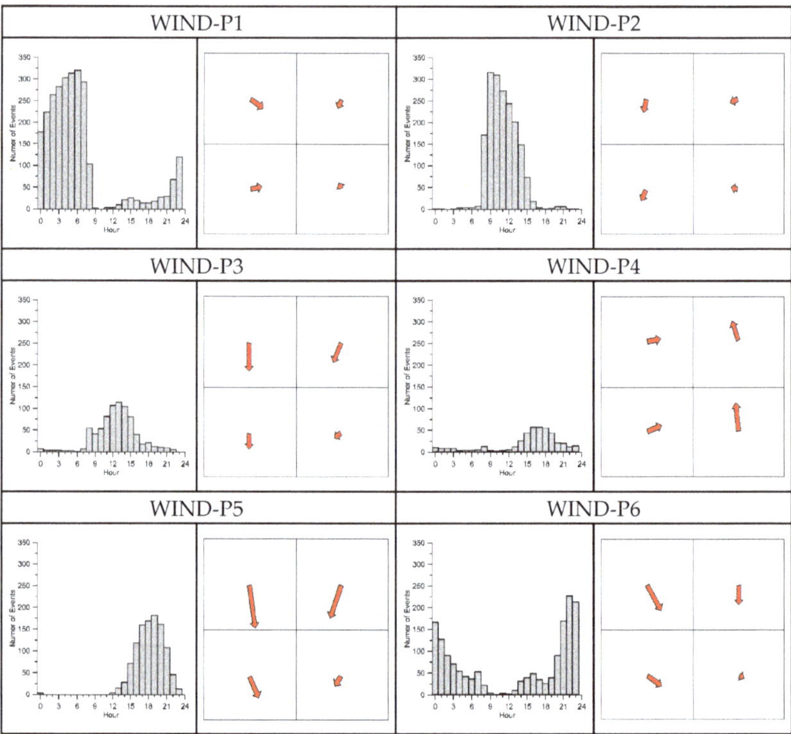

Figure 8. *Cont.*

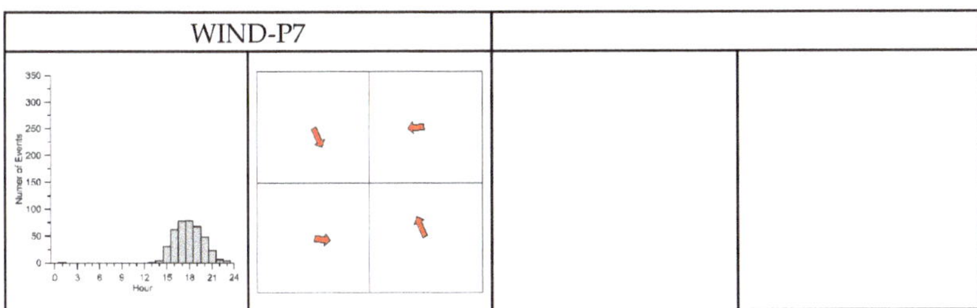

Figure 8. Wind circulation patterns in the Mexico City region during the period 2001–2010. For each pattern, we presented the hourly population (left) and the mean wind velocity (relative to the pattern) at the city quadrants (right). The size of red arrows that stand in for the velocity vectors indicates the velocity magnitude in a scale relative to the edge size of the squares that represent the quadrants.

The wind patterns with the most significant frequencies were WIND-P1 and WIND-P2, with annual values of 30% and 21%, respectively. These two patterns, however, correspond to small velocities in comparison with the other five wind patterns. An outstanding feature of WIND-P1 is that, although its events occurred mainly during the night with converging winds, it also comprises daylight events of winds with the same characteristic of convergence towards the city downtown. We cannot interpret these last events as katabatic winds produced by the mountain–valley system; instead, we must understand them as winds associated with convective upwards flows driven by the urban heat island (UHI) effect. The pattern WIND-P6 also occurred during the night, but it peaked at midnight. This wind pattern comprises the downslope winds of the first hours of the night and also reflects the effect of the late afternoon northerly gap winds. The patterns WIND-P3 and WIND-P5 reflect the well-known prevailing winds from N and NE of the Mexico City region, and WIND-P2 evidence the occurrence of the trade winds.

Our wind patterns agree quite well with the Mexico City wind patterns obtained in [25] for the period 2001–2006 with a slightly different methodology. In Table 3, we described the correspondence between these two sets of patterns. The first column contains the wind patterns WP1-WP7 reported in [25]; the second column presents the wind patterns obtained in the present work; and, the last column provides a brief description of the main characteristics of these patterns.

Table 3. Comparison against the wind patterns reported by Carreón-Sierra et al. [25].

Carreón-Sierra et al.	This Work	Description
WP1	WIND-P1	Early morning downslope winds
WP2	WIND-P2	Northeasterly and easterly winds
WP3	WIND-P3	Midday northerly winds
WP4	WIND-P4	Westerly winds at the western quadrants and
WP5		Southerly winds at the eastern quadrants
WP6	WIND-P5	Afternoon northerly winds
WP7	WIND-P6	Midnight downslope winds
—	WIND-P7	UHI-driven winds

The main differences between both sets of patterns were:

- The pattern WIND-P4 includes the wind patterns WP4 (southerly winds) and WP5 (westerly winds) reported in [25]. However, we detected the pattern WIND-P4 only during the winter season, while the pattern WP4 occurred throughout the year (although with tiny frequencies in spring, summer, and autumn), and the pattern WP5 was only observed during the first semester of the year (winter and spring seasons).

- We recognized a pattern (WIND-P7) associated with cyclonic winds strongly converging towards Mexico City downtown during the daylight hours. The winds in this pattern indicate an upwards flow driven by the UHI. No report exists in [25] about a similar pattern.

3.2. Pollutant Flow Patterns

In Tables 4–7, we enumerated the pollutant-flow patterns (NO_2, O_3, PM_{10}, and SO_2) that we identified for the Mexico City region, their seasonal and annual frequencies and flow intensities (the magnitudes of the flow vectors), and the associated wind circulation patterns. In Figures 9–12, for each pollutant-flow pattern, we presented the average population plot, the mean pollutant-flow vectors at the quadrants NW, NE, SW, and SE of the city, the pollutant surface distribution averaged over the pattern, and its mean spatial value. We denoted the pollution-flow patterns as POL-Pn, where POL indicates the pollutant (POL = NO2, O3, PM10, and SO2), and n is a consecutive integer number from 1 to 9. The cells with the value 0.00 in the tables indicate the seasonal periods where some flow patterns were not detected.

In general, although the pollutant concentrations modulate the magnitudes of the flow vectors, the pollutant-flow patterns reflected the directional characteristics inherited from the wind patterns.

3.2.1. The NO_2 Flow Patterns

We identified nine NO_2 flow patterns for the Mexico City region. Table 4 shows the seasonal and annual average frequencies and average flow intensities of the patterns, expressed in percent (%) and $\mu gm^{-2}s^{-1}$, respectively. As shown in Figure 9, these flow patterns closely resemble the wind circulation patterns, and Table 4 summarized the relationship between the pollutant-flow and wind circulation patterns.

Table 4. Frequencies (%) and flow intensities ($\mu g/m^2 s$) of Mexico City NO_2 flow patterns (2001–2010).

Pattern	Wind Pattern Inherited	JAN–MAR (Winter)		APR–JUN (Spring)		JUL–SEP (Summer)		OCT–DEC (Autumn)		ANNUAL	
		Freq	Flow	Freq	Flow	Freq	Flow	Freq	Flow	Freq	Flow
NO2-P1	WIND-P1	37.36	10.41	23.81	7.46	35.96	10.31	22.13	13.49	29.79	10.42
NO2-P2	WIND-P2	18.61	15.59	0.00	0.00	12.86	16.83	0.00	0.00	7.83	8.11
NO2-P3	WIND-P4	15.93	18.59	10.85	16.92	0.00	0.00	0.00	0.00	6.63	8.88
NO2-P4	WIND-P6	11.02	32.38	24.08	19.03	17.35	27.07	29.02	7.92	20.41	21.60
NO2-P5	WIND-P2	10.42	46.27	17.03	24.57	0.00	0.00	10.61	30.61	9.49	25.36
NO2-P6	WIND-P7	6.67	21.82	0.00	0.00	0.00	0.00	6.89	5.96	3.38	6.95
NO2-P7	WIND-P3	0.00	0.00	14.93	42.51	15.94	28.30	16.64	30.18	11.93	25.25
NO2-P8	WIND-P5	0.00	0.00	9.29	26.70	8.97	24.39	0.00	0.00	4.58	12.77
NO2-P9	WIND-P5	0.00	0.00	0.00	0.00	8.92	49.99	14.69	48.00	5.95	24.50

The NO_2 flow patterns with the most significant annual frequencies were NO2-P1 (30%), NO2-P4 (20%), and NO2-P7 (12%), but the patterns with the most substantial annual flow intensities were NO2-P5, NO2-P7, and NO2-P9 with 25.4, 25.3, and 24.5 $\mu gm^{-2}s^{-1}$, respectively.

The flow patterns NO2-P4, NO2-P7, NO2-P8, and NO2- P9 carry nitrogen dioxide from the north to the south quadrants of the city, particularly the pattern NO2-P9, although it occurred only during the second semester of the year. The events of these patterns take the ozone precursor to the SW and SE quadrants, contributing actively to the ozone formation in this area throughout the year.

The patterns NO2-P2 and NO2-P5 carry the pollutant from the eastern to the western quadrants of the city following the trade winds. The events of the NO2-P3 and NO2-P6 patterns, differently, take nitrogen dioxide from west to east on the west side of the city, but from south to north on the eastern side. However, while the pattern NO2-P3 reflects a coupling of the westerly winds and the afternoon southerly gap winds, the pattern NO2-P6 reflects cyclonic transport driven by the UHI.

During the night, the events of the pattern NO2-P1, follow the downslope winds from the surrounding mountains and carry NO_2 to the city downtown. In comparison with the patterns of other pollutants (O_3, PM_{10}, and SO_2), the NO2-P1 flow pattern is the only one that revealed considerable nocturnal transport due to the downslope winds. The main reason is that NO_2 accumulates during the night since there are no photochemical reactions that consume it, and some of the mobile sources, which are responsible for approximately 86% of nitrogen oxide emissions in the Mexico City region [3], remain active during the night.

The NO_2 surface distributions, which are averages over the events of the NO_2 flow patterns, reveal a North–South part of the city as the zone with the higher NO_2 concentrations. This zone extends around the N–S axis that separates the west from the east quadrants (although slightly shifted to the eastern quadrants). The highest NO_2 levels occurred, in general, in the south sector, close to the Huizachtepetl (Cerro de la Estrella).

The surface NO_2 distributions with the highest concentrations were in connection with the NO2-P5 and NO2-P6 flow patterns, with mean spatial levels of 46 and 41 ppb. These high levels of NO_2 were detected mainly during the winter and autumn. We note that, in these two cases, the high levels of NO_2 extended spatially, covering almost completely the city. In the first case, it was a consequence of blocked transport by the mountains of the Sierra las Cruces, while in the second case, the high levels were a consequence of the cyclonic wind convergence driven by the UHI.

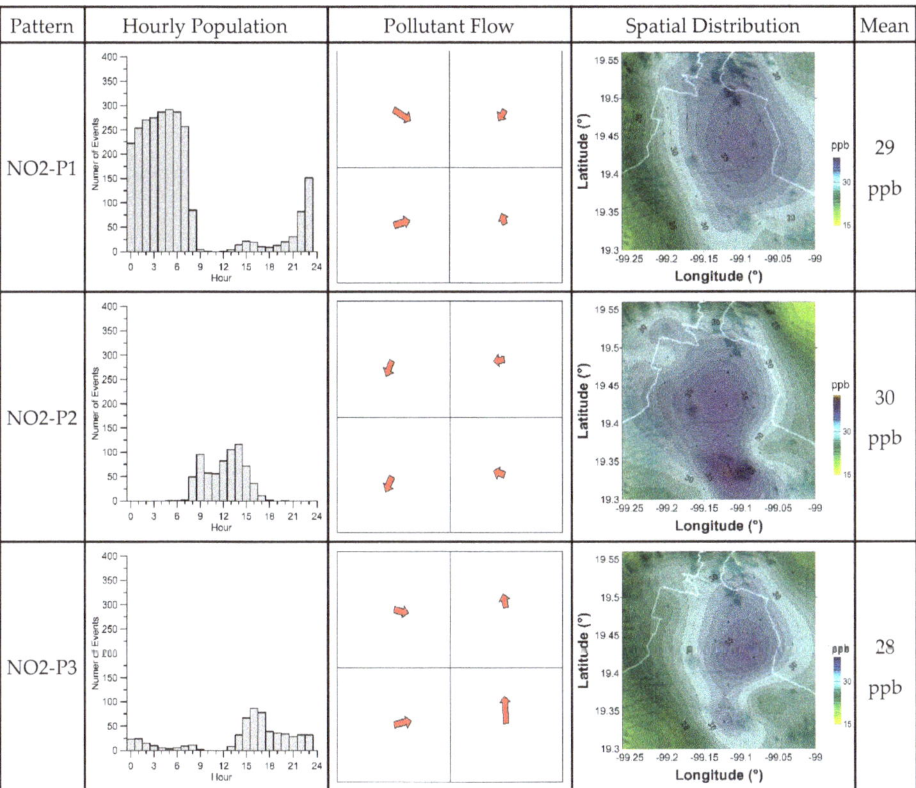

Figure 9. *Cont.*

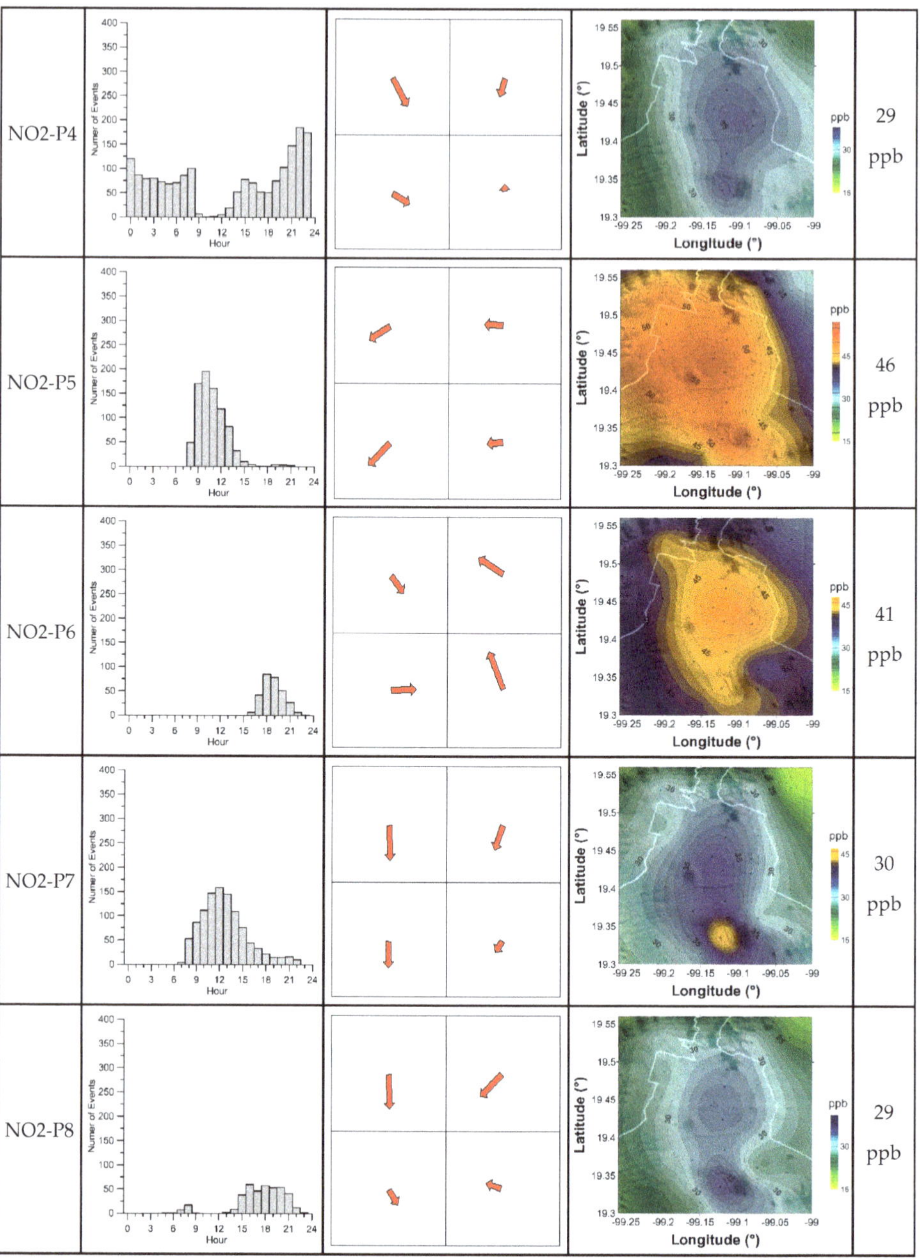

Figure 9. *Cont.*

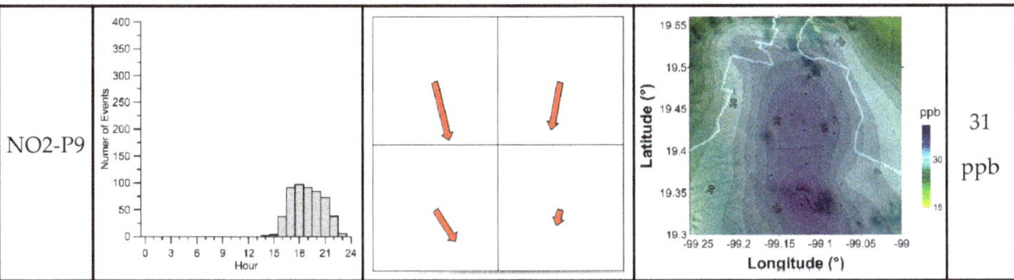

Figure 9. Flow patterns of nitrogen dioxide in the Mexico City region during the period 2001–2010. For each pattern, we included the hourly population (second column), the mean flow vectors (relative to the pattern) at the city quadrants (third column), the surface concentration distributions (fourth column) averaged over the events of the flow pattern, and the mean spatial concentration (last column). The size of red arrows that stand in for the NO_2 flow vectors indicates the flow intensity in a scale relative to the edge size of the squares that represent the quadrants.

3.2.2. The O_3 Flow Patterns

In Table 5 and Figure 10, we summarized the main characteristics of the O_3 flow patterns that we identified. In Table 5, we included the average seasonal and annual occurrence frequencies (%) and flow intensities (μg m^{-2}s^{-1}) of the flow patterns, and the wind patterns from which they inherited their circulation characteristics.

The flow vectors of the diurnal flow patterns O3-P6, O3-P8, and O3-P9 have considerable northerly components and convey ozone from the NW and NE to the SW and SE city quadrants, providing substantial contributions to the high ozone concentrations frequently observed in the south sectors (especially in SW) of the city throughout the year.

Table 5. Frequencies (%) and flow intensities (μg/m^{2}s) of the Mexico City O_3 flow patterns (2001–2010).

Pattern	Wind Pattern Inherited	JAN–MAR (Winter)		APR–JUN (Spring)		JUL–SEP (Summer)		OCT–DEC (Autumn)		ANNUAL	
		Freq	Flow	Freq	Flow	Freq	Flow	Freq	Flow	Freq	Flow
O3-P1	WIND-P1	54.54	6.47	46.57	6.66	41.35	7.89	52.20	4.41	48.64	6.36
O3-P2	WIND-P7	11.53	5.00	7.74	4.22	0.00	0.00	6.21	8.92	6.34	4.54
O3-P3	WIND-P2	11.48	13.84	0.00	0.00	0.00	0.00	10.79	29.21	5.55	10.76
O3-P4	WIND-P4	8.29	46.08	0.00	0.00	0.00	0.00	0.00	0.00	2.04	11.52
O3-P5	WIND-P2	7.50	43.99	0.00	0.00	6.52	53.26	0.00	0.00	3.49	24.31
O3-P6	WIND-P2	6.67	61.25	13.19	53.87	9.96	58.24	0.00	0.00	7.45	43.34
O3-P7	WIND-P6	0.00	0.00	18.45	32.28	27.49	20.02	16.51	22.15	15.69	18.62
O3-P8	WIND-P3	0.00	0.00	7.97	139.84	7.88	111.66	8.80	76.76	6.19	82.07
O3-P9	WIND-P5	0.00	0.00	6.09	116.44	6.79	78.33	5.49	99.39	4.61	73.54

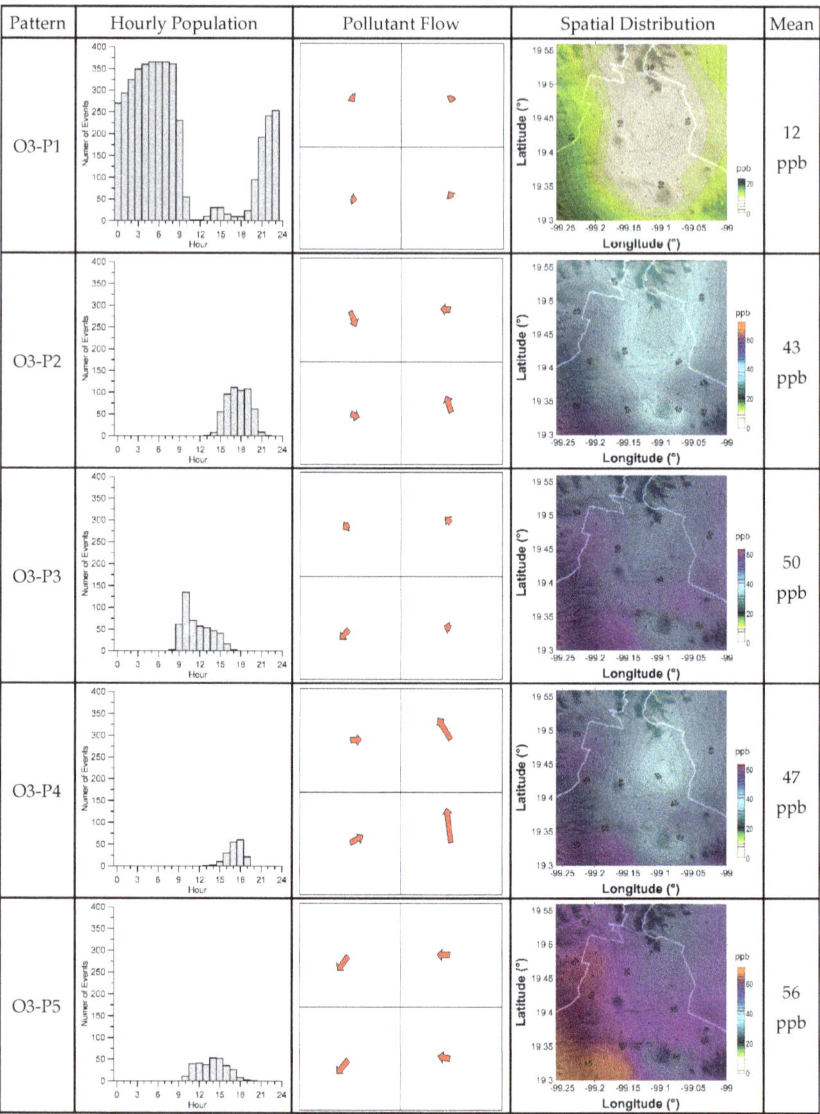

Figure 10. *Cont.*

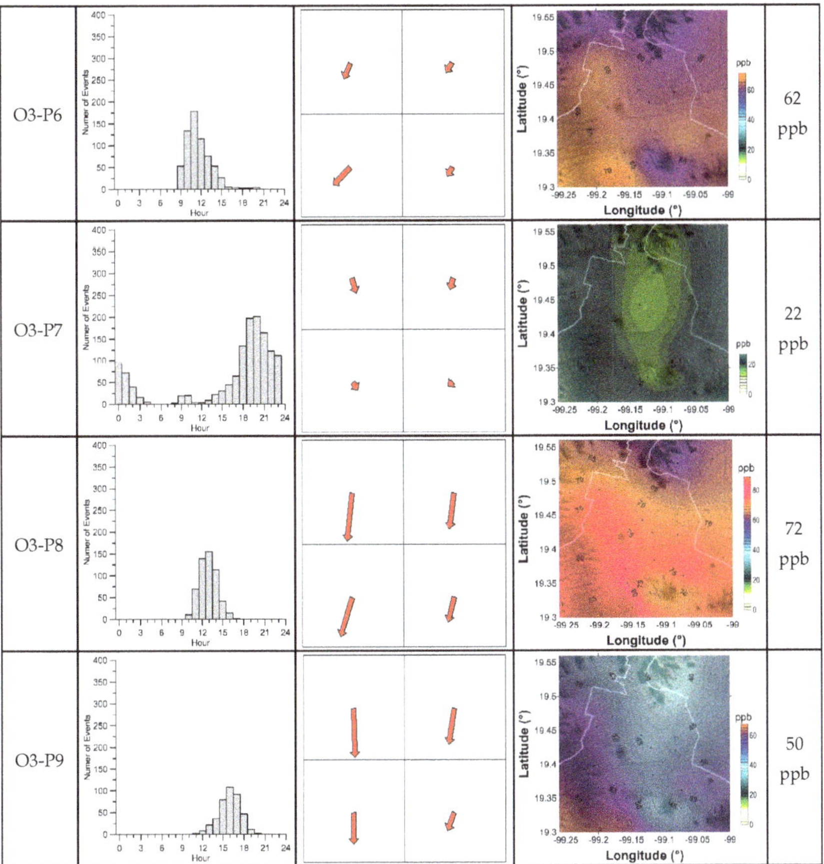

Figure 10. Flow patterns of ozone in the Mexico City region during the period 2001–2010. For each flow pattern, the figures show the hourly population (second column), the mean flow vectors (relative to the pattern) at the city quadrants (third column), the surface concentration distributions (fourth column) averaged over the events of the flow pattern, and the mean spatial concentration (last column). The size of red arrows that stand in for the O_3 flow vectors indicates the flow intensity in a scale relative to the edge size of the squares that represent the quadrants.

The flow patterns O3-P6, O3-P8, and O3-P9 revealed the highest flow intensities (43, 82, and 74 $\mu gm^{-2}s^{-1}$, on annual average). Seasonally, we observed the highest O_3 flow intensities during the spring, summer, and autumn.

The flow patterns O3-P3 and O3-P5 bring ozone from East to West in the city following the trade winds (pattern WIND-P2). These patterns reveal flow vectors from NE in the west quadrants of the city, and because of the mountains barrier (Sierra Las Cruces and Sierra Ajusco Chichinautzin), they contribute to the ozone accumulation at SW sector of the Mexico City region. The O_3 surface distributions, averaged over the events of these flow patterns, reveal, in fact, an evident and significant ozone accumulation (ranging from 70 to 100 ppb, approximately) at the SW quadrant.

The flow patterns O3-P1 and O3-P7 comprised nighttime flow events and had the first and second-largest occurrence frequencies. However, their flow intensities are tiny because of the low ozone concentrations (12 ppb and 22 ppb on average, respectively) in the nocturnal atmosphere (ozone production is of photochemical nature). It is interesting to note that the patterns O3-P1 and

O3-P7 and the patterns NO2-P1 and NO2-P4 reveal opposite behaviors (during the night, one observes NO_2 accumulation while O_3 decreases to its lowest levels) due to its relationship through the atmospheric photochemistry.

The O3-P2 is a flow pattern detected during the evening, driven by the UHI winds. It brings ozone from the surrounding parts of the city towards downtown, where an upwards convective flow takes ozone out there again, keeping the ozone levels relatively small in the city.

During winter, the flow events of the O3-P4 pattern bring ozone from west to east in the west side quadrants with small flow intensities, but the same flow pattern carries a considerable amount of ozone from south to north in the east side quadrants of the city, following the ventilation channel located at the west side of Sierra Nevada.

In general, the O_3 surface distributions show relative small concentrations where the NO_2 surface distributions show large emissions and vice versa. It is an evident behavior because of the NOx-O_3 photochemical interactions.

3.2.3. The PM_{10} Flow Patterns

We recognized nine PM_{10} flow patterns, which we enumerated in Table 6 and sketched out in Figure 11. In Table 6, we presented the seasonal and annual averages of the occurrence frequencies and the flow intensities of the PM_{10} flow patterns, expressed in % and $\mu g/m^2 s$, respectively. The directional characteristics of wind inherited by the PM_{10} patterns are also briefly indicated in this table.

Table 6. Frequencies (%) and flows ($\mu g/m^2 s$) of the Mexico City PM_{10} flow patterns (2001–2010).

Pattern	Wind Pattern Inherited	JAN–MAR Freq	JAN–MAR Flow	APR–JUN Freq	APR–JUN Flow	JUL–SEP Freq	JUL–SEP Flow	OCT–DEC Freq	OCT–DEC Flow	ANNUAL Freq	ANNUAL Flow
PM10-P1	WIND-P1	47.50	7.83	44.78	8.81	36.10	7.18	47.76	5.48	44.01	7.32
PM10-P2	WIND-P3	22.96	21.59	27.01	22.33	23.10	21.95	18.41	26.57	22.86	23.11
PM10-P3	WIND-P6	10.37	35.14	15.43	40.72	20.43	18.59	0.00	0.00	11.56	23.61
PM10-P4	WIND-P7	9.68	28.89	4.90	27.44	0.00	0.00	7.35	9.11	5.46	16.36
PM10-P5	WIND-P4	5.05	67.28	0.00	0.00	0.00	0.00	1.45	71.00	1.61	34.57
PM10-P6	WIND-P5	4.44	70.95	3.11	55.82	6.70	27.05	0.00	0.00	3.56	38.46
PM10-P7	WIND-P5	0.00	0.00	4.76	110.13	5.21	63.14	0.00	0.00	2.50	43.32
PM10-P8	WIND-P5	0.00	0.00	0.00	0.00	8.47	42.87	12.52	55.29	5.29	24.54
PM10-P9	WIND-P2	0.00	0.00	0.00	0.00	0.00	0.00	12.52	19.84	3.15	4.96

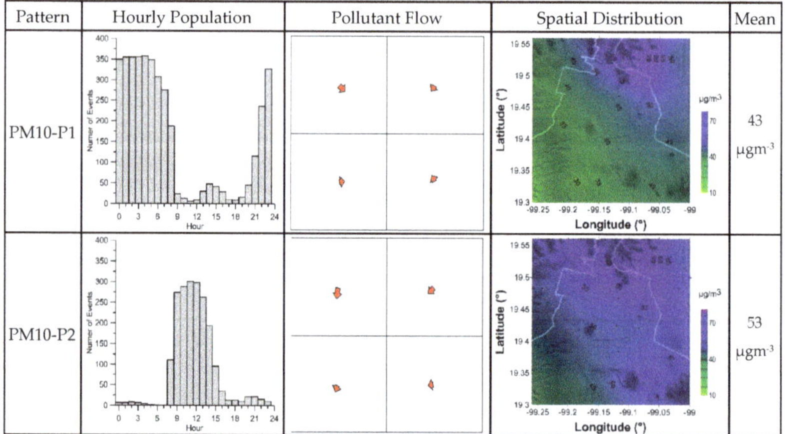

Figure 11. *Cont.*

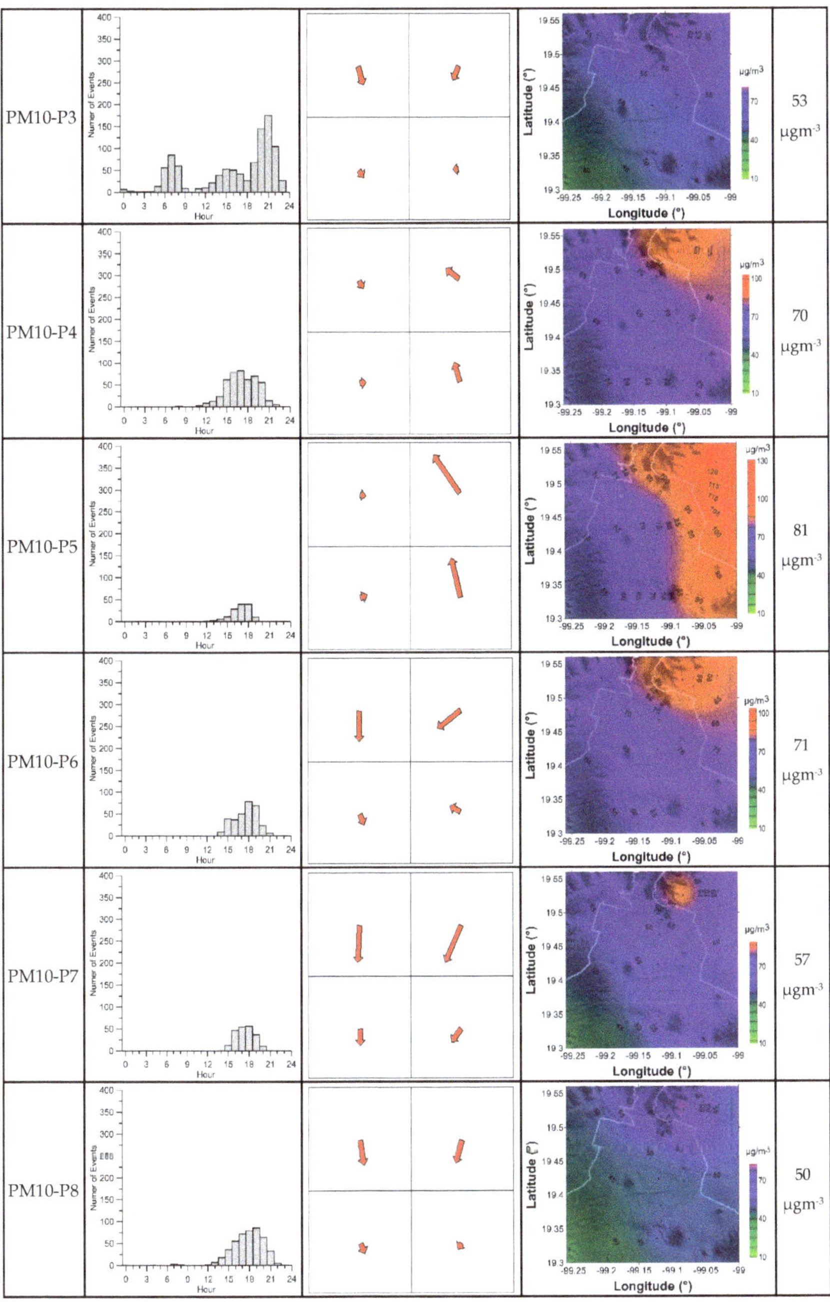

Figure 11. *Cont.*

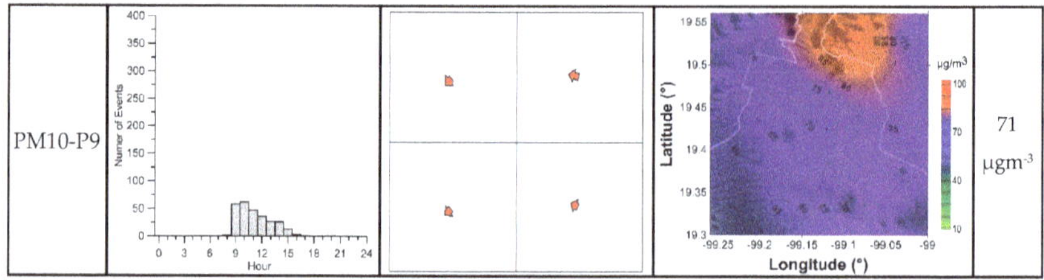

Figure 11. Flow patterns of PM_{10} in the Mexico City region during the period 2001–2010. For each flow pattern, the figures show the hourly population (second column), the mean flow vectors (relative to the pattern) at the city quadrants (third column), the surface concentration distributions (fourth column) averaged over the events of the flow pattern, and the mean spatial concentration (last column). The size of red arrows that stand in for the PM_{10} flow vectors indicates the flow intensity in a scale relative to the edge size of the squares that represent the quadrants.

The PM_{10} flow patterns with the highest annual flow intensities were PM10-P7 (43 $\mu gm^{-2}s^{-1}$), PM10-P6 (39 $\mu gm^{-2}s^{-1}$), PM10-P5 (35 $\mu gm^{-2}s^{-1}$), and PM10-P8 (26 $\mu gm^{-2}s^{-1}$); but their annual frequencies were of the smaller (3%, 4%, 2%, and 5%, respectively). All these four patterns reveal a remarkable flow intensity at the NE city quadrant throughout the year, although with different flow directions. The patterns PM10-P6, PM10-P7, and PM10-P8 exhibited also flow vectors with intense northerly components in the NW quadrant, which convey particle PM_{10} from the north to the south in the west side of the city. The events of these patterns also carried particulate matter from the northeast to the south sectors of the city, particularly in the NE quadrant. The flow intensities of the flow pattern PM10-P6 were 71, 56, and 27 $\mu gm^{-2}s^{-1}$ during the winter, spring, and summer seasons; for PM10-P7 were 110 and 63 $\mu gm^{-2}s^{-1}$ during the spring and summer, and for PM10-P8 were 43 and 55 $\mu gm^{-2}s^{-1}$ during the summer and autumn, respectively.

The events of the PM10-P4 and PM10-P5 flow patterns carried particles from south to northwest in the quadrants of the east side of the city: PM10-P4 with flow intensities of 29 and 27 $\mu gm^{-2}s^{-1}$ during the winter and spring, and PM10-P5 with flow intensities of 67 and 71 $\mu gm^{-2}s^{-1}$ during the winter and autumn, respectively.

The events of the flow patterns PM10-P1 and PM10-P2 occurred throughout the year with the highest frequencies of occurrence (44% and 22%, on an annual average). The flow intensity of the PM10-P1 (a nocturnal pattern) was too small all year, but the flow intensity of the PM10-P2 (a diurnal pattern) was similar to those of the other flow patterns. The patterns PM10-P1 and PM10-P3 reflect the influence of the downslope winds driven by the mountain-valley system. These patterns, furthermore, reflect the effect of the urban heat island, which is revealed in the hourly population by the small peak around the hour 15 (daylight winds converging to the downtown).

For all the PM_{10} flow patterns, the average surface distributions of the pollutant reveal the NE quadrant as the zone of the city with the highest PM_{10} concentrations, and the SW quadrant as the zone with the lowest levels of PM_{10}. It suggests that, at the NE of the town, in the surrounding area of Xalostoc on the east side of the Sierra de Guadalupe, there exists considerable sources of particulate matter, which release PM_{10} to the atmosphere all year long.

It is interesting to observe that the events of the pattern PM10-P5 revealed flow vectors at the NE and SE quadrants with intense southerly components, which produced, on average, the highest mean spatial concentration of PM_{10} (81 μgm^{-3}) in the city. This result displays the east side quadrants of the city as the most polluted areas by PM_{10} during the winter and autumn.

3.2.4. The SO_2 Flow Patterns

Table 7 and Figure 12 enumerate and sketch out the SO_2 flow patterns that we identified. Table 7 presents the average seasonal and annual occurrence frequencies and flow intensities of these patterns, including their relationship with the wind circulation patterns. Figure 12 shows the hourly population of the patterns, the corresponding flow vectors, and the surface distribution of the SO_2 emissions produced by the events of these flow patterns.

Table 7. Frequencies (%) and flows (μg/m^{2}s) of the Mexico City SO_2 flow patterns (2001–2010).

Pattern	Wind Pattern Inherited	JAN–MAR (Winter)		APR–JUN (Spring)		JUL–SEP (Summer)		OCT–DEC (Autumn)		ANNUAL	
		Freq	Flow	Freq	Flow	Freq	Flow	Freq	Flow	Freq	Flow
SO2-P1	WIND-P1	24.49	8.36	40.11	3.98	29.48	4.21	36.64	3.76	32.71	5.08
SO2-P2	WIND-P2	17.08	17.38	15.38	8.84	24.86	6.85	19.09	8.98	19.13	10.51
SO2-P3	WIND-P6	12.92	20.64	12.23	12.22	0.00	0.00	11.75	24.80	9.19	14.41
SO2-P4	WIND-P6	6.39	32.30	0.00	0.00	5.39	21.32	7.17	19.54	4.74	18.29
SO2-P5	WIND-P5	3.80	47.19	0.00	0.00	0.00	0.00	4.40	44.40	2.04	22.90
SO2-P6	WIND-P4	35.32	1.19	0.00	0.00	0.00	0.00	0.00	0.00	8.71	0.30
SO2-P7	WIND-P3	0.00	0.00	17.08	17.24	16.30	15.11	20.95	16.85	13.65	12.30
SO2-P8	WIND-P3	0.00	0.00	4.58	34.87	1.81	30.81	0.00	0.00	1.60	16.42
SO2-P9	WIND-P6	0.00	0.00	10.62	14.65	22.15	10.08	0.00	0.00	8.23	6.18

We observed, in general, that all the SO_2 flow patterns presented small flow intensities and produced small surface concentrations (we underline that the Mexican air quality standard for SO_2 is 110 ppb on a 24 h average, and 25 ppb on an annual average). Nevertheless, it is particularly interesting to observe that the flow patterns SO2-P3 (winter, spring, and autumn), SO2-P4 (winter, summer, and autumn), SO2-P5 (winter and autumn), and SO2-P9 (spring and summer), which were detected mainly during the night hours, exhibit the flow vector at the NW quadrant with a flow intensity relatively larger than in the other quadrants. However, the corresponding wind patterns show winds of similar magnitudes in all the quadrants.

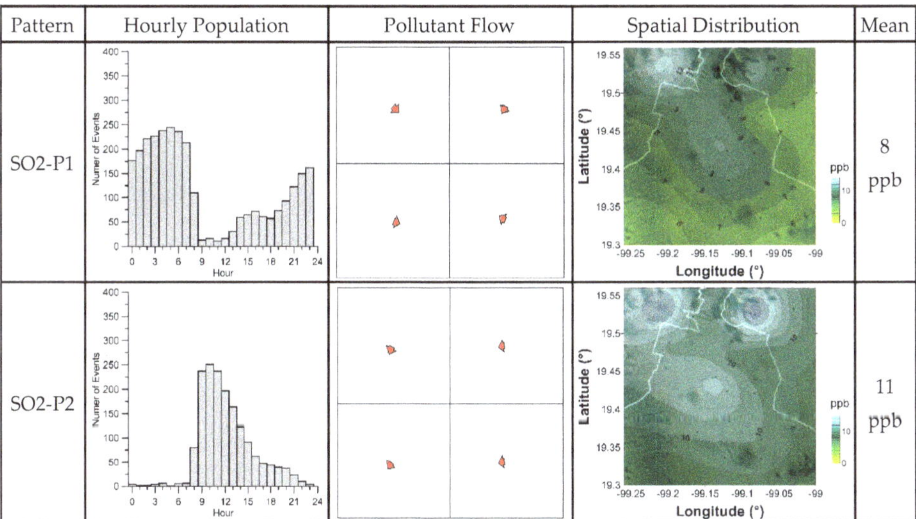

Figure 12. *Cont.*

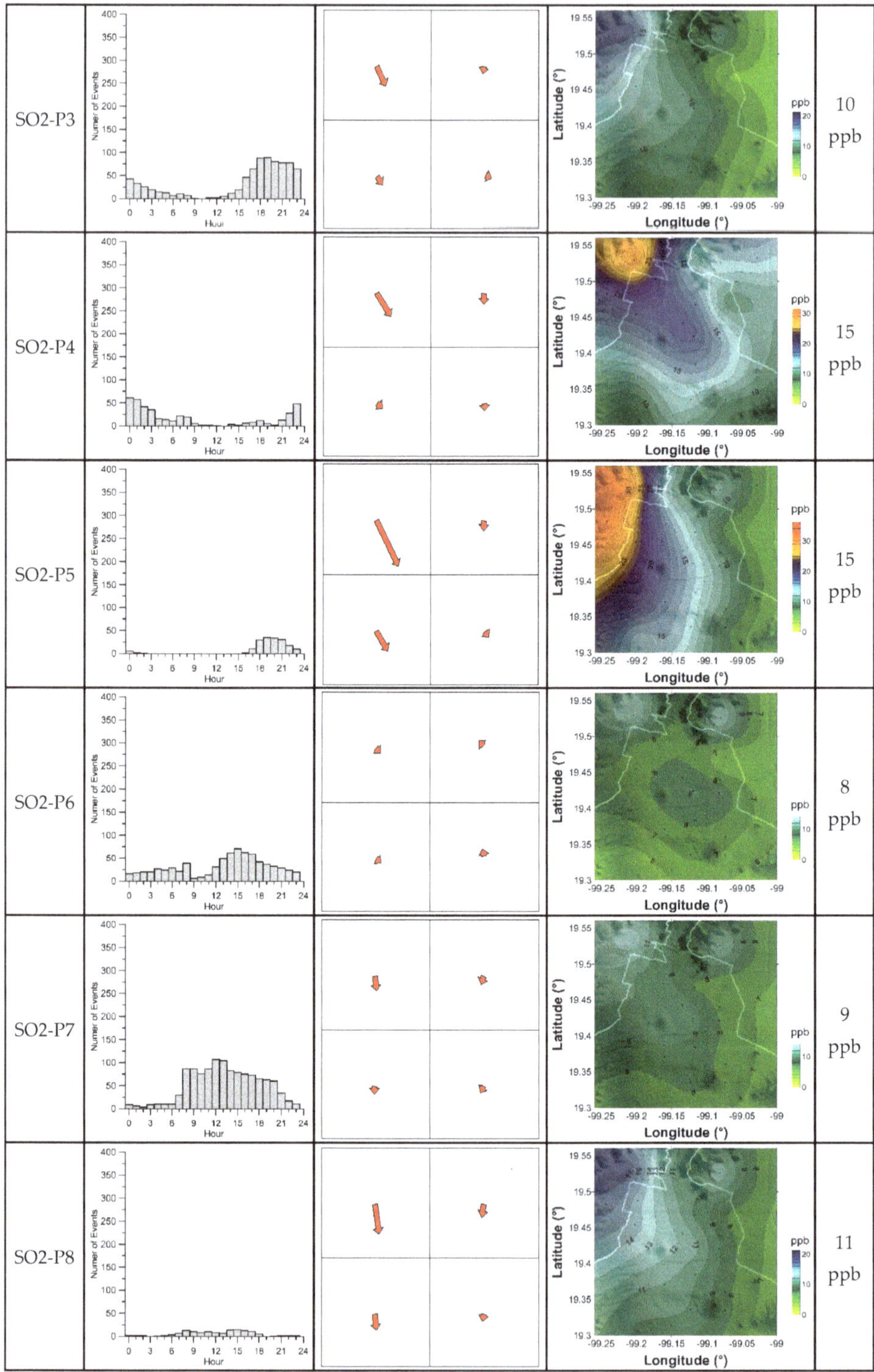

Figure 12. *Cont.*

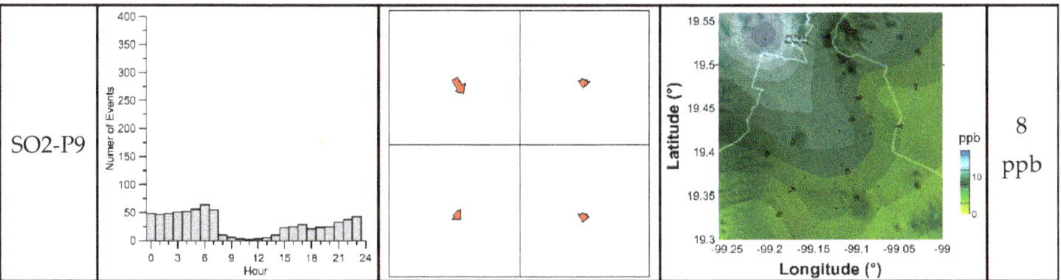

Figure 12. Flow patterns of sulfur dioxide in the Mexico City region during the period 2001–2010. For each flow pattern, the figures show the hourly population (second column), the mean flow vectors (relative to the pattern) at the city quadrants (third column), the surface concentration distributions (fourth column) averaged over the events of the flow pattern, and the mean spatial concentration (last column). The size of red arrows that stand in for the SO_2 flow vectors indicates the flow intensity in a scale relative to the edge size of the squares that represent the quadrants.

The events of the flow patterns SO2-P7 (spring, summer, and autumn) and SO2-P8 (spring and summer) occurred during daylight hours and show the NW-quadrant flow vector larger than in the other quadrants. Consistently with these observations, the mean spatial SO_2 surface distributions reveal the NW quadrant of the city as the zone with the highest levels of concentration, particularly in the case of the patterns SO2-P4 and SO2-P5. This situation seems to indicate that in the NW quadrant of the city, there are intense activities that release sulfur dioxide to the atmosphere during the night.

The SO2-P1 and SO2-P2 flow patterns, as was the case with the other pollutants, were the most frequent patterns (33% and 19%, respectively, on annual average) and the only ones detected all over the year; however, their flow intensities were tiny.

4. Conclusions

We used a simple model to define the pollutant-flow conditions in terms of the horizontal components of the flow vector and its gradients (divergence and curl). We applied this model to the hourly data of wind and pollutant concentration measured during the period 2001–2010 by the atmospheric monitoring network of Mexico City. We obtained the main flow patterns of NO_2, O_3, PM_{10}, and SO_2 using a 4-cell model of the city and the k-means algorithm of cluster analysis. Estimations of the seasonal and annual occurrence frequencies and flow intensities of the flow patterns were obtained.

The pollution flow patterns reflect some of the wind circulation conditions of the Mexico City wind patterns; however, since the pollutant-flow conditions also depend on the pollutant concentration, the flow patterns revealed situations of pollutant transport that cannot be inferred simply from the wind circulation modes. Different pollutants under the same wind patterns produced different pollutant-flow patterns. For example, the SO_2 flow patterns (SO2-P3, SO2-P4, and SO2-P5) display large flow intensities at the NW quadrant bringing this pollutant towards the Mexico City region, mainly during winter and autumn seasons; however, the corresponding wind patterns exhibit wind velocities with similar magnitudes in almost all the quadrants. The events of the flow patterns SO2-P8 occurred with similar conditions, but during the spring and summer periods. Similarly, the PM10-P4 and PM10-P5, and PM10-P6 and PM10-P7 flow patterns revealed large flow intensities in the NE quadrant, although winds have similar magnitudes in other quadrants.

The pollutant-flow patterns of NO_2, O_3, PM_{10}, and SO_2 represent the main scenarios of the pollutant transport in the Mexico City region. They provide information that can guide, enrich, and enlighten the results of Mexico City air quality studies based on modeling techniques.

Author Contributions: Conceptualization, A.S.; Formal analysis, A.S.; Investigation, S.C.-S. and A.-T.C.-M.; Methodology, A.S., S.C.-S. and A.-T.C.-M.; Software, A.S. and S.C.-S.; Supervision, A.S.; Writing—original draft, A.S., S.C.-S. and A.-T.C.-M.; Writing—review & editing, A.S. and A.-T.C.-M.

Funding: This research received no external funding.

Acknowledgments: We thank Ana Laura Colín Aguilar (Instituto Nacional de Electricidad y Energías Limpias) for beneficial comments to make the paper more comprehensible.

Conflicts of Interest: The authors declare no conflict of interest.

References

1. Pascal, M.; Corso, M.; Chanel, O.; Declercq, C.; Badaloni, C.; Cesaroni, G.; Henschel, S.; Meister, K.; Haluza, D.; Martin-Olmedo, P.; et al. Assessing the public health impacts of urban air pollution in 25 European cities: Results of the Aphekom project. *Sci. Total Environ.* **2013**, *449*, 390–400. [CrossRef] [PubMed]
2. Sicard, P.; Lesne, O.; Alexandre, N.; Mangin, A.; Collomp, R. Air quality trends and potential health effects —Development of an aggregate risk index. *Atmos. Environ.* **2011**, *45*, 1145–1153. [CrossRef]
3. Secretaría del Medio Ambiente. *Gobierno del Distrito Federal. Inventario de Emisiones de Contaminantes Criterio de la Zona Metropolitana del Valle de México*, 1st ed.; Secretaría del Medio Ambiente del DF: Mexico City, Mexico, 2008.
4. Bravo, H.A.; Torres, R.J. Air Pollution Levels and Trends in the Mexico City Metropolitan Area. In *Urban Air Pollution and Forests: Resources at Risk in the Mexico City Air Basin*; Springer: New York, NY, USA, 2002; pp. 121–159.
5. Bonner, J.C.; Rice, A.B.; Lindroos, P.M.; O'Brien, P.O.; Dreher, K.L.; Rosas, I.; Alfaro-Moreno, E.; Osornio-Vargas, A.R. Induction of the lung myofibroblast PDGF receptor system by urban ambient particles from Mexico City. *Am. J. Respir. Cell. Mol. Biol.* **1998**, *19*, 672–680. [CrossRef]
6. Osornio-Vargas, A.R.; Bonner, J.C.; Alfaro-Moreno, E.; Martinez, L.; Garcia-Cuellar, C.; Ponce-de-Leon-Rosales, S.; Miranda, J.; Rosas, I. Proinflammatory and cytotoxic effects of Mexico City air pollution particulate matter in vitro are dependent on particle size and composition. *Environ. Health Perspect.* **2003**, *111*, 1289–1293. [CrossRef]
7. Jáuregui, E. Local wind and air pollution interaction in the Mexico basin. *Atmósfera* **1988**, *1*, 131–140.
8. De Foy, B.; Caetano, E.; Magaña, V.; Zitácuaro, A.; Cárdenas, B.; Retama, A.; Ramos, R.; Molina, L.T.; Molina, M.J. Mexico City basin wind circulation during the MCMA-2003 field campaign. *Atmos. Chem. Phys.* **2005**, *5*, 2267–2288. [CrossRef]
9. Oke, T.R.; Zeuner, G.; Jauregui, E. The Surface Energy Balance in Mexico City. *Atmos. Environ. Part B Urban Atmos.* **1992**, *26*, 433–444. [CrossRef]
10. Jauregui, E.; Luyando, E. Global radiation attenuation by air pollution and its effects on the thermal climate in Mexico City. *Int. J. Clim.* **1999**, *19*, 683–694. [CrossRef]
11. Salcido, A.; Celada, A.T.; Villegas, R.; Salas, H.; Sozzi, R.; Georgiadis, T. A micrometeorological database for the Mexico City Metropolitan Area. *Nuovo Cim. C Geophys. Space Phys. C* **2003**, *26*, 317–355.
12. Castro, T.; Salcido, A. Influencia de la contaminación atmosférica de la Zona Metropolitana de la Ciudad de México en tres sitios perimetrales. In *Contaminación Atmosférica V*; El Colegio Nacional: Mexico City, Mexico, 2006.
13. Molina, L.T.; Madronich, S.; Gaffney, J.S.; Apel, E.; de Foy, B.; Fast, J.; Ferrare, R.; Herndon, S.; Jimenez, J.L.; Lamb, B.; et al. An overview of the MILAGRO 2006 Campaign: Mexico City emissions and their transport and transformation. *Atmos. Chem. Phys.* **2010**, *10*, 8697–8760. [CrossRef]
14. Bossert, J.E. An Investigation of Flow Regimes Affecting the Mexico City Region. *J. Appl. Meteorol.* **1997**, *36*, 119–140. [CrossRef]
15. Fast, J.D.; Zhong, S. Meteorological factors associated with inhomogeneous ozone concentrations within the Mexico City basin. *J. Geophys. Res.* **1998**, *103*, 18927–18946. [CrossRef]
16. Doran, J.C.; Abbott, S.; Archuleta, J.; Bian, X.; Chow, J.; Coulter, R.L.; de Wekker, S.F.J.; Edgerton, S.; Elliott, S.; Fernandez, A.; et al. The IMADA-AVER Boundary Layer Experiment in the Mexico City Area. *Bull. Am. Meteorol. Soc.* **1998**, *79*, 2497–2508. [CrossRef]
17. Mora, M.; Braun, R.A.; Shingler, T.; Sorooshian, A. Analysis of remotely sensed and surface data of aerosols and meteorology for the Mexico Megalopolis Area between 2003 and 2015. *J. Geophys. Res. Atmos. JGR* **2017**, *122*, 8705–8723. [CrossRef] [PubMed]

18. Fast, J.D.; de Foy, B.; Acevedo Rosas, F.; Caetano, E.; Carmichael, G.; Emmons, L.; McKenna, D.; Mena, M.; Skamarock, W.; Tie, X.; et al. A meteorological overview of the MILAGRO field campaigns. *Atmos. Chem. Phys.* **2007**, *7*, 2233–2257. [CrossRef]
19. Salcido, A.; Celada-Murillo, A.T.; Castro, T. A meso-β Scale Description of Surface Wind Events in Mexico City during the MILAGRO 2006 Campaign. In Proceedings of the IASTED Technology Conferences, Environmental Management and Engineering (EME), Banff, AB, Canada, 15–17 July 2010; Alhajj, R.S., Leung, V.C.M., Petela, R., Saif, M., Thring, R., Eds.; Acta Press: Calgary, AB, Canada, 2010. Track 699-041. [CrossRef]
20. Celada-Murillo, A.; Carreón-Sierra, S.; Salcido, A.; Castro, T.; Peralta, O.; Georgiadis, T. Main Characteristics of Mexico City Local Wind Events during the MILAGRO 2006 Campaign within a Meso-β Scale Lattice Wind Modeling Approach. *ISRN Meteorol.* **2013**, *2013*, 605210. [CrossRef]
21. Salcido, A.; Carreón-Sierra, S.; Celada-Murillo, A.T. A Brief Clustering Analysis of the Mexico City Local Wind States Occurred During the MILAGRO Campaign. In Proceedings of the IASTED International Conference on Environmental Management and Engineering (EME), Banff, AB, Canada, 16–17 July 2014. [CrossRef]
22. Klaus, D.; Poth, A.; Voss, M.; Jauregui, E. Ozone distributions in Mexico City using principal component analysis and its relation to meteorological parameters. *Atmósfera* **2001**, *14*, 171–188.
23. De Foy, B.; Fast, J.D.; Paech, S.J.; Phillips, D.; Walters, J.T.; Coulter, R.L.; Martin, T.J.; Pekour, M.S.; Shaw, W.J.; Kastendeuch, P.P.; et al. Basin-scale wind transport during the MILAGRO field campaign and comparison to climatology using cluster analysis. *Atmos. Chem. Phys.* **2008**, *8*, 1209–1224. [CrossRef]
24. Salcido, A.; Carreón-Sierra, S.; Georgiadis, T.; Celada-Murillo, A.T.; Castro, T. Lattice Wind Description and Characterization of Mexico City Local Wind Events. Period 2001–2006. *Climate* **2015**, *3*, 542–562. [CrossRef]
25. Carreón-Sierra, S.; Salcido, A.; Castro, T.; Celada-Murillo, A.T. Cluster Analysis of the Wind Events and Seasonal Wind Circulation Patterns in the Mexico City Region. *Atmosphere* **2015**, *6*, 1006–1031. [CrossRef]
26. Mexico City Air Quality Monitoring Network (Sistema de Monitoreo Atmosférico de la Ciudad de México). Available online: http://www.aire.cdmx.gob.mx/default.php (accessed on 23 July 2019).
27. Everitt, B.S.; Landau, S.; Leese, M. *Cluster Analysis*, 4th ed.; Arnold: London, UK, 2001; ISBN 9780340761199.
28. Hartigan, J.A.; Wong, M.A. Algorithm AS 136: A K-Means Clustering Algorithm. *Appl. Stat.* **1979**, *28*, 100–108. [CrossRef]
29. James, G.; Witten, D.; Hastie, T.; Tibshirani, R. *An Introduction to Statistical Learning with Applications in R*; Springer: New York, NY, USA; Heidelberg, Germany; Dordrecht, The Netherlands; London, UK, 2013; Corrected at Eighth Printing 2017. [CrossRef]
30. Shaefer, J.T.; Doswell, C.A. On the interpolation of a vector field. *Mon. Weather Rev.* **1979**, *107*, 458–476. [CrossRef]
31. Salcido, A. A Phenomenological Gradient Approach to Generalized Constitutive Equations for Isotropic Fluids. *J. Appl. Math. Phys.* **2018**, *6*, 1494–1506. [CrossRef]
32. Lohninger, H. *DataLab 3.911*; Epina GmbH: Gütersloh, Germany, 2019.
33. Klaus, D.; Jauregui, E.; Poth, A.; Stein, G.; Voss, M. Regular circulation structures in the tropical basin of Mexico City as a consequence of the urban heat island effect. *Erdkunde* **1999**, *53*, 231–243. [CrossRef]
34. Krishnamurti, T.N. The Subtropical Jet Stream of Winter. *J. Meteorol.* **1960**, *18*, 172–191. [CrossRef]
35. Doran, J.C.; Zhong, S. Thermally Driven Gap Winds into the Mexico City Basin. *J. Appl. Meteorol.* **2000**, *39*, 1330–1340. [CrossRef]

Article

The Role of Green Infrastructures in Urban Planning for Climate Change Adaptation

Luisa Sturiale [1,*] and Alessandro Scuderi [2]

[1] Department of Civil Engineering and Architecture, University of Catania, 95100 Catania, Italy
[2] Department of Agricultural, Food and Environment, University of Catania, 95100 Catania, Italy; scuderia@unict.it
* Correspondence: luisa.sturiale@unict.it

Received: 31 July 2019; Accepted: 28 September 2019; Published: 4 October 2019

Abstract: The population that lives in cities has surpassed the one that lives in the countryside. Cities are recognized as apriority source of pollution. The degradation of air quality and the phenomenon of Urban Heat Island (UHI) are some of the most well-known consequences of urban development. The adaptation of the cities is emerging as one of the greatest challenges that urban planners will face in this century. Urban Green Infrastructures (GIs) could help cities adapt to climate change, and the strategy of expansion of greening in urban planning could play an important role in enhancing the sustainability and resilience of cities and communities. Many studies have shown the benefits of GIs to climate change mitigation and adaptation in urban areas and their role as an important urban planning tool to satisfy environmental, social, and economic needs of urban areas. The objective of this article is to propose a methodological approach to evaluate the social perception of citizens regarding urban green areas. The proposed methodology, applied to the reality of the "urban green system" of Catania, is based on an integrated approach between participatory planning and the methods social multi criteria evaluation to guiding the city's government to realize a new urban resilient development.

Keywords: urban green system; ecosystem services; climate change benefits; resilient city; urban resilient development; green urban planning

---

1. Introduction

The actions against climate change and its effects on society and the environment are oriented in two directions: Mitigation, to progressively reduce the emissions of climate-changing gases responsible for global warming, and adaptation, to reduce the vulnerability of environmental, social, and economic systems and to increase their capacity for climate resilience." Many adaptation and mitigation options can help address climate change, but no single option is sufficient by itself. Effective implementation depends on policies and cooperation at all scales and can be enhanced through integrated responses that link mitigation and adaptation with other societal objectives" [1].

The evaluation of the integrated effects of planning and planning choices aimed at reducing climate-changing emissions is a priority theme in the document "Transforming our world: The 2030 Agenda for Sustainable Development [2]". The 2030 agenda establishes 17 sustainable development goals (SDGs) and 169 targets to be achieved within the next 15 years. Goal 11, sustainable cities and communities, is specifically dedicated to urban systems, and its ambitious goal is to "make cities and human settlements inclusive, safe, resilient, and sustainable."

The percentage of people living in urban areas will increase from 50.0% in 2010 to nearly 70.0% by 2150 [3]. For the first time in history, the population that lives in cities has surpassed the one that lives in the countryside or outside the inhabited centers. The cities are recognized as the priority source of pollution. Most energy consumption is connected to cities, which have to make the greatest efforts to manage sustainable resources under the social, environmental, and economic aspects and to improve the quality of life of their citizens. Heat waves in cities generate serious inconveniences for the most vulnerable citizens, especially the elderly and children. Urban socio-ecological systems are characterized by a high population density, an extensive change in land use, and the use of natural resources not directly present locally. In Europe, urbanization processes are progressing rapidly, causing soil sealing and the reduction of its functions and quality. One of the major consequences of urbanization, in terms of the impact on human health and environmental quality, is the "urban heat island" (UHI) effect (the phenomenon whereby cities appear to be warmer than the surrounding rural area). It is estimated that climate change will greatly aggravate the extent of the UHI, particularly in hot regions characterized by periods of summer dryness such as the Mediterranean Basin.

Protecting, upgrading, and increasing urban and peri-urban forests and street trees through the enhancement of the green infrastructures (GIs), is therefore fundamental for the sustainable development of urban areas that represent "demand areas for Ecosystem Services, "the goods and services provided to man by nature.

The maintenance of urban green spaces is also one of the approaches suggested by the IPCC (2014) [1]. For the management of climate change risk through adaptation, in particular through the reduction of vulnerability and exposure through development, planning and practices that include "low-regret" measures, i.e. those that produce benefits even in the absence of climate change and with which the adaptation costs are relatively low compared to the benefits of the action. Lastly, the maintenance of urban green spaces is one of the approaches suggested by SDGs 11 (Sustainable Cities and Communities) of the UN Agenda to 2030 [2].

In recent thinking, the GIs have been identified as 'best practices' in local governance when combined with traditional "grey" infrastructure to achieve greater urban sustainability and resilience. Moreover, GIs are recognized for their value for adapting to the emerging and irreversible impacts of climate change. In addition, some local governments have adopted GIs as a climate change adaptation measure, particularly if the strategies result in multiple other benefits. Indeed, adaptation to climate change is also seen as having ecological, economic, and social dimensions [3].

For three out of four European citizens, climate change is a very serious problem. The changes observed in the climate are already having far-reaching repercussions on ecosystems, economic sectors, human health, and welfare in Europe. Overall, the economic losses recorded in Europe in the period 1980–2016 caused by meteorological phenomena and other extreme climate-related events exceeded Euro 436 billion. The biggest losses would be in the industrial, transport, and energy sectors (CE, COM (2018), 738 final).

The cities will have an important role to adopt the law and the provisions that are necessary at the various levels but also to ensure the best quality of life in urban areas. Climate impact requires the use of innovative solutions and the rethinking of urban management and planning [4].

New urban and territorial structures, low energy consumption buildings and infrastructures, green areas and the adoption of advanced technologies mitigate global emissions and local pollution, promote adaptation to climate change, reduce the energy costs of families and businesses, and improve the climate of cities.

GIs and their integration in urban planning appear as one of the most appropriate and effective ways to improve the microclimate and face the impacts of climate change and mainly the UHI effect. GIs forms include green roofs, green walls, urban forest, bioswales, rain gardens, urban agriculture (urban gardens; community gardening; collective green; peri-urban agriculture, agricultural parks), river parks, local products markets, areas of constructed wetlands, alternative energy farms, and nature conservation areas, among the most common (Figure 1).

Figure 1. The different forms of Green Infrastructures (adapted from EU 2016) [5].

GIs provide a range of climate change services that can make both a substantial contribution towards adapting to climate change and a limited yet important contribution towards mitigating climate change. Such natural interventions are increasingly being recognized as a desirable 'win-win' approach to combating climate change, as they also help to deliver multiple other social, economic, and environmental benefits.

Green Infrastructures, according to the European Union (E.U.) definition, "... are networks of natural and semi-natural areas planned at strategic level with other environmental elements, designed and managed in such a way as to provide a wide spectrum of ecosystem services. This includes green (or blue, in the case of aquatic ecosystems) and other physical elements in areas on land (including coastal areas) and marine areas. On the mainland, green infrastructures are present in a rural and urban context" [6].

In the EU, the term GIs was first introduced in the 2009 commission white paper, "Adapting to Climate Change" [7]. In all the normative acts of the EU, the term "Green Infrastructures" is used in connection with landscape resources, with particular emphasis on ecological connectivity. In contrast, the European Environment Agency (EEA) and other European programs choose to use the term "green spaces," "green systems, "or "green structure" when referring to the urban environment or other related issues [8,9].

The objectives of the EU GIs Strategy (2013) [6] are:

- To enhance, conserve, and restore biodiversity by inter alia increasing of spatial and functional connectivity between natural and semi-natural areas and improving landscape permeability and mitigating fragmentation.
- To maintain, strengthen, and, where adequate, to restore the good functioning of ecosystems in order to ensure the delivery of multiple ecosystem and cultural services.
- To acknowledge the economic value of ecosystem services and to ncreasethe value itself, by strengthening their functionality.
- To enhance the social and cultural link with nature and biodiversity, to acknowledge and increase the economic value of ecosystem services and to create incentives for local stakeholders and communities to deliver them.
- To minimize urban sprawl and its negative effects on biodiversity, ecosystem services, and human living conditions.

- To mitigate and adapt to climate change, to increase resilience, and reduce vulnerability to natural disaster risks such as floods, water scarcity and droughts, coastal erosion, forest fires, mudslides, and avalanches as well as urban heat islands.
- To make the best use of limited land resources in Europe.
- To contribute to a healthy living, better places to live, provisioning open spaces and recreation opportunities, increasing urban-rural connections, contributing to sustainable transport systems, and strengthening the sense of community [10].

Many studies show the existence of important links between urban green and impacts on climatic conditions and the reduction of the heat island effects. Parks and trees offer shaded areas and help to cool the air, and they protect from solar radiation. The green surfaces also have a heat absorption effect and lower thermal inertia than when compared to concrete or asphalt surfaces. The integration of vegetation in the facades and on the roofs of buildings helps to balance the interior temperatures.

This article proposes a brief overview of the GIs and their benefits and their role as an important urban planning tool to satisfy different environmental, social, and economic needs of urban areas and to realize a resilient urban development. Urban resilience aims to increase the ability of the whole urban system (including physical, environmental, and socio-economic perspectives) to develop its adaptive capacity, to resist and recover from shocks and stresses, and, at the same time, to reduce its vulnerability. In this context, it becomes interesting to investigate the relationship between GIs and climate comfort within cities and the role of the urban green spaces as an essential component of the policies of mitigation and adaptation to climate change in the planning process. This is followed by the evaluation of green planning policies and adaptation measures in the city of Catania, to acquire useful information in order to guide the city government towards resilient urban planning. The methodology proposed and applied to the green areas of Catania and the planning of investments in GIs, is based on an integrated approach between participatory planning techniques (based on the establishment of focus groups with the various stakeholders) and the NAIADE method (Novel Approach to Imprecise Assessment and Decision Environments) [11] (for the Multi-Critical Social Assessment—SMCE) for the "complex" information collected (quantitative and qualitative data).

2. The Green Infrastructures as Tools for Climate Adaptation of Cities: Experiences and Evaluations

The relationship between city and climate change is now documented internationally by various studies, which highlight the negative effects on the well-being of the population and environmental ecosystems. The phenomena of pollution, heat islands and urban decay are now known, to name a few [12–14]. The E.U. has given a strong impetus to the fight against climate change and environmental protection. To this end, among various measures, it has promoted the European Strategy for Adaptation to Climate Change (2013) [15]. That has the objective of making Europe more resilient to climate change, promoting two types of interventions: Mitigation and adaptation.

The definition of the urban adaptation strategies actively involves citizens and other interested stakeholders to favor "non-regret" interventions that can remedy existing problems and bring immediate benefits and socio-economic benefits to increase adaptive capacity and actions based on an ecosystemic or "green" approach [4]. These actions envisaged are aimed to promote experimental interventions of climatic adaptation of public spaces and encourage the diffusion of different forms of GIs and the increase of public and private urban green areas, adopting the logic of green and blue infrastructures, and safeguarding biodiversity in urban areas.

Climate change requires the use of innovative solutions and new tools for urban management and planning. New urban structures, low energy consumption buildings and infrastructures, green areas, and the adoption of advanced technologies mitigate global emissions and local pollution, which promote adaptation to climate change. In the new vision of a city, sustainable and resilient, the green areas assume ever greater importance and become multifunctional resources for the city and its inhabitants.

The connotation of GIs includes the green space of the complex urban ecosystem, composed of various forms of non-built spaces, including gardens, parks, vertical plants, forestry, urban gardens, agricultural land, greenways, wetlands, and waterways (green and blue infrastructures) [16–18].

The green spaces in urban areas provide indirect and direct benefits to human health and well-being, which are defined as Ecosystem Services (ESs). The ESs "… consist of the flows of matter, energy, and information coming from the stocks of natural capital, which are combined with the services of anthropogenic artifacts to generate well-being and quality of life …" [19]; they perform the following functions: Environmental-regulator; hydrogeological protection; social, recreational and therapeutic; cultural and educational; aesthetic–architectural.

In particular, the ESs help adapt cities for climate change, through the air purification, global climate regulation, urban temperature regulation, noise reduction, and runoff mitigation. Several studies show the existence of important links between urban greening and impacts on climatic conditions: Parks and trees offer shaded areas and help to cool the air, they are places to find relief during heat waves, they offer plant cover, and they protect from solar radiation; the integration of vegetation in the facades and on the roofs of buildings helps to balance the interior temperatures as well as protect the structures [20–24].

The implementation of GIs promote an integrated approach to land management, determine the positive effects under the aspect: Economic, in the containment of some of the damages resulting from hydrogeological instability; environmental, in the fight against climate change and in restoring the quality of environmental matrices (air, water, soil); in promoting the well-being of citizens and their social relations and promoting social inclusion [4].

The benefits of the GIs are:

- *Health and well-being*: Increasing life expectancy and reducing health inequality; improving levels of physical activity and health; improving psychological health and mental well-being.
- *Climate change*: Heat amelioration; reducing flood risk; improving water quality; sustainable urban drainage; sustainable transport; improving air quality.
- *Land regeneration*: Regeneration of previously developed land; improving quality of the place; increasing environmental quality and aesthetics.
- *Wildlife and habitats*: Increasing habitat area; increasing populations of some protected species; increasing species movement.
- *Economic growth and investments*: Inward investments and job creation, land and property values; local economic regeneration.
- *Stronger communities*: Social interaction, inclusion, and cohesion; community engagement; education and participation; a sense of place; experiencing nature [4].

Worldwide, GIs have been spreading for at least a decade for their social and environmental benefits and as a tool for fighting climate change in urban areas. It is not possible to report the multiple experiences of projects already completed and underway, but some interesting examples will be recalled here.

In the USA, several cities have planned the development of specific GIs plans (New York, Chicago, Washington D.C.) or have foreseen their presence in the climate protection action plans (San Diego); the Greenworks Philadelphia aims to turn Philadelphia into "the greenest city in America" and includes an extensive list of GIs measures (Figure 2) [25,26].

Figure 2. Green waterstorm infrastructure in Philadelphia (USA).

From northern Europe to the Mediterranean area, climate adaptation plans and experimental projects have been adopted or are being drafted for the creation of sustainable eco-neighborhoods and various GIs, in urban areas and also around cities, playing an important role in regulating urban sprawl, regulating urbanization, and the growing senseless land consumption.

Just to name a few, this is the case of the British Green Belts (in urban planning in the UK), the Anella Verda of Barcelona, which includes a network of 12 protected areas around the city connected by increasingly enhanced ecological corridors; the vertical forest in Milan (Figure 3); the Green Belt in Turin; the green ring of the municipality of Mirandola (Modena); urban gardens in Catania [27]. Other projects and studies related to the progressive insertion of the different forms of GIs in the green planning of cities are found from Europe to Asia: In Copenaghen [28], Berlin [29] (Figure 4), Hong Kong [30], Beijing [31], and the Pukou District in Nanjing [32].

Figure 3. Vertical forest in Milan (Italy).

Figure 4. Germany's green roofs.

There is a large literature of specific researches that explore the important role that the GIs in urban areas play in adapting for climate change and as low-cost mitigation strategies [33–35]. Other researches place particular attention on models for evaluating ESs offered by urban GIs [4,27,31,36].

GIs practices such as trees, shrubs, lawns, green roofs, and green walls have been proven efficient in reducing the urban sprawl, the UHI, the harmful air pollutants, and regulating the Green House Gas (GHG) emissions within the cities [36–39].

Based on the urban specificities, on the socio-economic and environmental aspects of the cities considered and on the types of GIs studied, the methodological approaches found in the literature are different. It is possible to identify the application of the i-Tree eco model to present output from energy exchange and hydrological models showing surface temperature and surface runoff in relation to the green infrastructure under current and future climate scenarios (in greater Manchester [40]), or to quantify in biophysical and monetary terms the ecosystem services air purification, global climate regulation, and the ecosystem disservice air pollution associated with biogenic emissions (in Barcelona [36]). Several researches have used integrated approaches recognizing the multi-functionality of GIs, highlighting both the economic aspects and social and environmental aspects (in Nanjing China, morphological spatial pattern analysis MSPA method [29]; in Catania Italy, the *eco-social-green* model using Social Multi Criteria Evaluation SMCE [4]) and for planning and managing urban green infrastructure for urban sustainability and resilient cities [41,42].

Municipalities are called to respond to climate problems with new governance tools to distribute the risk of impacts, which aims to involve citizens to a greater extent in the project proposals for interventions and measures. Indeed, it is important for urban governance to assess the social perception of investments in GI and to estimate the benefits of related ecosystem services. The interactions between stakeholders involved in urban projects have proved useful in different experiences but havenot yet been widely applied to climate change adaptation actions in cities [4].

The good functionality and the correct use of the public green areas require the support of multi-criteria evaluation tools and specific government tools, able to guide the administrators in the choices of planning of the green investments and management as well as to provide citizens with elements of knowledge and respect towards this important common good [43,44]. The success of a particular public green space is not solely in the hands of the architect, urban designer or townplanner; it also relies on people adopting, using, and managing the space. People make places, more than places make people [45,46].

The implementation of GIs promotes an integrated approach to land management, determines positive effects under the economic, social, and environmental aspects. The GIs become an important tool of action for climate adaptation, for the enhancement of ESs and biodiversity, and social cohesion in the model of a sustainable and resilient city of the future.

3. The Perception of the GIs in Urban Green Planning: The Case Study of Catania

In Italy, the GIs are still few, limited to individual local initiatives and in any case, are not included in system logic, which essential for achieving the objectives. The attention to the presence of green in urban areas, in its various forms and functions, has been growing in recent years, so much so that even at the regulatory level several laws have been promulgated on the subject. Among them, the Law on the 14 January2013, n. 10 rules for the development of urban green spaces, in which the functions of urban green are recognized:

- *Ecosystem services*: Positive effects on the local climate, on air quality, on noise levels, and soil stability are evident, biodiversity conservation

- *Socio-economic aspects*: Meet the needs for recreation, social relations, cultural and healthy growth of its inhabitants.

Despite the multiple benefits associated with green, the Italian situation still shows some critical issues. Urban green is mainly managed on a technical and prescriptive level rather than as a strategic resource to orient local development policies to quality and resilience [4].

The city of Catania is the second city, in terms of importance, surfaces, and inhabitants of the Sicily Region (180,000 square meters, about 350,000 inhabitants, a density of 1.7 inhabitants per square meter). Under the urbanistic aspect, the regulatory plan, drawn up in 1964, was designed for the socio-economic needs of the city of the 60s, where the priorities and the vision were completely different from the current ones, mainly in terms of environmental protection, attention to climate change, and social inclusion. Concerning the socio-economic development of Catania, for the future potential of the urban green areas, the municipal administration has planned a program of interventions in line with climate adaptation measures [47–49]. The known negative impacts of climate change are also recorded in Catania, and some indicators are shown in the following figures (Figures 5 and 6)

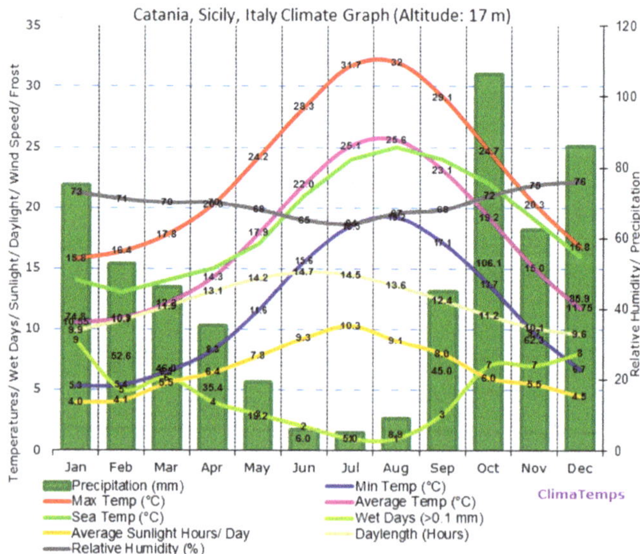

Figure 5. Trend of main climate indicators of the city of Catania (monthly averages).

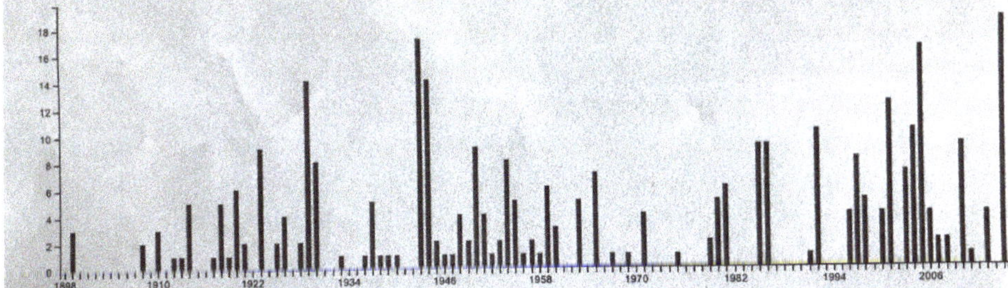

Figure 6. The hottest days in a year in Catania from 1900 to today.

The main climate indicators of Catania are indicated in Figure 5 and the climate trends from 1900 to the present in the area of Catania, analyzing the meteorological data (www.lab.gedidigital.it), it can be summarized as follows:

- The average temperature in the area between 2000 and 2017 was 0.6679 °C above the 20th-century average;
- The number of hot days (above 29 °C during 24 h on average) increased from 2.35 days per year in the 20th-century to 5.222 per year from 2000 onwards (Figure 6); the number of frost days (below −1 °C for 24 h on average 24) remained unchanged at zero days per year.

The municipality of Catania is in the process of defining their adaptation plan. However, the city has already started a series of adaptation measures, including the planning and implementation of GIs and oriented management of public green areas.

In the city of Catania (Kmsq 183), the extension of the urban green is equal to 4,843,660 square meters, and the urban green per inhabitant corresponds to 16.4 square meters.

The scenario of the urban green spaces in Cataniais shown in Table 1:

Table 1. The urban green spaces of the city of Catania (*).

Green Typology	Square Metres
Urban Parks (> 8,000 square metres)	513,577
Green equipment (< 8.000 square metres)	431,270
Urban Design Areas	715,500
Urban Forestation	-
School Gardens	350,000
Botanical Gardens and Vivai	20,000
Zoological Gardens	-
Cemetery	50,000
Urban Gardens (mainly managed by families)	2,500
Sports areas/Outdoor play	100,000
Bosch areas (> 500 square metres)	972,769
Uncultivated Green	1,668,044
Total Urban Green	**4,843,660**

(*) Source: Directorate for Environmental and Green Policies and Energy Management of the Autoparco—Service for the Protection and Management of Public Green, Giardino Bellini and Parchi, 2019.

The green areas play an important and specific role both as an urban component for the conservation and improvement of the landscape and the environment and as a means for aggregative purposes for social and cultural integration.

The GIs should pursue the following objectives among the ones envisaged:

- Improving and preserving the local landscape and environmental restoration;
- Favor urban climate control and reduction of albedo and heat islands;

- Increase the naturalness and biodiversity of the urban territory;
- To stimulate the aggregative, social and therapeutic functions of green areas (e.g., urban gardens, neighborhood parks, healing gardens, spaces for cultural events and shows) (Regulation of the public and private Green of the city of Catania, 2017).

The strategy for the GIs interventions includes the creation of green areas, green avenues, wetlands, and urban gardens according to the classification: Wedges, vortices, margins, strips, islands, hot points, hubs, fences, spurs, and vectors (Figure 7).

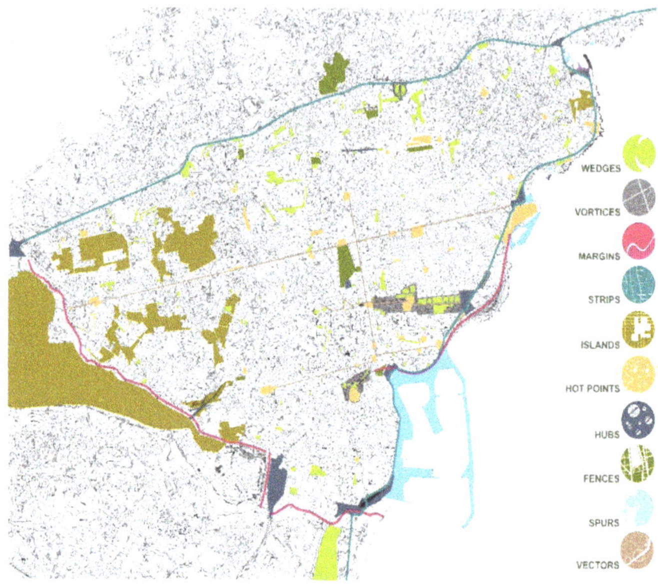

Figure 7. The types of green infrastructures foreseen in the municipality of Catania (2017).

Wedges: Urban areas with the aim of becoming green areas dedicated to developing the relationship between citizen and city. Vortices: Model of areas designed to interact with the city with an inaccurate destination (areas that can adapt to the different needs of different groups of society). Margins: Represent a bit of a breakthrough, as the boundary criterion of the city is redefined, moving it ever more towards the areas that assume a strategic significance under the aspect of livability (sea, river, park, etc.). In this vision of development, the synergies between the oriented reserve of the Simeto Oasis, Etna Park, and in general with the urban parks (Playa del Bosco, Gioeni Park, etc.) and the city are included. Strips: Areas dedicated to the connection between the different areas of Catania, also with the redefinition of the above-mentioned margins. Island areas allow the giving back to the territory its centrality, where man has included the buildings over the years. Hot spots are red areas, under the cultural, economic, and social aspects in which the interests of a large part of society are concentrated and in which a mitigating action is needed to improve the quality of the environment. This represents the areas where resident population flows, tourists, and populations that gravitates around the city with both structural and infrastructural needs. In these areas, the need for both congestion and pollution generated to create buffer areas to mitigate the effects of the hubs present in the area emerges. Spurs are the areas that correspond with the Waterfront project, which start from the port and end up in Ognina, where the city encourages the development of a tourist town on the sea. Fences: All the defined and delimited green parks are included. Vectors include the areas that represent the main routes of the city's traffic.

The municipality of Catania has realized some GIs and others are in the course of realization, as reported in Figures 8–13 that testify the situation of the interested area before and after the realization of the GI.

Figure 8. Piazza Galatea - Catania (before).

Figure 9. Piazza Galatea—Catania (today).

Figure 10. Tondo Gioeni—Catania (before).

Figure 11. Tondo Gioeni (today, after the realization of the vertical wall).

Figure 12. Corso Martiri della libertà—Catania (before).

Figure 13. Corso Martiri della Libertà (project in progress).

The urban green is a heritage of the complex city, which requires careful assessment that takes into account not only the economic variables but also the social, environmental, and institutional ones. The proposals for investments in GIs, their management, and the evaluation of benefits will have to be shared by the community for an effective pursuit of the objectives set. This is an assessment that considers the dimensions of sustainable development, which will contribute to providing useful elements for the promotion of a model of governance of the city *"eco-social-green"* [4].

Based on the *eco-social-green* model, it will be possible to carry out a social, environmental, and economic assessment of the proposed interventions, through a collaborative process and by sharing the objectives of valorization, transformation, and redefinition of the green spaces of the city of Catania. The model proposed is called *"eco-social-green"* because we have decided to integrate the aspects that we consider important for the new role attributed to cities. Both at a European level and in the scientific debate, there is increasing attention on the role that cities must have on climate change, biodiversity, conservation, and promotion of social cohesion. A multidimensional approach allows to make decisions in complex conditions, and it gives the possibility to consider economic and non-economic dimensions; the existing synergies and possible conflicts, and the different viewpoints of the subjects involved (public and private).

The attention of the citizens of Catania towards the environment and the presence of greening in the city has recently increased. In fact, the initiatives of "adoption of public green" by private individuals are multiplying (to date, the public green areas of Catania adopted by private individuals, for redevelopment and maintenance, are around 4000 square metres—with savings for municipal management of around 300,000 euros). Citizens are aware that the GIs perform important functions to improve the quality of urban life, both in environmental and social terms and in terms of adaptation to climate change.

4. Methodology

The analysis of the case study evaluated the GIs planning of the metropolitan city of Catania, to experiment with new approaches and opportunities for the definition of green strategies that have found concrete applications in the development of guidelines in politics and local planning tools, as a tool for climate mitigation in an urban environment.

The proposed approach was based on the integration between the participatory planning (based on the establishment of the Focus Groups with the various stakeholders) and the NAIADE method (Novel Approach to Imprecise Assessment and Decision Environments [11], for the Multi-Criteria Social Assessment—SMCE) as possible a methodological structure to acquire and evaluate the "complex" information collected (quantitative and qualitative data) on possible alternative scenarios in relation to the urban green spaces.

The methodology can be considered a social experiment, able to produce collective opinions, to detect communication barriers, to study conflictual behavior, to acquire local information, and to create acceptable options [50–53].

The innovative advantage is the interaction among the participants that highlight the fundamental tools to support a process of evaluation and reciprocal learning. This approach allows us to reveal new visions of the subjects involved in order to have a final participated decision.

The target is that of developing a methodological structure made by suitable tools to acquire first, and process second, qualitative and quantitative information concerning the possible alternative scenarios of the problem under study. Opinions were collected at specific meetings at the local level with stakeholders and sector's operators involved in the issue from environmental, social, climate, landscape, health safety, and economic points of view.

The opinions were collected through specific Focus Groups with local stakeholders, operators, and citizens interested in the issue in question from different economic, social, and environmental aspects [54]. This was in the presence of 2researchers, one with the role of moderator and the other with the detector of the responses of the individual subjects involved.

The adoption of this approach was limited to problems of territorial planning referable to SMCE [55–57]. There were numerous articles that employed SMCE for the resolution of problems related to the management of environmental resources, and in general, to valuations of sustainability, climate adaptation, energy policy, etc. [58–61].

The following picture (Figure 14) identifies the steps on which our SMCE was based, with some adaptation concerning the specificity of the context surveyed.

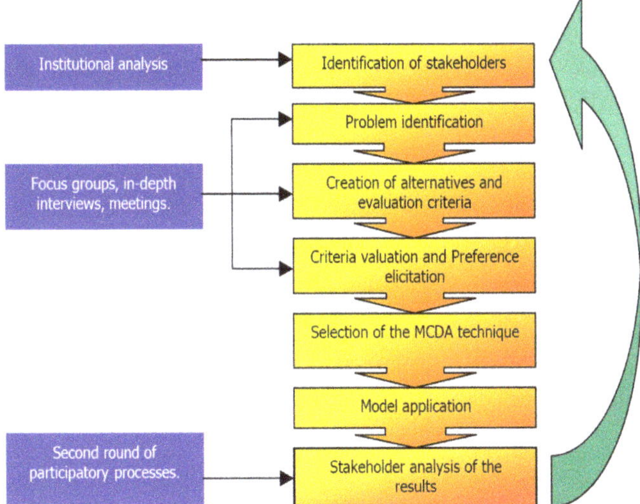

Figure 14. Structure of the model Multi-Critical Social Assessment (SMCE) [4].

In detail, the proposed model was based on:

- The individualization of the citizens and the stakeholders involved (100 questionnaires);
- The definition of the alternative scenarios (definition of the 3hypotheses of the scenario: Inclusive, resilient, and city);
- The definition of the context of evaluation, namely the decisional criteria (urban green spaces of Catania for the shared project);
- The evaluation of the impact of alternative scenarios relative to the criteria in question);
- The final creation of the impact matrix;

The structure used the focus groups as a social research methodology, aimed toacquire information on the opinions of stakeholders regarding a variety of scenarios for future development within the zone examined. The choice of focus groups and, therefore, of the interaction among the actors involved, aimed to support the phase of the choice and evaluation of the different aspects that would be included in the equity matrix. The matrices of impact and equity constituted the basis for the use of the discrete multicriteria evaluation NAIADE model (Novel Approach to Imprecise Assessment and Decision Environments) [62], able to manage qualitative and quantitative data in order to evaluate the measures of intervention. This instrument supported the classification of the alternative scenarios proposed based on the determined decisional criteria and considerations of possible "alliances" and "conflicts" between the groups of stakeholders on the proposed scenarios, thus measuring their acceptability. The NAIADE multicriteria evaluation method applied to this study constituted a discrete method of evaluation capable of managing qualitative and quantitative data. It was an appropriate tool for the planning of problems characterized by great "uncertainty" and "complexity" regarding existing territorial, social, and economic structures and their interrelations [63]. The basic input in the NAIADE method consisted of alternative scenarios to be analyzed, different decisional criteria for their evaluation, and different stakeholders who expressed opinions about the scenarios in question. One of the strengths of this tool in the application to the planning of interventions on green spaces was based on its ability to collect the conflicting perspectives of the stakeholders and to address the compromises among the environmental, social, and economic dimensions.

Concerning the objective of this study, the analysis will be applied to the principal priorities, the methodology used for the definition of the model of management for the green area of Catania, which is the area of investigation for this work.

The evaluation through the Focus Groups was divided into 3phases, referring in the specific case to the destination of urban areas in a degraded state to be valorized:

Phase 1—consisted of "planning" the meetings. During this phase, the following were established:

-   The number of sessions and the time dedicated to each of them (8, as an expression of the individual categories considered: Association of citizens, groups of pensioners, cultural associations, playrooms, trade unions, public institutions, scientific groups, tertiary sector's companies, with time varying from 4 to 8 h);
-   The creation of an interview guide to conduct the discussion (scientific and dissemination materials, research papers, photos, maps, relative the problems of urban green areas and on the social, climatic effects deriving from them);
-   Selection of participants (stratified selection for homogeneous groups: Age, gender, income).

The questionnaire used for the interviews was designed to explore the perception of environmental issues in the urban context and to evaluate the real needs of the population in terms of environmental quality, perception of climate change, and fruition of public green spaces. It was structured in 10 questions in order to collect information and opinions useful for the research, on the 3hypotheses proposed: Inclusive, resilient, and city.

Phase 2—consisted of "carrying out" the entire activity, based on the guide to the pre-established interview. It began with the presentation of the topic relating specifically to the action strategy for the management of degraded areas to be recovered, using the support material (articles, results, photographs), prepared specifically to introduce the issue under consideration and stimulate discussion and the interaction of the participants. During this phase, various ideas and opinions were acquired that represented the reactions of the participants involved in the issues raised.

Phase 3—consisted in the elaboration of the "qualitative results" and the production of the final report.

In this regard, various qualitative analysis tools were used, based on intentionally prepared inputs and specific rules. Overall, the Focus Groups can be considered a social experiment, able to produce collective opinions, reveal communication barriers, study conflicting behavior, acquire local information, create acceptable options, synthesize information, etc. The key advantage of the Focus Group dedicated to defining the intervention strategies in green urban areas to be enhanced, when compared to other participatory techniques, lay in the deep interaction between the participants, becoming a "social network."The participants became fundamental tools to support a "mutual learning process" on the questions examined. This participatory comparison technique made it possible to reveal new dimensions of the issue under discussion, thus underlining the possibility for the Focus Group to bring out the opinions in this regard rather than produce generalized results. The analysis phase of the results of the Focus Groups was followed by the multi-criteria analysis where the basic input of the NAIADE method consisted of alternative scenarios to be analyzed, different decision criteria for the relative evaluation, and different stakeholders that expressed opinions on the scenarios in question. Based on this method, 2 types of analysis can be performed:

-   A multi-criteria analysis based on the impact matrix, which leads to the definition of the priorities of alternative scenarios regarding certain decision-making criteria;
-   An analysis of equity based on the equity matrix, which analyzes possible "alliances" or "conflicts" among different interests in relation to the scenarios in question.

In this regard, the multicriteria analysis, according to the NAIADE methodology, aimed to classify alternative scenarios based on the preferences of individual groups based on certain decision criteria [64–67].

The basic input of the NAIADE method was constituted by the impact matrix (criteria/alternative matrix), including scores that can take the following forms: Crisp numbers, stochastic elements, fuzzy elements, and linguistic elements (such as "very poor", "poor", "good", "very good", "excellent") [18]. When comparing alternative scenarios, the concept of distance was introduced. In the presence of crisp numbers, the distance between the 2alternative scenarios with respect to the given evaluation criterion was calculated by subtracting the respective crisp numbers.

The classification of alternative scenarios was based on data from the impact matrix, used for:

- Comparison of everysingle pair of alternatives for all the evaluation criteria considered;
- Calculation of a credibility index for each of the aforementioned comparisons, that measured the credibility of one preference relation "... alternative scenario" a "is better/worse, etc., alternative scenario «b» ... "(preference relationships were used);
- Aggregation of the credibility indices produced during the previous stage leading to a preference intensity index $\mu * (a, b)$ of an alternative «a» with respect to another «b» for all the evaluation criteria, associated with the concept of entropy $H * (a, b)$, as an indication of the variation in the credibility indices;
- Classification of alternative scenarios based on previous information.

The final classification of the alternatives was the result (intersection) of 2 different classifications: the classification $\Phi + (a)$ based on the "best" and "decidedly better" preference relationships; and the classification $\Phi- (b)$, which was based on the "worst" and "decidedly worse" preference relationships.

In relation to the objective of the present study, the analysis will be applied to the main priorities, for the evaluation of the optimal management model for the enhancement of the green areas of Catania.

5. Results and Discussion

The results of the present work provide a further multidisciplinary contribution to research on the management and planning of green areas in cities. Specifically, the analysis was conducted on the basis of the following question:

What are the strategies for the enhancement of urban green spaces of the City of Catania for climate change adaptation?

Three hypotheses of green recreate strategies were envisaged (Table 1):

Hypothesis 1. *INCLUSIVE: Creation of green areas with inclusive and social functions (equipped parks, urban gardens, etc.).*

Hypothesis 2. *RESILIENT: Creation of urban green areas with non-usable landscape function but as a climate change adaptation measure.*

Hypothesis 3. *CITY: Conservative recovery, cleaning, and maintenance of the current green.*

In order to evaluate the three hypotheses mentioned above, evaluation criteria have been defined, which represent "... a measurable aspect of the judgment that can characterize a dimension of the various choices that are taken into consideration" [68]. In the present case study, twenty evaluation criteria or variables were used. These criteria were defined based on the purpose and objectives of the evaluation of the analyzed case, which can be considered representative of the reality of the City of Catania but overall very similar to other metropolitan areas.

The objectives of the evaluation activity were environmental, social, climate, economic, landscaping, and health and safety.

Specifically, for each objective, the related evaluation criteria are considered and are indicated in Table 2.

Table 2. The objectives and evaluation criteria adopted in the applied model for the urban green areas of Catania.

Goals	Evaluation Criteria
Environmental	air quality; human settlement
Social	usability, multi-functionality, agricultural production, employment commitment
Climate	reduction of temperatures, creation of accessible shade areas, thermal excursion
Economic	cost of realization, the value of the properties; productive exploitation
Landscape	quality of the landscape, the exaltation of the seasons, biodiversity
Health safety	pollution, pathogenic presence, use of pesticides and fertilizers

Source: Our elaboration, data collected direct survey.

According to the above-reported indicators, impact matrix results, as a whole, are reported in the following Table 3.

Table 3. Evaluation of the results of the impact matrix of the various alternatives.

Criteria of Evaluation	Scenario Inclusive	Scenario Resilient	Scenario City
Environmental			
Air quality	Good	Very good	Poor
Smell emanation	Good	Excellent	Good
Anthropization	Poor	Very good	Very good
Waste of water	Poor	Good	Excellent
Social			
Usability	Excellent	Medium	Poor
Multifunctional	Excellent	Very good	Good
Agricultural production	Very good	Poor	Poor
Occupational commitment	Very good	Good	Poor
Climate			
Temperature reduction	Good	Excellent	Very good
Creation of shaded areas	Good	Very good	Good
Temperature range	Good	Very good	Poor
Humidity	Very good	Good	Poor
Economic			
Cost of realization	Good	Poor	Very good
Value of real estate	Very good	Excellent	Good
Productive exploitation	Very good	Good	Poor
Landscape			
Landscape quality	Good	Excellent	Good
Exaltation of the seasons	Excellent	Very good	Poor
Biodiversity	Excellent	Very good	Poor
Health Safety			
Pollution	Good	Very good	Good
Presence of pathogens	Poor	Good	Very good
Use of pesticides and fertilizers	Poor	Good	Very good

Source: Our elaboration, data collected direct survey.

Hypothesis Inclusive as an option to be shared (highlighting the social, landscape, and economic aspects), followed at a short distance by hypothesis Resilient (highlighting environmental, climate, landscape results) and, then, hypothesis City (with a more negative evaluation).

Then the equity matrix was developed. It provided stakeholders' opinions on the three hypotheses suggested. The selection of stakeholders was based on their potentialities to influence the targets of the project. They represent citizens, socially vulnerable groups, and different interested associations and possible users of the interventions, with different qualifications, both into the private and public. In particular, eight typologies of stakeholders were involved (as listed in Table 4). It is important to underline that stakeholders' opinions in the NAIADE model can only be of a quality kind: Language expressions from very poor, poor, medium, good, very good, and excellent (Table 4).These results show that a big number of the stakeholders and operator groups selected agreed with the evaluation of the three hypotheses.

Table 4. The equity matrix—Stakeholder opinions on the three hypotheses

Typologies of Stakeholders	Scenario Inclusive	Scenario Resilient	Scenario City
A1. Associations of citizens	Very good	Excellent	Poor
A2. Groups of pensioners	Excellent	Poor	Poor
A3. Cultural associations	Very good	Good	Good
A4. Playroom	Excellent	Very good	Good
A5. Trade unions	Very good	Excellent	Good
A6. Public Institutions	Good	Very good	Poor
A7. Scientific groups	Very good	Excellent	Poor
A8. Tertiary sector's companies	Excellent	Excellent	Good

Source: Our elaboration.

The results of the multi-criteria analysis highlighted that our hypothesis was inclusive and was the predominant hypothesis, followed closely by hypothesis Resilient, while hypothesis City gained a marginal meaning.

The results obtained through the equity analysis were used to examine possible alliances or conflicts among the opinions of stakeholders about the decision of what hypotheses to adopt. Besides agreeing on the classification of the different hypotheses, the results (Table 5) show that a high number of stakeholders were in agreement with hypothesis Inclusive.

Table 5. Classification of the scenarios corresponding to the higher consent (0.8423).

	Typologies of Stakeholders	Scenario Inclusive	Scenario Resilient	Scenario City
A1	*Associations of citizens*	0.89	0.83	0.21
A2	*Groups of pensioners*	0.83	0.59	0.12
A3	*Cultural associations*	0.85	0.65	0.38
A4	*Playrooms*	0.74	0.38	0.11
A5	*Trade unions*	0.64	0.82	0.23
A6	*Public Institutions*	0.36	0.29	0.28
A7	*Scientific groups*	0.57	0.92	0.12
A8	*Tertiary sector's companies*	0.86	0.85	0.54

Source: Our elaboration, data collected direct survey.

The efficiency of this kind of approach relies on the possibility of establishing a "learning platform" that eases participation, information exchange, and reciprocal comprehension of participants, who stimulate each other towards a sharing of the territory. Results allowed the inclusion of several perspectives of the evaluation problem under study, as demonstrated by the different groups involved, increasing the perception of planners about the acceptability of the alternatives proposed that may lead to the improvement in strategic decisions and, therefore, create innovative ideas and new planning solutions based on the possibilities offered by the participated processes.

On the whole, the results obtained from this model of *collaborative governance, eco-social-green,* developed through the integration of a participative tool and a multi-criteria analysis, became strategic for the choices of urban green investments, in particular, related to climate change adaptation measures shared with the community.

6. Conclusions

In the last decade, literature has been enriched by a complex body of knowledge related to the evaluation of the benefits provided by GIs in an urban area to climate change mitigation and adaptation. Many pieces of research provide empirical evidence that can be used to design GIs to decrease the vulnerability of a city to climate change. In particular, the studies have shown the important role of

GIs in contributing to climate change mitigation and offsetting urban carbon emissions (especially with the employment of green roofs and green facades); the thermal comfort due to the cooling effect of green roofs in different types of buildings and in different seasons and others important positive effects. However, the GIs can offer social and psychological benefits to citizens, especially socially vulnerable ones.

From the experiences mentioned and from the researches cited (only some for brevity and even synthetically) it could highlight that various GIs could provide multiple benefits in urban areas and this should be taken into account in the planning and design of the urban landscape. It is important to consider the multi-functionality of the GIs because the focus on a single advantage could, in turn, be harmful from another point of view (trade-off) [45].

It is necessary to replace the expansive building processes with the virtuous ones of urban regeneration, in which the GIs take on an increasingly important role for their functions of social inclusion, environmental protection, and in the mitigation and adaptation for climate change. Adaptation strategies need to preserve and enhance exiting GIs, and increase them where possible, especially taking opportunities in restructuring and new developments to create significant new spaces.

The provision of GIs has been widely recognized as playing an important role in meeting the challenge of climate change adaptation. Integration of GIs into more ecosystem-based spatial planning makes the design of GI assets a crucial planning tool for building more sustainable urban environments, resistant to future challenges and adaptable to future needs [46].

The GIs are well established in climate adaptation strategies, but it is important that these strategic tools are encouraged through specific integrated territorial and urban planning; long-term investments, planning, and sustainability in decision making; models of GIs for climate change adaptation and for optimal multiple benefits; evaluating both the economic and social and environmental benefits; encouraging citizen participation in the planning of GIs for climate adaptation [4,42,43].

Due to the multifunctionality of GIs, there is no single science or reference discipline for its study [43]. They are based on the theories and practices of numerous scientific disciplines, such as conservation biology, landscape ecology, urban and regional planning, geographic analysis, information systems, and economists [69]. The research into GIs also needs to adjust to different spatial scales as its application can range from individual buildings to neighborhoods and cities to entire regions [70].

The expected impacts of climate change in urban settlements are very different: Impacts on health and quality of life (in particular of the weaker sections of the population), impacts on the buildings, on water, energy and transport infrastructures, on cultural heritage (due to landslides, floods and heatwaves), impacts on energy production and supply. Thus, to deal effectively with these impacts, we need the coordination of a very broad institutional network (multilevel governance) [43,71].

To date, however, the prospects for the development of the GIs are strictly dependent on its inclusion in planning policies, which local authorities and urban planning regulations must provide and support financially as well.

The ESs must be integrated into the planning and the choices of urban planning policies, making GIs and eco-innovation the fulcrum of an intelligent and sustainable urban transformation towards a new model of sustainable city enhancing biodiversity, environment and social inclusion [72].

The new guidelines on the protection of natural capital and biodiversity and attention to climate change are laying the foundations for outlining new ways of the government of the territory and the cities, with a proactive and engaging approach. Cities are ecosystems full of human presence, rich in knowledge and innovation, which welcome more than 50% of the world's population and about 70% of the Italian population. In cities, the conflict between artificiality and naturalness is maximum and causes loss of biodiversity, quality of ecosystem services, and resilience [73].

A direct consequence of the variety of expected impacts in urban settlements is the multiplicity of institutional actors who, together with citizens, must be involved in adaptation policies. Actors who will have a responsibility at different territorial scales (state, regions, provinces, municipalities) or responsibilities of certain sectors (basin authorities, energy service management bodies, water companies, etc.) and citizens, for which multilevel governance is required.

This research allowed us to point out that the methodological approach adopted, that was inspired by a model of city *"eco-social-green"* and based on the integration between the participated planning technique and the multi-criteria analysis, in the case of problems linked to the urban green spaces, represented a strategic tool.

The efficiency of this model of evaluation relies on the possibility of establishing a learning platform that facilitates the participation, information exchange, and reciprocal comprehension of the participants that support a strategy for the development and the fruition of the GIs of the city and to evaluate the perception of the importance of the GIs for climate change adaptation.

As other studies have already indicated [74–76], our study highlights an interesting potential for the wider use of SMCE in the governance of green spaces, for its ability to integrate ecological, social, and economic values, as well as the different stakeholder preferences among social groups, places, and temporal dynamics. Of course, policymakers have a broad ability to improve human well-being in cities through green space governance approaches that take into account their different components, economic, environmental, and social.

The value generated by the presence of GIs in the urban ecosystem, through the application of a vision of green-urban-planning, seems to be a valid instrument for achieving the objectives of realizing resilient and inclusive cities.

Author Contributions: All authors contributed to this work, in particular their individual contributions are: conceptualization L.S.; methodology L.S. and A.S.; validation and formal analysis A.S.; writing original draft and review and editing L.S. and A.S.

Acknowledgments: This work has been financed by the University of Catania within the project "Piano della Ricerca Dipartimentale 2016-2018 of the Department of Civil Engineering and Architecture".

References

1. Intergovernmental Panel on Climate Change (IPCC). *Climate Change 2014; Contribution of Working Groups I, II and III to the Fifth Assessment Report of the Intergovernmental Panel on Climate Change;* Core Writing Team, Pachauri, R.K., Meyer, L.A., Eds.; IPCC: Geneva, Switzerland, 2014; p. 151.

2. United Nations. *Transforming our World: The 2030 Agenda for Sustainable Development;* United Nations: New York, NY, USA, 2015.

3. United Nations. *Revision of the World Urbanization Prospects;* United Nations: New York, NY, USA, 2018.

4. Sturiale, L.; Scuderi, A. The evaluation of green investments in urban areas: A proposal of an eco-social-green model of the city. *Sustainability* **2018**, *10*, 4541. [CrossRef]

5. European Commission. *The Forms and the Functions of the Green Infrastructures;* European Commission/Envrinmental: Brussels, Belgium, 2016.

6. Environment Directorate-General for the Environment. *Communication from the commission to the European Parliament, the Council, the European Economic and Social Committee and the Committee of the Regions Green Infrastructure (GI) Enhancing Europe's Natural Capital,* Environment Directorate-General for the Environment: Bruxelles, Belgium, 2013.

7. EU. *2009 WHITE PAPER Adapting to Climate Change: Towards a EUROPEAN Framework for Action;* COM 2009 147/4; EU: Brussels, Belgium, 2009.

8. EEA. *Green Infrastructure and Territorial Cohesion. The Concept of Green Infrastructure and its Integration into Policies Using Monitoring Systems;* Technical Report no. 18/2011; European Environment Agency: Copenhagen, Denmark, 2011.

9. Werguin, A.C.; Duhem, B.; Lindholm, G.; Oppermann, B.; Pauleit, S.; Tjallingi, S. (Eds.) *Green Structure and Urban Planning;* Final Report, COST Action, No. C11; Office for Official Publications of the European Communities: Luxembourg, 2005.

10. DG Environmental New Alerts Services. *The Multifunctionality of the Green Infrastructures;* In-Depth Report; DG Environmental New Alerts Services: Bruxelles, Belgium, 2012.

11. Munda, G. Multicriteria Evaluationina Fuzzy Environment—Theory and Applications. In *Ecological Economics*; Physica-Verlag: Heidelberg, Germany, 1995.

12. OECD. *Competitive Cities and Climate Change*; OECD: Milan, Italy, 2008.

13. World Bank. *Cities and Climate Change: An Urgent Agenda*; The World Bank: Washington, DC, USA, 2010.

14. UN-HABITAT. *Cities and Climate Change: Global Report on Human Settlements 2011*; United Nations Human Settlements Programme; Earthscan: London, UK; Washington, DC, USA, 2011.

15. COM. *The European Strategy on Adaptation to Climate Change*; European Commission: Brussels, Belgium, 2013; p. 216.

16. Monclús, J. *From Park Systems and Green Belts to Green Infrastructures*; Visions, U., Díez Medina, C., Monclús, J., Eds.; Springer: Cham, Germany, 2018.

17. Anguluri, R.; Narayanan, P. Role of green space in urban planning: Outlook towards smart cities. *Urban. Urban Green.* **2017**, *25*, 58–65. [CrossRef]

18. Jim, C.Y. Green spaces preservation and allocation for sustainable greening of compact cities. *Cities* **2004**, *21*, 311–320. [CrossRef]

19. Costanza, R. (Ed.) *Ecological Economics: The Science and Management of Sustainability*; Columbia University Press: New York, NY, USA, 1991.

20. Foster, J.; Lowe, A.; Winkelman, S. *The Value of Green Infrastructures for Urban Climate Adaptation*; Center for Clean Air Policy: Washington, DC, USA, 2011.

21. Brink, E.; Aalders, T.; Ádám, D.; Feller, R.; Henselek, Y.; Hoffmann, A.; Ibe, K.; Matthey-Doret, A.; Meyer, M.; Negrut, N.L.; et al. Cascades of green: A review of ecosystem based adaptation in urban areas. *Glob. Environ. Chang.* **2016**, *36*, 111–123. [CrossRef]

22. Hunt, A.; Watkiss, P. Climate change impacts and adaptation in cities: A review of the literature. *Clim. Chang.* **2011**, *104*, 13–49. [CrossRef]

23. Escobedo, F.J.; Kroeger, T.; Wagner, J.E. Urban forests and pollution mitigation: Analyzing ecosystem services and disservices. *Environ. Pollut.* **2011**, *159*, 2078–2087. [CrossRef] [PubMed]

24. Gómez-Baggethun, E.; Barton, D.N. Classifying and valuing ecosystem services for urban planning. *Ecol. Econ.* **2013**, *86*, 235–245. [CrossRef]

25. Economides, C. *Green Infrastructure: Sustainable Solutions in 11 Cities across the United States*; Columbia University Water: New York, NY, USA, 2014.

26. Nowak, D.J.; Crane, D.N.; Stevens, J.C. Air pollution removal by urban trees and shrubs in the United States. *Urban For. Urban Green.* **2006**, *4*, 115–123. [CrossRef]

27. Sturiale, L.; Timpanaro, G.; Foti, V.T.; Scuderi, A.; Stella, G. *Social and Inclusive "Value" Generation in Metropolitan Area with the "Urban Gardens" Planning*; Green Energy and Technology; Springer: Cham, Switzerland, 2019; in press.

28. Caspersen, O.H.; Olafsson, A.S. Recreational mapping and planning for enlargement of the green structure in greater Copenhagen. *Urban Urban For. Green.* **2010**, *9*, 101–112. [CrossRef]

29. Kabisch, N. Ecosystem service implementation and governance challenges in "urban green spaces" planning-the case of Berlin, Germany. *Land Use Policy* **2015**, *42*, 557–567. [CrossRef]

30. Jim, C.Y. Planning strategies to overcome constraints on greenspace provision in urban Hong Kong Town. *Plan. Rev.* **2002**, *73*, 127–152.

31. Yang, J.; Mcbride, J.; Zhou, J.; Sun, Z. The urban forest in Beijing and its role in air pollution reduction. *Urban For. Urban Green.* **2005**, *3*, 65–78. [CrossRef]

32. Wei, J.; Qian, J.; Tao, Y.; Hu, F.; Ou, W. Evaluating Spatial Priority of Urban Green Infrastructure for Urban Sustainability in Areas of Rapid Urbanization: A Case Study of Pukou in China. *Sustainability* **2018**, *10*, 327. [CrossRef]

33. Beckett, K.P.; Freer-Smith, P.; Taylor, G. Urban woodlands: their role in reducing the effects of particulate pollution. *Environ. Pollut.* **1998**, *99*, 347–360. [CrossRef]

34. Nowak, D.J.; Mchale, P.J.; Ibarra, M.; Crane, D.; Stevens, D.J.; Luley, C.J. Modeling the Effects of Urban Vegetation on Air Pollution. In *Air Pollution Modeling and Its Application XII*; Plenum Press: New York, NY, USA, 1998; pp. 399–407.

35. Saunders, S.; Dade, E.; Van Niel, K. An Urban Forest Effects (UFORE) model study of the integrated effects of vegetation on local air pollution in the Western Suburbs of Perth WA. In Proceedings of the 19th International Congress on Modelling and Simulation, Cambridge, UK, 30 March–1 April 2011; pp. 1824–1830.

36. Baró, F.; Chaparro, L.; Gómez-Baggethun, E.; Langemeyer, J.; Nowak, D.J.; Terradas, J. Contribution of ecosystem services to air quality and climate change mitigation policies: the case of urban forests in Barcelona, Spain. *Ambio* **2014**, *43*, 466–479. [CrossRef]
37. Akbari, H.; Pomerantz, M.; Taha, H. Cool surfaces and shade trees to reduce energy use and improve air quality in urban areas. *Sol. Energy* **2001**, *70*, 295–310. [CrossRef]
38. Tzoulas, K.; Korpela, K.; Venn, S.; Yli-Pelkonen, V.; Kaźmierczak, A.; Niemela, J.; James, P. Promoting ecosystem and human health in urban areas using green infrastructure: a literature review. *Landsc. Urban Plann.* **2007**, *81*, 167–178. [CrossRef]
39. Jayasooriya, V.M.; Muthukumaran, A.W.M.N.G.S.; Perera, B.J.C. Green infrastructure practices for improvement of urban air quality. *Urban For. Urban Green.* **2017**, *21*, 34–47. [CrossRef]
40. Gill, S.E.; Handley, J.F.; Ennos, A.R.; Pauleit, S. Adapting Cities for Climate Change: The Role of the Green Infrastructure. *Built Environ.* **2007**, *1*, 115–133. [CrossRef]
41. Radovanović, M.; Lior, N. Sustainable economic–environmental planning in Southeast Europe—Beyond-GDP and climate change emphases. *Sustain. Dev.* **2017**, *25*, 580–594. [CrossRef]
42. Breuste, J.; Artmann, M.; Li, J.X.; Xie, M.M. Special issue on green infrastructure for urban sustainability. *J. Urban Plan. Dev.* **2015**, *141*. [CrossRef]
43. Scuderi, A.; Sturiale, L. *Evaluations of Social Media Strategy for Green Urban planning in Metropolitan Cities*; Smart Innovation, Systems and Technologies; Springer International Publishing AG, part of Springer Nature: New York, NY, USA, 2019; pp. 76–84.
44. Benedict, M.A.; McMahon, E.T. Green Infrastructure: Smart Conservation for the 21st Century. *Renew. Resour. J.* **2002**, *20*, 12–17.
45. Demuzerea, M.; Orrubc, K.; Heidrichd, O.; Olazabalej, E.; Genelettif, D.; Orrugh, H.; Bhavei, A.G.; Mittali, N.; Feliue, E. Mitigating and adapting to climate change: Multi-functional and multi-scale assessment of green urban infrastructure. *J. Environ. Manag.* **2014**, *146*, 107–115. [CrossRef] [PubMed]
46. Simonis, U.E. Greening urban development: on climate change and climate policy. *Int. J. Soc. Econ.* **2011**, *11*, 919–928. [CrossRef]
47. IPCC. *Fifth Assessment Report: Climate Change 2014: Mitigation of Climate Change*; IPCC: Geneva, Switzerland, 2014.
48. Morancho, A.B. A hedonic valuation of urban green areas. *Landsc. Urban Plan.* **2003**, *66*, 35–41. [CrossRef]
49. Haaland, C.; van den Bosch, C.K. Challenges and strategies for urban green-space planning in cities undergoing densiffication: A review. *Urban. Urban For. Green.* **2015**, *14*, 760–771. [CrossRef]
50. Miccoli, S.; Finucci, F.; Murro, R. Social evaluation approaches in landscape projects. *Sustainability* **2014**, *6*, 7906–7920. [CrossRef]
51. Lo, A.Y.; Jim, C.Y. Willingness of residents to pay and motives for conservation of urban green spaces in the compact city of Hong Kong. *Urban. Urban Green.* **2010**, *9*, 113–120. [CrossRef]
52. Maloutas, T.; Pantelidou, M. Debatsanddevelopments: The glassmenagerieofurbangovernanceandsocial cohesion concepts and stake/concepts as stakes. *Int. J. Urban Reg. Res.* **2004**, *28*, 449–465. [CrossRef]
53. Niemelä, J.; Saarela, S.-R.; Tarja Söderman, T.; Kopperoinen, L.; Yli-Pelkonen, V.; Väre, S.; Kotze, D.J. Using the ecosystem services approach for betterplanning and conservation of urban green spaces: A Finland case study. *Biodivers. Conserv.* **2010**, *19*, 3225–3243. [CrossRef]
54. Soderberg, H.; Karman, E. *MIKA: Methodologiesfor Integrationof Knowledge Areas—The Caseof Sustainable Urban Water Management*; Department of Built Environment & Sustainable Development, Chalmers Architecture, Chalmers University of Technology: Goeteborg, Sweden, 2003.
55. Panell, D.J.; Glenn, N.A. A framework for the economic evaluation and selection of sustainability indicators in agriculture. *Ecol. Econ.* **2000**, *33*, 135–149. [CrossRef]
56. Siciliano, G. Social multicriteria evaluation of farming practices in the presence of soil degradation. A case study in Southern Tuscany, Italy. Environ. *Dev. Sustain.* **2009**, *11*, 1107–1133. [CrossRef]
57. Vargas Isaza, O.L. La evaluación multicriterio social y su aporte a la conservación de los bosques social multicriteria. *Rev. Fac. Nac. De Agron. Medellín* **2005**, *58*, 2665–2683.
58. De Marchi, B.; Funtowicz, S.O.; Lo Cascio, S.; Munda, G. Combining participative and institutional approacheswith multicriteria evaluation. An empirical study for water issues in Troina, Sicily. *Ecol. Econ.* **2000**, *34*, 267–282. [CrossRef]

59. Greco, S.; Matarazzo, B.; Slowinski, R. Dominance based Rough Set Approach to decision under uncertainty and time preference. *Ann. Oper. Res.* **2010**, *176*, 41–75. [CrossRef]
60. Munaretto, S.; Siciliano, G.; Turvani, M. Integrating adaptive governance and participatory multicriteria methods: A framework for climate adaptation governance. *Ecol. Soc.* **2014**, *19*, 74. [CrossRef]
61. Scuderi, A.; Sturiale, L. Multi-criteria evaluation model to face phytosanitary emergencies: The case of citrus fruits farming in Italy. *Agric. Econ.* **2016**, *62*, 205–214. [CrossRef]
62. Munda, G. *Social Multicriteria Evaluation for a Sustainable Economy*; Springer: Berlin, Germany, 2008.
63. Munda, G. A NAIADE based Approach for Sustainability Benchmarking. *Int. J. Environ. Technol. Manag.* **2006**, *6*, 65–78. [CrossRef]
64. Shmelev, E.S.; Rodriguez-Labajos, B. Dynamic multidimensional assessment of sustainability at the macro level: The case of Austria. *Ecol. Econ.* **2009**, *68*, 2560–2573. [CrossRef]
65. Torrieri, F.; Concilio, G.; Nijkamp, P. Decision support tools for urban contingency policy. A scenario approach to risk management of the Vesuvio area in Naples—Italy. *J. Conting. Crisis Manag.* **2002**, *10*, 95–112. [CrossRef]
66. Tiwari, A.P. Choice and Preference of Water Supply Institutions—An Exploratory Study of Stakeholders' Preferences of Water Sector Reform in Metro City of Delhi, India. 2007. Available online: http://agua.isf.es/semana%E2%80%A6/Doc7_APTiwari_2pag_xcra_a_dobre%20cara.pdf (accessed on 24 May 2019).
67. Bekessy, S.A.; White, M.; Gordon, A.; Moilanen, A.; Mccarthy, M.A.; Wintle, B.A. Transparent planning for biodiversity and development in the urban fringe. *Landsc. Urban Plan.* **2012**, *108*, 140–149. [CrossRef]
68. Voogd, H. *Multiple Criteria Evaluation for Urban and Regional Planning*; Lion: London, UK, 1983.
69. European Commission. *The Multifuntionality of Green Infrastructures*; Science for Environment Policy/In-depth Reports; European Commission: Bruxelles, Belgium, 2012.
70. Naumann, S.; Davis, M.; Kaphengst, T.; Pieterse, M.; Rayment, M. *Design, Implementation and Cost Elements of Green Infrastructure Projects*; Final report; European Commission: Brussels, Belgium, 2011.
71. Matthews, T.; YLob, A.; AByrnec, J. Reconceptualizing green infrastructure for climate change adaptation: Barriers to adoption and drivers for uptake by spatial planners. *Landsc. Urban Plan.* **2015**, *138*, 155–163. [CrossRef]
72. Foti, V.T.; Scuderi, A.; Stella, G.; Sturiale, L.; Timpanaro, G.; Trovato, M.R. The integration of agriculture in the politics of social regeneration of degraded urban areas. In *Integrated Evaluation for the Management of Contemporary Cities*; Mondini, G., Fattinnanzi, E., Oppio, A., Bottero, M., Stanghellini, S., Eds.; Results of SIEV, Green Energy and Technology; Springer: Cham, Switzerland, 2018; pp. 99–111.
73. Forest Research. *Benefifits of Green Infrastructures*; Report by Forest Research; Forest Research: Farnham, UK, 2010.
74. Langemeyerab, J.; Gómez-Baggethuncf, E.; Haasede, D.; Scheuerd, S.; Elmqvistb, T. Bridging the gap between ecosystem service assessments and land-use planning through Multi-Criteria Decision Analysis (MCDA). *Environ. Sci. Policy* **2016**, *62*, 45–56. [CrossRef]
75. Srdjevic, Z.; Lakicevic, M.; Srdjevic, B. Approach of decision making based on the analytic hierarchy process for urban landscape management. *Environ. Manag.* **2013**, *51*, 777–785. [CrossRef]
76. Grêt-Regamey, A.; Celio, E.; Klein, T.M.; Wissen Hayek, U. Understanding ESs trade-offs with interactive procedural modeling for sustainable urban planning. *Landsc. Urban Plan.* **2013**, *109*, 107–111. [CrossRef]

Article

Negotiating Institutional Pathways for Sustaining Climate Change Resilience and Risk Governance in Indonesia

Jonatan A. Lassa

Emergency & Disaster Management, College of Indigenous Futures, Arts & Society, Charles Darwin University, Darwin, NT 0909, Australia; jonatan.lassa@cdu.edu.au; Tel.: +61-8-8946-6756

Received: 8 July 2019; Accepted: 23 July 2019; Published: 30 July 2019

Abstract: Institutions matter because they are instrumental in systematically adapting to global climate change, reducing disaster risks, and building resilience. Without institutionalised action, adapting to climatic change remains ad-hoc. Using exploratory research design and longitudinal observations, this research investigates how urban stakeholders and policy entrepreneurs negotiate institutional architecture and pathways for sustaining climate change adaptation and resilience implementation. This paper introduces hybrid institutionalism as a framework to understand how city administrators, local policy makers, and policy advocates navigate complex institutional landscapes that are characterised by volatility and uncertainties. Grounded in the experience from a recent experiment in Indonesia, this research suggests that institutionalisation of adaptation and resilience agenda involves different forms of institutionalisation and institutionalism through time. Future continuity of adaptation to climate change action depends on the dynamic nature of the institutionalism that leads to uncertainty in mainstreaming risk reduction. However, this research found that pathway-dependency theory emerges as a better predictor for institutionalising climate change adaptation and resilience in Indonesia.

Keywords: ACCCRN; Climate change adaptation; institutionalising adaptation; hybrid institutionalism; mainstreaming resilience; urban resilience and adaptation

---

1. Introduction

Many transformative adaptation projects have been exogenously driven by international donors with the aim to build and deepen resilience in both developed and developing worlds. They are often piloted in different ways to build institutional capacity and create institutional pathways that enable local actors worldwide to accelerate urban adaptation [1]. Some of the examples include Asian Cities Climate Change Resilience Network (ACCCRN), 100 Resilient Cities (both funded by Rockefeller Foundation), Making Cities Resilient campaign from United Nations International Strategy for Disaster Reduction (UNISDR), and the UN-Habitat's City Resilience Profiling Programme and many others. These initiatives have been serving as platforms to trigger local adaptation outcomes and disaster resilience.

Responding to the rise of climate risks and disaster vulnerabilities in Asian cities and the needs to build adaptive capacity of the city's governments in Asia, the Rockefeller Foundation, through the $59 million multiyear project, namely ACCCRN, has been supporting 50 secondary cities during 2009–2016 including its two pioneering cities in Indonesia—Semarang City and Bandar Lampung City—to help these cities develop a resilience strategy and build resilience [2].

Three specific adaptation outcomes of ACCCRN in Indonesia include (1) an improved capacity to plan, finance, coordinate, and implement climate change resilience strategies in the selected cities, (2) shared practical adaptation knowledge to address climate change and deepen the quality of awareness, engagement, demand, and implementation by the selected cities, and finally (3) expansion and/or replication of the ACCCRN models for urban resilience-building in other cities [3,4].

Institutions matter because they are instrumental in systematically adapting to global climate change, reducing disaster risks and building resilience [1]. Without institutionalised action, risk reduction and climatic change adaptation remains ad-hoc in many urban settings. This study investigates the experiments of institutionalising climate adaptation and resilience agendas initiated and implemented by ACCCRN project during 2009–2016 in Semarang City (Figure 1), Indonesia. The key questions include: how urban stakeholders and policy makers negotiate and come to terms with potential forms of institutional scenarios crafted to tackle climate change impacts and disaster risk reduction (DRR) in cities? And how urban stakeholders negotiate institutional pathways for sustaining climate risk governance and achieving resilience? This study contributes to the debate on how to make climate change resilience a local reality by understanding challenges and opportunities faced by local actors in mainstreaming urban adaptation.

Figure 1. City of Semarang Map.

As an institutionalist scholar, the author is motivated to show how institutionalism is used to institutionalise various agendas into institutionalised anticipatory adaptation and disaster risk reduction. This paper adopts the understanding of institutional change as a result of complex interplay between institutions and agents [4]. Institutions refer to formal and informal constraints while agents refer to actors such as local champions and organisations such as local governmental departments and NGOs.

2. Theoretical frameworks: Institutions, Institutionalism, and Institutionalisation

Institutions matter because they have become instrumental in making life and death decisions [5]. Institutions not only define what and who will be at risk from climate impacts, but also amend the way risks are defined, perceived, and acted upon [6]. Douglas North argued that "Institutions are the humanly devised constraints that structure human interaction, which are made of: formal constraints (i.e., rules, laws, constitutions), informal constraints (i.e., norms of behaviour, conventions, self-imposed codes of conduct), and their enforcement characteristics" [7]. North's vision of institutions suggests that institutions structure beyond human-to-human interactions as they also shape human-nature interactions. Unlike the view of behaviourists, institutionalists view institutions as the causality of the communities' behaviour and disaster risk outcomes [8]. Unmanageable risks and occurrence of preventable disasters indicates a lack of political commitments or an absence of public institutions [9]. Interestingly, while many have been working on institutionalisations of and/or mainstreaming climate change adaptation (CCA) and/or disaster risk reduction (DRR) [10–12], the author argues that there is still lack of critical discussion concerning institutions, institutionalism, and institutionalisation as both output/outcomes and process of adaptation and resilience experiments and interventions. In this paper, adaptation is understood as the human systems' adjustments and intervention to 'moderate or avoid harm or exploit beneficial opportunities' [1]. Despite not being synonymous, this paper uses the CCA/DRR as interchangeable as both have shared spaced in reducing risk, adapting to climatic extremes, and building resilience.

Institutionalism is a general approach to understanding institutions [13]. They matter because each type of institutionalism provides the lenses through which resilience initiatives and solutions are institutionalised. The author argues that institutions can be seen as outputs or outcomes of a long process of institutionalisations informed by institutionalism. Institutionalism is the rationality behind both institutions and the process of institutionalisation. For example, one legislative product, like a law, an act, or a bill, is a product of the long process of negotiation, debates and cooperation involving a wider range of actors as well as a process. Furthermore, one cannot study institutions and institutionalisation of CCA and DRR without clearly understanding the school of thought behind the change of and the formation of institutions and the institutionalisation processes that occur in the real world.

The theoretical approach to institutionalism is divided into "old" and "new" schools of thought. The old school emphasizes analysis of the formal-legal and administrative arrangements of government and the public sector [13]. Translating North's vision above to the context of DRR and CCA, institutions can be defined as an admixture of formal rules (e.g., climate-related bills, disaster management acts) [12]. However, the real world is too complex to be seen from a particular view of institutionalism. The "new" institutionalism deals with informal norms (including values, traditions, and beliefs). Both approaches deal with enforcement characteristics (e.g., coercive instruments that regulate land-use and building practices) that shape the landscape of adaptation and disaster risk reduction policy and implementation.

Figure 2 offers the general framework of institutionalisation of a development solution. It argues that an institutionalised action starts from the vision of resilience with a new discourse exercise regarding the change of status quo and the need for resilience and risk reduction. The translation of a resilience vision into institutionalised action depends very much on the types of institutionalism (which will be discuss in Sections 2.1 and 2.2) that will inform the process of mainstreaming and/or institutionalising adaptation and resilience.

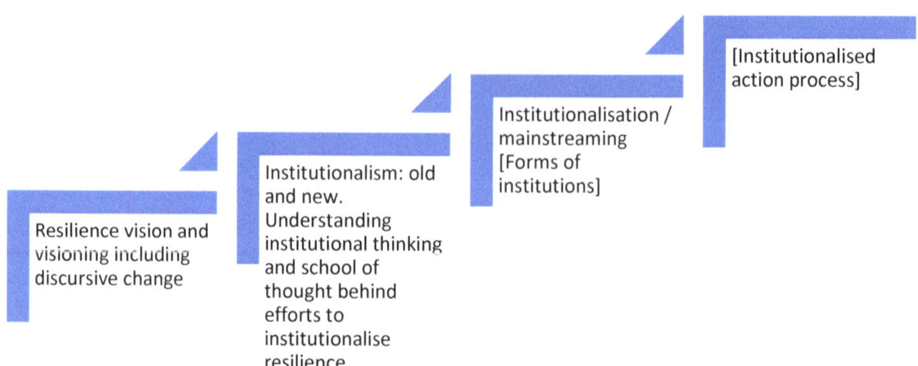

Figure 2. Resilience institutionalisation processes. Source: Author, 2019.

2.1. Old Institutionalism

A formal approach to resilience has been the main research agenda. For example, researchers have argued that the best instrument for addressing CCA and DRR is to work through the existing formal development process and mechanism [12] which can be defined as existing institutional machineries, ranging from formal bureaucratic processes and routines to existing political and social economic institutions and formal/informal processes that deal with the complexity of urban development. Nevertheless, institutionalisation also includes creating new and amending existing regulations, policies, codes, planning documents, and DRR/CCA-related support programmes [12].

DRR/CCA can be a routine development process as they can be embedded or nested inside existing local mechanism and institutions. Integration of risk and vulnerability information into development planning is an example of routinised adaptation and resilience building [14]. Anguelovski and others [15] defined institutionalisations as "linkage to existing urban planning, decision-making, and governance arrangement" [15].

One of the approaches of old institutionalism is often in favour of the roles of international institutions in shaping disaster and climate policy in developing world via the works of the United Nations and international non-governmental organisations (INGOs). Their initiatives can be seen as exogenous adaptation and resilience which can be transformed into endogenous adaptation through time. However, endogenous adaptation does not negate the need for external actors. Their interactions are structured in a way that it is impossible to understand continuity of DRR and CCA unless—as this paper argues—they are understood as an 'ecosystem of institutionalism' and/or where the real world of CCA/DRR operates according to complex interaction of institutions, institutionalisms, and institutionalisation. In this paper, the word 'institutionalisation' is used interchangeably with 'mainstreaming'.

The Hyogo Framework for Action (HFA) promoted the norm of institutionalisation that is based on a solid legal formal framework such as a specific legislation which eventually enables governments at different levels to develop comprehensive disaster resilience implementation [16]. UN-Habitat also sees the importance of specific climate-related legislation as a legitimate form of institutionalisation [17]. In the context of CCA, this can mean creating a specific administrative task put under existing environmental agencies [12,18]. HFA and Sendai Framework advocate for 'strong basis for implementation' as the institutionalisation process requires a constitutional basis, resource allocation, and the existence of multi-stakeholder platforms to ensure continued commitment and implementation of disaster resilience agenda [19].

2.2. New Institutionalism

The new institutionalism is divided into a few categories including the rational choice approach, the historical pathways approach, and the discursive approach [13,20,21].

2.2.1. Rational Choice and New Economics Institutionalism

Rational choice institutionalism (ROI) views policy response to climate change as an expression of pure rational choice of local actors to maximise their resilience and safety by adopting adaptation and risk reduction [13]. Regardless of the motivation of international donors, ROI also views governmental institutions and local actors in the developing world as rational agents as they approve any adaptation project based on the interest of their local affairs alone. The typical solution to the adaptation problem is therefore education and capacity building. Unfortunately, there has been mounting evidence that suggests human are not fully rational agents as everyone has limited rationality, and often make foolish decisions due to by-default challenges such as such as imperfect information, lack of motivation, limits of cognitive-ability, time-boundedness, and context [20].

North [7] is one of the key sources for new economic institutionalism (NEI) thinking. NEI views that institutional change from risk-ignorant to risk reduction occur because actors are motivated (or not) by incentives and/or disincentives provided by formal/informal institutions. Therefore, institutions incentivise or disincentivise actors' decisions and preferences to reduce CCA and DRR. This can manifest in the form of projects and resources or the lack of it. Future progress of CCA and DRR is heavily dependent on the institutional context that structures the enforcement of and/or the implementation of CCA + DRR agendas. This theory is often called new institutional economics theory [7] and in this paper, incentive is understood as more than monetary value to also include social, cultural, political, and symbolic incentives [20].

2.2.2. Pathway-Dependency Theory

Pathway-dependency often refers to the idea that institutional change occurs not according to rational choice but simply according to historical pathways [20,22]. Local actors' interests in adaptation and resilience is not simply based on knowledge informed by texts books and scientific papers. It is often known as historical regularities in the sense that future pathways (e.g., for urban adaptation and resilience) are simply built on the institutional paths from the past. Institutional pathways also point to the fact that climate disasters and urban crises and their impacts often create complex situations which pose difficulties for local actors to make strategic and rational decisions [22] about CCA and DRR. Consequently, their resilience strategies unfold as they interact with changes in the dynamic relationship between social dynamics and hazardous environments [20,22].

On the contrary, climate change adaptation policy and practice is likely to emerge incrementally in that it involves unpredictable institutional arrangements because ex-ante institutional design might be impossible to be developed by the climate resilience policy entrepreneurs. This implies that resilience strategies at each level of governance are more a result of historical interactions than of anything planned [22]. Such historical interactions include local and international interactions where endogenous climate change policy making can only be made if there is adequate exogenous support that trigger and empower endogenous responses [23].

2.2.3. Discursive Institutionalism

Discursive institutionalism aspires institutional changes through the roles of ideas and ideation. The roles of agents' "discursive abilities" [21] is critical to institutional change. This theory assumes a more dynamic and agent-centred approach to institutional change. The institutionalisation of new alternatives or approaches such as climate resilience within an already established institutional stream can be explained by discursive institutionalism. In the real world, roles of local champions can be seen as institutional solutions to unfamiliar agendas like climate change adaptation [24]. Without new ideas and new ideation exercise, the status quo remains. New ideas provide the opportunity to depart from the status quo. In trying to drive climate change adaptation agenda, discursive abilities of local actors are instrumental for change. The practical instruments for discursive exercises can manifest in the form of local champions [24], public relations and awareness, transmission of knowledge and ideas, training and capacity building, and so on.

2.3. Hybrid Institutionalism

Mainstreaming or institutionalising CCA/DRR in a modern urban context requires a multiplicity of efforts. Global champions such as Roberts [10] narrate long term evolution of the process of climate change institutionalisation in the development context of Durban, South Africa. Roberts offers a practical framework namely 'institutional marker' where she identifies institutionalisation or mainstreaming of resilience via a multipronged approach including: first, the existence of a local champion that serves as a messenger and climate policy entrepreneur, second, the adoption of certain climate change issues in municipal plans, third, resource allocation (human and financial) for climate related issues, and fourth, climate change becomes an important factor in both political and administrative decision making.

Table 1 offers a summary of mainstreaming adaptation options that are complementary in nature. These strategies are flexibly selected by local stakeholders as they see fit and necessary. This suggests that a mixture of formal and informal approaches is needed. Furthermore, a hybrid approach suggests that local reform occurs in the context of complex interaction of local and international actors, as well as state and non-state actors. This also suggests the reality in formal institutional settings are characterised by informalities. This can mean local champions adopt certain ideas or discourse that can be introduced informally. This kind of policy change tends to assume that actors and new ideas must come first (See Roberts [10]). This is later followed by processes (formal and informal) that lead to change in formal development plans and fiscal allocation that occurs in both political and formal administrative settings. Nevertheless, institutionalisation may also mean small changes such as adding new job descriptions of city administrators, training and guidance for local officials, and tweaking developing monitoring and evaluation tools that are sensitive to climate change [1]. Therefore, the author argues that institutionalisation of adaptation and resilience take place at the discretion of and interests of empowered local actors.

2.4. Climate Risk Governance Concept in Urban Context

Public governance means governing beyond the conventional governmental power to include different actors including all non-governmental organisation (NGOs) including civil society organisations, non-profit service providers, and business groups [13,20]. Therefore, climate risk governance and/or disaster risk governance concept suggests the polycentric nature of decision making in solving urban climate problem [20]. Furthermore, climate risk governance suggests that there is positive power exercised by external actors as they promote urban climate resilience to be endogenised [23]. ACCCRN is therefore treated here as an example of how urban climate risk governance is exercised where each player ranging from local (from government and NGOs) to international actors (international donors, international NGOs, think tanks) negotiate and shape the process of institutionalising resilience and adaptation.

Table 1. Hypothetical options for institutionalisation and mainstreaming adaptation. Source: Author, 2019.

Types of Institutions	Type of Institutionalisation	Type of Institutionalism	Timelines and Remarks
Formal approach	Legislation	Old institutionalism	Long-term implication—Potential stable budget allocation; Required political process and deliberative
	Mayor regulation	Old institutionalism	Mid-term; Required executive commitments
	Mid-term development plan	Old institutionalism	5-year period; Depending on drafting process
	Annual development plan	Old institutionalism	Short-term; Depending on context
	Establishment of specific department	Old institutionalism	Mid- to long-term; Depending on legislative mandates
	Fiscal allocation	Hybrid approach	Short-term to long-term: required stable political commitments
	Adding specific tasks and agendas to departmental plan	Hybrid approach	Short-term based on local actors' discretionary power
	Compliance to United Nations framework	Path-dependency and New economic institutionalism (NEI)	Timelines depending on the nature of the framework
Informal approach	Ideation through local network	Discursive	Short- to long-term; Depending on incentives
	Ideation through local champions	Discursive	Short-term; Depending on incentives
	Ideation through international networking	Discursive and NEI	Short- to long-term; Depending on incentives
	NGOs/CSCs driven initiatives	Path-dependency theory	Short- to long-term; Depending on incentives
	NGOnising formal process	Path-dependency theory	Short- to long-term; Depending on incentives
	International initiatives and projects	New economic institutionalism	Short- to long-term; Depending on incentives
Hybrid approach	Education and training	Rational choice theory	Capacity building
	Multi-stakeholder forums or platforms	Discursive	Short- to long-term; Depending on incentives to run the platforms
	Vulnerability analysis documents	Discursive	Short-term; Updating is needed—depending on incentives
	Shared-learning dialogues	Discursive	Project lifetime; Depending on incentives to run the platforms
	Resilience strategy document	Discursive	Short-term; Depending on incentives
	Systematic documentation	Discursive	Project lifetime; Depending on incentives
	Conference and seminars	Discursive	Project lifetime; Depending on incentives

Noted: NGO is non-governmental organisations; NGO-nising means, the process of institutionalisation of resilience using an NGO-like structure, such as foundations and/or associations.

3. Methods

The author used both exploratory research and longitudinal observations to understand the evolution of institutionalisation processes during the year 2009 to 2016 and post 2016. The participant observations and stakeholder interviews using open-ended interviews with 10 key informants recruited through snowball selection from June–December 2012 (face to face in Semarang). Participant-observation in several meetings including one conference in Surabaya in December 2015. Triangulations were made based on numerous approaches, including desk research, online observations from posted links, ACCCRN websites, published formal planning documents, and City of Semarang websites during 2012–2018. Qualitative analysis was applied to the analysis of the findings.

4. Findings: Mainstreaming Adaptation in Semarang City

Unless otherwise state, Section 4 is mainly informed by the field works during 2012 and personal observations including online observation completed during 2015–2018. To avoid confusion, all the personal interviews will be indicated as 'personal communication' when cited in the next sections.

4.1. Vulnerability and Risk Context

Semarang City has been increasingly affected by flood risks and severe coastal inundation. There are currently approximately 1.6 million people living in the city with average density of 4000 people/KM-2 in 2016, where some sub-districts, especially its vulnerable North Coast, host more than 11,000 people/KM-2. The city is also sinking due to the high rate of land subsidence which continues to occur at the rate from 1 mm/year to 10 cm/year. Some parts of the North Coast of the city already experience about 12–18 mm/year [25]. As of today, it is estimated that about 7 percent of the city's area have been inundated, significantly affecting a large population and many strategic assets such as the seaport infrastructure [26]. The main reason for the rise in risk level and vulnerabilities is the wide spread of urban settlements that has taken place over the last four decades [25]. The situation is likely to exacerbate in the future due to increases in mean sea level.

4.2. Asian Cities Climate Change Resilience Network (ACCCRN) Processes and Semarang City Team Formation

The ACCCRN project seeks to imprint new adaptation pathways within cities through the urban climate governance processes [27] which unfolded in four phases. The first phase (starting 2009—the introductory phase) involved city selection and early shared learning dialogues (SLDs) where city stakeholders were invited to participate and were able to learn and share climate change and other urban development issues [28]. SLDs is critical part of ACCCRN's urban governance framework as it emphasised on capacity building and shared learning [4].

During the second phase (2009–2011), ACCCRN worked through a multi-stakeholder forum, namely City Team, to complete a vulnerability assessment (VA) on the citywide scale. Following the VA, the City Team, in coordination with ACCCRN Indonesia (Mercy Corps) and the Rockefeller Foundation, conducted pilot projects and sector studies such as rainwater harvesting. The third phase (2011–2013) included the completion of the city resilience strategy (CRS) and the development of concept notes towards prioritised intervention (Figure 3). Physical project implementation occurred in the third phase. The final phase included engaging and influencing process at a national level. This included efforts to expand the approach to fifteen other cities that had shown strong interest in replicating resilience building processes [2].

During the early set-up (during the 2009–2011 period), a Semarang City Team (SCT) was formed, composed of an advisory team and a technical team. The advisory team consists of representatives from the city government, led by the city's executive secretary under the Mayor, and the technical team consists of representatives from municipal government agencies, local universities, and local NGOs. The role of the SCT was crucial, as the members were to monitor, control, organise, conduct studies, manage projects, and report on all activities, processes, and methodologies applied under ACCCRN. The City Team was mandated to lead, facilitate, and catalyse the development of the city resilience strategy document and to institutionalise the strategy for long-term development. SCT had been the backbone of the climate resilience initiatives and emerged as a collective decision-making body.

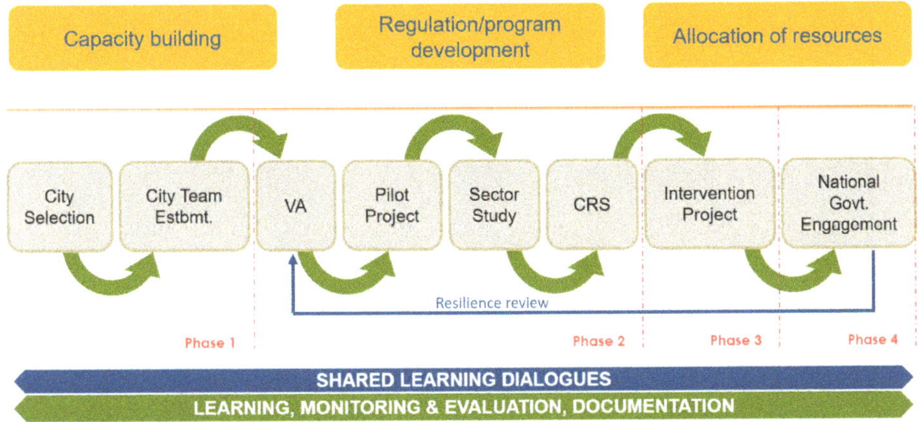

Figure 3. Asian Cities Climate Change Resilience Network (ACCCRN) Typical Process in Semarang City. (Source: Sutarto [29])

The ACCCRN Indonesia country strategy 2010–2013 focused on four key activities to ensure continuation of the interventions after the project. First, to manage the project implementation in the city, targeting the capacity-building of City Team members, transforming the City Team into a city climate change resource centre, and facilitating external support for the city government. Physical project implementation included the establishment of flood warning systems and the capacity development of the local communities to conserve the mangrove ecosystem in the coastal areas of the City [29,30]. Second, to disseminate the ideas and educational materials of urban climate change resilience through multiple means such as social networks, conferences, and workshops. Third, to link up with the national government ministries. Fourth, to scale-up projects through the national associations of cities governments and other national and international networks [29].

4.3. Kickstart of Project as First Level Institutionalisation

Governments are formal institutions. Therefore, driving new innovation with and by governments requires some formal basis. ACCCRN in Semarang City formally began with decision letters from the mayors. City's leadership was a key variable for this initial stage. The decision letters mandated the formation of the SCT (see Table 2). No other formal regulation was created at the city level to support the initiatives. In Indonesia's legal hierarchy, a decision letter from a mayor does not have good enough power to generate and mobilise both human resources as well as fiscal allocation to implement any innovative action.

Table 2. Evolution of climate risk governance in Semarang City. Source: Author, 2019.

Governance Variables	Agenda 21	Disaster Management	Urban Climate Change Resilience Arrangement Since 2009		
Timeframe	Semarang Agenda 21 2001	Existing structure since 2010	2009–2011	2011–2015	Since 2015 towards Post-ACCCRN
Nature of the organization	Multi-stakeholder forum	Government agency	Multi-stakeholder platform	Multi-stakeholder platform	NGO/association
Leadership	Single departmental leader	Single departmental leader	Collective leadership—Coordination team (17 head-of-city departments)	Appointed coordinator—Head of Development Planning Agency	Recruited executives
Decision making model	Top down	Command and control—top down	Shared decision making	Shared decision making	Board member
Chief executive	Head of Environmental Protection Unit	City Secretary	Head of City Environmental Protection Unit	Head of Planning Division of Development Planning Agency	Executive director
Executive secretary	Head of Environmental Protection Unit	Head of BPBD	None	Secretary of Development Planning Agency	NGO Manager
Membership	Loose membership	Single agency	20 members (city government staff, local NGOs and local universities)	15 members (city government staff, local NGOs and local universities)	Individual members of the present city team
Funders	World Bank	APBD	ACCCRN	ACCCRN	External funders
Quantity of public consultation	One-off workshop	n/a	Regular meeting—monthly	Regular meeting—monthly	Internal arrangement

Note: BPBD is the local disaster management office.

Therefore, in order to keep adaptation agenda on top of the city development plan, SCT must be able to create some spaces that allow them to work with limited resources. ACCCRN provided basic resources that can help by jump-starting the city to tackle climate change and urban risks. Two years after the launch of ACCCRN in Semarang, a small reform took place in the government of the City of Semarang where the Department of Development Planning (Bappeda) that used to be a less influential department in the past was revitalised to be a stronger planning institution. Bappeda has been mandated to not only coordinating city planning but also monitoring and evaluating city departments and all sectorial development. In reality, Bappeda had just been functioning as a positive advocate for any innovative policy including climate resilience ([31], personal communication)" Bappeda now is just like a sharp knife' because the agency has started to re-establish its mandates not only as a gatekeeper and quality controller for city development planning agenda but also as a sharpener of development ideas and proposals" ([32], personal communication).

4.4. City Resilience Strategy

ACCCRN facilitated the drafting of City Resilience Strategy (CRS)—a fundamental framework that aims at guiding the city to develop policy anticipating and addressing future impacts of climate change. The key features of CRS include a document containing broad adaptation agendas and guidance, prepared by local stakeholders and local government, vulnerability and risk context, organised evidence and analysis justifying adaptation interventions, priorities for resilience actions, consistency with existing planning documents and processes that are fit to local institutional settings, guidance for the private sector and civil society groups to design and implement their own adaptation actions, and linkage and coordination with complementary activities for donors and other funding [3,23,29].

The purpose of the CRS document was also to inform other development policy documents in Semarang City such as the Mid-Term Development Plan (RPJMD) documents. One of the reasons for this adoption is because the Chief Executive of the SCT and the Bappeda Head of Planning Unit happened to be the same person. The City Team was managed under the leadership of city development planning department (Bappeda). This coincidence allowed the CRS to inform the mid-term development plan (RPJMD) for the 2010–2015 period and Semarang Spatial Planning 2011–2030 ([31,33], personal communication). Prior to the CRS document, climate change-related discussion via shared learning dialogues (SLDs) have been directly 'fast tracked' into policy and practices ([34], personal communication). For example, the adoption of CRS recommendations such as rainwater harvesting, and a flood early warning system have been adopted as both policy and programmes in the city [30] and reference ([31,32], personal communication). The inclusion of adaptation agenda depends on the existence and commitments of local champions in the planning process as well as fiscal capability. One of the current issues is, VA and CRS documents might have been outdated and the question is how the actors allocate resources to update the documents?

4.5. Local Champion and Leadership

Effective progress towards building resilience in Semarang City has been associated with the roles of local champions. But local champions cannot be easily hand-picked and strategically planned. It took two years for the project to 'recruit' one of the most notable champions, Mr. Purnomo Dwi Sasongko (hereafter Purnomo), who was later elected as the Executive Secretary of the International Council for Local Environmental Initiatives (ICLEI) of Southeast Asia. He joined the SCT in Semarang in 2011 and was soon after promoted as the Head of the Planning and Infrastructure Unit of Bappeda. "Mr Purnomo was the key climate risk communicator and 'ACCCRN spokesperson' to the city's high officials, such as the head of Bappeda and the mayor" ([34], personal communication). He was aware that climate risk communication and strategic sense-making among the city departments are the necessary conditions to create critical awareness regarding the importance of climate integration into city development. Purnomo believed that Bappeda remains the city coordinating body overseeing annual development plans from 32 departments (a.k.a. local government units shorted as SKPDs) ([31], personal communication).

"Communication is key to all the cities departments. I have to find a proper message that is suitable for the respective department" ([31], personal communication). Purnomo proactively communicates CRS to cities' department focal points by making sense of the need to address climate change within the city departments (SKPDs) ([31,32], personal communication). "To the Public Works department, I could clearly articulate the linkage of drainage maintenance with climate adaptation. As a planner who only recently transformed myself from being a relative climate ignorant to be a climate advocate within the city planning departments, I believe that adaptation can be linked to many urban sectors, from public works to marine and fisheries, water resource management, disaster risk management, environmental services and protection, health, etc. Furthermore, Adaptation is not an extra task for city departments if the key staff in the departments understand how to integrate climate change adaptation into existing issues" ([31], personal communication). Both CRS and VA documents are considered 'academic inputs' to the local government. 'The city needs to deepen the climate adaptation details' ([31], personal communication). Regardless, both CRS and VA were noted as legitimate document that can inform mid- to long-term city planning ([31,34], personal communication).

4.6. Resilience Agendas in Development Planning Documents

There is solid evidence that the CRS document has been adopted into the mid-term development plan (RPJMD) 2010–2015, where it mentioned climate change six times including the fact that it recognised climate change impacts to city infrastructure such as road construction, as the rainfall drops become less unpredictable, road construction quality is compromised [35]. Furthermore, the document explicitly shows that climate change becomes a routine business of the Environmental Protection Department and formally budgeted about US$ 1 million a year during 2011–2015 with a focus on urban waste management ([35], Appendix 1 p. 13). Furthermore, there is a clear view that spatial planning is key to mitigate and adapt to climate change ([35] p.V-18 and VI-12).

Interestingly, climate change is still mentioned eight times in the RPJMD 2016–2021, mostly in the introduction. The document also briefly contains risk and vulnerability information. There is an earmarked budgeting for CCA (budget line 2.05.28), allocated for mangrove restoration each year about US$ 275,000 or about a 75 percent decline from previous RPJMD 2010–2015. Unfortunately, it is becoming more realistic than the previous mid-term development plan. Also, the government is aiming at creating 28 climate resilience villages (namely Kampung Proklim) by the end of 2021 ([36], p. VIII-14). The RPJMD 2016–2021 was later revised in 2017, where there is a new budget earmarked as "areas that have the capacity for adaptation and mitigation" with a budget plan from about US$ 425,000 in 2017 to about 625,000 in 2021 to cover all areas ([37], p. VI-14).

4.7. Fiscal Capability and Allocation

Cities can generate higher (fiscal) capacity and welfare that can assist their urban population to reduce existing risks [38]. In Indonesia, local opportunities for climate adaptation can be created through self-sustaining incentives that directly shape adaptation imperatives. The increase in cities' capacity to reform their local tax and retribution and curb corruption in tax collection, as demonstrated by Bandar Lampung City [18], is key to boosting fiscal capacity so cities can have greater freedom to invest in the sectors that challenge them most.

The issue is how local actors can sustain the CCA/DRR agenda, commitment, and discourse within the city. One transformative dimension of the ACCCRN initiative is that the Semarang City Planning Agency recently acknowledged the fact that almost all the coastal areas in Semarang are owned by private firms, making it particularly difficult to create mechanisms for coastal protection ([31], personal communication). The city faces serious governance challenges and financial burdens in buying back the coastlines. The SCT influenced the city officials to plan to buy back some of the coastal lands for public access ([31,32], personal communication).

The high visibility of urban risks such as the heavily inundated Northern parts of Semarang City has made it easier for the rest of the city departments to consider climate change adaptation in the annual budget allocation. Transformation is seen at the discursive level. There were adoptions of some proposed activities from the CRS document into annual programme and projects. For instance, at the earmarked budget line, there was no single indication that the Environmental Protection Office (BLHD) had made allocation for addressing the impact of climate change during 2005–2010 budgets. While for the 2011–2015 period, the city started to allocate US$10,000–12,500 annually to either host regular meetings of the SCT or to allocate annual budgets to scale-up rainwater harvesting and mangrove ecosystem development during 2011–2013. Climate change was budgeted under the 5th programme of BLHD, namely 'protection and conservation of natural resources' in 2011–2013. The amount seems trivial, but the discourse behind the allocation is an important first step towards bolder action.

Since the fiscal capacity of many cities in Indonesia has increased in the last 10 years [18], Semarang City has the capacity to fund its own climate adaptation activities in the future. During the 2012 fiscal period, under the flagship of 'climate adaptation', the city managed to allocate US$ 100,000 to buy back some hectares of coastal land ([31–33], personal communication). In the future, the plan to buy back land is expected to continue as long as the land value goes down as a result of nearly permanent inundation in the northern part of the city.

4.8. NGOnisation of Formal Platforms?

NGOnisation is defined by the author as a form of institutionalisation in that the multi-stakeholder platform comprises of governmental and non-governmental actors (including NGOs, academics, experts, private sectors and others) have been transformed into into an NGO-like structure. Such transformation, at least in theory, allows the actors to use it as a vehicle to carry on the mandate of adaptation and resilience in more flexible ways.

To ensure that adaptation outcomes can be sustained after the exit of ACCCRN in 2016, the SCT members negotiated to establish a working group or a unit outside the government that may be useful only for conducting objective studies to inform city planning ([31], personal communication). At least three scenarios have been discussed by members of the SCT concerning how to endogenously sustain climate change intervention in Semarang City beyond 2016, after ACCCRN.

The first scenario was a voluntary mechanism. This suggests that committed individuals from city departments who could influence from within using their discretionary roles in planning at sub-department and department levels. Their personal network with the higher authorities was vital to promote the idea of urban sustainability and adaptation. This option was limited. It requires committed leaders at all levels, and reality suggests that the city does not have this luxury all the time ([31,33], personal communication). In fact, there is only a small minority of positive deviant bureaucrats that function as climate goal-keepers in their own departments as well as city secretariat levels.

The second scenario involved mandatory mechanisms, which works through formalised processes with clear mandates and responsibilities as in the case of Surat [39]. Therefore, city resilience-building should be translated into formal rules that provide mandates for the relevant city institutions. This suggests that the city institutions will decide focal points (persons) that may (not) be fit for the task of being climate advocates. This requires two approaches that may complement each other. The first was having a strong formal scenario where there should be a regulation translated into annual operation (e.g., protocol or standard operational procedures) for planning and financing the city. This further requires changes at higher levels, e.g., a Ministry of Home Affairs (MoHA) regulation that mandates city governments to consider climate change as a cross-cutting issue. The second approach is a less formal scenario where local governments, at least at the mayor and legislative levels, can co-create climate regulation/legislation informed by the CRS document.

The later scenario can be achieved through three steps. The first is to create a mayoral regulation that could function for five years, suggesting that city departments work according to the mayor's interests as reflected in the regulation. This can be done at any time as long as the mayor is interested in and committed to the issue. Second, the mayor's five-year development agenda, reflected in the city's mid-term development plan (RPJMD), must cite or adopt a climate adaptation agenda from the CRS document. This scenario later materialised in the last two planning documents [35–37]. Third, a more long-term approach is that the city can create a local regulation, that can be drafted and endorsed by the mayor with or without the support from the city legislators. The later process involves a lengthy process including political lobbying with some degree of uncertainties.

The third Scenario involves transforming the present SCT into an NGO-like structure ([40], personal communication), where it can maintain its flexibility and interest in promoting climate change adaptation by working through personal contacts among the city's decision-makers. This option, of NGOnisation, in theory could work in the short term. It would already have been exercised in many places in Indonesia where strong environmental NGOs have been working over the last decades. However, strong environmental NGOs do not always succeed in the long run especially when the local political and administrative context change and funding mechanism for NGOs is not certain.

What is interesting is the view of the SCT that functions as a hub of knowledge (or rather as a discursive machine) and technology transfer for climate adaptation. Technology transfer is exemplified by the transfer of technology and knowledge of flood early warning system and appropriate breakwater technologies to protect mangroves [30] and reference ([34], personal communication). Interestingly, the successful technological transfer from ACCCRN can be seen as a direct outcome of its unique approach as it provided multiyear projects that guarantee deeper engagement with local administration and allowing climate change discourse to penetrate in the city structure in a rather informal fashion [30,41].

SCT members are aware of the high turnover of knowledgeable officials at middle and high ranks in public administration in many city/district departments have become a challenge in local governance. This will eventually lead to a lack of institutional memory within many local government offices. New mayors might mean new programmes and new ignorance ([31–34], personal communication). This awareness motivates the empowered officials and stakeholders to develop a structure that allow them to be drivers of CCA knowledge sharing beyond the City of Semarang.

Finally, as implied by all the scenarios above, institutional barriers are not easy to tackle. The SCT finally decided to transform itself into a CSO-like structure namely IUCCE (Initiative for Urban Climate Change and Environment) (Table 1). IUCCE [42] aims to "help achieve the objectives and foster urban resilience to climate change and changing environment". While this problem has been identified, it has become less clear how future climate governance in the city of Semarang will evolve under the UCCE regime.

A closer look at the work of IUCCE indicates that the organisation has been playing roles as a think tank that works beyond the city's jurisdiction. Some of the knowledge products include technical guide on community-based disaster risk reduction, documentation of knowledge concerning mangrove and flood early warning systems etc. While IUCCE can still be functional as an informal "City Team", it is safe to argue that investment of ACCCRN has been clearly successful in terms of knowledge and technological transfer.

4.9. 100 Resilience Cities Project in Semarang City

Climate change, including climate risk reduction, remains a marginal task under a few departments including environmental protection department where a few climate adaptation activities (especially mangrove planting) are budgeted for RPJMD 2016–2021 ([36], Section 3.6). While the DRR agenda remains at the discretion of local emergency management agency, for CCA however, based on the lessons during 2011–2015, the actual fiscal allocation does not reflect the budget as proposed by the planners.

After the ACCCRN, the formality of STC is not diminished even though incentives for regular activities, pilot projects, as well as adaptation projects such as flood warning systems ([31], personal communication) have literally come to an end, while at the same time, the Semarang City graduated from ACCCRN into the new initiatives, namely 100 Resilience Cities project funded by the Rockefeller Foundation [43].

5. Discussion

5.1. Institutionalisation: Formal and Informal

Old forms of institutionalisation that emphasise the formal approach for reform in formal policy settings remains the imperative of global framework and initiatives. The Hyogo Framework for Action and Sendai's norm of institutionalising resilience can be exemplified by the process of mainstreaming in Surat with the formation of the Climate Change Trust based on the Public Trust Act at state level [39]. However, context shapes the forms of institutionalisation as it can manifest as a general climate policy or specific DRR plan (case of Quito, Ecuador). Institutionalisation can also mean a shift from a resilience strategy that led to the endorsement of the Climate Change Trust (case of Surat, India) [15,39]. Such legal formal achievement in Surat remains to be seen in the City of Semarang as it requires a long local political process. It is also clear that some of the achievements such as resource allocation have been endogenously provided by the executive government in Semarang City, as also noted in the other study such as, Durban [10], Surat [39], and Bandar Lampung City [18,23].

The findings suggest that institutionalisation process take place in several domains, including formal development planning. In terms of institutionalised practice, CCA has become a key task of the local environmental agency while DRR is seen as the task for the local disaster management agency. Such an achievement is predictable, as informed by previous research in different context [12] as well as in Indonesia context [18,23].

In the context of Indonesia's national disaster risk reduction policy reform, institutional change started from legislation and followed by the creation of new administrative units to deal with broader disaster risk problems. Equivalent processes did not occur for climate change adaptation at a national level. In the absence of national guidelines for cities to be adaptive to climate change, secondary cities in the developing world often create their climate adaptation policies and practices through external influences, as exemplified by different ACCCRN cities in India and Indonesia [39,44].

5.2. Hybrid Approach to Institutionalisation

Local institutional uncertainty has made future adaptation in cities less predictable. Previous studies suggest that the negative outcomes are particularly due to little stability at public administration and bureaucratic levels, because the local government sector has been affected by dynamic political change and decentralisation. This has been quite clear from the other ACCCRN pioneering cities such as Bandar Lampung [23].

The process of adaptation in cities involved the complex process of exogenous and endogenous efforts in building resilience. While it seems almost impossible to fix the institutional mechanism under the project timeline, the local actors used a rather pragmatic approach to institutionalise climate action. The agenda of hybrid institutionalism has led to pragmatic institutionalisation as it goes beyond what was once seen as multi pronged approach [10].

Local champions exercised their discursive power. Their network and platform serve as guardians and gate-keepers of city planning, as seen in Western Cape, South Africa, where climate policy entrepreneurs have been the key to adaptation mainstreaming [45]. Their impact can help reduce institutional uncertainty temporarily. The challenge is, local champions and good leadership are often 'given' and cannot be easily planned or recruited in advance.

Responding to the challenges at local and national levels where climate adaptation agenda remains unclear and local capacity remains low, Sharma and Tomar [44] suggested pragmatic solutions namely 'entry points' including embedding adaptation and resilience through existing development and disaster management plans. The first ACCCRN process during 2009–2014 (Section 4.2) are the entry points. To gain quick wins, the 'entry points' approach has also been promoted by some researchers, such as in Reference [46]. The framing of 'entry points' indicates the nature of exogenous intervention. However, the challenge is how to win at the 'exit point' after the end of international projects remains an important issue for local actors.

The pragmatic approach also requires working with local proponents such as local champions as they can be seen as 'institutions' as they not only create their own rules of the game but play the roles as both goal-keepers and climate policy entrepreneurs. The champions have been the key officials from within existing institutions whose strong passion and interest in promoting innovation within the local government level were seen as vital [47–49] and Reference ([50], personal communication). These champions tend to have a balanced self-interest and public interest in promoting climate resilience agenda not only in Semarang but also in other ACCCRN cities such as Hat Yai City, Thailand [51]. This satisfies the rational choice theory approach as they acted based on their best interest. However, their inability to jumpstart adaptation without external aid suggests that their engagement is largely driven by their interests in incentives created by the projects. This justifies that NEI is the mechanism that helps local actors sustain CCA agendas.

The pragmatic approach to institutionalising CCA also includes the strategy of NGOnisation of the SCT platform. It has been partly used by the actors as a strategy for institutionalising climate change adaptation and resilience building in Semarang City. NGOnisation is an approach where urban stakeholders come to terms with the dynamic nature of complex realities of city development. This mechanism is used to solve institutional uncertainty in the city. While at the same time, the key actors continue to benefit from the existing formal mechanisms, such as continuing to use the platform of STC and others (e.g., 100 Resilient Cities Project funded by the Rockefeller Foundation) and existing departmental commitments related to climatic risks ([50], personal communication). Interestingly, the setup of IUCCE as a think tank in Semarang is favoured by most of the SCT members in Semarang City as they see that this 'NGO-like' structure can provide a balanced self-interest and public interest in promoting climate resilience.

5.3. Transforming Urban Adaptation Platforms into Permanent Institution?

Global disaster risk frameworks such as the Hyogo Framework for Action and its predecessor, the Sendai Framework for disaster reduction (SFDRR), have promoted the idea of multi-stakeholder engagement, namely DRR platforms (UNISDR 2005), that are supposed to exist at different levels from global to national to local levels of governance. ACCCRN's City Team in Semarang City can be seen as a CCA/DRR platform or forum. Forums can be seen as institutions as each forum has its own rules of the game. Small groups in any settings that meet regularly suggests that they are bounded by certain values and interests and their rules of the games function as the institutional avenue for them to continue to repeat their interactions [52] until they breakup and the game does not benefit the members.

Therefore, urban adaptation platforms including disaster management platforms are in themselves part of institutional development. Unfortunately, there are always costs associated with regular run of forums and platforms. Empirically speaking, local and national multi-stakeholder platforms have been established in many parts of the world including Indonesia [53], where some were more functional while others were simply function as institutional decoration.

5.4. Path-Dependency Theory as a Predictor for Urban Adaptation?

The challenge in the City of Semarang today is not entirely new as in the case of the Semarang Agenda 21 way back in 1997/1998. Semarang City was among the first Indonesian cities to adopt the sustainability framework of Local Agenda 21. The agenda was locally branded as 'Semarang Environmental Agenda: Toward a sustainable city 1998–2003' (hereinafter SEA21) [54]. Semarang City was selected for the pilot project initiated by the World Bank with the acceptance from the Semarang City government ([55], personal communication). Agenda 21 focused on process and trust building as the project conducted extensive consultation with sectoral experts, government officials, NGOs, academics, and others, which resulted in the 18 chapters of Agenda 21–Indonesia in 1997 [54]. The document identified high- and medium-priority programmes, to be completed in five years (1998–2003) and 10 years (1998–2008), respectively. These priorities included population management, self-resilient community, public transportation, coastal inundation, domestic waste reduction, treatment of human waste, waste management, clean production, healthy rivers, and clean air programmes.

In retrospect, the present institutional pathways for adaptation follow the historical path of the urban sustainability agendas stipulated by SEA21 developed 20 years ago in the same city. To the stakeholders, the most successful contribution of SEA21 to the city administration is the capacity-building ([55], personal communication). The knowledge transfers from the initiative facilitated new awareness for trained staff concerning environmental and urban sustainability. The programme may have been short-lived, but there was a discursive turn within the city's private sectors regarding environmental quality where Bapedalda (Local Environmental Protection Agency or now BLHD) was able to establish a stick-and-carrot approach [56] and reference ([55], personal communication).

Lessons from SEA21 suggested that the process of institutionalising international initiatives such as ACCCRN and others (through formal planning documentation, policy adoption, etc.) often hit the hard wall of local institutions. The issue is not that there was no innovation but innovation in ideas, policies, and practices are often short-lived because such initiatives relied more on persons than systems ([55,57], personal communication). Interestingly, the reliance on persons can be credited as a good start if the persons can play roles as champions with the capability to create adaptation discourse at different levels of governance in cities. The problem is, champions are also timebound. Champions today might not be champions tomorrow as external and internal incentives change through time.

6. Conclusions

This research investigates how local actors negotiate to ensure continuity of CCA and DRR as routine development agenda. The question is how urban stakeholders and policy entrepreneurs negotiate and come to terms with potential forms of institutional scenarios crafted to promote adaptation and deepen resilience in the City of Semarang. The whole arrangements and interactions of ACCCRN from the beginning have been about using different forms of institutionalisation to sustain adaptation and resilience agenda. The continuity of CCA and resilience-building depend on mechanisms where there is regular reproduction of resilience discourse including their urgency and importance at different levels and domains ranging from policy documents to the existence of CCA advocates or champions within the agencies in cities. This goes beyond the binary framework of endogenous versus exogenous initiatives for adaptation.

Hybrid institutionalism has merits to provide better understanding of the complexity around institutionalising urban adaptation in Semarang City. While projects such as ACCCRN have co-facilitated processes that aimed at promoting a more rational choice approach to establish a more permanent mechanism, it turned out that such adaptation initiatives have been trapped in the past institutional trajectory such as NGOnising the City Team structure. Lessons from the case of SA21 and the recent development of IUCCE suggests the model of institutionalisation as explained by the theory of historical institutionalism where the 'adaptation pathways' is skewed towards future institutional uncertainty, which makes it difficult for local actors such as the in the City of Semarang to make strategic decisions from within formal institutions [20]. As a result, the STC has transformed itself into an NGO-like structure and served as a think tank instead of policy makers. On the other hand, the emergence of new international initiatives such as 100 Resilient City provides new avenues for the local stakeholders to either restart again or to move forward to the next stages of a city resilient development strategy.

The Semarang City Team has a vision to drive and facilitate a permanent agenda for adaptation and resilience via formal mechanisms. Unfortunately, local dynamic process led these efforts to push the actors from shifting from formal into a more informal approach. The good news is that new exogenous initiatives remains available via different trajectories as exemplified by the shifts from Semarang Agenda 21, to ACCCRN and to 100 Resilience City programmes. And in between, there is often international frameworks (e.g., among others, the Hyogo Framework or the Sendai Framework) that can be used to ensure resilience discourse remains in the orbits of urban governance.

The author argues that that the (dis)continuity of urban sustainability initiatives, including climate change adaptation and resilience in Semarang city, do occur in the form of hybrid institutionalism but pathway-dependency theory emerges as the most dominant predictor as exemplified by the boom and burst of local platforms ranging from SEA21 to ACCCRN to 100 Resilience City and more into the future. Rational choice thinking and the effort to localise resilience and adaptation often ended up in path-dependency phenomenon where history repeats itself in the form of NGOnising the resilience platforms.

Funding: Initial funding was provided by Rockefeller Foundation via ACCCRN Indonesia project managed by Mercy Corps during 2012/2013. Since 2014 this study is self-funded.

Acknowledgments: Personal thanks to Ratri Sutarto, Aniessa Delima Sari, Paul Jeffery, Ninik Mulyawati, Omar Saracho (Mercy Crops) who have been supporting all the logistics of this research in Semarang in 2012. Thanks also to Purnomo (The Head of Planning Unit in Bappeda Semarang), Feri Prihantoro, Lilin Budiarti, Lutfi Muhamad, Gunawan Wicaksono, Raharjo Tjahyono and all colleagues at City Team of Semarang and Jawoto Sih Setyono (Diponegoro University) for kindly supporting this research. The author would like to thank the Indonesia Project at Australian National University for the opportunity to present this draft in late 2016. An earlier draft of this article was published as a Working Paper at the Resilience Development Institute. The author would like to thank the three anonymous reviewers who gave very valuable inputs to improve the previous draft. All the mistakes interpreting the information for the sources are the authors.

Conflicts of Interest: The author received research funds from ACCCRN Indonesia via Mercy Corps. This study is however independent and critical to satisfy scientific interest alone.

Abbreviations

ACCCRN	Asian Cities Climate Change Network
Bapedalda	Local Environmental Protection Agency (old)
Bappeda	Local development planning agency
BLHD	Local Environmental Protection Agency (new)
CCA	Climate change adaptation
HFA	Hyogo Framework for Action
IUCCE	Initiative for Urban Climate Change and Environment
SFDRR	Sendai Framework for disaster risk reduction
STC	Semarang City Team
UN	United Nations
UNISDR	United Nations International Strategy for Disaster Reduction

References

1. IPCC. *Climate Change 2014: Impacts, Adaptation, and Vulnerability. Part A: Global and Sectoral Aspects. Contribution of Working Group II to the Fifth Assessment Report of the Intergovernmental Panel on Climate Change*; Field, C.B., Barros, V.R., Dokken, D.J., Mach, K.J., Mastrandrea, M.D., Bilir, T.E., Chatterjee, M., Ebi, K.L., Estrada, Y.O., Genova, R.C., et al., Eds.; Cambridge University Press: Cambridge, UK; New York, NY, USA, 2014; 1132p.

2. Sari, A.D.; Prayoga, N.; Syam, D.; Theda, F.; Bielman, J.; Pradityo, D. *Climate Change Adaptation towards Urban Resilience: A Journey and Lesson Learnt from ACCCRN in Indonesia, 2009–2016*; ACCCRN Indonesia Project: Jakarta, Indonesia, 2018.

3. Brown, A.; Dayal, A.; Del Rio, C.R. From practice to theory: Emerging lessons from Asia for building urban climate change resilience. *Environ. Urban.* **2012**, *24*, 531–556. [CrossRef]

4. Tyler, S.; Moench, M. A framework for urban climate resilience. *Clim. Dev.* **2012**, *4*, 311–326. [CrossRef]

5. Douglas, M. *How Institutions Think*; Syracuse University Press: Syracuse, NY, USA, 1986.

6. Lebel, L.; Nikitina, E.; Kotov, V.; Manuta, J. Assessing institutionalised capacities and practices to reduce the risks of flood disaster. In *Measuring Vulnerability to Natural Hazards: Towards Disaster Resilient Societies*; Birkmann, J., Ed.; United Nations University Press: Tokyo, Japan; New York, NY, USA; Paris, France, 2006.
7. North, D.C. Economic Performance Through Time. In *The New Institutionalism in Sociology*; Brinton, M.C., Nee, V., Eds.; Stanford University Press: Stanford, CA, USA, 1998; pp. 247–257.
8. Underdal, A. *The Causal Significance of Institution*; The IDGEC Synthesis Conference—IHDP Newsletter #1; IHDP: Bali, Indonesia, 2007.
9. Lassa, J.; Surjan, A.; Caballero-Anthony, M.; Fisher, R. Measuring political will: An index of commitment to disaster risk reduction. *Int. J. Disaster Risk Reduct.* **2019**, *34*, 64–74. [CrossRef]
10. Roberts, D. Thinking globally, acting locally—Institutionalizing climate change at the local government level in Durban, South Africa. *Environ. Urban.* **2008**, *20*, 521–537. [CrossRef]
11. Román, M.; Linnér, B.-O.; Mickwitz, P. Development policies as a vehicle for addressing climate change. *Clim. Dev.* **2012**, *4*, 251–260. [CrossRef]
12. Anguelovski, I.; Carmin, J. Something borrowed, everything new: Innovation and institutionalization in urban climate governance. *Curr. Opin. Environ. Sustain.* **2011**, *3*, 169–175. [CrossRef]
13. Bevir, M. *Key Concept in Governance*; Sage: London, UK, 2009.
14. Heazle, M.; Tangney, P.; Burton, P.; Howes, M.; Grant-Smith, D.; Reis, K.; Bosomworth, K. Mainstreaming climate change adaptation: An incremental approach to disaster risk management in Australia. *Environ. Sci. Policy* **2013**, *33*, 162–170. [CrossRef]
15. Anguelovski, I.; Chu, E.; Carmin, J. Variations in approaches to urban climate adaptation: Experiences and experimentation from the global South. *Glob. Environ. Chang.* **2014**, *27*, 156–167. [CrossRef]
16. UNISDR. *Hyogo Framework for Action—Disaster Risk Reduction*; United Nations: Geneva, Switzerland, 2005.
17. Bulkeley, H.; Tuts, R. Understanding urban vulnerability, adaptation and resilience in the context of climate change. *Local Environ.* **2013**, *18*, 646–662. [CrossRef]
18. Lassa, J.A.; Nugraha, E. From shared learning to shared action in building resilience in the city of Bandar Lampung, Indonesia. *Environ. Urban.* **2015**, *27*, 161–180. [CrossRef]
19. UNISDR. *Sendai Framework for Disaster Risk Reduction*; United Nations: Geneva, Switzerland, 2015.
20. Lassa, J.A. Institutional Vulnerability and Governance of Disaster Risk Reduction: Macro, Meso and Micro Scale Assessment. PhD. Thesis, University of Bonn, Bonn, Germany, 2011.
21. Schmidt, V.A. Discursive Institutionalism: The Explanatory Power of Ideas and Discourse. *Annu. Rev. Political Sci.* **2008**, *11*, 303–326. [CrossRef]
22. Kaag, M.M.A.; Brons, J.; de Bruijn, M.E.; van Dijk, J.W.M.; de Haan, L.J.; Nooteboom, G.; Zoomers, A. *Poverty is Bad. Ways Forward in Livelihood Research*; CERES 'Pathways of Development' Seminar Paper; Universiteit Utrecht: Utrecht, The Netherlands, 6 February 2003.
23. Nugraha, E.; Lassa, J.A. Towards endogenous disasters and climate adaptation policy making in Indonesia. *Disaster Prev. Manag.* **2018**, *27*, 228–242. [CrossRef]
24. Leck, H.; Roberts, D. What lies beneath: Understanding the invisible aspects of municipal climate change governance. *Curr. Opin. Environ. Sustain.* **2015**, *13*, 61–67. [CrossRef]
25. Yastika, P.E.; Shimizu, N.; Abidin, H.Z. Monitoring of long-term land subsidence from 2003 to 2017 in coastal area of Semarang, Indonesia by SBAS DInSAR analyses using Envisat-ASAR, ALOS-PALSAR, and Sentinel-1A SAR data. *Adv. Space Res.* **2009**, *63*, 1719–1736. [CrossRef]
26. Husnayaen, A.; Rimba, B.; Osawa, T.; Parwata, I.N.S.; Astarini, I.A. Physical assessment of coastal vulnerability under enhanced land subsidence in Semarang, Indonesia, using multi-sensor satellite data. *Adv. Space Res.* **2018**, *61*, 2159–2179. [CrossRef]
27. Archer, D.; Dodman, D. Making capacity building critical: Power and justice in building urban climate resilience in Indonesia and Thailand. *Urban Clim.* **2015**, *14*, 68–78. [CrossRef]
28. Reed, S.O.; Friend, R.; Toan, V.C.; Thinphanga, P.; Sutarto, R.; Singh, D. "Shared learning" for building urban climate resilience—Experiences from Asian cities 393. *Environ. Urban.* **2013**, *25*, 393–412. [CrossRef]
29. Lindsay, J.; Rogers, B.; Church, E.; Gunn, A.; Hammer, K.; Dean, A.J.; Fielding, K. The role of community champion in long-term sustainable urban water planning. *Water* **2019**, *11*, 476. [CrossRef]

30. Sari, A.D.; Prayoga, N. Enhancing citizen engagement in the face of climate change risks: A case study of the flood early warning system and health information system in Semarang City, Indonesia. In *Climate Change in Cities Innovations in Multi-Level Governance*; Hughes, S., Chu, E.K., Mason, S.G., Eds.; Springer: New York, NY, USA, 2017; pp. 1121–1137.

31. Purnomo, D.S. (Bappeda, Semarang City, Jawa Tengah). Personal communication, 24–25 June 2012; 2 October 2012; 5 December 2012.

32. Lutfi, M. (Bappeda, Semarang City, Jawa Tengah). Personal communication, 15 December 2012.

33. Gunawan, W. (BLHD, Semarang City, Jawa Tengah). Personal communication, 6 June 2012.

34. Delima, S.A. (Mercy Corps/ACCCRN Indonesia Project, Semarang City, Jawa Tengah). Personal communication. 28 June 2012; 11 December 2012.

35. Semarang City. *Mid-Term Development Plan [Rencana Pembangunan Jangka Menengah] 2011–2015*; Semarang City Government: Semarang, Indonesia, 2011.

36. Semarang City. *Mid-Term Development Plan [Rencana Pembangunan Jangka Menengah] 2016–2021*; Semarang City Government: Semarang, Indonesia, 2016.

37. Semarang City. *Revision of Mid-Term Development Plan [Rencana Pembangunan Jangka Menengah] 2016–2021*; Semarang City Government: Semarang, Indonesia, 2017.

38. Lall, S.V.; Deichmann, U. *Density and Disasters: Economics of Urban Hazard Risk*; Policy Research Working Paper 5161; World Bank: Washington, DC, USA, 2009.

39. Karanth, A.; Archer, D. Institutionalising mechanisms for building urban climate resilience: Experiences from India. *Dev. Pract.* **2014**, *24*, 514–526. [CrossRef]

40. Jawoto, S. (Diponegoro University, Semarang City, Jawa Tengah). Personal communication, 18 December 2012.

41. Taylor, J.; Lassa, J. *How can Climate Change Vulnerability Assessments Best Impact Policy and Planning? Lessons from Indonesia*; IIED Asian Cities Climate Resilience WP Series, No 22; IIED: London, UK, 2015; pp. 1–33.

42. IUCCE. About Initiative for Urban Climate Change and Environment. 2019. Available online: iucce.org (accessed on 10 January 2019).

43. Semarang City. Resilient Semarang Moving Together Towards a Resilient Semarang. 2016. Available online: www.100resilientcities.org/wp-content/uploads/2016/05/Semarang20Resilience20Strategy20-202016. pdf (accessed on 10 January 2019).

44. Sharma, D.; Tomar, S. Mainstreaming climate change adaptation in Indian cities. *Urban. Dev.* **2010**, *22*, 451–465. [CrossRef]

45. Pasquini, L.; Ziervogel, G.; Cowling, R.M.; Shearing, C. What enableslocal governments to mainstream climate change adaptation? Lessons learned from two municipal case studies in the Western Cape, South Africa. *Clim. Dev.* **2015**, *7*, 60–70. [CrossRef]

46. Bahadur, A.; Tanner, T.; Pichon, F. *Enhancing Urban Climate Change Resilience: Seven Entry Points for Action*; Sustainable Development Working Paper Series 47; Asian Development Bank: Manila, Philippines, 2016.

47. Sutarto, R.; Jarvie, J. *Integrating Climate Resilience Strategy Into City Planning In Semarang, Indonesia*; Urban Climate Resilience Working Paper Series #1; Institute for Social and Environmental Transition ISET: Boulder, CO, USA, 2012.

48. Carmin, J.; Anguelovski, I.; Roberts, D. Urban Climate Adaptation in the Global South: Planning in an Emerging Policy Domain. *J. Plan. Educ. Res.* **2012**, *32*, 18–32. [CrossRef]

49. Sutarto, R. *Building Resilience in Indonesian Cities Case Study: ACCCRN Program*; Charles Darwin University: Darwin, Australia, 2017.

50. Prihantoro, F. (Bintari Foundation, Semarang City, Jawa Tengah). Personal communication, 13 December 2012.

51. Siriporananon, S.; Visuthismajarn, P. Key success factors of disaster management policy: A case study of the Asian cities climate change resilience network in Hat Yai city, Thailand. *Kasetsart J. Soc. Sci.* **2018**, *39*, 269–276. [CrossRef]

52. Mehta, L.; Leach, M.; Newell, P.; Scoones, I.; Sivaramakrishnan, K.; Way, S.A. *Exploring Understandings of Institutions and Uncertainty: New Directions in Natural Resource Management*; IDS Discussion Paper 372; University of Sussex: Sussex, UK, 1999.

53. Djalante, R. Adaptive governance and resilience: The role of multi-stakeholder platforms in disaster risk reduction. *Nat. Hazards Earth Syst. Sci.* **2012**, *12*, 2923–2942. [CrossRef]

54. Tjahjono, R. The Semarang Environmental Agenda: A stimulus to targeted capacity building among the stakeholders. *Habitat Int.* **2000**, *24*, 443–453.

55.	Budiarti, L. (Training and Education/DIKLAT Propinsi Jawa Tengah, Semarang, Jawa Tengah Province). Personal communication, 13 December 2012.
56.	Aden, J.; Rock, M. *"What Is Driving the Environmental Behavior of Manufacturing Plants in Semarang? Implications for Policy-Makers"*; Research Report for USAEP Richard Sheppard Region: Asia; USAID Natural Resources Management and Development Portal: Washington, DC, USA, 1998.
57.	Tjahjono, R. (Soegijapranata University, Semarang City, Jawa Tengah). Personal communication, 6 December 2012.

Review

Meteorological and Ancillary Data Resources for Climate Research in Urban Areas

Sorin Cheval [1,2,3,*], Dana Micu [4], Alexandru Dumitrescu [2,5], Anișoara Irimescu [2], Maria Frighenciu [6], Cristian Iojă [7], Nicu Constantin Tudose [3], Șerban Davidescu [3] and Bogdan Antonescu [8,9]

[1] Department of Aviation, "Henri Coandă" Air Force Academy, 500187 Brașov, Romania
[2] National Meteorological Administration, 013686 Bucharest, Romania; dumitrescu@meteoromania.ro (A.D.); anisoara.irimescu@meteoromania.ro (A.I.)
[3] Torrential Watershed Management Unit, Brașov Station, National Research and Development Institute in Forestry "Marin Drăcea", 500040 Brașov, Romania; cntudose@yahoo.com (N.C.T.); s_davidescu@icas.ro (S.D.)
[4] Department of Physical Geography, Institute of Geography, Romanian Academy, 023993 Bucharest, Romania; micudanamagdalena@gmail.com
[5] Research Institute of the University of Bucharest, 030018 Bucharest, Romania
[6] Doctoral School of Sciences, University of Craiova, 200585 Craiova, Romania; maria.frighenciu@gmail.com
[7] Center for Environmental Research and Impact Studies, University of Bucharest, 1 Bd. Nicolae Balcescu, 010041 Bucharest, Romania; cristian.ioja@geo.unibuc.ro
[8] Remote Sensing Department, National Institute of Research and Development for Optoelectronics INOE2000, 77125 Măgurele, Romania; bogdan.antonescu@inoe.ro
[9] Faculty of Physics, University of Bucharest, 77125, Măgurele, Romania
[*] Correspondence: sorin.cheval@afahc.ro or sorin.cheval@meteoromania.ro

Received: 15 January 2020; Accepted: 20 February 2020; Published: 25 February 2020

Abstract: An increasing plethora of both meteorological and ancillary data are presently available for climate research and applications in urban areas. The data are often held by local or national institutions (i.e., meteorological services, universities or environmental agencies). This paper outlines a total number of 33 datasets, organized into three main categories of meteorological data resources (14 datasets) and four categories of ancillary data resources (19 datasets), selected for their potential to support urban climate studies, but also for their free accessibility. Such a collection cannot be exhaustive, but we aim to draw the attention of the scientific community to relevant datasets, freely available at temporal and spatial resolutions appropriate for urban climatology. Each dataset contains information about its availability, limitations, and examples of research in urban areas.

Keywords: urban climate; open data; data sources; urban climate monitoring

1. Introduction

Fast natural and anthropogenic variations have been observed at global, regional and local scale in the recent decades, triggering complex impacts on environment and society, such as biodiversity loss, land degradation, reduced water resources and migration. The world population has experienced continuous growth, and urbanization has brought the people living in cities to 55.2% of total world population [1]. At the same time, climate change is a high-priority topic for the public agenda, and huge efforts have been devoted to understanding the temporal and spatial variability of our atmosphere in order to enhance the accuracy of climate predictions. Numerous studies documented climate change hotspots from various perspectives, e.g., climate modelling [2], hydrology [3], or ecosystems [4]. As they are particularly vulnerable to both present and future climate impacts, with considerable risks for human security [5], cities are hotspots both for the present climate and climate change.

Cities are territorial structures with the highest heterogeneity among the spatial systems of the Earth. Cities are also a unique combination of more or less prevalent anthropogenic and natural patches, very diverse landscape fragmentation, land use and land cover, dynamic energy and material fluxes. The availability of consistent information represents a key condition for the daily functionality and long-term development of the urban areas. Climate services address data provisions, statistic indicators, overview or products useful for different applications. In situ observations and gridded data are extensively used for climate services, satellite and aerial remote sensing, crowdsourcing, big data and artificial intelligence have produced a true revolution in climate data assimilation, storage and modelling.

The need of meteorological input for the decision making in urban development has been acknowledged for a long time [6]. Meteorological data are primary resources supporting the climate services designed mainly for present necessities, and continuous information from many other fields, such as land cover, demography, economy, are needed for the efficient adaptation of urban communities to future climate. The combined use of meteorological and ancillary data requires comparable quality, continuity and fine temporal and spatial resolution for efficient applications in urban applications.

There is an extensive body of literature of urban climate studies, which employed a wide range of datasets derived through different algorithms or modelling assumptions from primary data resources (e.g., in situ measurements or satellite). However, these data products have never been overviewed to highlight their advantages and disadvantages and applicability for urban climatological studies. It is worth mentioning some examples of systematic reviews with topics related to urban climatology, which tackled the issue of urban climate/meteorological datasets indirectly, within assessments of methodologies for urban heat island (UHI) [7,8] or of development status of urban meteorological networks [9].

This paper aims at providing a review of the meteorological and ancillary data resources available for urban areas, with emphasis on the possible applications which facilitate the current use of climate resources and adaptation to climate change challenges in urban areas. The review presents relevant meteorological and ancillary data resources currently used in urban climatology, derived from ground-based measurements, gridded datasets, crowdsourcing, remote sensing and airborne sensors, and highlights the benefits and specific limitations of the datasets.

2. Methodology

This research relied on authors' experience in the field of climate research related to urban environment, but also on Internet search technique to find semantically important entities, by querying the Web of Science database of academic literature. The search targeted the English-written scientific articles disseminated within the scientific community as journal papers, conference proceedings papers or book chapters, published after 2000. Some examples of searching terms and syntaxes are: "urban climate", "urban climate reviews", "urban climate datasets", "meteorological and climate datasets", and "urban meteorological networks".

Acknowledging the high importance of other datasets and variables (i.e., anthropogenic heat flux or radiation fluxes), the review targeted three main categories of meteorological resources, namely (1) ground-based measurements, (2) gridded datasets, (3) remote sensing, and four categories of ancillary data. This review tackles four categories, namely (1) societal information, (2) land cover, (3) urban morphology and (4) climate change, adaptation and urban resilience. It has to be stressed that the approach is not aimed to be exhaustive and we consider it as a fundament for more comprehensive review reports.

3. Data Resources

3.1. Meteorological Data Resources

Based on the method of retrieval, the meteorological information can be classified in ground observations, model outputs, and remote sensing data, with different characteristics and utility. Ideally, the meteorological data should simultaneously address the following issues:

(a) Temporal and spatial stability and homogeneity. Ideally, the observations should be performed in constant locations, quasi-continuously over time, with limited and isolated gaps. The shortcomings related to missing data or changes in station locations can be successfully secured by homogenization procedures [10,11]. The period covered by satellite images can be too short and contain too many gaps for developing climatic studies, but the remote sensing products are valuable for meteorological applications as much as they are consistent temporally and spatially. While any meteorological data retrieved from urban sensors may bring valuable information, the data stability and homogeneity are sometimes difficult to address due to inherent spatial heterogeneity and to the intense changes of the urban morphology and land cover–land use categories.

(b) Reliability. The observations should comply with the WMO standards for the stations monitoring the regional climate or other known standards for monitoring the local climate [12,13]. Many synoptic stations worldwide are placed within the administrative limits of a city, but it is very likely that more sensors will capture more relevant information about of the multifaceted urban climate even if they are not placed in standard conditions [14–16].

(c) Metadata. Geographical coordinates, instrument specifications, information about the working procedures, spatial and temporal resolution and any changes which have eventually occurred along time must be associated to any meteorological observations and remote sensing products. Besides, information about the proximities are particularly important for sensors placed in an urban environment.

In the recent decades, urban areas have faced a rapid population increase and dynamic transformation of infrastructure and functions, generating a high demand for environmental data. The access to meteorological information has been significantly improved by higher instrumental accuracy, larger storage possibilities and faster transfer capabilities.

3.1.1. In Situ Meteorological Data

The in situ or ground-based observations represent the oldest method for monitoring the urban atmosphere. The Medici Network was the first meteorological network in the world, operated between 1654 and 1670, and included 11 stations, from which 9 were placed in European cities [17]. The longest-running meteorological records in Europe are registered in cities, such as Uppsala 1722, Padua 1725, Milan 1763, Stockholm 1754, Basel 1755, or Prague 1775 [18]. At present, the ground-based observations in urban areas are performed by (a) WMO stations, (b) Urban Meteorological Networks or, more recently, by (c) citizen observatories. While the data collected by WMO stations are available from national meteorological services, this paper focusses on data freely available, usually with a transnational coverage. The main benefits of using in situ meteorological data are the potential continuity of records, flexibility of locations and relatively good accessibility in term of cost and maintenance. The main shortcomings derive from the limited spatial coverage and relevance for urban conditions. The integrated use of in situ and remote sensing data results in more complex products for urban climate studies [19].

The **Integrated Surface Database (ISD)** consists of global hourly and synoptic observations compiled from more than 100 sources into a single common ASCII format and common data model. ISD has a global coverage and it contains a variety of hourly meteorological data from more than 20,000 stations worldwide [20], ranging from 1891 to present. Based on the ISD, the HadISD is a global sub-daily station dataset, homogenized and quality controlled, including extremes of temperature, pressure and humidity from about 7600 stations from 1931 to present [21].

The **Global Summary of the Day (GSOD)** dataset produced by the National Climatic Data Center (NCDC). These data are derived from the Integrated Surface Hourly (ISH) dataset, which includes global data obtained from the USAF Climatology Center (synoptic/hourly data). This dataset comprises daily averages computed from global hourly station data, providing access to daily weather elements including mean values of: temperature, dew point temperature, sea level pressure, station pressure, visibility, and wind speed plus maximum and minimum temperature, maximum sustained wind speed and maximum gust, precipitation amount, snow depth, and weather indicators (e.g., fog, rain, drizzle, snow, hail, thunder). GSOD data are updated on a real-time basis with a lag of 1–2 days relative to the date–time of the observations, based on Greenwich Mean Time, for about 9000 stations worldwide. Historical data are accessible from 1929 to the present, but the records since 1973 are more complete and consistent. This dataset was used to analyze the observed changes in climate extremes (temperature, precipitation and wind) in 217 urban areas and 142 paired non-urban and urban stations across the globe, over the 1972–2012 period [22].

The **European Climate Assessment and Dataset (ECA&D)** is a platform aggregating a daily series of observations provided by the climatological divisions of national meteorological and hydrological services of 63 participating European countries, as well as by station time series of the observatories and research centers throughout Europe and the Mediterranean [23]. The input data are provided by a total number of 69 participants, and they are tested for homogeneity and quality control by the ECA&D team. The platform provides access to a blended series, for which it has applied an automated updating procedure relying on the daily data extracted from SYNOP messages distributed in near real-time through the Global Telecommunication System (GTS) (including a procedure of gap filling based on a daily series of nearby observations located within a 12.5 km distance and at a height difference of less than 25 m), and a non-blended series, accessible for public use as provided by the participants. A predefined set of aggregated indices for climate extremes derived from the ECA&D is also available. Gridded daily temperature, precipitation and pressure fields constructed from the ECA&D create the E-OBS daily gridded database [24], available at 0.1° to 0.25° regular grids, covering the same geographic area from 1950 to present. E-OBS also allows grid box average comparisons with the results of Regional Climate Models for validation purposes [25–27]. These datasets were used in various urban climatology studies relying on observations addressing the challenges of climate change in European cities, e.g., assessment of the impact of climate change on thermal performance of residential buildings [28]; and change in extreme heat/cold stress [29–31].

Urban Meteorological Networks (UMNs) have been implemented in several cities around the world in order to supplement and detail the observations provided by standard WMO stations. It is extremely useful information for assessing the urban heat island (UHI), as it can capture the urban thermal behavior at better resolution than the WMO stations. [9] overview the scientific and logistical issues in a study of 24 UMNs from the USA, Europe and Asia. The urban climate of Bucharest (Romania) is currently monitored by 3 long-term WMO standard meteorological stations, and 6 sensors placed in urban conditions, fully operational since November 2014 [32]. The city of Ghent (Belgium) has implemented since July 2016 the MOCCA network (Monitoring the City's Climate and Atmosphere) to monitor the canopy layer UHI. [14] demonstrated the MOCCA network importance and its complementarity with two modelling approaches (i.e., the SURFEX land surface model and UrbClim boundary layer model). The Birmingham Urban Climate Laboratory (BUCL) is another high-density urban meteorological network (https://www.birmingham.ac.uk/schools/gees/centres/bucl/maps-data/index.aspx), comprising 25 weather stations and more than 100 air temperature low-cost wireless sensors, which provided hourly air temperature records in near real-time for the city of Birmingham (UK) between June 2012 and December 2014 [33,34]. The Berlin city (Germany) monitoring network of the Freie University, Institute for Meteorology (FUMINET) is measuring meteorological data every five minutes for microclimate and human thermal comfort investigations. Novi Sad (Serbia) has also implemented an automatic microclimatic urban monitoring network [35,36], comprising 25 urban stations and two stations located in non-urbanized environments, collecting air temperature and humidity data every 10 minute. In its turn, the climate of Szeged (Hungary) is monitored by 23 weather stations placed in urban conditions [37] and available at http://en.urban-path.hu/monitoring-system. html. UNM data can be accessed by contacting the corresponding authors/organizations.

Air temperature in the Barrow city (Alaska) and its surroundings has been monitored since 2001 through 70 temperature data loggers, for highlighting winter urban heat island [38] and urban-suburban soil temperature contrasts [39]. The winter heat urban effects were also investigated in the arctic city of Norilsk (Russia), using data provided by automatic weather stations, iButton sensors, combined with MODIS remote sensing data [40]. The lack of a dense network of air temperature measurement points across the Eurasian arctic region was the reason for the establishing in 2015 of the Urban Heat Island Arctic Research Campaigns (UHIARC) network, which provides access to accurate, spatially dense and interconnected climate information about temperature anomalies at city scale (http://urbanreanalysis.ru/uhiarc.html). This network was deployed in several mid-sized cities of the region and combines data provided by air temperature data loggers (iButton) and automatic weather stations located in each target city. Its importance and applicability were already proven in some studies focusing on the winter urban heat island in the cities of Apatity, Vorkuta, Salekhard, Nadym and NovyUrengoy of Russia [41,42]. In Asia, a Community Weather Information Network (CWIN) was established in Hong Kong in 2007 in order to improve the data coverage of the Hong Kong Observatory (HKO) by extending meteorological measurement in schools and at community levels. Using the data provided by CWIN, HKO provides improved impact-based forecasts at multi-time scales and warnings/advisories for several natural phenomena and processes (tropical cyclones, thunderstorms, heavy rains, landslides, flooding, cold/very hot weather episodes) [43].

There is still an inadequate number of networks providing data with an appropriate high spatial density for high resolution modelling of urban climate. This situation is mostly determined by security, associated high costs, and difficulty in finding appropriate measurements sites [9]. However, under the future changing climate projections with expected increases of weather extremes by the end of the 21st century in most regions of the globe [44], the need for denser measurement networks and high resolution meteorological data is likely to increase, especially in densely populated areas.

The potential of **crowdsourcing** to provide useful data for urban climatology has been carefully considered in the recent years, once the number of sensors held by citizen extensively increased [44–47]. Data collected either by volunteers running personal weather stations, such as Weather Underground or Netatmo networks [48], or by smartphones [49,50] are now publicly accessible and allow individuals to share real-time weather information. Such data are collected in non-standard conditions and quality control is mandatory, but they can be retained as a valuable data resource, with a continuous expansion, which could provide real-time and high temporal and spatial resolution meteorological information over areas with heterogeneous environments lacking in dense traditional meteorological networks such as cities.

Ref [47] provided a systematic review of crowdsourcing for climate and atmospheric research, identifying initiatives, projects and programs based on citizen science and amateur weather stations (e.g., UK Met Office Weather Observation Website in the UK; Meteoclimatic in the Iberian Peninsula; CoCoRaHS in the US; Birmingham snow depth; Air Quality Egg), mobile app (e.g., WeatherSignal; iCelsius), moving platforms (e.g., OpenSense), which have been implemented in many areas of regions around the world. The applicability of these data resources was demonstrated in several studies (e.g., [51–54]). A further example of applicability is offered by [55], who integrated air temperature measurements from the Weather Underground network of Atlanta and Chicago in the analysis of the performance of the National Weather Service Heat Warning System against ground observations and satellite imagery, by assigning them to the pixels of LST Aqua and Terra MODIS satellite retrievals. In this study, the crowdsource climate information was found to be reliable in the description of the general patterns described by the National Weather Service and weather station measurements. [56]

provide evidence of crowdsource meteorological data from citizen weather stations to improve weather forecasting with the Weather Research and Forecasting (WRF) model for two cities in Russia (Saint Petersburg and Moscow).

3.1.2. Gridded Datasets

Gridded climate datasets represent an alternative to instrumental measurements, especially for areas with spatially scarce distribution of WMO stations or poor quality measurements. While better spatial coverage provided at low cost is an important advantage, the gridded meteorological data are associated with a lower accuracy than the measurements, depending on the quality and density of input data.

The **E-OBS** dataset of the ECA&D provides free access to a long time gridded series of daily climate data for Europe, available at 0.1° spatial resolution, covering the period from 1950 to present [24]. E-OBS was also designed to allow grid box average comparisons with the results of Regional Climate Models for validation purposes [25–27]. Both ECA&D and E-OBS datasets were used in various urban climatology studies relying on observations addressing the challenges of climate change in the European urban areas, e.g., assessment of the impact of climate change on the thermal performance of residential buildings [29]; change in extreme heat/cold and heat/cold stress [29,30].

The **Climatologies at High Resolution for the Earth's Land Surface Areas (CHELSA)** dataset is hosted by the Swiss Federal Institute for Forest, Snow and Landscape Research WSL and is based on a statistical downscaling of the ERA interim global circulation model [57]. CHELSA has a global coverage, and it contains monthly mean temperature and precipitation for the time period 1979–2013 at a 30-arc sec–spatial resolution. Recently, [58] employed the CHELSA dataset, validated with independent station data provided by the Global Historical Climatology Network (https://www.ncdc.noaa.gov/data-access/land-based-station-data/land-based-datasets/ global-historical-climatology-network-ghcn), to analyze the trend in urban heat island intensity from 1992 to 2012 for eight megacities in Asia. The study revealed the ability of the dataset to capture the characteristics of the urban heat island and a good match between the increase in air temperature and urban sprawl.

Berkeley Earth provides open access to historical temperature data products that allow statistical diagnostics of local urban climatologies and trend magnitudes based on the estimates of the monthly means of average, maximum and minimum surface air temperature anomaly over land areas [59]. The land temperature dataset has an extensive time coverage from 1753 (or 1833 for minimum and maximum temperatures) until present, relying on a large inventory of weather station observations (from over 30,000 weather stations). The dataset provides access to gridded temperatures, regional averages (available for the northern and southern hemisphere, country, state, city and individual station scale) and bias-corrected station data based on 'breakpoint' detection (using a similar method used in the Global Historical Climatology Network dataset—[60]) in station records, which resulted from changes in station location or measurement equipment. With regards to the water impact in metropolitan areas, the Service for Water Indicators in Climate Change Adaptation (SWICCA) offers open source climate impact information necessary for the sustainable management of watersheds, including urban areas, across Europe. The time-series of the future river flow and water-related indicators, bias-corrected data associated with three climate change projections (RCP 2.6; 4.5 and 8.5), socio-economic scenarios (0.1 deg. grid) and land use projections (10 km grid) are available through SWICCA, a service run by the Swedish Meteorological and Hydrological Institute (SMHI).

Other observational platforms may supplement the ground sensors with very useful information for urban climate research (i.e., airborne sensors or different vehicles). Here, we refer only to airborne sensors, which can be used for temperature profiling over the city and rural areas, infra-red imaging or the development of digital surface and elevation models. More than five decades ago, sensors placed in a helicopter were used to retrieve temperature and pressure information for studying the urban heat island effect in New York City [61]. Unmanned Aerial Vehicles (UAVs) or drones are conveying unprecedented advantages for urban climate monitoring, as they can retrieve the variability of temperatures across urban land surfaces, at high resolution and low cost [62,63]. The Natural Environment Research Council's Data Repository for Atmospheric Science and Earth Observation (**CEDA**) **data collection archive** provides access to several datasets of airborne observations by aircraft available from ongoing or ended collaborative projects, for assessing atmospheric composition and air quality in urban areas e.g., urban emissions (QUEFA), urban visibility (VISURB), biomonitoring of urban habitat (BIOHYPE), analysis of atmospheric chemical species and meteorological parameters (GASPOL) and airborne PM10 pollutants (PHYTOX), street-level air circulation and pollutants mix within the urban canopy (URBMET).

3.1.3. Remote Sensing Data

The need for reliable information with good spatial coverage in urban areas is also addressed by **remote sensing applications** aiming to retrieve data relevant for climate studies. Products delivered by satellite (Table 1), radar and airborne sensors represent a powerful tool set for urban climate studies, since the cost at user's desk and technical developments enhanced their availability for scientific research in the last two decades. [64] and [65] document the remote sensing of urban climates and differentiate the UHI, which refers to UHI effects in the canopy or boundary layer, by the surface urban heat island (SUHI), representing the radiative temperature difference between urban and non-urban areas at the level of the subjacent surface. The use of satellite imagery in urban climate studies may be impended by factors like cloudiness, technical limitation and various time constraints (i.e., short range or temporal discontinuities).

Table 1. Characteristics of satellite remote sensing products delivering Land Surface Temperature data.

Sensor	Satellite	Spatial Resolution	Temporal Resolution	Time Span
SEVIRI	MSG	3 to 5 km	15 min	2004 to date
AVHRR	NOAA	1.1 km	2 images/24 h	1981 to date
MODIS	Terra/Aqua	1 km	4 images/24 h	2000/2002 to date
SLSTR	Copernicus Sentinel-3	1 km	1 image/24 h	2017 to date
TM, ETM+, OLI, TIRS	Landsat 4, 5, 7, 8	60–120 m (30 m resampled)	1 image/8 or16 days	1982 to date

The World Meteorological Organization acknowledges 105 satellite instruments which have provided LST over time, categorized from primary to marginal relevance [66]. The first LST products were available at fair quality in the 1960s, but the first high-relevance products were delivered by the NASA satellite Nimbus-5 in 1973. However, the operation for urban climate research was prohibited by the 30 to 32 km spatial resolution.

The **Moderate-Resolution Imaging Spectro-Radiometer (MODIS)** instruments aboard NASA Terra/Aqua satellites have become a popular tool for urban climate studies due to a few major advantages: (1) spatial and temporal resolutions (1-km and 4 images daily) appropriate for urban climate applications; (2) significant time span, suitable for climatology, as the monitoring has been regularly performed since 2000 (Terra) and 2002 (Aqua). LST is a key parameter in the estimation of urban heat fluxes and highlighting the presence and magnitude of UHI (Figure 1) and heat health risk [67,68] and has been widely used in climatological studies at urban scale (e.g., [69–75]).

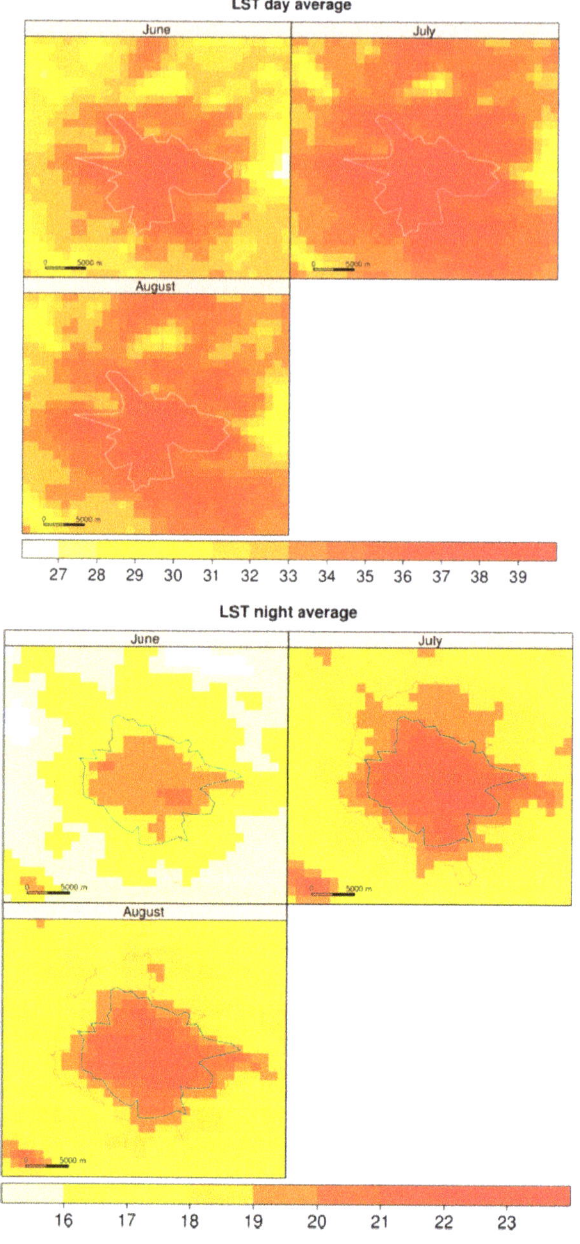

Figure 1. Average LST values and Bucharest's UHI limit (white contour line for daytime and blue contour line for nighttime), as retrieved from MODIS images (2000–2012). The administrative limit is marked with a light grey line [71] (Material from: 'Cheval, S.; Dumitrescu, A. The summer surface urban heat island of Bucharest (Romania) retrieved from MODIS images, Theor. Appl. Climatol., published [2015], [Springer]'.)

The **Spinning Enhanced Visible Infra-Red Imager (SEVIRI)** sensors placed on the Meteosat Second Generation (MSG) satellites operated by EUMETSAT provide LST information since 2004. The spatial resolution (3 to 5 km) is reasonable for investigating the urban environment, especially in cities with large extent covering dozen of kilometres, while the 15-min full disk coverage make SEVIRI a very robust tool for operational purposes, such as forecast and near-real-time meteorological evaluation.

NOAA's Advanced Very High-Resolution Radiometer (AVHRR) two and three instruments have retrieved LST at 1.1 km resolution since 1981. Early studies of the SUHI were based on AVHRR data [76–78], and the datasets are still used [79].

The Landsat missions jointly operated by NASA and the U.S. Geological Survey have provided a series of Earth Observation satellites since 1972. The spatial resolution of the **Landsat's Thematic Mapper (TM), Enhanced Thematic Mapper Plus (ETM+), Operational Land Imager (OLI) and Thermal Infrared Sensor (TIRS)** reached 30 m and contributed significantly to implementation of urban climate research [80–82]. Landsat is among the most widely used satellite resources in the detection and characterization of SUHI [83], due to its main advantages residing from the free availability (since 2008), long time series (more than 30 years), high spatial resolution (30 to 120 m) and the swath coverage of 185x185 km (Landsat 5, 7 and 8). However, the long revisiting cycle of 16 days, only daytime data availability, the strong dependence of weather conditions (e.g., cloud cover or cloud proximity which could result in missing data) and the lack of an operational Landsat LST product combining data from the Landsat sensors have to be noted as the main drawbacks impeding the detection and monitoring of inter-annual variability of SUHI [84,85]. In response to the need of an operational land surface temperature product for Landsat thermal data, [86] developed an operational algorithm for (emissivity correction and retrieval methodology) Landsat LST for all sensors, which have been successfully validated with in situ observations from four surface radiation budget network sites and two inland water bodies (Salton Sea and Lake Tahoe) in the US. The proposed algorithm will be implemented by the United States Geological Survey/The National Aeronautics and Space Administration and made available through the Land Processes Distributed Active Archive Center portal, which will provide access, for the first time, to consistent LST records dating back to Landsat 4 (1982 to present).

The **Sea and Land Surface Temperature Radiometer (SLSTR)** sensor, on-board on Sentinel-3 satellite, have retrieved LST at 1 km spatial resolution and daily temporal resolution since 2016. UHI research is based on LST retrieved by SLSTR sensors [87].

The **Copernicus Atmosphere Monitoring Service (CAMS)** has been implemented by the European Centre for Medium-Range Weather Forecasts (ECMWF), and it offers reliable information for improving life quality and urban planning. CAMS combines satellite and non-satellite observations with computer-based forecast models to provide near-real-time analysis and forecast of air quality and atmospheric composition related to pollutants and greenhouse gases (e.g., particulate matter, pollen, nitrogen dioxide and sulphur dioxide). The CAMS service data portfolio comprises a broad range of CAMS products available for urban agglomerations, such as (1) solar radiation and UV, (2) air quality and atmospheric composition, (3) emissions and surface fluxes of pollutants and (4) greenhouse gases and radiative climate forcing.

The urban heat island (UHI) is a key challenge for climate change adaptation in urban environments worldwide. Exploiting the benefits of remote sensing resources, the Center for International Earth Science Information Network (CIESIN), University of Columbia released in 2013 the first version of the **Global Urban Heat Island Dataset**, providing access to global data of average summer daytime maximum (1:30 p.m. overpass) and nighttime minimum (1:30 a.m. overpass) land surface temperature (LST) for urban areas, as well as to data referring to LST difference between the urban extents and their rural surroundings within a 10 km buffer. This dataset relies on the urban extents resulting from the SEDAC's Global Rural-Urban Mapping Project, Version 1 (GRUMPv1) and land surface temperatures are from SEDAC's Global Summer Land Surface Temperature (LST) Grids (2013), which are derived from the Aqua Level-3 Moderate Resolution Imaging Spectroradiometer (MODIS) Version 5 global daytime and nighttime LST 8-day composite data (MYD11A2). LST grid data are available for a 40-day window from July to August (Julian dates 185 to 224) for the urban areas located in the northern hemisphere, and from January to February (Julian dates 001 to 040) for those in the southern hemisphere. This dataset allowed the analysis of surface temperature anomalies in more than 30,000 cities, which have been further used to model the links between population, background climate and UHI intensity [88].

Light detecting and ranging (LIDAR) is an active type of remote sensing technology that ensures collection of accurate elevation-based information about anthropogenic features of urban environments and land cover, with a great potential to assist smart cities and support climate change resilience at city scale [89]. This technology proved a good applicability in urban flood modelling related to storm water flow and accumulation [89,90], solar potential at building roof level and city parcels for identifying optimal solar panel parameters [89,91]; Figure 2 and emergency response planning using building footprints, building heights, travel and congestion times [89]. [92] reviewed the results of climatic and vegetation surveys in urban environments relying on static, mobile or aerial laser scanning. We identified few examples of microclimate urban studies employing the LIDAR technique for showing the vegetation effects on urban climate [93,94]. Such results highlighted the relevance of LIDAR acquired data in urban green space planning under an increasing need for urban climate change adaptation.

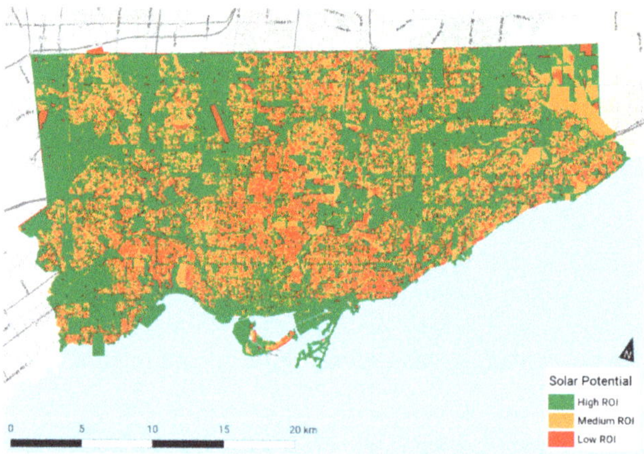

Figure 2. Estimation of solar potential in Toronto City using Lidar-derived DTM [89]

Table 2 summarizes the categories of meteorological data for urban climate studies and applications included in this review.

3.2. Ancillary Data Resources

Climate research always needs to be sustained by non-meteorological information, such as topography, roughness or anthropogenic influence, which can explain the formation and dynamics of certain atmospheric phenomena. Ancillary data are contextual information incorporated in urban climate studies together with meteorological input. [95] classify the input data for both mesoscale and microscale urban modelling in five categories, as follows: (1) land cover, (2) building morphology, (3) building design and architecture, (4) building use, anthropogenic heat and socio-economic data, and (5) urban vegetation data. This paper describes several data sources referring to air quality, land cover / land use, and urban morphology.

Table 2. Meteorological data sources for urban climate studies and applications.

Category	Dataset	Data Format	Spatial Resolution	Temporal Resolution	Temporal Coverage	Source
In situ	Integrated Surface Database (ISD)	ASCII	Data from 35,000 weather stations worldwide	Hourly, daily	1901 to present	https://www.ncdc.noaa.gov/isd
	Global Summary of the Day (GSOD)	ASCII	Data from over 9000 weather stations		1929 to present (since 1973 data are the most complete)	http://www.climate.gov/global-summary-day-gsod
	European Climate Assessment and Dataset (ECA&D)	ASCII	Data from 18,909 weather stations across Europe and the Mediterranean	Daily	1900 to present (blended series) and 1900 to a certain year (depending on the ECA&D participant)	https://www.ecad.eu/
	Urban Meteorological Networks (UMNs)	ASCII	Various	Hourly or sub-hourly	Various	http://en.urban-path.hu/monitoring-system.html https://www.birmingham.ac.uk/schools/gees/centres/bucl/maps-data/index.aspx Data could be accessed by contacting the owners
	Crowdsourcing	ASCII	Various	Sub-hourly	Various	https://dev.netatmo.com/ https://www.wunderground.com/pws/overview
Gridded Datasets	E-OBS	Grid (NetCDF-4)	0.1 to 0.25° regular grids	Daily	1950-01-01 to 2019-07-31 (v20.0e the latest version —released on October 2019)	https://www.ecad.eu/download/ensembles/download.php
	Climatologies at High Resolution for the Earth's Land Surface Areas (CHELSA)	Grid (tif)	30 arcsec, ~1 km	Monthly	1979–2013	http://chelsa-climate.org/
	Berkeley Earth	ASCII, graphs	-	Monthly and annual summaries	1750-present (land only) 1850-present (land and ocean)	http://berkeleyearth.org/
	Service for Water Indicators in Climate Change Adaptation (SWICCA)					http://swicca.eu/
Remote Sensing Data	Moderate-resolution Imaging Spectro-radiometer (MODIS)	HDF4	1 km	Daily eight-day mean	2000-present	https://pdaac.usgs.gov/tools/data-pool/ https://pdaac.usgs.gov/tools/earthdata-search/
	Spinning Enhanced Visible Infra-Red Imager (SEVIRI)	HDF5	3 km	15 min Hourly Daily Weekly Monthly Seasonal Yearly	1991-present	https://landsaf.ipma.pt/ChangeSystemProdLong.do?system=LandSAF+MSG&algo=LST https://wui.cmsaf.eu/safira/action/viewProduktSearch?menuName=PRODUKT_SUCHE, https://and.copernicus.eu/global/products/lst
	NOAA's Advanced Very High Resolution Radiometer (AVHRR)	NetCDF	4 km	Daily	1978-present	https://www.bou.class.noaa.gov/saa/products/search?sub_id=0&datatype_family=AVHRR&submit_x=15&submit.y=6
	the Landsat's Thematic Mapper (TM), Enhanced Thematic Mapper Plus (ETM+), Operational Land Imager (OLI) and Thermal Infrared Sensor (TIRS)		30 m		1982-present	http://rslab.gr/downloads_LandsatLST.html
	Sea and Land Surface Temperature Radiometer (SLSTR)	NC	1 km	Daily	2016-present	https://scihub.copernicus.eu/dhus/#/home https://search.earthdata.nasa.gov/
	Copernicus Atmosphere Monitoring Service (CAMS)	Grid (NetCDF, Grib Edition2)	-	Daily and hourly	Near real-time data and hourly forecast	http://copernicus-atmosphere.eu

3.2.1. Societal Information

Urban Audit Data Collection of EUROSTAT is a valuable resource of indicators providing relevant information about the quality of life in individual European cities and their commuting zones (Functional Urban Areas). The key topics covered within the database include demography, housing, health, labor market, education, environment, transport and tourism. Data are collected by the National Statistical Institutes, the Directorate-General for Regional and Urban Policy and Eurostat. The statistics have been used in several climate change vulnerability assessments in European urban areas (e.g., [96–98]).

LandScan High Resolution global Population Dataset, created by the Oak Ridge National Laboratory, is a high-resolution population distribution dataset (30 arc seconds or 1km at Equator) available in a GIS raster (ESRI Grid) format. The dataset has a global coverage and comprises sub-national census records provided by the International Program Center Bureau of Census and is a valuable resource allowing assessments, estimations and visualizations of population at risk. This dataset was employed to identify major cities across the globe (administrative capitals or cities that account more than 1,000,000 inhabitants) to investigate the extent to which 520 iconic cities are likely to experience a shift in response to climate change by 2050 (future cities) relative to their current climate conditions (current cities), in relation to the changes in climate variability and seasonality [99].

3.2.2. Land Cover

Ref [100] published a comprehensive inventory of land cover (LC) products available at global and regional level, including high spatial resolution products useful for urban climate applications, used as a basis for this review.

The **Global Land Survey (GLS)** is a 30 m global land cover dataset, based on Landsat images. The U.S. Geological Survey (USGS) and the National Aeronautics and Space Administration (NASA) have collaborated to develop the Global Land Surveys (GLS) datasets. This collection contains images acquired from 1972 to 2012, centered on 1975 (based on images acquired from 1972 to 1983 and from 1982 to 1987), 1990 (based on images acquired from 1987 to 1997), 2000 (based on images acquired from 1999 to 2003), 2005 (based on images acquired from 2003 to 2008) and 2010 (based on images acquired from 2008 to 2012). The GLS datasets were used to estimate the intensity of urbanization [101].

The **Global High-Resolution Urban Data from Landsat** are produced by NASA Goddard Space Flight Center and the Department of Geographical Sciences at the University of Maryland from Landsat data and include:

(a) The **Global Man-Made Impervious Surface (GMIS)** is one of the first 30m global dataset estimates of fractional impervious cover derived from the Global Land Survey (GLS) for 2010. The GMIS dataset includes: the global percent of impervious cover and the uncertainty for the global impervious cover. The urban studies are related to urban extend [102];

(b) The **Global Human Built-Up and Settlement Extent (HBASE)** is one of the first 30 m global datasets and estimates the urban extend cover derived from the Global Land Survey (GLS) for 2010. The urban studies are related to urban extend [102].

(c) The **Urban Landsat: Cities from Space** (1999–2003) is one of the first 30m global datasets providing composite Landsat images and raw data for urban areas that can be used in interdisciplinary studies of remote sensing and the environment.

The GMIS and HBASE datasets are complementary. The built-up and settlement extend mask (in HBASE dataset) was created for the post-processing of the GMIS dataset.

All the datasets can be used for local modelling in order to study the urban impacts on the energy, water, and carbon cycles or to analyse at country level.

The **GlobeLand30 (GLOB)** is a high-resolution dataset (30 m) of global land use/cover, based on TM5 and ETM + of America Land Resources Satellite (Landsat) and the multispectral images of China Environmental Disaster Alleviation Satellite (HJ-1) from 2010) [103]. The LC product includes information on land cover data, and other information highly relevant for urban climate analysis, such as artificial surfaces (habitation, industrial and mining area, transportation facilities, and interior urban green zones and water bodies, etc.). GlobeLand30 was exploited by [104] for investigating the urban expansion impact on a global scale.

At European level most of the LC products at local level are produced under **ESA CCI LC** and **Copernicus Land Monitoring Service (CLMS)** initiatives. The **ESA CCI LC** products have global coverage at 300 m spatial resolution. The datasets were derived based on a multi-year and multi-sensor strategy of MERIS and SPOT-Vegetation time series, continued under the Copernicus Climate Change Service (C3S), implemented by the European Centre for Medium-Range Weather Forecasts (ECMWF) on behalf of the European Commission, in order to assure continuity. At C3S Climate Data Store (CDS), 27 land cover datasets (from 1992 to 2018) are available for download, used for climate modelling, urban expansion [105,106] or UHI [73]. Figure 3 illustrates the relationship between LST and land cover based on the CCI-LC dataset.

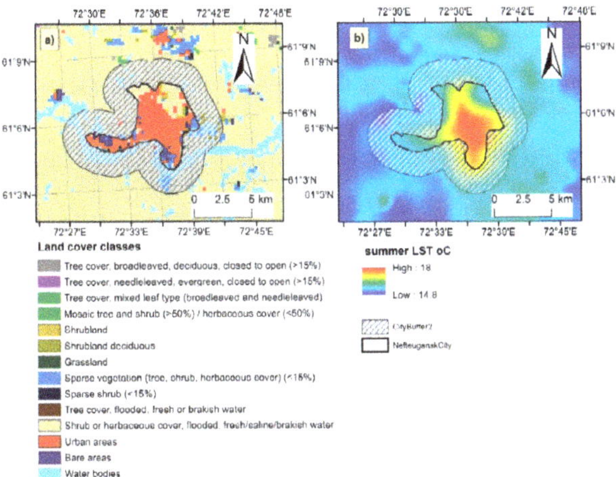

Figure 3. Example of CCI-LC and LST use in the urban study by [73].

The CLMS offers LC products at global, regional (Pan-European) and local level, as follows:

(a) The **global LC product** is available only for 2015 based on Proba-V at 100 m spatial resolution. It uses the FAO Land Cover Classification System (LCCS), totaling 23 classes [107].

(b) The Pan-European LC products include:

    i.   **Corine Land Cover (CLC)** is a land cover inventory (in 44 classes) project initiated in 1980s and updated in 1990, 2000, 2006, 2012 and 2018 with 100 m spatial resolution.;

    ii.   **Pan-European High-Resolution Layers (HRL)** provide information on specific land cover characteristics (imperviousness density, forest, grassland, wetland and water bodies and the new small woody features [108], and are complementary to the CLC dataset. The imperviousness data have a spatial resolution of 20 m and 100 m for 2006, 2009, 2012 and 2015.

    iii.   The **European Settlement Map (EMS)** is a very high-resolution dataset with spatial resolutions of 2.5 m, 10 m and 100 m. The first dataset, released in 2014, mapped the settlements in Europe based on 2010–2013 images. During the last year, different improvements have been developed: benchmarking process with population data; increases the number of classes to 13 (buildings, green, streets, water, railways, airports, open space inside the built-up area and same categories outside the built-up area) [109].

(c) The **Local LC Products** include the **Urban Atlas (UA)** which provide reliable, inter-comparable, high-resolution land use maps with 17 classes for large urban zones and their surroundings (more than 100,000 inhabitants), at 10 m spatial resolution for 2006 and 2012. The UA is also including the Building Height at 10 m spatial resolution for 2012.

The **MODIS Land Cover product** is a 500 m global land cover dataset based on MODIS Terra and Aqua data from 2001 to 2018, extensively used in urban climate research [110].

The **GlobCover Portal** provides access to the results of the GlobCover project, an ESA initiative started in 2005 in partnership with JRC, EEA, FAO, UNEP, GOFC-GOLD and IGBP. The aim of the project was to develop a service capable of delivering global composites and land cover maps (with 22 land cover classes) using as input observations from the 300m MERIS sensor on board the ENVISAT satellite mission. ESA makes the land cover maps available, which cover two periods: December 2004–June 2006 and January–December 2009.

The **Open-ECOCLIMAP** dataset has been available since 30 June 2014 at The ECOCLIMAP program, and is a dual database at 1 km resolution, that includes an ecosystem classification and a coherent set of land surface parameters, that are primarily mandatory in meteorological modelling (notably leaf area index and albedo). Hence, the aim of this innovative physiography is to enhance the quality of initialization and impose some surface attributes within the scope of weather forecasting and climate related studies.

3.2.3. Urban Morphology

The World Urban Database and Access Portal Tool (WUDAPT) project is a community-based initiative of the International Association of Urban Climate (IAUC) which supports urban climate research and various forecasting applications on surface energy budget, meso- to urban-scale weather and atmospheric chemical composition and air quality. The tool adopted a common framework for acquisition, storage and dissemination of relevant data about urban form (e.g., surface cover, properties of construction materials, surface geometry) and function (e.g., transportation, energy use), which are relevant for a coherent description of heterogeneous urban landscapes, as a prerequisite for understanding the urban climate characteristics.

WUDAPT rely on the local climate zone (LCZ) typology scheme proposed by [111], which categorizes 17 distinct landscapes (out of which 10 are urban), in relation to a number of surface parameters related to the main built and land cover types. The procedure employs geospatial data (multi-year series of multispectral and SAR satellite observations) as input in the random forest classifier to detect specific features for both urban and natural LCZ types [112].

The WUDAPT portal provides free access to maps (in GeoTIFF and KML format) and metadata of classified local climate zones for more than 150 cities worldwide. Data are structured in a hierarchical format, with increasing details from level zero (local climate zones along with parameter ranges) to level one (more precise parameter values for each LCZs) and level two (detailed description of urban landscape parameters at refined scales suited for boundary-layer modelling).

Alternatively, urban morphology can be obtained from OpenStreetMap [113] or from the building-resolving data provided by the local authorities.

3.2.4. Climate Change, Adaptation and Urban Resilience

Urban Audit dataset through the Geographic Information System of the Commission (GISCO) supplies the users with geographical information concerning the boundaries of cities and the functional urban areas, defined according to the EC-OECD city internationally harmonized definition [114]. The coverage is the EU-28 plus Iceland, Norway and Switzerland, and the dataset is available in ESRI format (shapefile and geostatistical database) at scales ranging from 1:100,000 (2011-2014, 2018) to 1:3,000,000 (2001, 2004). [29] employed the Urban Audit GISCO dataset and the urban morphological zones from the European Environment Agency to analyze future change in heatwaves, droughts and floods in all 571 European cities targeted by Urban Audit, and to elaborate low, medium and high hazard impact scenarios in relation to climate projections of all ensemble members of CMIP5 for the RCP8.5 emission scenario. The study provides evidence of an increasing exposure to heatwave days in all cities, an intensification of drought in the southern European cities and increasing frequency of river floods in the northern European cities.

Urban Adaptation Map Viewer (UAMV) is a web platform available at <https://climate-adapt.eea.europa.eu/knowledge/tools/urban-adaptation/Urban-Adaptation-datasets>, aggregating information from various sources (e.g., Met Office, EURO-CORDEX, E-OBS dataset, C3S, JRC, Urban Audit) and providing free access to five types of datasets and metadata from European cities, as follows: (1) climate and climate related hazards (e.g., hot summer days, cooling degree days, extreme heat waves, winter heavy precipitation, meteorological drought, forest fire danger, sea level rise); (2) exposure of cities to climate-related hazards (e.g., areas and population affected by wildfires, river flooding, coastal flooding); (3) physical characteristics of urban areas (urban morphological zone—UMZ, share of green and impervious areas in city core and UMZs); (4) socio-economic characteristics of cities (e.g., population, share of elderly and children, share of lone-pensioner households, unemployed people); and (5) adaptation activities.

Lobellia Earth is a newly resealed (2018) web platform for visualizing and exploring the annual and monthly climatologies in user specific locations for the period 1981–2010. The visualization tool (as maps and graphs) of the past climate uses ERA5 data from ECMWF that allow the analysis of spatial distribution of air temperature (average, maximum and minimum air temperature), precipitation, average wind speed, wind gust and cloud cover and several climate extreme indices (e.g., frost days, warm nights, heavy and very heavy precipitation days). In this platform, the user could only visualize the climatological information (maps and graphs) about different target urban areas. The platform supports the UN Habitat's City Resilience Profiling Programme (CRPT) using downscaling and bias correction techniques for providing high-resolution climate projection data of EUROCORDEX project and tailored climate information about future climate change threats and vulnerabilities for five pilots of cities sensitive to climate change (i.e., Asuncion, Yakutsk, Maputo, Dakar and Port Vila). Climate projection data for the target urban areas are available only by request from the Lobelia Earth developers. Exploring the benefits of Earth observation, Lobelia Earth, with the contribution of Royal Netherlands Meteorological Institute, supplies air quality information for three European pilot cities (i.e., NO_2 and PM10 concentrations in Barcelona, Madrid and Amsterdam). Table 3 summarizes the categories of ancillary data for urban climate studies and applications described in this review.

Table 3. Ancillary data sources for urban climate studies and applications.

	Dataset	Details (e.g., Data Format, Resolution, Year of Compliance)	Source
Societal information	Urban Audit data collection of EUROSTAT	Data format: ASCII Temporal coverage: 2010–2019 (depending on the indicator)	https://ec.europa.eu/eurostat/web/cities/data/database
	LandScan High Resolution global Population Dataset	Data format: ESRI Grid Spatial resolution: 30 arc seconds or ~1km at equator	https://www.eastview.com/resources/e-collections/landscan/
	Global Land Survey (GLS) The Global Man-made Impervious Surface (GMIS) Global Human Built-up And Settlement Extent (HBASE) Urban Landsat: Cities from Space	Data format: GeoTIFF Spatial resolution: 30 m Temporal coverage: 2010	http://earthexplorer.usgs.gov/ https://sedac.ciesin.columbia.edu/data/set/ulandsat-gmis-v1/data-download https://sedac.ciesin.columbia.edu/data/set/ulandsat-hbase-v1/data-download https://sedac.ciesin.columbia.edu/data/set/ulandsat-cities-from-space/data-download
Land cover	GlobeLand30 (GLOB)	Data format: GeoTIFF Spatial resolution: 30 m Temporal coverage: 2010 (2009–2011)	http://www.globallandcover.com/GLC30Download/index.aspx
	Copernicus Land Monitoring Service (CLMS)	Data format: GeoTIFF Spatial resolution: 100 m Temporal coverage: 2015	https://lcviewer.vito.be/download
	Corine Land Cover (CLC)	Data format: GeoTIFF, ESRI Geodatabase, SQLite Database Spatial resolution: 100 m Temporal coverage: 1990 (1986–1998), 2000 (+/- 1 year), 2006 (+/- 1 year), 2012 (2011–2012), 2018 (2017–2018)	https://land.copernicus.eu/pan-european/corine-land-cover/clc2018?tab=download
	Pan-European High-Resolution Layers (HRL)	Data format: GeoTIFF Spatial resolution: 20 m and 100 m Temporal coverage: 2006 (2005–2007), 2009 (2008–2010), 2012 (2011–2013), 2015 (2014–2016)	https://land.copernicus.eu/pan-european/high-resolution-layers/imperviousness/status-maps/2015?tab=download
	European Settlement Map (EMS)	Data format: GeoTIFF Spatial resolution: 2.5 m, 10 m and 100 m Temporal coverage: 2012 (2010–2013) Release: 2014, 2016, 2017	https://land.copernicus.eu/pan-european/GHSL/european-settlement-map/esm-2012-release-2017-urban-green?tab=download
	Urban Atlas (UA)	Data format: vector file ESRI format Spatial resolution: 10 m Temporal coverage: 2006 (2005–2007), 2012 (2011–2013)	https://land.copernicus.eu/local/urban-atlas/urban-atlas-2012?tab=download
	Building Height	Data format: GeoTIFF Spatial resolution: 10 m Temporal coverage: 2012 (2011–2014)	https://land.copernicus.eu/local/urban-atlas/building-height-2012?tab=download
	MODIS Land Cover	Data format: HDF4 Spatial resolution: 500 m Temporal coverage: 2001-2018	https://lpdaac.usgs.gov/tools/earthdata-search/, https://lpdaac.usgs.gov/tools/usgs-earthexplorer/, https://lpdaac.usgs.gov/tools/data-pool
	GlobCover Portal	Data format: GeoTIFF Spatial resolution: 300 m Temporal coverage: 2005, 2009	http://due.esrin.esa.int/page_globcover.php
	Open-ECOCLIMAP	Data format: ASCII Spatial resolution: 1 km Temporal coverage: 2005, 2009	https://opensource.umr-cnrm.tr/projects/ecoclimap/files

Table 3. *Cont.*

	Dataset	Details (e.g., Data Format, Resolution, Year of Compliance)	Source
Urban morphology	World Urban Database and Access Portal Tool (WUDAPT)		http://www.wudapt.org/
Climate change, adaptation and urban resilience	Urban Audit dataset	Data format: ESRI shapefile and geodatabase Scale 1:100000 (UA2018, 2011–2014), scale 1:3 Million (UA2004, 2001) Temporal availability: 2015–2018 (UA2018), 2011–2014 (UA2011–2014), 2004 (UA2004), 2001 (UA2001) Available only for visualization (e.g., in ArcGIS JavaScript, ArcGIS Online Map Viewer, ArcGis Earth)	https://ec.europa.eu/eurostat/web/gisco/geodata/reference-data/administrative-units-statistical-units/urban-audit
	Urban Adaptation Map Viewer (UAMV)	Dataset comprises 15 indicators of climate and climate-related hazards (e.g., heat waves, heavy precipitation, meteorological drought, fire danger); 7 exposure indicators of cities to climate-related hazards (e.g., wildfires, river flooding, sea level rise); 4 indicators of physical characteristics of urban areas (e.g., urban morphological zone; percentage of impervious area in core city); 10 socio-economic indicators for cities (e.g., population, total use of water); adaptation activities of cities.	https://climate-adapt.eea.europa.eu/knowledge/tools/urban-adaptation/Urban-Adaptation-datasets
	LOBELIA EARTH	Available only for visualization as distribution maps and graphs using Climate Explorer (past climate)	https://www.lobelia.earth/

4. Conclusions

The rapid development of urban areas has derived complex challenges for environmental sciences, demanding better data, in terms of accuracy, resolution and easy accessibility. At present, there are several data sources providing valuable meteorological and ancillary information for urban climate research, and this paper presents an outline based on criteria like data availability, geographical coverage and, above all, the demonstrated utility for urban climate. We are confident that, far from being comprehensive, such an outline is a starting point and a useful tool for urban climate researchers and practitioners.

Five categories of meteorological data are identified and presented in this review paper, i.e., ground-based data and remote sensing, while the ancillary data can be classified in air quality, land cover/land use, and urban topography. One can remark the following conclusions: 1) the information sources providing meteorological and ancillary data for urban climate research are numerous; 2) spatial and temporal resolutions support a variety of applications, from weather forecast to climate modelling. It has to be mentioned that the combined use of data from various sources have a significant potential to support the improvement of data and products, as the shortcomings may be addressed with keeping the advantages associated to each dataset.

Author Contributions: Conceptualization, S.C., D.M., A.D., A.I.; methodology, S.C., D.M., A.D., A.I.; software, S.C., D.M., A.I.; validation, M.F., C.I., N.C.T., S.D. and B.A.; formal analysis, S.C., D.M., A.I. All authors have read and agreed to the published version of the manuscript.

Funding: This research was partially funded by Ministry of Research and Innovation, Romania, grant number PN-III-P1-1.1-PD-2016-1579.

Acknowledgments: This work was partially conducted in the framework of the research theme "Climate hazards in the Romanian Plain" of the Institute of Geography, Romanian Academy.

Conflicts of Interest: The authors declare no conflict of interest.

References

1. The World Bank Group. 2019. Available online: https://data.worldbank.org/indicator/sp.urb.totl.in.zs (accessed on 10 January 2020).
2. Diffenbaugh, N.S.; Giorgi, F. Climate change hotspots in the CMIP5 global climate model ensemble. *Clim. Chang.* **2012**, *114*, 813–822. [CrossRef]
3. Prudhomme, C.; Giuntoli, I.; Robinson, E.L.; Clark, D.B.; Arnell, N.W.; Dankers, R.; Fekete, B.; Franssen, W.; Gerten, D.; Gosling, S.; et al. Hydrological Droughts in the 21st Century, Hotspots and Uncertainties from a Global Multimodel Ensemble Experiment. *Proc. Natl. Acad. Sci. USA* **2013**, *111*. [CrossRef]
4. Turco, M.; Palazzi, E.; von Hardenberg, J.; Provenzale, A. Observed climate change hot-spots. *Geophys. Res. Lett.* **2015**, *42*. [CrossRef]
5. de Sherbinin, A. Climate Change Hotspots Mapping: What Have We Learned? *Clim. Chang.* **2013**, *123*, 23–37. [CrossRef]
6. Chandler, T.J. *Urban Climatology and Its Relevance to Urban Design*; WMO No 438, Technical Note No 149; WHO: Geneva, Switzerland, 1976; 80p.
7. Stewart, I.D. A systematic review and scientific critique literature of methodology in modern urban heat island literature. *Int. J. Climatol.* **2011**, *31*. [CrossRef]
8. Deilami, K.; Kamruzzaman, M.D.; Liu, Y. Urban heat island effect: A systematic review of spatio-temporal factors, data, methods and mitigation measures. *Int. J. Appl. Earth Obs. Geoinf.* **2018**. [CrossRef]
9. Muller, C.; Chapman, L.; Grimmond, C.; Young, D.; Cai, X. Sensors and the city: A review of urban meteorological networks. *Int. J. Clim.* **2013**, *33*, 1585–1600. [CrossRef]
10. Savić, S.; Milošević, D.; Marković, V.; Kujundžić-Dačović, R. Homogenisation of mean air temperature time series from Vojvodina (North Serbia). *Geogr. Pannonica* **2012**, *16*, 38–43. [CrossRef]
11. Dumitrescu, A.; Cheval, S.; Guijarro, J.A. Homogenization of a combined hourly air temperature dataset over Romania. *Int. J. Climatol.* **2019**. [CrossRef]

12. Oke, T.R. *Initial Guidance to Obtain Representative Meteorological Observations at Urban Sites*; WMO/TD No. 1250; WMO: Geneva, Switzerland, 2006; 51p.

13. WMO. *Guide to Meteorological Instruments and Methods of Observations*; WMO-no.8; WMO: Geneva, Switzerland, 2017; 1177p.

14. Caluwaerts, S.; Hamdi, R.; Top, S.; Lauwaet, D.; Berckmans, J.; Degrauwe, D.; Dejonghe, H.; De Ridder, K.; De Troch, R.; Duchêne, F.; et al. The urban climate of Ghent, Belgium: A case study combining a high-accuracy monitoring network with numerical simulations. *Urban Clim.* **2020**, *31*, 100565. [CrossRef]

15. Bilandžija, D. Spatio-Temporal Climate and Agroclimate Diversities over the Zagreb City Area. *Geogr. Pannonica* **2019**, *23*, 385–397. [CrossRef]

16. Dian, C.; Pongrácz, R.; Incze, D.; Bartholy, J.; Talamon, A. Analysis of the Urban Heat Island Intensity Based on air Temperature Measurements in a Renovated Part of Budapest (Hungary). *Geogr. Pannonica* **2019**, *23*, 277–288. [CrossRef]

17. Camuffo, D.; Bertolin, C. The earliest temperature observations in the world: The Medici Network (1654-1670). *Clim. Chang.* **2012**, *111*, 335–363. [CrossRef]

18. Burt, S. *The Weather Observer's Handbook*; Cambridge University Press: Cambridge, UK, 2012; 458p.

19. Sheng, L.; Tang, X.; You, H.; Gu, Q.; Hu, H. Comparison of the urban heat island intensity quantified by using air temperature and Landsat land surface temperature in Hangzhou, China. *Ecol. Indic.* **2017**, *72*, 738–746. [CrossRef]

20. Smith, A.; Lott, N.; Vose, R. The integrated surface database: Recent developments and partnerships. *Bull. Am. Meteorol. Soc.* **2011**, *92*, 704–708. [CrossRef]

21. Dunn, R.; Willett, K.; Parker, E.D.; Mitchell, L. Expanding HadISD: Quality-controlled, sub-daily station data from 1931. *Geosci. Instrum. Methods Data Syst.* **2016**, *5*, 473–491. [CrossRef]

22. Mishra, V.; Ganguly, A.R.; Nijssen, B.; Lattenmaier, D.P. Changes in observed climate extremes in global urban areas. *Environ. Res. Lett.* **2015**, *10*, 024005. [CrossRef]

23. Klein Tank, A.M.G.; Wijngaard, J.B.; Können, G.P.; Böhm, R.; Demarée, G.; Gocheva, A.; Heino, R. Daily dataset of 20th-century surface air temperature and precipitation series for the European Climate Assessment. *Int. J. Climatol.* **2002**, *22*, 1441–1453. [CrossRef]

24. Cornes, R.; van der Schrier, G.; van den Besselaar, E.J.M.; Jones, P.D. An Ensemble Version of the E-OBS Temperature and Precipitation Datasets. *J. Geophys. Res. Atmos.* **2018**, *123*, 9391–9409. [CrossRef]

25. Haylock, M.R.; Hofstra, N.; Klein Tank, A.M.G.; Klok, E.J.; Jones, P.D.; New, M. A European daily high-resolution gridded dataset of surface temperature and precipitation. *J. Geophys. Res.* **2008**, *113*, D20119. [CrossRef]

26. Pereira, S.C.; Marta-Almeida, M.; Carvalho, A.C.; Rocha, A. Heat waves and cold spells changes in Iberia for a future climate scenario. *Int. J. Clim.* **2017**, *37*, 5192–5205. [CrossRef]

27. Fallmann, J.; Wagner, S.; Emeis, S. High resolution climate projections to asses the future vulnerability of European urban areas to climatological extreme events. *Theor. Appl. Climatol.* **2017**, *127*, 667–683. [CrossRef]

28. Soutullo, S.; Giancola, E.; Jiménez, M.J.; Ferrer, J.A.; Sánchez, M.N. How climate trends impact on the thermal performance of a typical residential building in Madrid. *Energies* **2019**, *13*, 237. [CrossRef]

29. Guerreiro, S.B.; Dawson, R.J.; Kilsby, C.; Lewis, E.; Ford, A. Future heat-waves, droughts and floods in 571 European cities. *Environ. Res. Lett.* **2018**, *13*, 034009. [CrossRef]

30. Lorenz, R.; Stalhandske, Z.; Fischer, E.M. Detection of a climate change signal in extreme heat, heat stress d cold in Europe from observations. *Geophys. Res. Lett.* **2019**, *46*, 8363–8374. [CrossRef]

31. Founda, D.; Oierros, F.; Katavoutas, G.; Keramitsoglou, I. Observed trends in thermal stress at European Cities with different background climates. *Atmosphere* **2019**, *10*, 436. [CrossRef]

32. Cheval, S.; Dumitrescu, A. Rapid daily and sub-daily temperature variations in an urban environment. *Clim. Res.* **2017**, *73*, 233–246. [CrossRef]

33. Chapman, L.; Muller, C.L.; Young, D.T.; Cai, X.-M.; Grimmond, C.S.B. An introduction to the Birmingham urban climate laboratory. In Proceedings of the 8th International Conference on Urban Climates, Dublin, Ireland, 6–10 August 2012.

34. Azevedo, J.A.; Chapman, L.; Muller, C.L. Quantifying the daytime and night-time urban heat island in Birmingham, UK: A comparison of satellite derived land surface temperature and high resolution air temperature observations. *Remote Sens.* **2016**, *8*, 153. [CrossRef]

35. Šećerov, I.; Savić, S.; Milošević, D.; Marković, V.; Bajšanski, I. Development of an automated urban climate monitoring system in Novi Sad (Serbia). *Geogr. Pannonica* **2015**, *19*, 174–183. [CrossRef]
36. Šećerov, I.; Savić, S.; Milošević, D.; Arsenović, D.M.; Dolinaj, D.M.; Popov, S.B. Progressing urban climate research using a high-density monitoring network system. *Environ. Monit Assess.* **2019**, *191*, 89. [CrossRef]
37. Skarbit, N.; Stewart, I.D.; Unger, J.; Gál, T. Employing an urban meteorological network to monitor air temperature conditions in the "local climate zones" of Szeged, Hungary. *Int. J. Climatol.* **2017**, *37*, 582–596. [CrossRef]
38. Hinkel, K.M.; Nelson, F.E. Anthropogenic heat island at Barrow, Alaska, during winter (2001-2005). *J. Geophys. Res. Atmos.* **2007**, *112*, D6. [CrossRef]
39. Klene, A.E.; Nelson, F.E.; Hinkel, K.M. Urban–rural contrasts in summer soil-surface temperature and active-layer thickness, Barrow, Alaska, USA. *Polar Geogr.* **2013**, *36*, 183–201. [CrossRef]
40. Varentsov, M.I.; Konstantinov, P.I.; Samsonov, T.E.; Repina, I.A. Investigation of the urban heat island phenomenon during polar night based on experimental measurements and remote sensing of Norilsk city (in Russian). *Sovrem. Probl. Distantsionnogo Zondirovaniya Zemliiz Kosm.* **2014**, *11*, 329–337.
41. Konstantinov, P.I.; Grishchenko, M.Y.; Varentsov, M.I. Mapping urban heat islands of arctic cities using combined data on field measurements and satellite images based on the example of the city of Apatity (Murmansk Oblast) Izv. *Atmos. Ocean Phy.* **2015**, *51*, 992–998. [CrossRef]
42. Konstantinov, P.; Varentsov, M.; Esau, I. A high density urban temperature network deployed for several cities of Euroasian Arctic. *Environ. Res. Lett.* **2018**, *13*, 075007. [CrossRef]
43. Lee, T.; Wong, W.; Tam, K. Urban-focused weather and climate services in Hong Kong. *Geosci. Lett.* **2018**, *5*, 18. [CrossRef]
44. IPCC. *2012: Managing the Risks of Extreme Events and Disasters to Advance Climate Change Adaptation. A Special Report of Working Groups I and II of the Intergovernmental Panel on Climate Change*; Field, C.B., V. Barros, T.F., Stocker, D., Qin, D.J., Dokken, K.L., Ebi, M.D., Mastrandrea, K.J., Mach, G.-K., Plattner, S.K., Allen, M.T., et al., Eds.; Cambridge University Press: Cambridge, UK; New York, NY, USA, 2012; 582p.
45. Wolters, D.; Brandsma, T. Estimating the urban heat island in residential areas in The Netherlands using observations by weather amateurs. *J. Appl. Meteorol. Climatol.* **2012**, *51*, 711–721. [CrossRef]
46. Bell, S.; Cornford, D.; Bastin, L. How good are citizen weather stations? Addressing a biased opinion. *Weather* **2015**, *70*, 75–84. [CrossRef]
47. Muller, C.L.; Chapman, L.; Johnston, S.; Kidd, C.; Illingworth, S.; Foody, G.; Overreem, A.; Leigh, R.R. Crowdsourcing for climate and atmospheric sciences: Current status and future potential. *Int. J. Climatol.* **2015**, *35*, 3185–3203. [CrossRef]
48. Meier, F.; Fenner, D.; Grassmann, T.; Marco, O.; Scherer, D. Crowdsourcing air temperature from citizen weather stations for urban climate research. *Urban Clim.* **2017**, *19*, 170–191. [CrossRef]
49. Overeem, A.; Robinson, J.; Leijnse, H.; SteeneveldGert-Jan Horn, B.; Uijlenhoet, R. Crowdsourcing urban air temperatures from smartphone battery temperatures. *Geophys. Res. Lett.* **2013**, *40*, 4081–4085. [CrossRef]
50. Droste, A.; Pape, J.J.; Overeem, A.; Leijnse, H.; SteeneveldGert-Jan van Delden, A.; Uijlenhoet, R. Crowdsourcing urban air temperatures through smartphone battery temperatures in São Paulo, Brazil. *J. Atmos. Ocean. Technol.* **2017**, *34*, 1853–1866. [CrossRef]
51. Cifelli, R.; Doesken, N.; Kennedy, P.; Carey, L.D.; Rutledge, S.A.; Gimmestad, C.; Depue, T. The community collaborative rain, hail, and snow network: Informal education for scientists and citizens. *Bull. Am. Meteorol. Soc.* **2005**, *86*, 1069–1077. [CrossRef]
52. Shan, Q.; Brown, D. Wireless temperature sensor using bluetooth. In Proceedings of the IWAN 2005: International Workshop on Wireless Ad Hoc Networks, London, UK, 23–26 May 2005.
53. Anderson, A.R.S.; Chapman, M.; Drobot, S.D.; Tadesse, A.; Lambi, B.; Wiener, G.; Pisano, P. Quality of mobile air temperature and atmospheric pressure observations from the 2010 development test environment experiment. *J. Appl. Meteorol. Clim.* **2012**, *51*, 691–701. [CrossRef]
54. Devarakonda, S.; Sevusu, P.; Liu, H.; Liu, R.; Iftode, L.; Nath, B. Real-time Air Quality Monitoring through Mobile Sensing in Metropolitan Areas. In Proceedings of the Conference UrbComp'13, Chicago, IL, USA, 11–14 August 2013.
55. Vargo, J.; Xiao, Q.; Liu, Y. The performance of the National Weather Service Heat Warning System against ground observations and satellite imagery. *Adv. Meteorol.* **2015**. [CrossRef]

56. Uteov, A.; Kalyuzhnaya, A.; Boukhanovsky, A. The cities weather forecast by crowdsourced atmospheric data. *Procedia Comput. Sci.* **2019**, *156*, 347–356. [CrossRef]

57. Karger, D.N.; Conrad, O.; Böhner, J.; Kawohl, T.; Kreft, H.; Soria-Auza, R.W.; Zimmermann, N.; Linder, H.P.; Kessler, M. Climatologies at high resolution for the Earth land surface areas. *Sci. Data* **2016**, *4*, 170122. [CrossRef] [PubMed]

58. Lee, K.; Kim, Y.; Chan Sung, H.; Ryu, J.; Woo Jeon, S. Trend analysis of urban island intensity according to urban area change in Asian Mega Cities. *Sustainability* **2020**, *12*, 112. [CrossRef]

59. Rohde, R.; Muller, R.; Jacobsen, R.; Perlmutter, S.; Rosenfeld, A.; Wurtele, J.; Curry, J.; Wickham, C.; Mosher, S. Berkeley earth temperature averaging process. *Geoinfor. Geostat. An Overv.* **2013**, *1*, 2. [CrossRef]

60. Menne, M.J.; Williams, C.N., Jr. Homogenization of Temperature Series via Pairwise Comparisons. *J. Clim.* **2009**, *22*, 1700–1717. [CrossRef]

61. Bornstein, R.D. Observations of the Urban Heat Island Effect in New York City. *J. Appl. Meteorol.* **1968**, *7*, 575–582. [CrossRef]

62. Gaitani, N.; Burud, I.; Thiis, T.; Santamouris, M. High-resolution spectral mapping of urban thermal propertieswith Unmanned Aerial Vehicles. *Build. Environ.* **2017**, *121*, 215–224. [CrossRef]

63. Naughton, J.; Mcdonald, W. Evaluating the Variability of Urban Land Surface Temperatures Using Drone Observations. *Remote Sens.* **2019**, *11*, 1722. [CrossRef]

64. Voogt, J.; Oke, T. Thermal remote sensing of urban climates. *Remote Sens. Environ.* **2003**, *86*, 370–384. [CrossRef]

65. Sun, H.; Chen, Y.; Zhan, W. Comparing surface- and canopy-layer urban heat islands over Beijing using MODIS data. *Int. J. Remote Sens.* **2015**, *36*, 5448–5465. [CrossRef]

66. OSCAR. Observing Systems Capability Analysis and Review Tool. 2019. Available online: https://www.wmo-sat.info/oscar/gapanalyses?variable=96 (accessed on 20 October 2019).

67. Tomlinson, C.J.; Chapman, L.; Thornes, J.E.; Baker, C.J. Remote sensing land surface temperature for meteorology and climatology: A review. *Meteorol. Appl.* **2011**, *18*, 296–306. [CrossRef]

68. Ward, K.; Lauf, S.; Kleinschmit, B.; Endlicher, W. Heat waves and urban heat islands in Europe: A review of relevant drivers. *Sci. Total Environ.* **2016**, *569–570*, 527–539. [CrossRef]

69. Jin, M.; Dickinson, R.E.; Zhang, D.-L. The footprint of urban areas on global climate as characterized by MODIS. *J. Clim.* **2005**, *18*, 1551–1565. [CrossRef]

70. Tomlinson, C.J.; Chapman, L.; Thornes, J.E.; Baker, C.J. Derivation of Birmingham's summer surface urban heat island from MODIS satellite images. *Int. J. Climatol.* **2012**, *32*, 214–224. [CrossRef]

71. Cheval, S.; Dumitrescu, A. The summer surface urban heat island of Bucharest (Romania) retrieved from MODIS images. *Theor. Appl. Climatol.* **2015**, *121*, 631–640. [CrossRef]

72. Morabito, M.; Crisci, A.; Gioli, B.; Gualtieri, G.; Toscano, P.; Di Stefano, V.; Orlandini, S.; Dalal, K. Urban-hazard risk analysis: Mapping of heat-related risks in the elderly in major Italian cities. *PLoS ONE* **2015**, *10*, e0127277. [CrossRef] [PubMed]

73. Miles, V.; Esau, I. Seasonal and spatial characteristics of urban heat islands (UHIs) in Northern West Siberian cities. *Remote Sens.* **2017**, *9*, 989. [CrossRef]

74. Polydoros, A.; Mavrakou, T.; Cartalis, C. Quantifying the trends in land surface temperature and surface urban heat island intensity in Mediterranean cities in view of smart urbanization. *Urban Sci.* **2018**, *2*, 16. [CrossRef]

75. Chen, Q.; Ding, M.Y.X.; Hu, K.; Qi, J. Spatially explicit assessment of heat health risk using multi-sensor remote sensing images and socioeconomic data in Yangtze River Delta, China. *Int. J. Health Geogr.* **2018**, *17*, 15. [CrossRef]

76. Gutman, G.G. Multi-annual time series of AVHRR-derived land surface temperature. *Adv. Space Res.* **1994**, *14*, 27–30. [CrossRef]

77. Streutker, D.R. Satellite-measured growth of the urban heat island of Houston, Texas. *Remote Sens. Environ.* **2003**, *85*, 282–289. [CrossRef]

78. Weng, Q.; Lu, D.; Schubring, J. Estimation of land surface temperature-vegetation abundance relationship for urban heat island studies. *Remote Sens. Environ.* **2004**, *89*, 467–483. [CrossRef]

79. Che, Y.; Guang, J.; de Leeuw, G.; Xue, Y.; Sun, L.; Che, H. Investigations into the development of a satellite-based aerosol climate data record using ATSR-2, AATSR and AVHRR data over north-eastern China from 1987 to 2012. *Atmos. Meas. Tech.* **2019**, *12*, 4091–4112. [CrossRef]

80. Liu, L.; Zhang, Y. Urban heat island analysis using the landsat TM data and ASTER Data: A case study in Hong Kong. *Remote Sens.* **2011**, *3*, 1535–1552. [CrossRef]

81. Kaplan, G.; Avdan, U.; Avdan, Z.Y. 2018. Urban heat island analysis using the landsat 8 satellite data: A case study in Skopje, Macedonia. *Multidiscip. Digit. Publ. Inst. Proc.* **2018**, *2*, 358.

82. Parastatidis, D.; Mitraka, Z.; Chrysoulakis, N.; Abrams, M. Online Global Land Surface Temperature Estimation from Landsat. *Remote Sens.* **2019**, *9*, 1208. [CrossRef]

83. Zhou, D.; Xiao, J.; Bonafoni, S.; Berger, C.; Deilami, K.; Zhou, Y.; Frolking, S.; Yao, R.; Qiao, Z.; Sobrino, J.A. Satellite remote sensing of surface urban heat islands: Progress, challenges and perspectives. *Remote Sens.* **2019**, *11*, 48. [CrossRef]

84. Quan, J.; Zhan, W.; Chen, Y.; Wang, M.; Wang, J. Time series decomposition of remotely sensed land surface temperature and investigation of trends and seasonal variations in surface urban heat islands. *J. Geophys. Res. Atmos.* **2016**, *121*, 2638–2657. [CrossRef]

85. Shen, H.; Huang, L.; Zhang, L.; Wu, P.; Zeng, C. Long-term and fine-scale satellite monitoring of the urban heat island effect by the fusion of multi-temporal and multi-sensor remote sensed data: A 26-year case study of the city of Wuhan in China. *Remote Sens. Environ.* **2016**, *172*, 109–125. [CrossRef]

86. Malakar, N.K.; Hulley, G.C.; Hook, S.J.; Laraby, K.; Cook, M.; Schott, J.R. An operational land surface temperature product for Landsat thermal data: Methodology and validation. *IEEE Trans. Geosci. Remote Sens.* **2018**, *56*, 5717–5735. [CrossRef]

87. Shumilo, L.; Kussul, N.; Shelestov, A.; Korsunska, Y.; Yailymov, B. Sentinel-3 Urban Heat Island Monitoring and analysis for Kyiv Based on Vector Data. In Proceedings of the 10th International Conference on Dependable Systems, Services and Technologies (DESSERT), Leeds, UK, 5–7 June 2019; pp. 131–135.

88. Manoli, G.; Fatichi, S.; Schläpfer, M.; Yu, K.; Crowther, T.W.; Meili, N.; Burlando, P.; Katul, G.G.; Bou-Zeid, E. Magnitude of urban heat islands largely explained by climate and population. *Nature* **2019**, *573*, 55–60. [CrossRef] [PubMed]

89. Garnett, R.; Adams, M.D. LIDAR—A technology to assist with smart cities and climate change resilience: A case study in an urban metropolis. *ISPRS Int. J. Geo-Inf.* **2018**, *7*, 161. [CrossRef]

90. Haile, A.; Rientjes, T. Effects of LiDAR DEM resolution in flood modelling: A model sensitivity study for the city of Tegucigalpa, Honduras. In Proceedings of the ISPRS Wg Iii/3, Iii/4, Vienna, Austria, 29–30 August 2005; pp. 168–173.

91. Rowlands, I.H.; Kemery, B.P.; Beausoleil-Morrison, I. Optimal solar-PV tilt angle and azimuth: An Ontario (Canada) case-study. *Energy Policy* **2011**, *39*, 1397–1409. [CrossRef]

92. Szabó, Z.; Schlosser, A.; Túri, Z.; Szabó, S. A Review of Climatic and Vegetation Surveys in Urban Environment with Laser Scanning: A Literature-based Analysis. *Geogr. Pannonica* **2019**, *23*, 411–421. [CrossRef]

93. Kong, F.; Yan, W.; Zheng, G.; Yin, H.; Cavan, G.; Zhan, W.; Zhang, N.; Cheng, L. Retrieval of threedimensional tree canopy and shade using terrestrial laser scanning (TLS) data to analyze the cooling effect of vegetation. *Agric. For. Meteorol.* **2016**, *217*, 22–34. [CrossRef]

94. Bournez, E.; Landes, T.; Saudreau, M.; Kastendeuch, P.; Najjar, G. Impact of level of details in the 3D reconstruction of trees for microclimate modeling. *Int. Arch. Photogramm. Remote Sens. Spat. Inf. Sci. ISPRS Arch.* **2016**, *41*, 257–264. [CrossRef]

95. Masson, V.; Heldens, W.; Bocher, E.; Bonhomme, M.; Bucher, B.; Burmeister, C.; de Munck, C.; Esch, T.; Hidalgo, J.; Kanani-Sühring, F.; et al. City-descriptive input data for urban climate models: Model requirements, data sources and challenges. *Urban Clim.* **2019**, *31*, 100536. [CrossRef]

96. EEA. *Urban Adaptation to Climate Change in Europe. Challenges and Opportunities for Cities Together with Supportive National and European Policies*; EEA Report 2/2012; EEA: Luxemburg, 2012; 148p.

97. Reckien, D.; Flacke, J.; Olazabal, M.; Heidrich, O. The influence of drivers and barriers in urban adaptation and mitigation plans—Empirical analysis of European cities. *PLoS ONE* **2015**, *10*, e0135597. [CrossRef] [PubMed]

98. Tapia, C.; Abajo-Alda, B.; Feliu, E.; Mendizabal, M.; Martínez-Sáenz, J.; Fernández, G.J.; Laburu, T.; Lejarazu, A. Profiling urban vulnerability to climate change: An indicator-based vulnerability assessment for European cities. *Ecol. Indic.* **2017**, *78*, 142–155. [CrossRef]

99. Bastin, J.-F.; Clark, E.; Elliott, T.; Hart, S.; van den Hoogen, J.; Hordijk, I.; Ma, H.; Majumder, S.; Manoli, G.; Maschler, J.; et al. Understanding climate change from a global analysis of city analogues. *PLoS ONE* **2019**, *14*, e0217592. [CrossRef]

100. Grekousis, G.; Mountrakis, G.; Kavouras, M. An overview of 21 global and 43 regional land-cover mapping products. *Int. J. Remote Sens.* **2015**, *36*, 5309–5335. [CrossRef]
101. Wang, P.H. Mapping 2000–2010 Impervious Surface Change in India Using Global Land Survey Landsat Data. *Remote Sens.* **2017**, *9*, 366. [CrossRef]
102. Liu, X.; Pei, F.; Wen, Y.; Li, X.; Wang, S.; Wu, C.; Cai, Y.; Wu, J.; Chen, J.; Feng, K.; et al. Global urban expansion offsets climate-driven increases in terrestrial net primary productivity. *Nat. Commun.* **2019**, *10*, 5558. [CrossRef]
103. Chen, J.; Chen, J.; Liao, A.; Cao, X.; Chen, L.; Chen, X.; He, C.; Han, G.; Peng, S.; Lu, M.; et al. Global land cover mapping at 30 m resolution: A POK-based operational approach. *ISPRS J. Photogramm. Remote Sens.* **2015**, *103*, 7–27. [CrossRef]
104. Liu, X.; de Sherbinin, A.; Zhan, Y. Mapping Urban Extent at Large Spatial Scales Using Machine Learning Methods with VIIRS Nighttime Light and MODIS Daytime NDVI Data. *Remote Sens.* **2019**, *11*, 1247. [CrossRef]
105. Radwan, T.M.; Blackburn, G.A.; Whyatt, J.D.; Atkinson, P.M. Dramatic Loss of Agricultural Land Due to Urban Expansion Threatens Food Security in the Nile Delta, Egypt. *Remote Sens.* **2019**, *11*, 332. [CrossRef]
106. van Vliet, J. Direct and indirect loss of natural area from urban expansion. *Nat. Sustain.* **2019**, *2*, 755–763. [CrossRef]
107. Buchhorn, M.; Smets, B.; Bertels, L.; Lesiv, M.; Tsendbazar, N.-E.; Herold, M.; Fritz, S. Copernicus Global Land Service: Land Cover 100m: Epoch 2015: Globe. *Zenodo* **2019**. [CrossRef]
108. Congedo, L.; Sallustio, L.; Munafo, M.; Ottaviano, M.; Tonti, D.; Marchetti, M. Copernicus high-resolution layers for land cover classification in Italy. *J. Maps* **2016**, *12*, 1195–1205. [CrossRef]
109. Ferri, S.; Siragusa, A.; Sabo, F.; Pafi, M.; Halkia, M. *The European settlement map 2017 release. Methodology and output of the European settlement map (ESM2p5m)*; JRC Technical reports; UR 28644 EN; European Commission: Brussels, Belgium, 2017. [CrossRef]
110. Schneider, A.; Friedl, M.A.; Potere, D. Mapping global urban areas using MODIS 500-m data: New methods and datasets based on 'urban ecoregions'. *Remote Sens. Environ.* **2010**, *114*, 1733–1746. [CrossRef]
111. Stewart, I.D.; Oke, T. Local Climate Zones for Urban Temperature Studies. *Bull. Amer. Meteor. Soc.* **2012**, *93*, 879–1900. [CrossRef]
112. Demuzere, M.; Bechtel, B.; Middel, A.; Mills, G. Mapping Europe into local climate zones. *PLoS ONE* **2019**. [CrossRef]
113. Fan, H.; Zipf, A.; Fu, Q.; Neis, P. Quality assessment for building footprints data on OpenStreetMap. *Int. J. Geogr. Inf. Sci.* **2014**, *28*, 4. [CrossRef]
114. Dijkstra, L.; Poelman, H. Cities in Europe the New OECD-EC Definition. European Commision. RF 01/2012. Available online: https://ec.europa.eu/regional_policy/sources/docgener/focus/2012_01_city.pdf (accessed on 3 December 2019).

Review

The Maturing Interdisciplinary Relationship between Human Biometeorological Aspects and Local Adaptation Processes: An Encompassing Overview

Andre Santos Nouri [1,2] **and Andreas Matzarakis** [3,4,*]

[1] Faculty of Architecture, University of Lisbon, Rua Sá Nogeuira, Pólo Universitário do Alto da Ajuda, 1349-063 Lisbon, Portugal; andrenouri@fa.ulisboa.pt
[2] Department of Interior Architecture and Environmental Design, Faculty of Art, Design and Architecture, Bilkent University, 06800 Bilkent, Turkey
[3] Research Centre Human Biometeorology, German Meteorological Service, D-79104 Freiburg, Germany
[4] Chair of Environmental Meteorology, Faculty of Environment and Natural Resources, Albert-Ludwigs-University, D-79085 Freiburg, Germany
[*] Correspondence: andreas.matzarakis@dwd.de; Tel.: +49-69-8062-9610

Received: 10 October 2019; Accepted: 22 November 2019; Published: 25 November 2019

Abstract: To date, top-down approaches have played a fundamental role in expanding the comprehension of both existing, and future, climatological patterns. In liaison, the focus attributed to climatic mitigation has shifted towards the identification of how climatic adaptation can specifically prepare for an era prone to further climatological aggravations. Within this review study, the progress and growing opportunities for the interdisciplinary integration of human biometeorological aspects within existing and future local adaptation efforts are assessed. This encompassing assessment of the existing literature likewise scrutinises existing scientific hurdles in approaching existing/future human thermal wellbeing in local urban contexts. The respective hurdles are subsequently framed into new research opportunities concerning human biometeorology and its increasing interdisciplinary significance in multifaceted urban thermal adaptation processes. It is here where the assembly and solidification of 'scientific bridges' are acknowledged within the multifaceted ambition to ensuring a responsive, safe and thermally comfortable urban environment. Amongst other aspects, this review study deliberates upon numerous scientific interferences that must be strengthened, inclusively between the: (i) climatic assessments of both top-down and bottom-up approaches to local human thermal wellbeing; (ii) rooted associations between qualitative and quantitative aspects of thermal comfort in both outdoor and indoor environments; and (iii) efficiency and easy-to-understand communication with non-climatic experts that play an equally fundamental role in consolidating effective adaptation responses in an era of climate change.

Keywords: human biometeorology; thermal comfort; interdisciplinarity; climate change adaptation; thermal sensitive design

---

1. Introduction

Before the turn of the century, the limited local specificity of global top-down approaches to climatic risk factors within urban environments was already well known by the international scientific community. In particular, such fragility in applicative know-how led to the growing interest in identifying how local bottom-up approaches could be instigated. As a result, and likely associated to the contiguous maturing Climate Change Adaptation (CCA), there has been a rapidly growing interest in how adaptation tools can be locally instigated to improve the climatic responsiveness of the urban public realm, e.g., [1–14].

Moreover, when considering the urban climate condition and human wellbeing within public realm, the scientific community has already recognised the growing importance of bottom-up approaches to climatic risk factors that are already presenting aggravations associated to climate change impacts [15–19]. For this reason, and as exemplified by studies undertaken by [12,20–24] local scales are becoming an arena in which both decision makers and designers are seeking means to address physiological and psychological factors pertaining to human thermal comfort within the public realm in an era of climate change.

Although examinations pertaining to the characteristics of the urban climate date well back to the previous century, e.g., [25–31], the practical application upon contemporary practices of urban design and planning has been limited [3,17,32,33]. Such a desire on behalf of the scientific community to further develop climatic tools can be intertwined with the earlier encompassing perspective of Wilbanks and Kates [34] who suggested that "the bulk of the research relating to local places to global climate change has been top-down, from global toward local, concentrating on methods of impact analysis that use as a starting point climate change scenarios derived from global models, even though these have little regional or local specificity. There is a growing interest, however, in considering a bottom-up approach, asking such questions as (…) how efforts at mitigation and adaptation can be locally initiated and adapted" (p. 1).

Such scientific interrogations chronologically coincided with the international recognition that mitigation efforts alone were no longer sufficient to address the potential impacts of climate change. Resultantly, the turn of the century witnessed an exponential leap for CCA efforts. Almost twenty years onwards, the demand for local application orientated approaches and tools are at an all-time high, both at assessment and at design levels. Within local scales, these assessments are predominantly focused upon the concrete symbiotic relationship between that of built form and encircling atmospheric conditions beneath the Urban Canopy Layer (UCL). Subsequently, it is here where the planning and design of the public realm and indoor environments can serve as a niche for interdisciplinary bottom-up approaches that question how efforts at adaptation can be locally initiated and adopted.

2. Review Structure

Within the scientific community, the balance between research articles and review articles discloses the encompassing ambition to develop existing knowledge, and at the same time, continually organise, review and structure the state-of-the-art. The objective of this study falls within the latter category, and aims to present an encompassing overview of the growing interdisciplinary relationships and interrogations concerning human biometeorological aspects associated to the practice of local thermal adaptation efforts. Divided into three predominant sections, the study situates: (i) the investigative prospectus for human biometeorology within the ever consolidating CCA agenda; (ii) the opportunities to further develop existing thermo–physiological approaches to identify both existing and future thermal risk factors that jeopardise urban wellbeing in indoor/outdoor settings; and, (iii) existing approaches and creative means to improve the thermal responsiveness of the urban environment through thermal sensitive urban design and planning efforts within local scales. Aiming to transition from the broader to the more specific facets of disclosed interdisciplinary relationship between human biometeorology and that of thermal adaptation processes, a summary of this division is presented in Figure 1.

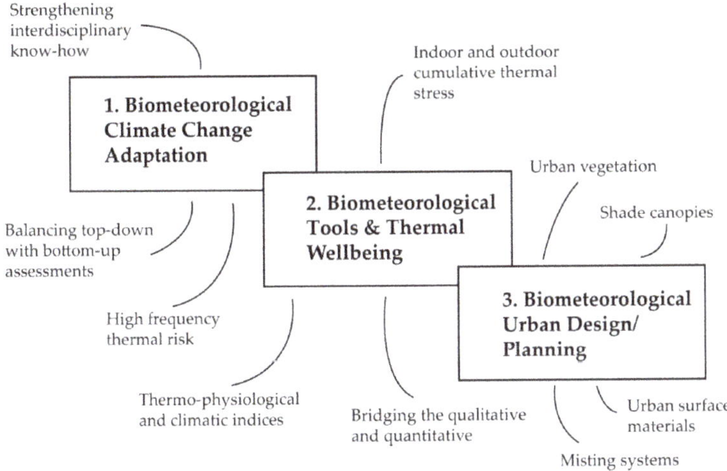

Figure 1. Structure schematic of review throughout the three sequential sections presented in the study.

Within each of the sections, the opportunities presented by human biometeorology to enhance local bottom-up adaptation processes are identified. In Section 2, and within the scope international CCA, based upon the unequivocal and direct thermo–physiological affects climate change shall have upon the human body, current methods, warning systems and impact projection assessments are discussed. Based upon local scales and recognising the increased need to go beyond 'high-impact but low frequency' impact projections; biometeorological tools to address the high frequency thermal risk factors are adjacently reviewed. Subsequently, and again with an emphasis on strengthening interdisciplinary bridges, this section moreover discusses the better integration and recognition of qualitative and quantitative aspects of thermal comfort in wholesome thermal comfort evaluations. Interconnected with the previous two points, the review study deliberates upon the fundamental transient associations between comfort thresholds and urban indoor and outdoor contexts. Lastly, and in direct association to local scales and how they can be feasibly modified through creative and flexible thermal sensitive adaptation processes, different existing measures to address local thermo–physiological risk factors in the urban public realm are reviewed. These disclosed measures are considered to be a part of a newly emerging scope for practices such as urban design and planning as a result of the effective bridging with local human biometeorology.

Overall, and throughout this review study, it is argued that the successful approach towards these factors highlights the growing significance of interdisciplinarity between different fields of practice. Ultimately, it is this assembly and fortification of collaborative scientific bridges that will bring different professionals together to tackle the same pressing issues within the urban environment. Naturally, while both scientific existing outcomes and obstacles are identified within the existing state-art-of-the-art, such obstacles are correspondingly framed into scientific opportunities to expanding interdisciplinary mentality/know-how regarding human biometeorology, and moreover, its unequivocally growing prominence in urban adaptation processes.

3. Biometeorological Climate Change Adaptation

Well before the turn of the century, and inclusively prior to the arrival of the CCA, Oke [35] identified that "relatively little of the large body of knowledge concerning urban climate has permeated through to working planners (...) the reasons for this state of affairs are many, but amongst those most cited are the inherent complexity of the subject, its interdisciplinary nature and lack of meaningful dialogue between planners and the climatological research community." (p. 1). Today, and even with growing consolidation of the CCA, the adjacent enclosure of pertinent biometeorological data and information within municipal and policy documents has been a complicated and a slow process.

3.1. Strengthening Interdisciplinary Know-How

In an earlier study conducted by Alcoforado and Vieira [36] that identified that within the Portuguese context, many cities presented a significant lack of pertinent meteorological data that could otherwise inform such local thermal adaptation efforts. Through the analysis of 15 master plans of urban municipalities, the respective study identified that although climatic information was considered in almost all of them, the information often proved either unreliable, or of little use for local adaptation efforts. Such a discrepancy was later argued by Alcoforado, Andrade, Lopes and Vasconcelos [17] to be attributable to numerous causes, including that the meteorological data from typical stations used in such plans were not applicable for microclimatic studies. Such a conclusion goes back to the prior conclusions of Oke [35], who additionally pointed at the difficulty in translating such information into robust tools for concrete local urban planning. Naturally, such difficulty is further increased at municipal policy and guideline levels.

Nevertheless almost two decades into the twenty-first century, and still within the Portuguese context, although further interdisciplinary strengthening is still considered essential [16], promising interest and integration with municipal entities are starting to be established [37]. This establishment can be inclusively associated to the scientific disseminations of the 'Climate Change and Environmental Systems Research' (CEG/CliMA) group. A group that has thus far conducted research into numerous topics, including overall bioclimatic conditions within Lisbon [38,39], causalities and intensities of Urban Heat Island (UHI) effects [18,40,41], urban wind patterns [42,43] and potential climatic integration within planning policy [15,44].

The exemplified disseminations focused on Lisbon mark a clear progression in addressing Okes [35] early outlook. Nonetheless, and with CCA serving as a continually growing catalyst for interdisciplinary thermal adaptation efforts, the international growing interaction with non-climatic experts (e.g., urban planners/designers, architects and landscape architects) must be upheld to address local thermo–physiological risk factors, as identified in [2,3,11,33,45–53],

3.2. Balancing Top-Down with Bottom-Up Assessments

As already mentioned, the international scientific community has already recognised the crucial role of bottom-up approaches that focus upon the importance of local scales. Although top-down approaches and disseminations have presented an imperative emerging international co-operative understanding of the existing and future global climatic system, such outcomes are rarely capable (nor so intended) to provide guidance at local scales. Resultantly, the amount of disseminated studies on this topic has increased dramatically since the turn of the century. In accordance, both the limitations and means to improve local scale analysis tools have grown across different disciplines such as urban climatology, and urban planning/design.

So far, and in accordance with Global Circulation Models (GCMs), global temperatures shall continue to rise throughout the 21st century. Yet, it has adjacently been recognised that such top-down climatic assessments are often less useful for local scale analysis tools and adaptation. For instance, within the assessments reports of international entities such as the Intergovernmental Panel on Climate Change (IPCC), the effects of weather are often described with a simple index based upon amalgamations of air temperature (T_a) and Relative Humidity (RH). Although it is indispensable to recognise the value of such descriptions within the maturing CCA agenda, when pondering upon bottom-up approaches to climatic vulnerability, the exclusion of vital non-temperature factors (i.e., radiation fluxes, wind speed (V) and human thermo–physiological factors) have been argued to decrease their usefulness for local thermal decision making and design [2,16,19,54,55].

In the study conducted by Matzarakis and Amelung [54], through the use of synoptic global radiation estimations retrieved from monthly sunshine fractions (extracted from the Hadley Centre's

HadCM3 model—one of the predominant models utilized by the IPCC in its third assessment report in 2001), clear underestimations of global climate change impacts on human thermal comfort thresholds were identified. As an example, Western European areas could witness changes in thermo–physiological indices by up to 15 °C based upon worst case scenarios. The synoptic projections sharply differ from the IPCC projections established upon singular climatic variables such as T_a [56,57]. Retrospectively, the significance of the study was twofold, it: (1) presented how the inclusion of non-temperature variables (i.e., radiation fluxes) could dramatically amplify the gravity of climate change projections; and, (2) showed an initial approach to running climate change scenario variables through a biometeorological model to understand how such variables would interact with the human body, and subsequently, obtain an estimation of thermo–physiological stress levels by the end of the century. Derivative from GCMs, and recognising the analogous limitations of standard climatic variables in weather forecasting activities, a recent study undertaken by Giannaros et al. [58] also emphasised the: (1) significance of human biometeorology in not only assessing present-day meteorological conditions, but warning provisions for both heat waves and cold outbreaks; and, (2) crucial role of effectively and accessibly communicating easy-to-understand to the general public.

Processed from GCMs, both studies conducted by Matzarakis and Amelung [54] and Giannaros, Lagouvardos, Kotroni and Matzarakis [58] marked clear strides in further consolidating the imperative role of human biometeorology in identifying and managing existing/future thermo–physiological risk factors. Subsequently, these strides also validate the continual importance of top-down assessments, even when specific local urban characteristics were not variables considered either study. More specifically, this was accomplished by the on-going robust emphasis upon: (i) frequently overlooked variables such as radiation fluxes; (ii) the symbiotic relationship with the human thermo–physiological system; and lastly, (iii) the critical role of the ease-of-assess, transmission and comprehensibility of the results for non-climatic experts and general public.

3.3. High Frequency Thermal Risk Factors

Contrastingly to the former studies, many top-down climatic disseminations (especially from international bodies), while fundamental, remain frequently "focussed [on] the exposure of cities to hazards that have a huge impact but low frequency. [They] have little to say about the high-frequency and microscale climatic phenomena created within the anthropogenic environment of the city" [59]. Contiguously, and as identified by numerous authors that address human thermal comfort through the elaboration of creative measures through urban planning and design, it is within the anthropogenic environment of the city where human wellbeing becomes crucial [13,60,61].

For this reason, it is here where "landscape architects and urban designers strive to design places that encourage [urban] activities, places where people will want to spend their time (…) however unless people are thermally comfortable in the space, they simply won't use it. Although few people are even aware of the effects that design can have on the sun, wind, humidity and air temperature in a space, a thermally comfortable microclimate is the very foundation of well-loved and well-used outdoor places." [23]. Analogous inferences were reached by the earlier study conducted by Whyte [29] who advocated that "by asking the right questions in sun and wind studies, by experimentation, we can find better ways to board the sun, to double its light, or to obscure it, or to cut down breezes in winter and induce them in the summer" (p. 45).

Respectively, and based upon the overarching principal that actual adaptation measures take place at finer scales, it is the concrete bond within specific localities which can substantiate such bioclimatic adaptation initiatives and tools in cities [1,13,55,62,63]. It is at this scale where the encircling microclimatic under the UCL that has direct 'in-situ' influences upon pedestrian comfort thresholds.

Undoubtedly, such principals enforce the fundamental relationship between resulting local climatic variables beneath the UCL, with that of human biometeorology. Under a more encompassing perspective, this suggests that urban form, layout and design have an enormous capability to enhance (or reduce) human wellbeing standards in cities. In this way, the interdisciplinary spheres of human biometeorology with that of climatic adaptation measures/tools must continue to be explored further.

4. Biometeorological Tools and Thermal Wellbeing

In accordance with the previously discussed scope of Oke [35], and the 'climate-comfort' rational discussed by Olgyay [30], this segment discusses the potentiality of interdisciplinarity in linking human biometeorology tools and assessments with local urban thermal wellbeing. The term 'locality' is again approached as the physical niche in which creative interdisciplinary practices such as Public Space Design (PSD) can render relevant, yet direct, thermal modifications of pedestrian thermo–physiological stress thresholds. Specifying this rational a little further in the greater context of local decision making and design, this catalyses two predominant perspectives, as suggested by Nouri, Costa, Santamouris and Matzarakis [3]: (i) the requirement to improve and facilitate the bioclimatic design guidelines within such environmental perspectives for local action and adaptation; and, (ii) given the growth of the CCA agenda, the accompanying cogency for local, thermal and pre-emptive climate-sensitive action and tools.

4.1. Thermo–Physiological and Climatic Indices

To undertake such an exercise, the direct effects of the thermal environment must be evaluated against the human biometeorological system. Such a multifaceted bond can be examined through the use of thermal indices that are centred on the energy balance of the human body [64]. Thus far, within the international community, a vast amount of thermal indices have been developed, and moreover, reviewed against one another. Examples of these studies are presented in Table 1.

Table 1. Example of studies that review and compare the application efficiency of different thermal indices in different settings.

No. of Investigated Indices	Dominant Focused Context	Region Specified	Year	Source
5	Not Stipulated	No	1988	[65]
2	Outdoor	Taiwan	2012	[66]
40	Outdoor	Mediterranean Zones	2014	[67]
162	Indoor/Outdoor	No	2015	[68]
3	Outdoor	Doha, Qatar	2015	[69]
24	Outdoor	Polar, Cold, Temperate, Arid and Tropical	2016	[70]
165	Indoor/Outdoor	No	2016	[71]
4 specific (from 165)	Outdoor	No	2018	[47]
6	Outdoor	No	2018	[72]
6	Outdoor	Mediterranean Zones	2019	[73]
6	Outdoor	Mediterranean Zones	2019	[74]
4	Indoor/Outdoor	No	2019	[75]
1 (MRT [*1])	Indoor/Outdoor	No	2019	[76]
- (SVF [*2])	Outdoor	No	2019	[77]

[*1] MRT—Mean Radiant Temperature, [*2] SVF—Sky View Factor.

Since the emergence of thermal indices well before the turn of the century, the international scientific community has since developed hundreds of different indices. Again, such an occurrence naturally leads to review and comparative studies of the indices themselves through different analytical methodologies and within different climatic contexts. Furthermore, and as exemplified by the studies undertaken by Golasi, Salata, Vollaro and Coppi [72], there still remain scientists that pursue the further standardisation of a global outdoor standardisation thermal indices. Although met with some resistance due to the already extensive amount and versatility of existing indices, such studies still salient the continual and important scientific desires to further develop additional approaches to human biometeorology. Adjacently, from the large identified sample of indices, many studies have suggested that only between 6 and 4 thermal indices can provide wholesome local human thermo–physiological evaluations [47,69,73–75]. In addition, and as a distinguished example from many related studies (discussed later in this section), the work undertaken by Lin, Tsai, Hwang and Matzarakis [66] presented important outputs pertaining to crucial relationships with microclimatic variables such as Mean Radiant Temperature (MRT) and Sky View Factor (SVF) ratios. Regarding these two aspects, and although the former two studies in Table 1 do not categorically refer to the comparison of thermal indices, both present noteworthy contemporary reviews regarding the calculation methods of: (i) MRT in indoor and outdoor environments through different applicative algorithms and models [76]; (ii) SVF through a diverse range of reviewed methodologies and software packages, moreover highlighting their respective weaknesses and strengthens within local microclimatic assessments and linkage with urban planning processes and decision making [77].

To illustrate a sample of the inherently different utilised thermal indices within the scientific community, and based upon the typological division suggested by Freitas and Grigorieva [68], of the eight, four typologies were included in Table 2, these being: (i) B–singular parameter model; (ii) C–climatic index based upon algebraic or statistical model; (iii) F–energy balance strain model; (iv) G–energy balance stress model.

Table 2. Illustration of selected thermal indices and their respective index typologies as defined by Freitas and Grigorieva [68].

Index	Acronym	Typology	Source
Perceived Temperature	(PT)	(G)–Energy balance stress model	[78]
Standard Effective Temperature	(SET *)	(G)–Energy balance stress model	[79,80]
Outdoor Standard Effective Temperature	(OUT_SET *)	(G)–Energy balance stress model	[63,81]
Thermal Humidity Index	(THI)	(C)–Algebraic/statistical model	[82]
Predicted Mean Vote	(PMV)	(G)–Energy balance stress model	[28,83]
Predicted Percentage of Dissatisfied	(PPD)	(G)–Energy balance stress model	[28]
Humidex	(HD)	(C)–Algebraic/statistical model	[84]
Index of Thermal Stress	(ITS)	(F)–Energy balance strain model	[31]
Outdoor thermal comfort model	(COMFA)	(G)–Energy balance stress model	[85,86]
Universal Thermal Climate Index	(UTCI)	(G)–Energy balance stress model	[87–89]
Wet Bulb Temperature	(WBGT)	(B)–Single-parameter model	[90,91]
Predicted Heat Strain	(PHS)	(F)–Energy balance strain model	[92]
Physiologically Equivalent Temperature	(PET)	(G)–Energy balance stress model	[26,93,94]
modified Physiologically Equivalent Temperature	(mPET)	(G)–Energy balance stress model [*1]	[95]

[*1] New modified physiologically equivalent temperature (mPET) index included in (G) typology due to its close proximity to the original Munich energy-balance model for Individuals (MEMI).

Based upon the studies disclosed in Table 1, of the four typologies presented in Table 2, the predominantly utilized indices for outdoor studies have been those constructed upon the energy balance stress models, in particular, the Physiologically Equivalent Temperature (PET), Predicted Mean Vote (PMV), Universal Thermal Climate Index (UTCI) and Standard Effective Temperature (SET*) indices [96], especially for the climatic evaluation for urban planning and design [75].

In the case of the latter two examples in Table 2, of all of the thermo–physiological indices, PET has been one of the most commonly used steady-state model in human biometeorological studies [67]. Constructed upon the Munich Energy-balance Model for Individuals (MEMI) [97], it is designated as the T_a at which, in a typical indoor setting, the human energy budget is maintained by the skin temperature (T_{skn}), core temperature (T_{core}) and perspiration rate that are equivalent to those under the conditions to be investigated [93]. Retrospectively, the likely reason for its higher application can be attributable to: (i) its feasibility in being calibrated on easily obtainable microclimatic elements, and (ii) its measuring unit being (°C), which in turn, simplifies its comprehension by non-climatic experts, including urban designers/planners and architects. This being said, synonymous to the equally maturing body of knowledge in human biometeorology, numerous studies have already made headway in the development of the PET index as well. Directed specifically towards improving the calibration of the integrated thermoregulation and clothing models utilised by the PET index, Chen and Matzarakis [95] launched the new modified Physiologically Equivalent Temperature (mPET) index. As discussed in the study, the main modifications of the mPET are the integrated thermoregulation model (modified from a single double-node body model to a multiple-segment model) and updated the clothing model, resulting in more accurate evaluations of the human bio-heat transfer mechanism, particularly during periods of higher thermal stimuli. Such increased accuracy of the modified index was subsequently verified by numerous studies in different countries and climatic contexts [5,19,73,98,99].

4.2. Bridging the Qualitative with the Quantitative

As mentioned in the introduction, in addition to physiological aspects, there is also a demand in accompanying the associated call for investigating psychological factors of human thermal comfort. Although located predominantly within the qualitative spectrum, the 'intangible' attributes of human psychology have also been recognised to play a crucial role in diurnal human thermal comfort investigations. Such recognition has arisen at both in indoor contexts, e.g., [100–106], and outdoor contexts, e.g., [24,29,32,48,107–114].

Based upon a bottom-up perspective that focuses upon the role of local scales in ensuring human wellbeing during an era of climate change, Figure 2 illustrates the required interactions between that of: (1) Physiological aspects, which consider the direct quantitative influences of encircling microclimates upon the human-biometeorological system; (2) Psychological aspects, which prompts the adjacently important value of qualitative aspects of thermal comfort thresholds, including assessments of human behaviour patterns, and that of thermal adaptability; and lastly, (3) the interaction with further climate change impacts during the unravelling of the twenty-first century, that are already aggravating existing human thermal comfort standards. From the interaction of these three aspects, originates the requirement for further interdisciplinary biometeorological tools that can aid local assessment and design practices, both now, and in the future through informed CCA efforts.

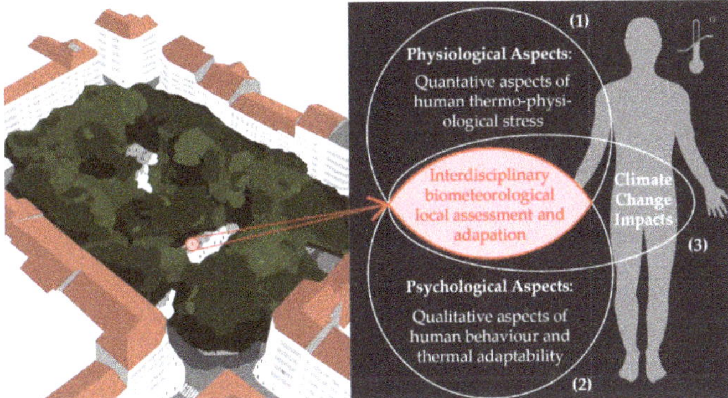

Figure 2. Illustrative division of human biometeorological facets within local scales based upon a bottom-up approach in century prone to climate change impacts.

Although the comparative significance between qualitative and quantitative aspects of thermal comfort is still debated within different studies (i.e., where some authors methodically favour one aspect more than the other), numerous veracities are concomitant to both schools of thought. To start with, it is consensus that there is an unmistakable opportunity to explore how specific qualitative aspects of thermal comfort can build upon quantitative assessments. Such an opportunity can be allied to a few simple premises, that: (i) predominantly in outdoor environments, human beings rarely pursue microclimatic monotony [109,115], reversely, it is the very desire of climatic diversity and stimulation (and even overstimulation beyond stipulated thermal comfort levels [108,111]) that also lures pedestrians outdoors [29]; (ii) human beings are by default peripatetic, meaning that their movement patterns are based upon complex behaviour and decision making processes associated to 'intangible' attributes (e.g., expectations, past experience, perceived control and time of exposure) [32,116].

As a result, improving this integration between these two aspects could potentially present means to better predict and account for human psychological attributes for local thermal sensitive design and planning. Of the attributes previously mentioned, it is suggested that these main attributes can open up new interdisciplinary lines of research, which by default, coerce the bridging with quantitative aspects of thermal comfort. In addition, such a bridging can also entice further considerations also interrelated to indoor conditions, as also suggested by past review studies exemplified by the prominent example disseminated by Brager and de-Dear [117]. As part of their review, they inclusively referenced an entire issue from *Energy and Buildings* [118] that focused upon the variation amongst the human psychological 'perceived need' or 'desire' for indoor mechanical air conditioning.

4.3. Indoor and Outdoor Cumulative Thermal Stress

In accordance with the human biometeorological evaluation study undertaken by Charalampopoulos, Tsiros, Chronopoulou-Sereli and Matzarakis [11] who utilised the PET index, two preliminary factors were adjoined, these being: (1) the PET Load (PETL), i.e., the amount of variation from the optimal physiological stress range (between PET values of 18–23 °C) as defined by [119,120]; and, (2) the cumulative PET Load (cPETL), i.e., the sum of the PETL for an X amount of hours which can be configured to represent a portion, or the full 24 hours of a respective day. Such an approach enables a preliminary understanding of cumulative human thermal stress loads beyond 'neutral' (or background) conditions. Although intended for outdoor assessments, principals of cumulative human thermal stress can also be transposed to methodically approach human psychological attributes, particularly during periods and/or events of accentuated thermal stress, and even climate change [19]. Subsequently, such accentuation periods with higher stimuli can be unambiguously associated to numerous urban events, particularly heatwaves. Alarmingly, and beyond the early consensus that increases in heatwaves are 'very likely' throughout the twenty-first century [121]; the subsequent fifth assessment report moreover stipulated that the influences of climate change upon heatwaves shall be more significant than the impacts upon global average temperatures [122].

Taking the European heatwave of 2003 as an extreme example which explicitly amplified the need for additional measures to warn, cope and prevent the recurrence of such events upon public health and welling [58,123,124]. Within Western Europe, the data provided by Nogueira et al. [125] identified that between the 29 July and 13 August 2003 within the district of Lisbon there was/were: (i) 15 days with a maximum T_a above 32 °C; (ii) a noteworthy consecutive run of 10 days with T_a above 32 °C; and, (iii) a 5 day period consecutively experiencing T_a above 35 °C. This extreme heat event led to severe implications on urban health, resulting in an estimated mortality rate increase of 37.7% in comparison to what would be expected under normal conditions.

Key lessons for human biometeorology can continue to be extracted from this type climatic event that has serious implications for human health and wellbeing in urban contexts. Such teachings, in turn, again call for more sophisticated integration and analytical tools between the quantitative and qualitative aspects of thermal comfort, both for outdoor and indoor environments. More specifically, and considering the early principals of the urban energy balance as defined by Oke [126] the reciprocal dynamics of indoor environments also play an essential role resultant of the: (i) increased heat storage within urban materials and buildings [18,22,40,127–131]; and the cause-and-effect of, (ii) anthropogenic emissions resultant of urban cooling energy loads associated to interior air conditioning [117,132–135], which by the end of the century can potentially increase by 166% (in energy demand) as a result of climate change [136].

In the case of naturally ventilated residential indoor environments during periods of extreme and extended heat stress, the principals of cumulative human thermal stress load can be strongly connected to psychological aspects. Although previously observed by Givoni [137] that "during periods of rising outdoor temperatures, e.g., a heat wave lasting for several days, the rate of rise of the indoor temperature is lower than that of the outdoors (. . .). As a result, the indoor temperatures during the heat-wave period will be somewhat lower" (p. 22), it is important to note that during extreme events, this 'somewhat' reduction while significant, is indicative of continued cumulative human thermal stress load during the night period. Such an extension, invariably, results in disruptions in human sleep cycles as a result of higher nocturnal indoor T_a levels [138,139]. These conclusions were also extended by a more recent review study conducted by Lan et al. [140], who also depicted upon the 2003 heatwave in Europe, and moreover, the associated future risk factors associated to human sleep disruptions as a result of climate change.

With regards to specific implications upon the human biometeorological system, the preceding study by Haskell et al. [141] indicated that T_a above the thermo–neutral thresholds increased wakefulness and decreases Slow Wave Sleep (SWS) which takes place in the late stages of non-Rapid Eye Movement ((n)REM). Such a stage is where energy restoration occurs, including the regulation of body glycogen levels, that are subsequently heavily consumed during active brain function [142]. Up until the more crucial and profound REM stage of sleep, the human body continues to thermo-regulate, and perspiration is proportional to the encircling thermal load [143]. During the latter stage of the sleep cycle, perspiration does not take place [144], and the human hypothalamic thermostat (in control of the body's T_{core}) becomes sedentary as a result of the poikilothermic state during the REM stage [145]. Resultantly, the successfulness in reaching REM sleep strongly depends upon the adequate down-regulation of T_{core} beforehand. If not accomplished, inclusively in circumstances with high thermal loads, both SWS and REM will likely be replaced by wakefulness to maintain bodily homeothermic conditions [146,147]. Such homeothermy can be backtracked to the functioning principals of the previously mentioned MEMI.

This being said, and in addition to heat stress, exposure to elevated nocturnal RH also plays a pivotal role in thermal stress. More specifically, increased RH levels impedes sweat to evaporate, thus impeding T_{skn} to dissipate heat and remain wet, thereby, suppressing adequate down-regulation of the body's T_{core}, and similarly decreasing the likelihood of REM sleep [147]. This influence of encircling nocturnal RH upon sleep quality has moreover been identified by other comparable studies, e.g., [140,148,149].

While suggested by prominent thermal comfort studies that people living in naturally ventilated buildings become accustomed to, and moreover grow to accept higher T_a and RH, [137], human biometeorological investigations have come to respectfully rebut such acclimatization easement (especially during periods of higher thermo–physiological loads). Respectively, and as identified by the early analysis undertaken by Libert et al. [150], heat-related sleep disruptions do not adapt even after five days of continuous diurnal and nocturnal heat exposure. Likewise, it was also later documented that the cerebral dynamics of SWS does not change after partial sleep deprivation (SD), where 'sleep pressure' would inevitably be augmented [151].

Subsequently, such results were also evidenced by the more recent study conducted by Nastos and Matzarakis [152] who analysed the daily records of SD against the frequency of daily weather conditions (with PET > 35 °C) and nocturnal conditions (with minimum T_a > 23 °C). The recorded events/admissions for SD were obtained from the psychiatric emergency unit of Eginition Hosptial of the Athens University Medical School during the years of 1989 and 1994. It is important to note that in this particular study, the SD admissions dataset did not include cases which were associated to specific organic disturbances. Such an inclusion would very likely increase admission data numbers; but invariably, excessively extend the investigation parameters due to the inherent intricacies of specific human organic disturbances in relationship with SDs (e.g., pertinent to the respiratory system [153], and in oncological cases [154]). Irrespectively, the study identified that during continued periods of both diurnal and nocturnal thermal load, there was a substantial increase in SD, which moreover, did not seem to placate, nor adapt, to the respective conditions over time.

Overall, the studies in this section depict upon the significance of the Circadian Rhythm Cycle (CRC) in human wellbeing, which by definition, also extends to the human biometeorological thermoregulation dynamics during the night. Naturally, the circumstances during the CRC influence wellbeing standards, whereby if one part of the cycle inept, there will be a cause-and-effect relationship upon the following stage. In other words, if the cumulative thermal loads do not fluctuate adequately to allow the human-biometeorological system to regulate, replenish and restore attributes of the human physiology (including during different sleep stages), then this shall have direct implications upon human psychology as well. In this way, the physiological and the psychological attributes pertaining to thermal comfort can be directly related to one another. It is here where central intangible aspects as of human psychology as presented by, e.g., [32] can be further explored, including for urban sensitive planning and design.

Inarguably, there still remain other noteworthy impromptu influences upon these intangible characteristics that influence human behaviour. However, it is argued that such qualitative thermal comfort aspects (e.g., expectations, past experience, perceived control and time of exposure) can be rendered less subjective by more efficiently cross-examining human behaviour patterns and decision making against CRC dynamics and cumulative thermal stress.

Evaluating the specific case-by-case peripatetic behaviour of individual human beings is very complex. Yet it is here reasoned that further studies on this interdisciplinary topic can be undertaken based upon the unambiguous certainties that are already held by the scientific community, including the: (1) universal conduct of the human biometeorological system to thermal stimulus (including in cumulative terms); and, (2) impacts that extreme urban events can have upon both indoor and outdoor environments upon urban human wellbeing, including those associated to future climatic aggravation. For this reason, when one considers the urban populace as whole, it feasible to acknowledge that pedestrians shall, in general, show higher psychological predispositions under certain climatic conditions, particularly under prolonged extreme events.

As represented in Figure 3, this shall not only affect the peripatetic transitioning between indoor and outdoor movement patterns/durations, but the individual psychological aspects that catalyse such human behaviour. More precisely, during periods of extended thermo–physiological stress, elicited from 'past experience' of thermal discomfort, there shall be a greater pursuit (i.e., 'expectation') to address cumulative discomfort. Since this is associated with the CRC, it cannot be assumed that thermal stress simply resets at the end of the day. Naturally, the longer the susceptibility to cumulative load (including throughout the night) the greater the 'expectation' and reduced willingness for more 'time of exposure'. Subsequently, and as developed throughout this section, it is suggested that there are opportunities for future concrete investigations to better link this symbiotic physiological and psychological relationship.

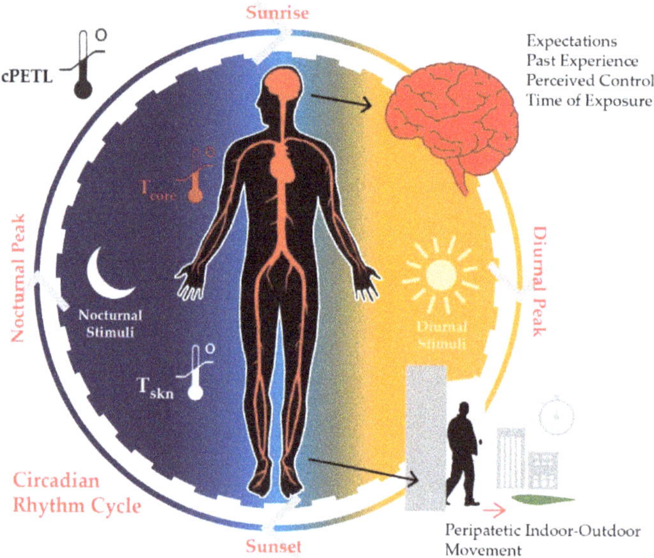

Figure 3. The relationship of the circadian rhythm cycle, cumulative stress and general psychological characteristics.

5. Biometeorological Urban Design/Planning

5.1. Urban Vegetation

So far within the existing literature, numerous review articles have already discussed the state-of-the-art of various aspects pertaining to the influences of vegetation upon urban climates. Of these review articles thus far, e.g., [3,20,128,155–159], the two predominant influences of vegetation pertaining to urban thermal comfort aspects have thus far been the: (i) direct reductions of urban T_a; and moreover the (ii) associated interrelating reduction of UHI intensities. Other disseminated review studies have moreover deliberated on further positive attributes that vegetation can have upon indoor/outdoor human wellbeing standards in urban contexts. Within Table 3, these review studies are divided into five summarised topics that also play an important role in ensuring urban environmental health and welfare.

Table 3. Selected review studies concerning further positive attributes of vegetation within urban environments.

No.	Predominant Review Topic Summary	Icon	Study Year	Example Review Studies
(i)	Specific effects of green roofs, including indoor thermal behaviour, cooling loads and performance		2014	[160]
			2014	[161]
			2018	[162]
(ii)	Specific quantitative influences and performance of urban green walls/facades		2014	[163]
			2014	[164]
			2017	[165]
(iii)	Air quality and particles dispersion/abatement through the presence of vegetation		2015	[166]
			2015	[167]
			2017	[168]
(iv)	Overall socio-economic benefits, and challenges, of growing urban vegetation in the public realm		2011	[169]
			2015	[170]
(v)	Wider social impacts of street vegetation upon urban ecosystems and communities		2016	[171]

(i)   As suggested by the comprehensive review undertaken by Berardi, GhaffarianHoseini and GhaffarianHoseini [160] there is a very tactile opportunity to continue the exploration into the further quantification and assessments of interdisciplinary approaches regarding urban landscaping, plantations, construction and that of mechanical/environmental engineering. Moreover and in addition to stipulating the different classification of green roofs, the authors also cross-examined the typologies against their ability in mitigating UHI/air pollution, improve stormwater management, reduce urban noise and augment urban diversity. From the same year, and focused at the city scale, Santamouris [161] identified four categories to determine the particular efficiency of green roofs, namely through: (i) climatological variables, including radiation fluctuations; (ii) optical variables, including changes in albedo and absorptivity of the roof's vegetation; (iii) thermal variables, including thermal capacity and heat storage; and lastly, (iv) hydrological variables, including the dynamics of latent heat loss due to evaporation of the water vapour from the vegetative material (or in other words, evapotranspiration). Within the more recent study conducted by Shafique, Kim and Rafiq [162], it was revealed how green roofs can aid simulating urban natural hydrology systems, and also reduce factors such as UHI effects. Still within this recent study, the prominence of further interdisciplinary research was recognised, including in accompanying the demand for such technology through economically sustainable methods.

(ii)   With regards to the application of green walls and facades, the review study conducted by Hunter, Williams, Rayner, Aye, Hes and Livesley [163] reported that their efficiency must be based on multiple microclimatic factors, including G_{rad}, T_a and V (both adjacent to the structure, and in-between the gap with the respective wall). In the summary of the study, while the significant potential of green facades were recognised in urban contexts, it was adjacently argued that: (i) they are unlikely mechanisms to modulate internal buildings in all types of construction typologies and climatic contexts; and, (ii) its associated engineering terminology is often too specific to be readily understood across design and planning disciplines. Similarly, and also relating the application of these vegetation structures to different climates, and moreover the influences of different vegetative species, Perez, Coma, Martorell and Cabeza [164] came to similar conclusions. Finally, and within the more recent review study conducted by Medl, Stangl and Florineth [165] (and in addition to the recognised positive attributes mentioned above), the authors argued that there still remains a clear need for further interdisciplinary and standardized measurement approaches to guarantee the better application and erection of effective urban green facades.

(iii) While the aforementioned studies also discussed issues of urban air quality and pollution dispersion through urban vegetation, Gallagher, Baldauf, Fuller, Kumar, Gill and McNabola [166] and Abhijith, Kumar, Gallagher, McNabola, Baldauf, Pilla, Broderick, Sabatino and Pulvirenti [168] took this analysis a step further. More specifically, it was identified that wind-tunnel and modelling results provide adequate evaluations, yet further real-world studies are still required to validate such findings. Similarly, and still in line with the aforementioned perspective of Oke [35], both studies moreover suggest that to develop clear guidelines for urban planners with regards to air quality and pollution dispersal; better interdisciplinary 'channels' must be fortified to enable such knowledge to be translated into practical guidelines to ensure their effective urban implementation. Convergent conclusions pertaining to the associated translation into urban planning and design tools/guidelines were also met by Janhall [167].

(iv) Undertaking a more socio-economic approach, the review study launched by Soares, Rego, McPherson, Simpson, Peper and Xiao [169] described the application of the Street Tree Resource Analysis Tool for Urban forest Managers (STRATUM) within Lisbon. The results of the study disclosed a clear quantitative breakdown of economic maintenance/managerial costs of urban vegetation species which was subsequently crossed examined with urban 'energy savings', air purification, increased property values, reduced stormwater runoff and CO2 emissions. Still predominantly within the socio-economic spectrum, the later review study undertaken by Mullaney, Lucke and Trueman [170] also provided an investigation into financial aspects of urban vegetation. More specifically, beyond also disclosing environmental and socio-economic benefits, the costs/management of detailed characteristics such as pavement damage from tree roots were also case-studied.

(v) In the last segment, the study conducted by Salmond, Tadaki, Vardoulakis, Arbuthnott, Coutts, Demuzere, Dirks, Heaviside, Lim, Macintyre, McInnes, and Wheeler [171] undertook a more encompassing perspective, which suggested that based upon the existing literature, there needs to be a locally based bottom-up decision making process. Such a process was argued to be innately better associated with local community engagement to better determine 'what matters to them', and not just constructed upon the technical scientific aspects of ecological interventions. As a result, a matured interdisciplinary relationship between these cultural and scientific approaches was suggested to be essential to further exploit the disclosed societal and wider benefits provided by urban vegetation.

Parting from review studies, and focussing henceforth on individual investigations regarding the specific relationship of human thermo–physiological thresholds with urban vegetation, two distinct types of studies can be established, those: (1) which focus upon the direct 'In-Situ' (IS) influences of vegetation directly upon the encircling area (such as beneath the vegetative crown); and, (2) which investigate the effects of Park Cooling Islands (PCI) resultant of urban vegetation amid different spaces (where normally one is labelled as an urban 'green space').

Both within the IS and PCI types of study, the methical approach towards human thermal comfort thresholds have been different. Most prominently, there is a clear distinction between studies which have concentrated more upon singular variables (such as T_a), and those which have applied thermo–physiological indices that account for non-temperature variables, including radiation fluxes. Thus far, significant IS effects of urban vegetation specifically upon T_a and its associated connotations upon human thermal comfort have been well documented as exemplified in the studies in Table 4. Adjacently, studies focussing on the effects PCI upon T_a are successively presented in Table 5. Within these tables, the maximum thermal result obtained by the study, year, city and climatic context (through the Köppen Geiger (KG) [172] climatic classification system) are presented.

Table 4. Studies concerning in-situ (IS) changes in T_a resultant of urban vegetation.

Thermal Result (T_a Max)	Location	KG	Study Year	Source
−1.5 °C	California	*'Csa'*	1988	[173]
−0.7 °C	Tokyo	*'Cfa'*	2008	[174]
−2.2 °C	Athens	*'Csa'*	2010	[175]
−0.5 °C	Singapore	*'Af'*	2010	[176]
−1.0 °C	Melbourne	*'Cfb'*	2013	[177]
−1.0 °C	Manchester	*'Cfb'*	2014	[178]
Table Result Avg. = −1.2 °C				

Table 5. Studies concerning changes in T_a as a result of urban park cooling islands (PCI) effects.

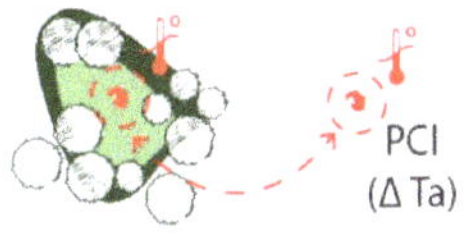

Thermal Result (T_a Max)	Location	KG	Study Year	Source
−4.0 °C	Mexico City	*'Cwb'*	1990-1	[179]
−2.5 °C	Dehli	*'BSh'*	1990-1	[180]
−3.0 °C	Kumanoto	*'Cfa'*	1991	[181]
−2.5 °C	Fukuoka	*'Cfa'*	1993	[182]
−2.0 °C	Tokyo	*'Cfa'*	1998	[132]
−4.0 °C	(Scaled model)	-	1999	[183]
−4.0 °C	Tel Aviv	*'Csa'*	2000	[184]
−4.0 °C	Botswana	*'BSh'*	2004	[185]
−3.5 °C	Tel Aviv	*'Csa'*	2006	[186]
−4.4 °C	Taipei	*'Cfa'*	2007	[187]
−2.5 °C	Taipei	*'Cfa'*	2010	[188]
−5.0 °C	Athens	*'Csa'*	2014	[189]
−5.0 °C	Chania	*'Csa'*	2014	[190]
−7.4 °C	Lisbon	*'Csa'*	2019	[191]
Table Result Avg. = −3.9 °C				

Between Tables 4 and 5, it is possible to identify that the studies have predominantly been undertaken within 'Temperate' climates, with variations mostly being discernable within the subcategories pertaining to both annual precipitation levels, and average temperature levels during the summer. Additionally, when considering the thermal effects in both Tables, the summarised averages of 'maximum thermal effects' present by the studies show clear thermal differences between IS and PCI typologies. As expected, this divergence can be attributable to the: (i) increased amount of vegetative mass within PCI studies that were able to have a greater impact upon elements such as atmospheric T_a and RH, which in IS studies were easier to dissipate in the atmosphere before allowing the utilised apparatus to record such modifications; and (ii) distances between the study's measurement locations, which in the case of the some PCI studies could vary up to hundreds of metres, thus entailing other potential microclimatic influences (e.g., sufficiently notable morphological and topographical disparities, rendering urban corridor cooling effects through channelled V acceleration and V gusts).

Adjacently to these T_a studies, aspects of urban vegetation as a tool address and regulate human thermal comfort in urban environments through non-temperature factors have also been discussed extensively. Such factors included the direct role urban tree crowns reducing the amount of radiation reaching pedestrian levels [192–196]. Subsequently, this was followed by studies which focused upon assessments that included non-temperature dynamics through the use of human biometeorological models and indices in both IS studies (Table 6) and PCI studies (Table 7).

Table 6. Studies concerning IS changes in physiologically equivalent temperature (PET)/ mean radiant temperature (MRT) resultant of urban vegetation.

IS
(Δ PET/
Δ MRT)

Thermal Result (PET/MRT Max)	Location	KG	Study Year	Source
−11.2 °C (PET)	Szeged	'*Cfb*'	2006	[197]
−12.0 °C (PET)	São Paulo	'*Cfa*'	2008	[198]
≈−12.0 °C (MRT)	Huwei	'*Cfa*'	2010	[199]
−20.0 °C (PET)	Shanghai	'*Cfa*'	2011	[200]
−8.0 °C (PET)	Campinas	'*Cwa*'	2012	[201]
−8.3 °C (PET)	Athens	'*Csa*'	2012	[33]
−16.6 °C (PET)	Campinas	'*Cwa*'	2015	[202]
−27.0 °C (MRT)	Manchester	'*Cfb*'	2016	[203]
−4.6 °C (PET)	Toronto	'*Dfb*'	2016	[22]
−3.4 °C (PET)	Hong Kong	'*Cwa*'	2017	[204]
−9.9 °C (PET)	Lisbon	'*Csa*'	2017	[108]
−15.6 °C (PET)	Lisbon	'*Csa*'	2018	[99]
Table Result Avg. = −11 °C (PET)/−19.5 °C (MRT)				

Table 7. Studies concerning changes in PET/MRT as a result of urban PCI effects.

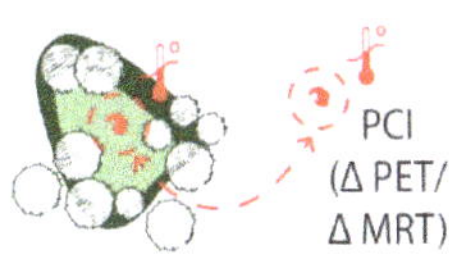

Thermal Result (PET/MRT Max)	Location	KG	Study Year	Source
−17.6 °C (PET)	Freiburg	*'Cfb'*	2003	[205]
−9.0 °C (PET)	Freiburg	*'Cfb'*	2006	[206]
−33.0 °C (MRT)	Lisbon	*'Csa'*	2007	[207]
−10.7 °C (PET)	Tel Aviv	*'Csa'*	2010	[208]
−39.2 °C (MRT)	Lisbon	*'Csa'*	2011	[209]
−12.0 °C (PET)	Tel Aviv	*'Csa'*	2012	[210]
−20.0 °C (MRT)	Milan, Genoa, Rome	*'Cfa'*, *'Csa'*, *'Csa'*	2014	[211]
−10.0 °C (PET)	Toulouse	*'Cfb'*	2016	[212]
−18.0 °C (PET)	Tel Aviv, Beer Sheva, Eilat	*'Csa'*, *'BSh'*, *'BWh'*	2017	[4]

Table Result Avg. = −12.3 °C (PET)/−30.7 °C (MRT)

Both IS and PCI studies that included non-temperature variables revealed very important differences, that can be related back to the outcomes also obtained by Matzarakis and Amelung [54] due to the crucial importance of radiation within thermal comfort studies and projections. In comparison with the former maximum T_a averages from Table 4 (−1.2 °C) and Table 5 (−3.9 °C), PET calculations revealed average maximum reductions of −11.0 °C for IS studies and −12.3 °C for PCI studies. Furthermore, it is worth noting the elevated changes in MRT with a maximum reduction of −39.2 °C obtained by Oliveira, Andrade and Vaz [209]. Such a measurement was undertaken within a small urban park in Lisbon and compared with values presented by the local meteorological station during August 2007. It is worth noting that during the same assessment period/day, the identified difference of T_a was of −3.2 °C.

When considering the differences of obtained PET/MRT measurements between the IS and PCI studies, the changes were more modest. However (and unlike in the case of T_a studies) it is suggested that such difference are less associated to IS/PCI study typology, and more attributable to the type of tree species used in each study. As identified by numerous studies, e.g., [213], the biggest influence upon thermo–physiological impacts from urban trees is associated to tree species, rather than planting layout.

This being said, the type and quantification of the influences on human biometeorology resultant of urban vegetation also greatly depends upon the proposed evaluation methods and assessment scale. Naturally, between the disclosed studies in Table 4 through to Table 7, the applied methodologies on behalf of the authors varied. Yet general trends amongst these studies are clearly identifiable. Such trends in summary demonstrate that PCI effect studies render greater reductions in singular variables due to the larger cluster and/or arrangement of vegetation mass within the designated 'green space'. Although much more modest, the same can be recognised for studies which utilised thermo–physiological variables in PCI assessments. While both types of studies have rendered important outcomes, it is suggested that those which considered reductions in radiation fluxes upon the human biometeorological system are able to present more wholesome evaluations of human thermal wellbeing.

5.2. Shade Canopies

Within the existing literature, numerous studies have also acknowledged the critical role between urban canyons, radiation fluxes and human biometeorological thresholds [5,12,214–218]. Resultantly, when considering the application of shade canopies within local scales, the main microclimatic factor that must be investigated is the structures ability to attenuate solar radiation. In IS terms, while recorded T_a beneath a canopy may be the same as a recording fully exposed to the sun, the amount of solar radiation can vary significantly [16,60,108].

Unlike in the aforementioned studies, the majority of the application of (either ephemeral or permanent) shade canopies in the urban realm has not considered the specific impacts upon the human biometeorological system. Inversely, they more frequently originate from artistic influence (particularly ephemeral solutions) and more-often-then-not thermally impassive urban amenity placement. Resultantly, the current application of shade canopies as an effective thermal attenuation measure typology must be reconsidered, especially in the case of permanent solutions due to risks of over-shading during colder months.

5.3. Urban Surface Materials

Returning again to the principals of the urban energy balance, elements such as local surface materials play a large role in urban heat storage patterns. As a result, and due to the vast amount of paving within the public realm, its heat flux and implications on human biometeorology has been extensively discussed within the existing literature. More specifically, and due to the consensually recognised poor thermal performance of urban materials such as asphalt and concrete, means to augment surface albedos are continually being explored. As suggested by the study undertaken by Gaitani et al. [219] both researchers and manufacturers are already been developing 'cool' materials with higher reflectance values compared to the conventionally pigmented materials of the matching colour. In addition, such materials have been considered in local sites where the use of light colours could lead to solar glare issues, thus rendering another type of human discomfort within the public realm. Such efforts fall within a strong growing body of existing literature which consider divergent methods (including the implementation of 'infrared reflective cool paint' and 'photocatalytic compounds') to increase the thermal behaviour of pavement materials [129,130,220–226].

Retrospectively, considering the existing state-of-the-art, the relationship between human thermal comfort thresholds and surface materials should be approached as a two-step sequential approach. Whereby the assessment and design of pavements through interdisciplinary urban planning/design processes should: (1) be integrated with other measures that can, beforehand, reduce the energy load upon street materials; and subsequently, (2) ensure an effective thermal balance of the material itself by considering factors such as absorbed radiation, emitted infrared radiation, heat storage/convection and the effects of anthropogenic heat caused by urban activities such as vehicular traffic.

5.4. Misting Systems

The application of water and misting systems within the urban public realm has been a measure typology which was witnessed an increasing change in paradigm. In other words, cooling purposes behind that of pure aesthetics are increasingly becoming more obvious in urban environments. As a result, they are moreover growing in applicative meaning within interdisciplinary CCA efforts to improve local thermal conditions at local scales [227,228].

So far within the state-of-the-art, there is a fairly observable division within existing projects that use water or misting systems to improve thermal comfort levels during the hotter periods of the year. The first cluster, more often orientated towards bioclimatic design approaches [227,229,230], often lack the mechanical background in concretely attaining the correct balance between RH and T_a to cool microclimates without exacerbating acceptable atmospheric moisture levels. On the other hand, the second cluster can be linked back to the rudimentary, yet effective, Japanese cultural cooling method called 'Uchimizu' [231]. Correspondingly, with time, the simple action of scattering of water upon the entrances of residential buildings has evolved, and rendered Japan as one of the frontrunner countries in this cluster. Congruently with the Japanese nocturnal thermal comfort studies previously mentioned in this article, the focus upon diurnal/nocturnal equilibrium between T_a and RH can largely be attributable to Japan having simultaneously elevated humidity levels and temperatures, including during the summer [172]. As a result, the application of 'dry-mist' technology both within indoor and outdoor environments has been extensively researched within Japanese cities as demonstrated in Table 8.

Table 8. Exemplification of Japanese misting system studies originating from ancient 'Uchimizu' concepts to specifically reduce encircling T_a levels.

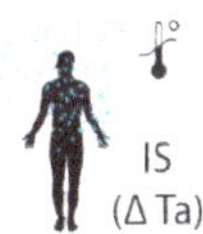

Thermal Result (Ta Max)	Location	KG	Study Method.	Study Year	Source
−2.0 °C	Nagoya	'Cfa'	Field Study	2008	[232]
−1.5 °C	Tokyo	'Cfa'	Field Study	2008	[233]
−2.5 °C	Tokyo	'Cfa'	CFD * Study	2008	[234]
−2.0 °C	Yohohama	'Cfa'	CFDStudy	2009	[231]
−0.8 °C	Osaka	'Cfa'	Field + CFD Study	2011	[235]
Table Result Avg. = −1.8 °C			* CFD—Computational Fluid Dynamic		

Based upon KG classification climates concomitant with 'temperate' yet with hot and wet summers, the outputs from the Japanese studies have marked an important step forwards in addressing how misting systems can be specifically configured to reduce T_a, without exacerbating local atmospheric moisture content in the air. Similar to the average maximum reductions in T_a as presented in Table 4 discussing the IS reductions as a result of urban vegetation, Table 8 presented a comparable result of −1.8 °C. This being said, this second cluster of measures has the problem of being extensively orientated towards engineering approach, with limited or no design connotations. Such a lapse between the engineering spectrum with design applications reinforce the aforementioned conclusions reached by the review study conducted by Hunter, Williams, Rayner, Aye, Hes and Livesley [163] regarding urban green walls and façades.

Subsequently, there needs to be a better interdisciplinary integration of these two clusters, where design and engineering approaches combine to tackle issues of human thermal comfort. Encouragingly, there have already numerous bioclimatic studies which have provided both conceptual, e.g., [108] and constructed, e.g., [115,236] examples of how water can be used to cool the human biometeorological system without exacerbating encircling atmospheric moisture levels. Nevertheless, there still remains a large opportunity for future studies to continue to dilute the disparity between these two clusters, even if initially based on rudimentary atmospheric principals.

6. Concluding Remarks

Within the contemporary city, the interdisciplinary relationship between biometeorology and local adaptation has already made important progress in addressing urban thermal wellbeing and safety. This being said, and as recognised within the three interconnecting sections of this encompassing review study, there remains the opportunity to further communicate climatic information into urban thermal planning/design tools and assessments. As further advocated by the adjacent review studies discussed in this study, the interdisciplinary transposal of technical know-how into concrete adaptation tools needs to be enforced. Pertaining specifically to urban human biometeorological aspects with that of urban wellbeing, an encompassing schematic summary is illustrated in Figure 4.

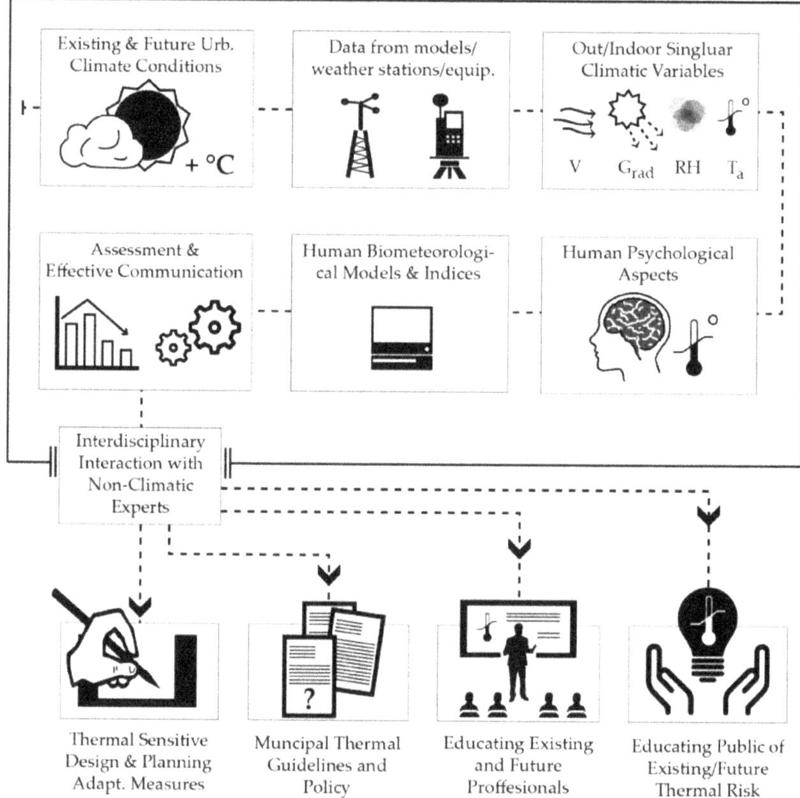

Figure 4. Schematic overview of human biometeorological aspects and augmenting associated interdisciplinary communication with non-climatic experts.

As discussed within the first section of the study, human biometeorology must play an increased role within existing and future thermo–physiological assessments of urban climate conditions. More specifically:

1. Methods of approaching climatic data from climatic models and meteorological stations/equipment should not rely solely upon singular climatic variables to obtain wholesome evaluations of existing or future human thermal comfort conditions as a result of climate change.
2. The information retrieved from such assessments must, unequivocally, be translatable through easy-to-understand guidance for non-climatic experts, and that of the general public. Such interdisciplinary communication channels shall become moreover significant given the expected increase of extreme heat/cold events within urban contexts. Eventually, the sequential multifaceted process of going from identified risk factors, to establishing better thermal response measures and transposing these into municipal climatic policy and guidelines can be strengthened.
3. Due to the inherent nature of thermo–physiological risk factors, undertaken assessments and projections must reach a better equilibrium between 'huge-impact-but-low-frequency' with that of 'high-frequency-yet-continuous' stimulus within the built environment, particularly during

summer/winter periods. In this way, thermal sensitive urban planning and design can better tackle both of these different, yet, decisive facets of urban climatology.

As discussed within the second segment of the study, further interdisciplinary scientific bridging must be accomplished between:

1. The remarkable and continual evolution of different thermo–physiological indices (including those arising from energy balance stress models, energy balance stain models, statistical/algebraic models and single-parameter models) with that of psychological factors. As mentioned, as this remains a less explored characteristic, this originates the respective opportunity to decrease 'qualitative subjectivity' through further research. Such research outlines can be launched through the association between continued physiological cumulative stimulus, circadian rhythm cycles, and anticipated triggers of human psychological behaviour patterns.

2. In association to the previous point, such future lines of research shall also diminish the often over-powering differentiation between evaluation methods between outdoor and indoor environments. Although clear why thermal evaluation methods must be different between these environments, the analytical relationship between the two types of environments must be strengthened. As an example, the effects existing/future extreme heat events shall influence both outdoor and indoor environments; meaning that the daily peripatetic transient relationship between the two environments can be better explored in future thermal comfort research. As result, this shall once again present better means of establishing better thermal response measures, both in indoor and in outdoor contexts.

Within the final section of the study, an overview of existing biometeorological aspects has already been integrated within different typologies of local thermal responsive measures. Such efforts are expected to propagate as the climate change adaptation agenda further matures its approaches and tools to present bottom-up means to attenuate thermo–physiological factors during the twenty-first-century. This being said, the final section of the study also discussed means in which human biometeorological aspects can more concretely aid local bottom-up adaptation processes through thermal sensitive design and planning. Beyond the recurring recognition from the disclosed review and research articles calling for further interdisciplinary communication amongst engineering/technical climatic know-how and urban planning-design tools, it was moreover acknowledged that:

1. While singular-variable evaluations pertaining to the thermal benefits of urban vegetation have been vital for thermal comfort studies (including in both IS and PCI typologies), the exclusion of non-temperature variables limit thermo–physiological interface comprehension with the human biometeorological system.

2. Given that the biggest potential of shade canopies is to limit the amount of global radiation projected upon the human body, the very limited amount of existing studies examining this aspect needs to be addressed by future research. Here, material types, structure size and distribution can all serve as analytical variables for addressing thermal comfort in open spaces that are particularly susceptible to high amounts of radiation. Ironically, the scientific community has produced a very strong body of research concerning the relationship amongst different urban morphological compositions and that of solar radiation. As a result, a rich body of research into which types of street configurations and orientations will serve as an excellent platform to guide such future research.

3. Due to the predominant use of thermally poor preforming materials, (such as urban concrete and asphalt) within cities, while there has been a considerable amount of research into surface materials, there is the opportunity to further explore the application of thermally efficient pavement materials. While existing studies have made clear strides in examining individual material, aggregate and finishing performances, there needs to be further studies that link such materials with other measure typologies within urban fabrics. As an example, there is the prospect

to further examine the affiliation with shading patterns resulting from both tree crowns and shading amenities. Since both of these measures can be utilised to reduce the amount of absorbed solar radiation upon a specific type of investigated pavement, the resulting inferences upon emitted infrared radiation, heat storage and convections can be further explored under specific urban conditions/layouts. Such studies will propagate means to address high surface temperatures and inefficient albedos values in local thermal urban planning/design and decision making.

4. As a result of the remaining disparity between engineering and design approaches to misting systems within the urban realm, similar to the case of shade canopies, there needs to be additional studies that consider wholesome projects which consider actual influences upon the human biometeorological system. Although commonly found within cities, there is a large opportunity for future studies to continue to dilute the segregation between engineering and thermal sensitive design approaches, even if based upon simple atmospheric principals to accomplish their full potential in attenuating thermal comfort stress without exacerbating atmospheric moisture content levels.

Overall, while the scientific community has made large strides in thermal comfort studies which were arguably catalysed by the maturing CCA agenda, there still remains a fertile opportunity for more scientific investigation. Inclusively, this review article suggests that this scientific advancement shall always be synonymously concomitant with interdisciplinarity. In other words, while the deepening of knowledge within a specific field is a quintessential cornerstone to science, the same can be said for the bridging between different disciplines that, are ultimately, set to face similar challenges throughout the twenty-first century. As it stands, the interdisciplinary interaction with non-climatic experts is still not strong enough, or at least, nowhere as strong as the existing thermal assessment capacity of climatological related experts. Again summarised in Figure 4, this fragility arguably risks the weakening of: (i) the concrete know-how of architects, landscape architects, urban designers and planners to more efficiently address human biometeorological constraints, and suggest local bioclimatic measures to address both indoor and outdoor human wellbeing standards; (ii) municipal policy and guidelines that could otherwise aid thermal sensitive design and planning; (iii) the transposition of knowledge to other non-climatic experts such as students to catalyse this interdisciplinary bridging upon upcoming generations of scientists and professionals; and finally, (iv) the production and dissemination of easy-to-use, easy-to-understand information and warning mechanisms for the general public regarding urban climatic risks on wellbeing standards, and just as importantly, the augmentation of such risk factors as a result of climate change.

Author Contributions: The authors declare that each has contributed equally to the production of this document. Whereby, (i) A.S.N. consolidated the overall results obtained within the study; and (ii) A.M. equally provided both general and specific contributions to the scientific content elaborated within this review study.

Funding: This research received no external funding.

Conflicts of Interest: The authors declare no conflict of interest.

References

1. Matzarakis, A. Chapter 14—Climate change and adaptation at regional and local scale. In *Tourism and the Implications of Climate Change: Issues and Actions (Bridging Tourism Theory and Practice)*; Schott, C., Jafari, J., Cai, L., Eds.; Emerald Group Publishing Limited: Bingley, UK, 2010; Volume 3, pp. 237–259.
2. Matzarakis, A.; Endler, C. Climate change and thermal bioclimate in cities: Impacts and options for adaptation in Freiburg, Germany. *Int. J. Biometeorol.* **2010**, *54*, 479–483. [CrossRef]
3. Nouri, A.S.; Costa, J.P.; Santamouris, M.; Matzarakis, A. Approaches to outdoor thermal comfort thresholds through public space design: A review. *Atmosphere* **2018**, *9*, 108. [CrossRef]
4. Potchter, O.; Shashua-Bar, L. Urban Greenery as a tool for city cooling: The Israeli experience in a variety of climatic zones. In Proceedings of the PLEA 2017—Design to Thrive, Edinburgh, UK, 3–5 July 2017; pp. 2–9.

5. Nouri, A.S.; Costa, J.P.; Matzarakis, A. Examining default urban-aspect-ratios and sky-view-factors to identify priorities for thermal-sesnsitive public space design in hot-summer Mediterranean climates: The Lisbon case. *Build. Environ.* **2017**, *126*, 442–456. [CrossRef]

6. Fröhlich, D. Development of a microscale model for the thermal environment in complex areas. In *Fakultät für Umwelt und Natürliche Ressourcen*; der Albert-Ludwigs-Universität: Freiburg im Breisgau, Germany, 2017.

7. Chen, Y.-C.; Fröhlich, D.; Matzarakis, A.; Lin, T.-P. Urban Roughness Estimation Based on Digital Building Models for Urban Wind and Thermal Condition Estimation—Application of the SkyHelios Model. *Atmosphere* **2017**, *8*, 247. [CrossRef]

8. Santamouris, M.; Kolokotsa, D. *Urban. Climate Mitigation Techniques*; Earthscan, Routledge: London, UK, 2016.

9. Santamouris, M. Cooling the buildings—Past, present and future. *Energy Build.* **2016**, *128*, 617–638. [CrossRef]

10. Matos Silva, M. *Public Space Design for Flooding: Facing Challenges Presented by Climate Change Adaptation*; Universitat de Barcelona: Barcelona, Spain, 2016.

11. Charalampopoulos, I.; Tsiros, I.; Chronopoulou-Sereli, A.; Matzarakis, A. A methodology for the evaluation fo the human-biolcimatic performance of open spaces. *Theor. Appl. Climatol.* **2016**, 1–10.

12. Ketterer, C.; Matzarakis, A. Human-biometeorlogical assessment of heat stress reduction by replanning measures in Stuttgart, Germany. *Landsc. Urban Plan.* **2014**, *122*, 78–88. [CrossRef]

13. Matzarakis, A.; Fröhlich, D.; Ketterer, C.; Martinelli, L. Urban bioclimate and micro climate—How to construct cities in the era of climate change. In *Climate Change and Sustainable Heritage*; Hofbauer, C., Kandjani, E.M., Meuwissen, J., Eds.; Cambridge Scholars Publishing: Cambridge, UK, 2018; pp. 38–61.

14. Nouri, A.S. A Framework of Thermal Sensitive Urban Design Benchmarks: Potentiating the Longevity of Auckland's Public Realm. *Buildings* **2015**, *5*, 252–281. [CrossRef]

15. Alcoforado, M.J.; Matzarakis, A. Planning with urban cilmate in different climatic zones. *Geographicalia* **2010**, *57*, 5–39.

16. Nouri, A.S.; Lopes, A.; Costa, J.P.; Matzarakis, A. Confronting potential future augmentations of the physiologically equivalent temperature through public space design: The case of Rossio, Lisbon. *Sustain. Cities Soc.* **2018**, *37*, 7–25. [CrossRef]

17. Alcoforado, M.-J.; Andrade, H.; Lopes, A.; Vasconcelos, J. Application of climatic guidelines to urban planning—The example of Lisbon (Portugal). *Landsc. Urban Plan.* **2009**, *90*, 56–65. [CrossRef]

18. Lopes, A.; Alves, E.; Alcoforado, M.J.; Machete, R. Lisbon Urban Heat Island Updated: New Highlights about the Relationships between Thermal Patterns and Wind Regimes. *J. Adv. Meteorol.* **2013**, *2013*, 1–11. [CrossRef]

19. Nouri, A.S.; Charalampopoulos, I.; Matzarakis, A. Beyond Singular Climatic Variables—Identifying the Dynamics of Wholesome Thermo-Physiological Factors for Existing/Future Human Thermal Comfort during Hot Dry Mediterranean Summers. *Int. J. Environ. Res. Public Health* **2018**, *15*, 2362. [CrossRef] [PubMed]

20. Santamouris, M.; Ding, L.; Fiorito, F.; Oldfield, P.; Osmond, P.; Paolini, R.; Prasad, D.; Synnefa, A. Passive and active cooling for the built environment—Analysis and assessment of the cooling potential of mitigation technologies using performance data form 220 large scale projects. *Sol. Energy* **2016**, *154*, 14–33. [CrossRef]

21. Alchapar, N.; Pezzuto, C.; Correa, E.; Labaki, L. The impact of different cooling strategies on urban air temperatures: The cases of Campinas, Brazil and Mendoza, Argentina. *Theor. Appl. Climatol.* **2017**, *130*, 33. [CrossRef]

22. Wang, Y.; Berardi, U.; Akbari, H. Comparing the effects of urban heat island mitigation strategies for Toronto, Canada. *Energy Build.* **2016**, *114*, 2 19. [CrossRef]

23. Brown, R. *Design with Microclimate—The Secret to Comfortable Outdoor Space*; Island Press: Washington, DC, USA, 2010.

24. Nouri, A.S.; Costa, J.P. Placemaking and climate change adaptation: New qualitative and quantitative considerations for the "Place Diagram". *J. Urban Int. Res. Placemaking Urban Sustain.* **2017**, *10*, 1–27. [CrossRef]

25. Oke, T. The distinction between canopy and boundary layer urban heat islands. *J. Atmos.* **1976**, *14*, 268–277. [CrossRef]

26. Mayer, H.; Höppe, P. Thermal comfort of man in different urban environments. *Theor. Appl. Climatol.* **1987**, *38*, 43–49. [CrossRef]

27. Brown, R.; Gillespie, T. *Microclimatic Landscape Design: Creating Thermal Comfort and Energy Efficiency*; John Wiley and Sons: Hoboken, NJ, USA, 1995.

28. Fanger, P. *Thermal Comfort: Analysis and Applications in Environmental Engineering*; McGraw-Hill Book Company: New York, NY, USA, 1972; p. 244.
29. Whyte, W.H. *The Social Life of Small Urban. Spaces*; Project for Public Spaces Inc.: New York, NY, USA, 1980.
30. Olgyay, V. *Design with Climate, Bioclimatic Approach to Architectural Regionalism*; Princeton University Press: Princeton, NJ, USA, 1963.
31. Givoni, B. *Man, Climate and Architecture*; Applied Science Publishers: London, UK, 1976.
32. Nikolopoulou, M.; Steemers, K. Thermal comfort and psychological adaptation as a guide for designing urban spaces. *Energy Build.* **2003**, *35*, 95–101. [CrossRef]
33. Shashua-Bar, L.; Tsiros, I.X.; Hoffman, M. Passive cooling design options to ameliorate thermal comfort in urban streets of a Mediterranean climate (Athens) under hot summer conditions. *Build. Environ.* **2012**, *57*, 110–119. [CrossRef]
34. Wilbanks, T.J.; Kates, R.W. *Global Change in Local Places: How Scale Matters*; Kluwer Academic Publishers: Dordrecht, The Netherlands, 1999; pp. 601–628.
35. Oke, T. Towards a prescription for the greater use of climatic principles in settlement planning. *Energy Build.* **1984**, *7*, 1–10. [CrossRef]
36. Alcoforado, M.J.; Vieira, H. Urban climate in Portuguese management plans (in Portuguese with abstract in English). *Soc. E Territ.* **2004**, *37*, 101–116.
37. Alcoforado, M.J.; Lopes, A.; Andrade, H. Urban climatic map studies in Portugal. In *The Urban Climatic Map: A Methodology for Sustainable Urban Planning*; Ng, E., Ren, C., Eds.; Routledge: Abingdon, UK, 2015.
38. Andrade, H. *Bioclima Humano E Temperatura Do Ar Em Lisboa*; Universidade de Lisboa: Lisbon, Portugal, 2003.
39. Oliveira, S.; Andrade, H. An initial assessment of the bioclimatic comfort in an outdoor public space in Lisbon. *Int. J. Biometeorol.* **2007**, 69–84. [CrossRef]
40. Alcoforado, M.J.; Andrade, H. Nocturnal urban heat island in Lisbon (Portugal): Main features and modelling attempts. *Theor. Appl. Climatol.* **2006**, *84*, 151–159. [CrossRef]
41. Alcoforado, M.J.; Lopes, A.; Alves, E.; Canário, P. Lisbon Heat Island; Statistical Study. *Finisterra* **2014**, *98*, 61–80.
42. Lopes, A. *Modificações No Clima Da Lisboa Como Consequência Do Crescimento Urbano*; University of Lisbon: Lisbon, Portugal, 2003.
43. Alcoforado, M.J.; Lopes, A.; Andrade, H.; Vasconcelos, J.; Vieira, R. *Observational Studies on Summer Winds in Lisbon (Portugal) and Their Influence on Daytime Regional and Urban Thermal Patterns*; Tel Aviv University Tel Aviv: Merhavim, Israel, 2006; pp. 88–112.
44. Alcoforado, M.J.; Lopes, A.; Andrade, H.; Vasconcelos, J. *Orientações Climáticas Para O Ordenamento Em Lisboa (Relatório 4)*; Centro de Estudos Geográficos da Universidade de Lisboa: Lisboa, Portugal, 2005; p. 83, ISBN 978-972-636-165-7.
45. Matzarakis, A.; Endler, C. Climate change and urban bioclimate: Adaptation possibilities. In Proceedings of the Seventh International Conference on Urban Climate, Yokohama, Japan, 29 June–3 July 2009.
46. Algeciras, J.A.R.; Matzarakis, A. Quantification of thermal bioclimate for the management of urban design in Mediterranean climate of Barcelona, Spain. *Int. J. Biometeorol.* **2015**, *8*, 1261–1270.
47. Potchter, O.; Cohen, P.; Lin, T.; Matzarakis, A. Outdoor human thermal perception in various climates: A comprehensive review of approaches, methods and quantification. *Sci. Total Environ.* **2018**, *631–632*, 390–406. [CrossRef]
48. Lin, T.-P. Thermal perception, adaptation and attendance in a public square in hot and humid regions. *Build. Environ.* **2009**, *44*, 2017–2026. [CrossRef]
49. Andreou, E. Thermal comfort in outdoor spaces and urban canyon microclimate. *Renew. Energy* **2013**, *55*, 182–188. [CrossRef]
50. Santamouris, M.; Xirafi, F.; Gaitani, N.; Saliari, M.; Vassilakopoulou, K. Improving the microclimate in a dense urban area using experimental and theoretical techniques—The case of Marousi, Athens. *Int. J. Vent.* **2012**, *11*, 1–16. [CrossRef]
51. Labaki, L.C.; Fontes, M.S.G.d.C.; Bueno-Bartholomei, C.L.; Dacanal, C. Thermal comfort in public open spaces: Studies in pedestrian streets in São Paulo state, Brazil. *Ambiente Construído* **2012**, *12*, 167–183. [CrossRef]
52. Fröhlich, D.; Gangwisch, M.; Matzarakis, A. Effect of radiation and wind on thermal comfort in urban environments—Applications of the RayMan and SkyHelios Model. *Urban Clim.* **2019**, *27*, 1–7. [CrossRef]

53. Katzschner, L. *Microclimatic Thermal Comfort Analysis in Cities for Urban Planning and Open Space Design Network for Comfort and Energy Use in Buildings*; NCUB: London, UK, 2006.
54. Matzarakis, A.; Amelung, B. Physiological Equivalent Temperature as Indicator for Impacts of Climate Change on Thermal Comfort of Humans. In *Seasonal Forecasts, Climatic Change and Human Health. Advances in Global Research 30*; Thomson, M.C., Garcia-Herrera, R., Beniston, M., Eds.; Springer: Berlin, Germany, 2008; pp. 161–172.
55. Matzarakis, A.; Georiadis, T.; Rossi, F. Thermal bioclimate analysis for Europe and Italy. *IL Nuovo Cim.* **2007**, *30*, 623–631.
56. IPCC. *Climate Change 2014: Synthesis Report. Contribution of Working Groups I, II and III to the Fifth Assessment Report of the Intergovernmental Panel on Climate Change*; IPCC: Geneva, Switzerland, 2014; p. 151.
57. IPCC. *IPCC Special Report Emission Scenarios, A Special Report of Working Group III of the Intergovernmental Panel on Climate Change*; IPCC: Cambridge, UK, 2000; p. 27.
58. Giannaros, T.; Lagouvardos, K.; Kotroni, V.; Matzarakis, A. Operational forecasting of human-biometeorological conditions. *Int. J. Biometeorol.* **2018**, 1–5. [CrossRef]
59. Hebbert, M.; Webb, B. Towards a Liveable Urban Climate: Lessons from Stuttgart. In *Liveable Cities: Urbanising World: Isocarp 07*; Routledge: Manchester, UK, 2007.
60. Kántor, N.; Chen, L.; Gal, C.V. Human-biometeorological significance of shading in urban public spaces—Summertime measurements in Pécs, Hungary. *Landsc. Urban Plan.* **2018**, *170*, 241–255. [CrossRef]
61. Nouri, A.S. *Addressing Urban Outdoor Thermal Comfort Thresholds through Public Space Design—A Bottom-Up Interdisciplinary Research Approach for Thermal Sensitive Urban Design in An Era of Climate Change: The Lisbon Case*; University of Lisbon: Lisbon, Portugal, 2018.
62. Nouri, A.S. A bottom-up perspective upon climate change—Approaches towards the local scale and microclimatic assessment. In *Green Design, Materials and Manufacturing Processes*; Bártolo, H., Ed.; Taylor & Francis: Lisbon, Portugal, 2013; pp. 119–124.
63. Spagnolo, J.; de-Dear, R. A field study of thermal comfort in outdoor and semi-outdoor environments in subtropical Sydney, Australia. *Build. Environ.* **2003**, *38*, 721–738. [CrossRef]
64. VDI. Part I: Environmental meteorology, methods for the human-biometeorological evaluation of climate and air quality for the urban and regional planning at regional level. Part I: Climate. In *VDI/DIN-Handbuch Reinhaltung der Luft*; Verein Deutscher Ingenieure: Düsseldorf, Germany, 1998; p. 29.
65. Beshir, M.; Ramsey, J. Heat Stress Indicies: A Review Paper. *Int. J. Ind. Ergon.* **1988**, *3*, 89–102. [CrossRef]
66. Lin, T.-P.; Tsai, K.-T.; Hwang, R.-L.; Matzarakis, A. Quantification of the effect of thermal indices and sky view factor on park attendance. *Landsc. Urban Plan.* **2012**, *107*, 137–146. [CrossRef]
67. Pantavou, K.; Santamouris, M.; Asimakopoulos, D.; Theoharatos, G. Empirical calibtation of thermal indices in an urban outdoor Mediterranean environment. *Build. Environ.* **2014**, *80*, 283–292. [CrossRef]
68. Freitas, C.; Grigorieva, E. A comprehensive catalogue and classification of human thermal climate indices. *Int. J. Biometeorol.* **2015**, *59*, 1–12. [CrossRef] [PubMed]
69. Fröhlich, D.; Matzarakis, A. A quantitative sensitivity analysis on the behaviour of common thermal indices under hot and windy conditions in Doha, Qatar. *Theor. Appl. Climatol.* **2015**, *124*, 179–187. [CrossRef]
70. Coccolo, S.; Kämpf, J.; Scartezzini, J.; Pearlmutter, D. Outdoor thermal comfort and thermal stress: A comprehensive review on models and standards. *Urban Clim.* **2016**, *18*, 33–57. [CrossRef]
71. Freitas, C.; Grigorieva, E. A comparison and appraisal of a comprehensive range of human thermal climate indices. *Int. J. Biometeorol.* **2016**, *61*, 1–26. [CrossRef] [PubMed]
72. Golasi, I.; Salata, F.; Vollaro, E.; Coppi, M. Complying with the demand of standarization in outdoor thermal comfort: A first approach to the Global Outdoor Comfort Index (GOCI). *Build. Environ.* **2018**, *130*, 104–119. [CrossRef]
73. Charalampopoulos, I. A comparative sensitivity analysis of human thermal comfort indices with generalized additive models. *Theor. Appl. Climatol.* **2019**, 1–18. [CrossRef]
74. Charalampopoulos, I.; Nouri, A.S. Investigating the behaviour of human thermal indices under divergent atmospheric conditions: A sensitivity analysis approach. *Atmosphere* **2019**, *10*, 580. [CrossRef]
75. Staiger, H.; Laschewski, G.; Matzarakis, A. Selection of appropriate thermal indices for applications in human biometeorological studies. *Atmosphere* **2019**, *10*, 18. [CrossRef]

76. Guo, H.; Aviv, D.; Loyola, M.; Teitelbaum, E.; Houchois, N.; Meggers, F. On the understanding of the mean radiant temperature within both the indoor and outdoor environment, a critical review. *Renew. Sustain. Energy Rev.* **2019**, in press. [CrossRef]

77. Miao, C.; Yu, S.; Hu, Y.; Zhang, H.; He, X.; Chen, W. Review of methods used to estimate the sky view factor in urban street canyons. *Build. Environ.* **2019**, *168*, 6497. [CrossRef]

78. Tinz, B.; Jendrizky, G. Europa- und Weltkarten der gefühlten Temperatur. In *Beiträge zur Klima- und Meeresforschung*; Chmielewski, F., Foken, T., Eds.; Deutschland: Berlin/Bayreuth, Germany, 2003; pp. 111–123.

79. Gagge, A.; Fobelets, P.; Bergland, L. A standard predictive index of human responce to thermal environment. *Ashrae Trans.* **1986**, *92*, 709–731.

80. Gonzalez, R.; Nishi, Y.; Gagge, A. Experimental evaluation of standard effective temperature: A new biometeorological index of man's thermal discomfort. *Int. J. Biometeorol.* **1974**, *18*, 1–15. [CrossRef] [PubMed]

81. de-Dear, R.; Pickup, R. An outdoor thermal comfort index (OUT_SET*)—Part I—The model and its assumptions. In Proceedings of the International Conference on Urban Climatology Macquarie University, Sydney, Australia, 8–12 November 1999.

82. Unger, J. Comparisons of urban and rural bioclimatological conditions in the case of a Central-European city. *Int. J. Biometeorol.* **1999**, *43*, 139–144. [CrossRef]

83. Alfano, D.A.; Olesen, B.; Palella, B. Polv Ole Fanger's impact ten years later. *Energy Build.* **2017**, *152*, 243–249. [CrossRef]

84. Masterton, J.; Richardson, F. *Humidex: A Method of Quantifying Human Discomfort Due to Excessive Heat and humidty*; Environment Canada: Downsview, ON, Canada, 1979.

85. Kenny, A.; Warland, S.; Brown, R. Part A: Assessing the performance of the COMFA outdoor thermal comfort model on subjects performing physical activity. *Int. J. Biometeorol.* **2009**, *53*, 415–428. [CrossRef] [PubMed]

86. Brown, R.; Gillespie, T. Estimating outdoor thermal comfort using a cylindrical radiation thermometer and an energy budget model. *Int. J. Biometeorol.* **1986**, *30*, 43–52. [CrossRef]

87. Jendritzky, G.; Maarouf, A.; Fiala, D.; Staiger, H. An update on the development of a Universal Thermal Climate Index. In Proceedings of the 15th Conference Biometeorology/Aerobiology and 16th International Congress of Biometeorology, Kansas City, KS, USA, 27 October 2002; pp. 129–133.

88. Jendritzky, G.; de-Dear, R.; Havenith, G. UTCI—Why another thermal index? *Int. J. Biometeorol.* **2012**, *56*, 421–428. [CrossRef]

89. Havenith, G.; Fiala, D.; Błazejczyk, K.; Richards, M.; Bröde, P.; Holmér, I.; Rintamaki, H.; Benshabat, Y.; Jendritzky, G. The UTCI clothing model. *Int. J. Biometeorol.* **2012**, *56*, 461–470. [CrossRef]

90. Yaglou, C.; Minard, D. Control of heat casualties at military training centers. *AMA Arch. Ind. Health* **1957**, *16*, 302–316.

91. Alfano, F.; Malchaire, J.; Palella, B.; Riccio, G. WBGT index revisited after 60 years of use. *Ann. Occup. Hyg.* **2014**, *58*, 955–970.

92. Malchaire, J.; Piette, A.; Kampmann, B.; Mehnerts, P.; Gebhardt, H.; Havenith, G.; Hartog, E.; Holmer, I.; Parsons, K.; Alfanoss, G.; et al. Development and Validation of the Predicted Heat Strain Model. *Ann. Occup. Hyg.* **2001**, *45*, 123–135. [CrossRef]

93. Höppe, P. The physiological equivalent temperature—A universal index for the biometeorological assessment of the thermal environment. *Int. J. Biometeorol.* **1999**, *43*, 71–75. [CrossRef] [PubMed]

94. Matzarakis, A.; Mayer, H.; Iziomon, G.M. Applications of a universal thermal index: Physiological equivalent temperature. *Int. J. Biometeorol.* **1999**, *42*, 76–84. [CrossRef] [PubMed]

95. Chen, Y.-C.; Matzarakis, A. Modified physiologically equivalent temperature—Basics and applications for western European climate. *Theor. Appl. Climatol.* **2017**, *132*, 1–15. [CrossRef]

96. Binarti, F.; Koerniawan, M.; Triyadi, S.; Utami, S.; Matzarakis, A. A review of outdoor thermal comfort indices and neutral ranges for hot-humid regions. *Urban Clim.* **2020**, *31*, 100531. [CrossRef]

97. Höppe, P. *The Energy Balance in Humans (Original Title—Die Energiebilanz des Menschen)*; Universitat Munchen, Meteorologisches Institut: Munich, Germany, 1984.

98. Lin, T.-P.; Yang, S.-R.; Chen, Y.-C.; Matzarakis, A. The potential of a modified physiologically equivalent temperature (mPET) based on local thermal comfort perception in hot and humid regions. *Theor. Appl. Climatol.* **2018**, *135*, 873–876. [CrossRef]

99. Nouri, A.S.; Fröhlich, D.; Silva, M.M.; Matzarakis, A. The Impact of Tipuana tipu Species on Local Human Thermal Comfort Thresholds in Different Urban Canyon Cases in Mediterranean Climates: Lisbon, Portugal. *Atmosphere* **2018**, *9*, 2–28.

100. de-Dear, R.; Foldvary, V.; Zhang, H.; Arens, E.; Luo, M.; Parkison, T.; Du, X.; Zhang, W.; Chun, C.; Liu, S. Comfort is in the mind of the beholder: A review of progress in adaptative thermal comfort research over the past two decades. In Proceedings of the 5th International Conference on Human-Environmental System, Nagoya, Japan, 29 October–2 November 2016.

101. Kim, J.; de-Dear, R. Thermal comfort expectations and adaptative behavioural characteristics of primary and secondary school students. *Build. Environ.* **2017**, *127*, 13–22. [CrossRef]

102. de-Dear, R.; Kim, J.; Candido, C.; Deuble, M. Adaptive thermal comfort in Australian school classrooms. *Build. Res. Inf.* **2015**, *43*, 383–398. [CrossRef]

103. Humphreys, M.; Nicol, F.; Raja, I. Field studies of indoor thermal comfort and the progress of the adaptive approach. *J. Adv. Build. Energy Res.* **2007**, *1*, 55–88. [CrossRef]

104. de-Dear, R. Thermal comfort in air-conditioned office buildings in the tropics. In *Standards for Thermal Comfort: Indoor Air Temperature Standards for the 21st Century*; Nicol, F., Humphreys, M., Sykes, O., Roaf, S., Eds.; Chapman and Hall: London, UK, 1995.

105. Heijs, W.; Stringer, P. Research on residential thermal comfort: Some contributions from environmental psychology. *J. Environ. Psychol.* **1988**, *8*, 235–247. [CrossRef]

106. Song, X.; Yang, L.; Zheng, W.; Ren, Y.; Lin, Y. Analysis on human adaptive levels in different kinds of indoor thermal environment. *Procedia Eng.* **2015**, *121*, 151–157. [CrossRef]

107. Chen, L.; Ng, E. Outdoor thermal comfort and outdoor activities: A Review of research in the past decade. *Cities* **2012**, 118–125. [CrossRef]

108. Nouri, A.S.; Costa, J.P. Addressing thermophysiological thresholds and psychological aspects during hot and dry mediterranean summers through public space design: The case of Rossio. *Build. Environ.* **2017**, *118*, 67–90. [CrossRef]

109. Nikolopoulou, M.; Baker, N.; Steemers, K. Thermal Comfort in Outdoor Urban Spaces: Understanding the Human Parameter. *Sol. Energy* **2001**, *70*, 227–235. [CrossRef]

110. Thorsson, S.; Lindqvist, M.; Lindqvist, S. Thermal bioclimatic conditions and patterns of behaviour in an urbn park in Göteborg, Sweden. *Int. J. Biometeorol.* **2004**, *48*, 149–156. [CrossRef]

111. Katzschner, L. Behaviour of people in open spaces in dependence of thermal comfort conditions. In Proceedings of the 23rd Conference on Passive and Low Energy Architecture, Geneva, Switzerland, 6–8 September 2006.

112. Nikolopoulou, M.; Steemers, K. Thermal comfort and psychological adaptation as a guide for designing urban spaces. *Energy Build.* **2003**, *35*, 95–101. [CrossRef]

113. Thorsson, S.; Honjo, T.; Lindberg, F.; Eliasson, I.; Lim, E.-M. Thermal comfort and outdoor activity in Japanese urban public places. *Environ. Behav.* **2007**, *39*, 661–684. [CrossRef]

114. Zacharias, J.; Stathopoulos, T.; Wu, H. Spatial behavior in San Francisco's plazas: The effects of microclimate, other people, and environmental design. *Environ. Behav.* **2004**, *36*, 639–658. [CrossRef]

115. Velazquez, R.; Alvarez, S.; Guerra, J. *Climatic Control. of the Open Spaces in Expo. 1992*; College of Industrial Engineering of Seville: Seville, Spain, 1992.

116. Hirashima, S.; Assis, E.; Nikolopoulou, M. Daytime thermal comfort in urban spaces: A field study in Brazil. *Build. Environ.* **2016**, *107*, 243–253. [CrossRef]

117. Brager, G.; de-Dear, R. Thermal adaptation in the built environment; A literature review. *Energy Build.* **1998**, *27*, 83–96. [CrossRef]

118. National Benefits Assessment. *Energy Build.* **1992**, *18*, 171–291.

119. Matzarakis, A.; Mayer, H. Heat Stress in Greece. *Int. J. Biometeorol.* **1997**, *41*, 34–39. [CrossRef] [PubMed]

120. Matzarakis, A.; Mayer, H. Another Kind of Environmental Stress: Thermal Stress. WHO Colloborating Centre for Air Quality Management and Air Pollution Control. *News Lett.* **1996**, *18*, 7–10.

121. IPCC. *Synthesis Report—An Assessment of the Intergovermental Panel on Climate Change—Adopted Section by Section at IPCC Plenary XXVII (Valencia, Spain, 12–17 November 2007)*; IPCC: Geneva, Switzerland, 2007; p. 104.

122. IPCC. *Climate Change 2013: The Physical Science Basis. Working Group Contribution to the IPCC 5th Assessment Report*; IPCC: Cambridge, UK, 2013.

123. Kovats, R.; Ebi, K. Heatwaves and public health in Europe. *J. Public Health* **2006**, *16*, 592–599. [CrossRef] [PubMed]

124. Matzarakis, A. The Heat Health Warning System of DWD—Concept and Lessons Learned. In *Perspectives on Atmospheroc Sciences*; Karacostas, T., Bais, A., Nastos, P., Eds.; Springer Atmospheric Sciences: Cham, Switzerland, 2016.

125. Nogueira, P.; Falcão, J.; Contreiras, M.; Paixão, E.; Brandão, J.; Batista, I. Mortality in Portugal associated with the heat wave of August 2003: Early estimation of effect, using a rapid method. *Eurosurveillance* **2005**, *10*, 5–6. [CrossRef]

126. Oke, T. The urban energy balance. *J. Prog. Phys. Geogr.* **1988**, *12*, 38. [CrossRef]

127. Stathopoulou, M.; Synnefa, A.; Cartalis, C.; Santamouris, M.; Karlessi, T.; Akbari, H. A surface heat island study of Athens using high-resolution satellite imagery and measurements of the optical and thermal properties of commonly used building and paving materials. *J. Sustain. Energy* **2009**, *28*, 59–76. [CrossRef]

128. Akbari, H.; Cartalis, C.; Kolokotsa, D.; Muscio, A.; Pisello, A.L.; Rossi, F.; Santamouris, M.; Synnefa, A.; Wong, N.H.; Zinzi, M. Local climate change and urban heat island mitigation techniques—The state of the art. *J. Civ. Eng. Manag.* **2016**, *22*, 1–16. [CrossRef]

129. Yang, J.; Wang, Z.; Kaloush, K. *Unintended Consequences—A Research Synthesis Examining the Use of Reflective Pavements to Mitigate the Urban Heat Island Effect (Revised April 2014)*; Arizona State University National Center for Excellence for SMART Innovations: Phoenix, Arizona, 2013.

130. Santamouris, M. Using cool pavements as a mitigation strategy to fight urban heat islands—A review of the actual developments. *J. Renew. Sustain. Energy Rev.* **2013**, *26*, 224–240. [CrossRef]

131. Dimoudi, A.; Zoras, S.; Kantzioura, A.; Stogiannou, X.; Kosmopoulos, P.; Pallas, C. Use of cool materials and other bioclimatic interventions in outdoor places in order to mitigate the urban heat island in a medium size city in Greece. *Sustain. Cities Soc.* **2014**, *13*, 89–96. [CrossRef]

132. Ca, T.; Asaeda, T.; Abu, M. Reductions in air conditioning energy caused by a nearby park. *Energy Build.* **1998**, *29*, 83–92. [CrossRef]

133. Elhelw, M. Analysis of energy management for heating, ventilating and air-conditioning systems. *Alex. Eng. J.* **2016**, *55*, 811–818. [CrossRef]

134. Song, L.; Zhou, X.; Zhang, J.; Zheng, S.; Yan, S. Air-conditioning usage pattern and energy consumption for residential space heating in Shanghai China. *Procedia Eng.* **2017**, *205*, 3138–3145. [CrossRef]

135. Lundgren-Kownacki, K.; Hornyanszky, E.; Chu, T.; Olsson, J.; Becker, P. Challenges of using air conditioning in an increasingly hot climate. *Int. J. Biometeorol.* **2017**, *62*, 401–412. [CrossRef]

136. Isaac, M.; Vuuren, D.P.V. Modeling global residential sector energy demand for heating and air conditioning in the contect of climate change. *Energy Policy* **2009**, *37*, 507–521. [CrossRef]

137. Givoni, B. Comfort, climate analysis and building design guidelines. *Energy Build.* **1991**, *18*, 11–23. [CrossRef]

138. Muzet, A.; Libert, J.; Candas, V. Ambient temperature and human sleep. *Experientia* **1984**, *40*, 425–429. [CrossRef]

139. Glotzbach, S.; Heller, H. Temperature regulation. In *Principles and Practice of Sleep Medicine*; Kryger, M., Roth, T., Dement, W., Eds.; Saunders: New York, NY, USA, 1999; pp. 289–304.

140. Lan, L.; Tsuzuki, K.; Liu, Y.; Lian, Z. Thermal environment and sleep quality: A review. *Energy Build.* **2017**, *149*, 101–113. [CrossRef]

141. Haskell, E.; Palca, J.; Walker, J.; Berger, R.; Heller, H. The effects of high and low ambient temperatures on human sleep stages. *Electro-Clin. Neurophysiol.* **1981**, *51*, 494–501. [CrossRef]

142. Sapolsky, R. *Why Zebras Don't Get Ulcers*, 3rd ed.; Holt Paperbacks: New York, NY, USA, 2004.

143. Law, T. *The Future of Thermal Comfort in An Energy-Contrained World*; University of Tasmania Australia: Hobart, Australia, 2013.

144. Gisolfi, C.; Mora, M. *The Hot Brain: Survival, Temperature, and the Human Body*; MIT Press: Cambridge, MA, USA, 2000.

145. Parmeggiani, P. Thermoregulation and sleep. *Front. Biosci.* **2003**, *8*, 557–567. [CrossRef] [PubMed]

146. Parmeggiani, P. Interaction between sleep and thermoregulation: An aspect of the control of behavioural states. *Sleep* **1987**, *10*, 426–435. [CrossRef] [PubMed]

147. Okamoto-Mizuno, K.; Mizuno, K. Effects of thermal environment on sleep and circadian rhythm. *J. Physiol. Anthropol.* **2012**, *31*, 1–9. [CrossRef] [PubMed]

148. Tsuzuki, K.; Okamoto-Mizuno, K.; Mizuno, K. Effects of humid heat exposure on sleep, thermoregulation, melatonin, and microclimate. *J. Therm. Biol.* **2004**, *29*, 31–36. [CrossRef]
149. Okamoto-Mizuno, K.; Mizuno, K.; Michie, S.; Maeda, A.; Lizuka, S. Effects of humid heat exposure on human sleep stages and body temperature. *Sleep* **1999**, *22*, 767–773.
150. Libert, J.; Di, J.; Fukuda, H.; Muzet, A.; Ehrhart, J.; Amoros, C. Effect of continuous heat exposure on sleep stages in humans. *Sleep* **1988**, *11*, 195–209. [CrossRef]
151. Bach, V.; Maingourd, Y.; Libert, J.; Oudart, H.; Muzet, A.; Lenzi, P.; Johnson, L. Effect of continuous heat exposure on sleep during partial sleep deprivation. *Sleep* **1994**, *17*, 1–10. [CrossRef]
152. Nastos, P.; Matzarakis, A. Human-biometeorological effects on sleep in Athens, Greece: A Preliminary Evaluation. *Indoor Built Environ.* **2008**, *17*, 535–542. [CrossRef]
153. Douglas, J.; Krieger, J.; Peter, J.; Rauscher, H.; Stradling, J. *Sleep Related Breathing Disorders*; Springer-Verlag Wien: New York, NY, USA, 1992.
154. Roscoe, J.; Kaufman, M.; Matteson-Rusby, S.; Palesh, O.; Ryan, J.; Kohli, S.; Perlis, M.; Morrow, G. Cancer-Related Fatigue and Sleep Disorders. *Oncologist* **2007**, *12*, 35–42. [CrossRef]
155. Givoni, B. Impact of planted areas on urban environmental quality: A review. *J. Atmos. Environ.* **1991**, *25B*, 289–299. [CrossRef]
156. Santamouris, M. Heat Island Research in Europe: The State of the Art. *Adv. Build. Energy Res.* **2007**, *1*, 123–150. [CrossRef]
157. Bowler, D.E.; Buyung-Ali, L.; Knight, T.M.; Pullin, A.S. Urban greening to cool towns and cities: A systematic review of the empirical evidence. *Landsc. Urban Plan.* **2010**, *97*, 147–155. [CrossRef]
158. Gago, E.; Roldan, J.; Pacheco-Torres, R.; Ordonez, J. The city and urban heat islands: A review of strategies to mitigate adverse effects. *Renew. Sustain. Energy Rev.* **2013**, *25*, 749–758. [CrossRef]
159. Qiu, G.-Y.; Li, H.-Y.; Zhang, Q.-T.; Chen, W.; Liang, X.-J.; Li, X.-Z. Effects of Evapotranspiration on Mitigation of Urban Temperature by Vegetation and Urban Agriculture. *J. Integr. Agric.* **2013**, *12*, 1307–1315. [CrossRef]
160. Berardi, U.; GhaffarianHoseini, A.; GhaffarianHoseini, A. State-of-the-art analysis of the environmental benefits of green roofs. *Appl. Energy* **2014**, *115*, 411–428. [CrossRef]
161. Santamouris, M. Cooling the cities—A review of reflective and green roof mitigation technologies to fight heat island and improve comfort in urban environments. *Sol. Energy* **2014**, *103*, 682–703. [CrossRef]
162. Shafique, M.; Kim, R.; Rafiq, M. Green roof benefits, opportunities and challenges—A review. *Renew. Sustain. Energy Rev.* **2018**, *90*, 757–773. [CrossRef]
163. Hunter, A.; Williams, N.; Rayner, J.; Aye, L.; Hes, D.; Livesley, S. Quantifying the thermal performance of green facades: A critical review. *Ecol. Eng.* **2014**, *63*, 102–113. [CrossRef]
164. Perez, G.; Coma, J.; Martorell, I.; Cabeza, L. Vertical Greenery Systems (VGS) for energy saving in buildings: A review. *Renew. Sustain. Energy Rev.* **2014**, *39*, 139–165. [CrossRef]
165. Medl, A.; Stangl, R.; Florineth, F. Vertical greening systems—A review on recent technologies and research advancement. *Build. Environ.* **2017**, *125*, 227–239. [CrossRef]
166. Gallagher, J.; Baldauf, R.; Fuller, C.; Kumar, P.; Gill, L.; McNabola, A. Passive methods for improving air quality in the built environment; A review of porous and solid barriers. *Atmos. Environ.* **2015**, *120*, 61–70. [CrossRef]
167. Janhall, S. Review on urban vegetation and particle air pollution—Deposition and dispersion. *J. Atmos. Environ.* **2005**, *105*, 130–137. [CrossRef]
168. Abhijith, K.; Kumar, P.; Gallagher, J.; McNabola, A.; Baldauf, R.; Pilla, F.; Broderick, B.; Sabatino, S.; Pulvirenti, B. Air pollution abatement performances of green infrastructure in open road and built-up street canyon environments—A review. *Atmos. Environ.* **2017**, *162*, 71–86. [CrossRef]
169. Soares, A.; Rego, F.; McPherson, E.; Simpson, J.; Peper, P.; Xiao, Q. Benefits and costs of street tree in Lisbon, Portugal. *Urban. For. Urban Green.* **2011**, *10*, 69–78. [CrossRef]
170. Mullaney, J.; Lucke, T.; Trueman, S. A review of benefits and challenges in growing street trees in paved urban environments. *Landsc. Urban Plan.* **2015**, *134*, 157–166. [CrossRef]
171. Salmond, J.; Tadaki, M.; Vardoulakis, S.; Arbuthnott, K.; Coutts, A.; Demuzere, M.; Dirks, K.; Heaviside, C.; Lim, S.; Macintyre, H.; et al. Health and climate related ecosystem services provided by street trees in the urban environment. *Environ. Health* **2016**, *15*, 96–171. [CrossRef]
172. Peel, M.; Finlayson, B.; McMahon, T. Updated world map of the Koppen-Geiger climate classification. *J. Hydrol. Earth Syst. Sci.* **2007**, *11*, 1633–1644. [CrossRef]

173. Taha, H.; Akbari, H.; Rosenfeld, A. *Vegetation Canopy Micro-Climate: A Field Project in Davis, California*; Lawrence Berkeley Laboratory, University of California: Berkley, CA, USA, 1988.

174. Narita, K.; Sugawara, H.; Honjo, T. Effects of roadside trees on the thermal environment within a street canyon. *Geogr. Rep. Tokyo Metrop. Univ.* **2008**, *43*, 41–48.

175. Tsiros, I. Assessment and energy implications of street air temperature cooling by shade trees in Athens (Greece) under extremely hot weather conditions. *J. Renew. Energy* **2010**, *35*, 1866–1869. [CrossRef]

176. Wong, N.; Jusuf, S. Study on the microclimatic condition along a green pedestrian canyon in Singapore. *Archit. Sci. Rev.* **2010**, *53*, 196–212. [CrossRef]

177. Berry, R.; Livesley, S.; Aye, L. Tree canopy shade impacts on solar irradiance received by building walls and their surface temperature. *Build. Environ.* **2013**, *69*, 91–100. [CrossRef]

178. Skelhorn, C.; Lindley, S.; Levemore, G. The impact of vegetation types on air and surface temperatures in a temperate city: A fine scale assessment in Manchester, UK. *Landsc. Urban Plan.* **2014**, *121*, 129–140. [CrossRef]

179. Jauregui, E. Influence of a large urban park on temperature and convective precipitation in a tropical city. *Energy Build.* **1990**, *15–16*, 457–463. [CrossRef]

180. Padmanabhamurty, B. Microclimates in Tropical Urban Complexes. *Energy Build.* **1990**, *15*, 83–92. [CrossRef]

181. Saito, I.; Ishihara, O.; Katayama, T. Study of the effect of green areas on the thermal environment in an urban area. *Energy Build.* **1991**, *15*, 2624–2631. [CrossRef]

182. Katayama, T.; Ishii, A.; Hayashi, T.; Tsutsumi, J. Field surveys on cooling effects of vegetation in an urban area. *J. Therm. Biol.* **1993**, *18*, 571–576. [CrossRef]

183. Spronken-Smith, R.; Oke, T. Scale modelling of nocturnal cooling in urban parks. *Bound. Layer Meteorol.* **1999**, *93*, 287–312. [CrossRef]

184. Shashua-Bar, L.; Hoffman, M. Vegetation as a climatic component in the design of an urban street; An empirical model for predicting the cooling effect of urban green areas with trees. *Energy Build.* **2000**, *31*, 221–235. [CrossRef]

185. Jonsson, P. Vegetation as an urban climate control in the subtropical city of Gaborone, Botswana. *Int. J. Climatol.* **2004**, *24*, 1307–1322. [CrossRef]

186. Potchter, O.; Cohen, P.; Bitan, A. Climatic behaviour of various urban parks during hot and humid summer in the Mediterranean city of Tel Aviv, Israel. *Int. J. Climatol.* **2006**, *26*, 1695–1711. [CrossRef]

187. Chang, C.; Li, M.; Chang, S. A preliminary study on the local cool-island intensity of Taipei city parks. *Landsc. Urban Plan.* **2007**, *80*, 386–395. [CrossRef]

188. Lin, B.; Lin, Y. Cooling effect of shade trees with different characteristics in a subtropical urban park. *HortScience* **2010**, *45*, 83–86. [CrossRef]

189. Skoulika, F.; Santamouris, M.; Kolokotsa, D.; Boemi, N. On the thermal characteristics and the mitigation potential of a medium size urban park in Athens, Greece. *Landsc. Urban Plan.* **2014**, *123*, 73–86. [CrossRef]

190. Tsilini, V.; Papantoniou, S.; Kolokotsa, D.; Maria, E. Urban gardens as a solution to energy poverty and urban heat island. *Sustain. Cities Soc.* **2014**, *14*, 323–333. [CrossRef]

191. Reis, C.; Lopes, A. Evaluating the cooling potential of urban green spaces to tackle urban climate change in Lisbon. *Sustainability* **2019**, *11*, 2480. [CrossRef]

192. Viñas, F.; Solanich, J.; Vilardaga, X.; Montilo, L. *El Árbol En Jardinería Y Paisajismo—Guía de Aplicación Para ESPAÑA Y Países de Clima Mediterráneo Y Templado*, 2nd ed.; Ediciones Omega: Barcelona, Spain, 1995.

193. McPherson, E. Planting design for solar control. In *Energy-Conserving Site Design*; McPherson, E., Ed.; American Soceity of Landscape Architects: Washington, DC, USA, 1984; pp. 141–164.

194. Brown, R.; Gillespie, T. Estimating radiation received by a person under different species of shade trees. *J. Aboric.* **1990**, *16*, 158–161.

195. Brown, R.; Cherkezoff, L. Of what comfort value, a tree? *J. Arboric.* **1989**, *15*, 158–162.

196. Takács, Á.; Kiss, M.; Gulyás, Á.; Tanács, E.; Kántor, N. Solar Permeability of Different Tree Species in Szeged, Hungary. *Geogr. Pannonica* **2016**, *20*, 32–41. [CrossRef]

197. Gulyas, A.; Unger, J.; Matzarakis, A. Assessment of the microclimatic and human comfort conditions in a complex urban environment: Modelling and measurements. *Build. Environ.* **2006**, *41*, 1713–1722. [CrossRef]

198. Spangenberg, J.; Shinzato, P.; Johansson, E.; Duarte, D. Simulation of the influence of vegetation on microclimate and thermal comfort in the city of São Paulo. *Revista da Sociedade Brasileira de Arborização Urbana* **2008**, *3*, 1–19. [CrossRef]

199. Lin, T.-P.; Matzarakis, A.; Hwang, R.-L. Shading effect on long-term outdoor thermal comfort. *Build. Environ.* **2010**, *45*, 213–221. [CrossRef]

200. Yang, F.; Lau, S.; Qian, F. Thermal comfort effects of urban design strategies in high-rise urban environments in a sub-tropical climate. *Archit. Sci. Rev.* **2011**, *54*, 285–304. [CrossRef]

201. Abreu-Harbich, L.V.d.; Labaki, L.; Matzarakis, A. Reduction of mean radiant temperature by cluster of trees in urban and architectural planning in tropical climate. In Proceedings of the PLEA2012—28th Conference: Opportunities, Limits & Needs Towards an Environmentally Responsible Architecture, Lima, Perú, 7–9 November 2012.

202. Abreu-Harbich, L.V.d.; Labaki, L.; Matzarakis, A. Effect of tree planting design and tree species on human thermal comfort in the tropics. *Landsc. Urban Plan.* **2015**, *138*, 99–109. [CrossRef]

203. Tan, Z.; Lau, K.K.-L.; Ng, E. Urban tree design approaches for mitigating daytime urban heat island effects in a high-density urban environment. *Energy Build.* **2016**, *114*, 265–274. [CrossRef]

204. Kong, L.; Lau, K.; Yuan, C.; Chen, Y.; Xu, Y.; Ren, C.; Ng, E. Regulation of outdoor thermal comfort by trees in Hong Kong. *Sustain. Cities Soc.* **2017**, *31*, 12–25. [CrossRef]

205. Streiling, S.; Matzarakis, A. Influence of single and small clusters of trees on the bioclimate of a city: A case study. *J. Arboric.* **2003**, *29*, 309–317.

206. Mayer, H.; Matzarakis, A. Impact of street trees on the thermal comfort of people in summer: A case study in Freiburg (Germany). *Merchavim* **2006**, *6*, 285–300.

207. Andrade, H.; Vieira, R. A climatic study of an urban green space: The Gulbenkian Park in Lisbon (Portugal). *Finisterra* **2007**, *XLII*, 27–46. [CrossRef]

208. Cohen, P.; Potchter, O. Daily and Seasonal Air Quality Characteristics of Urban Parks in the Mediterranean City of Tel Aviv. In Proceedings of the CLIMAQS Workshop 'Local Air Quality and its Interactions with Vegetation', Antwerp, Belgium, 21–22 January 2010.

209. Oliveira, S.; Andrade, H.; Vaz, T. The cooling effect of green spaces as a contribution to the mitigation of urban heat: A case study in Lisbon. *Build. Environ.* **2011**, *46*, 2186–2194. [CrossRef]

210. Cohen, P.; Potchter, O.; Matzarakis, A. Daily and seasonal climatic conditions of green urban open spaces in the Mediterranean climate and their impact on human comfort. *Build. Environ.* **2012**, *51*, 285–295. [CrossRef]

211. Perini, K.; Magliocco, A. Effects of vegetation, urban density, building height, and atmospheric conditions on local temperatures and thermal comfort. *Urban For. Urban Green.* **2014**, *13*, 495–506. [CrossRef]

212. Martins, T.; Adolphe, L.; Bonhomme, M.; Faraut, S.; Ginestet, S.; Michel, C.; Guyard, W. Impact of Urban Cool Island measures on outdoor climate and pedestrian comfort: Solutions for a new district of Toulouse, France. *Sustain. Cities Soc.* **2016**, *26*, 2–26. [CrossRef]

213. Abreu-Harbich, L.V.d.; Labaki, L.C.; Bueno-Bartholomei, C.L. How much does the shade provided by different trees collaborate to control the urban heat island in tropical climates?—A study in Campinas, Brazil. In Proceedings of the Conference: IC2UHI—Third International Conference on Countermeasures to Urban Heat Islands, Venice, Italy, 13–15 October 2014; pp. 838–849.

214. Ali-Toudert, F.; Mayer, H. Numerical study on the effects of aspect ratio and orientation of an urban street canyon on outdoor thermal comfort in hot and dry climate. *Build. Environ.* **2006**, *41*, 94–108. [CrossRef]

215. Herrmann, J.; Matzarakis, A. Mean radiant temperature in idealised urban canyons—Examples from Freiburg, Germany. *Int. J. Biometeorol.* **2012**, *56*, 199–203. [CrossRef]

216. Qaid, A.; Ossen, D. Effect of asymmetrical street aspect ratios on microclimates in hot, humid regions. *Int. J. Biometeorol.* **2015**, 1–21. [CrossRef]

217. Algeciras, J.A.R.; Consuegra, L.G.; Matzarakis, A. Spatial-temporal study on the effect of urban street configurations on human thermal comfort in the world heritage city of Camagüey-Cuba. *Build. Environ.* **2016**, *101*, 85–101. [CrossRef]

218. Algeciras, J.A.R.; Tablada, A.; Matzarakis, A. Effect of asymmetrical street canyons on pedestrian thermal comfort in warm-humid climate of Cuba. *Theor. Appl. Climatol.* **2017**, 1–17.

219. Gaitani, N.; Spanou, A.; Saliari, M.; Vassilakopoulou, K.; Papadopoulou, K.; Pavlou, K.; Santamouris, M.; Lagoudaki, A. Improving the microclimate in urban areas: A case study in the centre of Athens. *Build. Serv. Eng. Res. Technol.* **2011**, *32*, 53–71. [CrossRef]

220. Santamouris, M.; Gaitani, N.; Spanou, A.; Salirai, M.; Giannopoulou, K.; Vasilakopoulou, K.; Kardomateas, T. Using cool paving materials to improve microclimate of urban areas—Design realization and results of the flisvos project. *J. Build. Environ.* **2012**, *53*, 128–136. [CrossRef]

221. Fintikakis, N.; Gaitani, N.; Santamouris, M.; Assimakopoulos, M.; Assimakopoulos, D.; Fintikaki, M.; Albanis, G.; Papadimitriou, K.; Chryssochoides, E.; Katopodi, K.; et al. Bioclimatic design of open public spaces in the historic centre of Tirana, Albania. *Sustain. Cities Soc.* **2011**, *1*, 54–62. [CrossRef]

222. Erell, E.; Pearlmutter, D.; Boneh, D.; Kutiel, P. Effect of high-albedo materials on pedestrian heat stress in urban street canyons. *Urban Clim.* **2014**, *10*, 367–386. [CrossRef]

223. Boriboonsomsin, K.; Reza, F. Mix Design and Benefit Evaluation of High Solar Reflectance Concrete for Pavements. *Transp. Res. Rec. J.* **2011**, *361*, 11–20. [CrossRef]

224. NCAT. *Strategies for Design and Construction of High-Reflectance Asphalt Pavements*; National Centre for Asphalt Technology: Auburn, AL, USA, 2009; p. 28.

225. Gui, J.; Phelan, P.; Kaloush, K.; Golden, J. Impact of Pavement Thermophysical Properties on Surface Temperatures. *J. Mater. Civ. Eng.* **2007**, *19*, 683–688. [CrossRef]

226. Synnefa, A.; Karlessi, T.; Gaitani, N.; Santamouris, M.; Assimakopoulos, D.; Papakatsikas, C. Experimental testing of cool colored thin layer asphalt and estimation of its potential to improve the urban microclimate. *Build. Environ.* **2011**, *46*, 38–48. [CrossRef]

227. Nunes, J.; Zolio, I.; Jacinto, N.; Nunes, A.; Campos, T.; Pacheco, M.; Fonseca, D. *Misting-Cooling Systems for Microclimatic Control in Public Space*; PROAP Landscape Architects: Lisbon, Portugal, 2013; pp. 1–16.

228. Nouri, A.S. A Framework of Thermal Sensitive Urban Design Benchmarks: Potentiating the Longevity of Auckland's Public Realm. In Proceedings of the Building A Better New Zealand, Auckland, New Zealand, 12 March 2015.

229. TVK. *Place de la Republique*; Trevelo & Viger-Kohler Architectes Urbanistes: Paris, France, 2013; p. 21.

230. Knuijt, M. One Step Beyond. *Open Space* **2013**, *85*, 60–67.

231. Ishii, T.; Tsujimoto, M.; Yoon, G.; Okumiya, M. Cooling system with water mist sprayers for mitigation of heat-island. In Proceedings of the Seventh International Conference on Urban Climate, Yokohama, Japan, 1 July 2009.

232. Ishii, T.; Tsujimoto, M.; Yamanishi, A. The experiment at the platform of dry-mist atomization. In Proceedings of the Summaries of Technical Papers of the Annual Meeting of the Architectural Institute of Japan, Yamagata, Japan, 6 September 2018.

233. Yamada, H.; Yoon, G.; Okumiya, M.; Okuyama, H. Study of cooling system with water mist sprayers: Fundemental examination of particle size distribution and cooling effects. *J. Build. Simul.* **2008**, *1*, 214–222. [CrossRef]

234. Yoon, G.; Yamada, H.; Okumiya, M. Study on a cooling system using water mist sprayers; System control considering outdoor environment. In Proceedings of the Korea-Japan Joint Symposium on Human-Environment Systems, Cheju, Korea, 30 November 2008.

235. Farnham, C.; Nakao, M.; Nishioka, M.; Nabeshima, M.; Mizuno, T. Study of mist-cooling for semi-enclosed spaces in Osaka, Japan. *Urban Environ. Pollut.* **2011**, *4*, 228–238. [CrossRef]

236. Alvarez, S.; Rodriguez, E.; Martin, R. Direct air cooling from water drop evaporation. In Proceedings of the PLEA 91—Passive and Low Energy Architecture, Seville, Spain, 24–27 September 1991.

Perspective

Urban Transformation: From Single-Point Solutions to Systems Innovation

Eleanor Tonks [1],* and Sean Lockie [2]

[1]   EIT Climate-KIC, 1018 JA Amsterdam, The Netherlands
[2]   EIT Climate-KIC, Fellow RICS, 40129 Bologna, Italy; sean.lockie@climate-kic.org
*   Correspondence: ellie.tonks@climate-kic.org

Received: 20 December 2019; Accepted: 15 January 2020; Published: 20 January 2020

Abstract: Adapting our cities to the new climate regime is critical to ensure that human development is not jeopardized and that the world's citizens can thrive where they live. Faced as we are with the imperative to act, we now need to accept that the challenges we face are not technical in nature—they are systemic. Traditionally, investments in low-carbon city solutions have suffered from being small and disaggregated, with a focus on single-point solutions. To truly enable city transformation at scale, we need to completely rewire our approach to urban innovation and implementation. To face our new reality, EIT Climate-KIC works on catalysing systems change through innovation in areas of human activity that have a critical impact on greenhouse gas emissions—cities, land use, materials, and finance—and to create climate-resilient communities. In this paper, EIT Climate-KIC reflects on its key learnings, as an innovation community, on how to apply innovation in service of urban transformation through the application of nature-based solutions.

Keywords: systems change; innovation; nature-based solutions; cities

1. Introduction

Adapting our cities to the new climate regime is critical to ensure that human development is not jeopardized and that the world's citizens can thrive where they live. The number of weather- and climate-related loss events has been increasing rapidly since 1980, causing a similar increase in the economic damages from these events. Today, the annual number of events approximate 700 as compared to 200–300 in the 1980s [1]. There is no doubt that climate change is exacerbating the vulnerability of cities. Despite increasing awareness of short-term climate hazards, the medium- and long-term hazards of climate change are currently being under-reported and actioned on by cities [2]. By 2050, eight times as many city dwellers will be exposed to high temperatures (amounting to 1.6 billion citizens), and over 800 million people will be at risk from the impacts of rising seas and storm surges [3]. Cities cannot afford inaction

Global targets for emissions reductions set forth in the Paris Climate Agreement have put low-carbon climate-resilient cities high on the political, financial, and social agenda. They are an essential response to climate change and fundamental to achieving the goals of the Paris Agreement. However, urban transformations into low-carbon societies are not only about saving our climate. Multiple benefits arise from this change, including improved health and well-being, cleaner air, employment creation, an opportunity for community renewal, and positively refreshed state–society relationships.

However, so far, the vast majority of action toward sustainable change in cities has arguably been in words—through the creation of policies, strategies, and targets for the slightly too-distant future, established through too many meetings, workshops, conferences, and talking shops, with Climate Action Plans often missing the detail or the financial backing to attract the investment needed to result in change on the ground. Many cities have made ambitious 2050 targets, but the reality is that these will come too late. With the acknowledgement that the EU should "concentrate on immediate and urgent" [4] climate policies for 2030, we argue that 2050 is moving too slowly.

Faced as we are with the imperative to act, we now need to accept that the challenges we face are not technical in nature—they are systemic. For too long, investments in low-carbon city solutions have suffered from being small and disaggregated, with a focus on single-point solutions. Understandably, most cities around the world are interested in similar, well-tested, and well-proven projects in climate adaptation and mitigation. However, the cities and projects themselves are regularly too small to justify the involvement of financiers, and often lack the technical knowledge and network to develop and implement their projects at scale.

To truly enable city transformation at scale, we need to completely rewire our approach to urban innovation and implementation, working toward 2030 rather than 2050 and implementing system-orientated approaches in preference to point solutions. According to the IPCC's 1.5 Degree Special Report [5], the world requires rapid and unprecedented transformations not just in energy supply and consumption, but in land-use, urban, infrastructure, and industrial systems, in order to avoid the most perilous effects of global warming. To face this new reality, EIT Climate-KIC works on catalysing systems change through innovation in areas of human activity that have a critical impact on greenhouse gas emissions—cities, land use, materials, and finance—and to create climate-resilient communities.

In the context of systems innovation for urban adaptation, we view nature-based solutions (NBS) as an opportunity to progress toward a low-carbon and more resilient city: one that puts citizens' health and well-being at the centre of urban development and design. NBS are actions which use or are supported by nature and its restorative system processes to address societal challenges, while enhancing citizen well-being and socially inclusive green growth [6]. Furthermore, NBS can play a crucial role in increasing the resilience of urban environments in the new climate regime. Despite the opportunities presented by NBS for citizen health, urban liveability, and climate adaptation and mitigation efforts, the deployment of solutions within our cities remains limited. Integrating long-term considerations into planning requires thinking beyond infrastructure solutions and focusing on systemic changes.

In this paper, we reflect on the key learnings from EIT Climate-KIC, as an innovation community, on how to apply innovation in service of urban transformation. First, we present the origins and initial purpose of EIT Climate-KIC when it was established nine years ago. Then, we discuss four key learnings on the need for a systems approach. Finally, we present three NBS case studies in which we see elements of these learnings being implemented in a real-world context.

2. An Innovation Community, Nine Years On

EIT Climate-KIC is a knowledge and innovation community established and funded by the European Institute of Innovation and Technology (EIT) in 2010. Our purpose is to tackle climate change through innovation. We draw our purpose from a context in which climate change is advancing fast and its damaging effects are beginning to take hold. We are Europe's largest public–private partnership with this purpose—a growing pan-European community of diverse organisations united by a commitment to direct the power of creativity and human ingenuity at the climate crisis. We bring together large and small companies, scientific institutions, and universities, city authorities, and other public bodies, start-ups, and students. With nearly 400 formal organisational partners from across 25 countries, we work on innovation to mitigate climate change and to adapt to its unavoidable impacts.

To understand our systems innovation approach, it is first vital to reflect on our nine years of experience in tackling climate change through innovation.

In 2009, the context in which we worked was that research leadership in Europe was not being translated into business growth and job creation, and that such leadership could be orientated to focus on tackling some of Europe's most pressing challenges (of which climate change was one). This research leadership was then targeted toward three key areas:

I.   Tackling climate change through innovation;

II.   Developing impactful start-ups and new generations of environmental entrepreneurs;

III.   Working with universities to educate future decision makers.

We have been reasonably successful in this, as we have achieved the following: supported over 1400 innovative start-ups, attracted plus €900 million to those start-ups, leveraged €3.4 billion climate funding, created over 2000 jobs, launched 367 new products and services, and empowered 17,000 participants through our education activities. Furthermore, the innovations we have supported and nurtured are now starting to support climate action. However, a large question remains, did we tackle climate change at the speed and scale that we need? We acknowledge that the answer is probably not. Europe is not on track to its 2050 targets.

3. A Systems Innovation Approach

We understand that most systems we need to transform behave in complex adaptive ways. This means that there are no *right solutions* and that deterministic interventions are bound to fail [7]. Instead, we believe that the best way to shift these systems is through exploration and experimentation, testing ourselves forward through real-world experiences [8]. Now, in 2020, the context in which we work is set out in the IPCC Summary for Policy Makers: "Limiting the risks from global warming of 1.5 °C in the context of sustainable development and poverty eradication implies **system transitions** that can be enabled by an increase of adaptation and mitigation investments, policy instruments, the acceleration of technological innovation and behaviour changes" [5].

Our new strategy (Transformation, in time [9]) is underpinned by four core lessons from our past experiences, which can also be contextualised further through the lens of urban-resilience.

3.1. We Cannot Treat Innovation as Techno-Centric

Over the past nine years, we have realised that a supply-driven innovation approach leads to a techno-centric pipeline characterised by incremental improvements of single-point solutions—and rarely to systemic solutions. EIT Climate-KIC is not alone in treating climate change as a complicated problem [7]. Ever since climate change entered the political stage [10] in the 1970s efforts to stem global warming have mostly focused on developing technical solutions through research and engineering. Furthermore, adaptation has often been framed as a purely technical issue that can be addressed through climate-proofing interventions rather than something integral to how city systems function. This means that we continue to see escalating risk and exposure within urban developments, for example, new developments in high risk locations, insurance priced in ways that does not account for positive adaptation actions, or mortgages that are not systematically pricing in future climate risk. This techno-centric approach has produced many important building blocks of a sustainable future (such as renewable-energy technologies or advanced batteries); however, what we must learn now is how to weave these technological advances into our societal systems, along with other cultural, institutional, social, and economic innovations. At EIT Climate-KIC, we define systems innovation as integrated and coordinated interventions in economic, political, technological, and social systems, and along whole value chains.

3.2. We Cannot Work on a Project-by-Project Level

Instead we must orchestrate the development and deployment of interventions in portfolios. Portfolios are collections of deliberately chosen innovation experiments. These experiments—representing the supply side of innovation—come in the form of diverse and coordinated innovation projects, education programs, start-ups, ecosystem building activities, citizen engagement strategies, and communication initiatives. They focus on different parts of the same problem, on different opportunities to leverage change. On considering a city's adaptive capacity, we need to shift the locus of interest from individual interventions to the aggregate level, to consider, for example, density and risk exposure, infrastructure vulnerability and resilience, governance, and institutional capacity. One of the distinguishing features of portfolio-based innovation is that, rather than focusing on individual experiments, we seek to understand the aggregate impact of the portfolio.

3.3. We Need to Demonstrate What Is Possible

To both inspire and legitimise climate action, we must work with the demand side: governments, corporations, and other challenge owners that share our ambition for transformative action. Ambitious demand-side actors have a strong appetite for change and appreciate the full complexity of the problem, trade-offs or synergies, and the necessary scale for intervention. They are also well-positioned to identify a broad and diverse set of points in the systems they aim to transform, whilst holding the networks to pull through the connected, system-wide solutions we need. EIT Climate-KIC is working at the whole city, region, and supply-chain scale through our Deep Demonstrations programme on systems transformation [11].

3.4. We Need to Broaden the Innovation Tent

No single individual, organisation, or government can address the climate crisis alone. Furthermore, the climate risks urban populations face are not only environmental; they raise equity, social justice, and sustainable development issues. To ensure the transition is socially just, diverse communities need to be engaged in decisions made about the future direction of resilience in their towns and cities. For example, involving trade unions, using tools of citizen monitoring and science, and engaging social movements, youth or community organisations, on specific issues, can embed adaptation activities in a socially just resilient transition and contribute to building a politics for transformation. EIT Climate-KIC orchestrates close to 400 innovation partners—from the private, public, and academic sectors—operating as a diverse community that understands the specific needs of different places and contexts. As a community, they bring their own expertise and experience to the systems in which we intervene.

4. Discussion

In the case of the built environment, we see nothing but opportunity, but the scale of change is not happening fast enough, nor is it being framed correctly. With political will and the right policies and incentives in place, NBS can be integrated in new developments from the outset (seen through the Sponge City initiatives in China). Retrospectively including them in existing, densely built-up city districts on public and private land and connecting them on the level of the whole city, however, requires working at scale in a long-term mindset (seen in the City of Copenhagen Cloudburst Plan, launched in 2012 and Philadelphia's Green City, Clean Waters programme, launched in 2011). Below, we explore three cities where nature is helping to catalyse change at the scale needed.

The Wild West End Project [12] in London (UK), a collaboration between public authorities, property owners, NGOs, and the private sector, seeks to create a network of green corridors that can sustain biodiversity and improve ecological connectivity. Taking an aggregate view, this diverse consortium is implementing a portfolio of solutions, including green and biodiversity roofs, living walls, sustainable drainage systems, pocket parks, street trees, terrace planters, and pop-up green spaces, to not only address biodiversity, climate, and micro-climate challenges, but also societal well-being. Between 2016 and 2018, the project saw a 17.5% increase in documented multi-functional green infrastructure measures (54 interventions) across the West End. The project builds on a previous Ecology Masterplan, adopted by The Crown Estate, developed to connect two major parks (Regent's Park and St James's Park) in London's West End, by adopting a long-term, estate-wide approach of installing NBS.

Building on the success of the "My tree—My city" public donation campaign [13] (launched as part of the European Green Capital program in 2011), Hamburg has developed a Green Power Strategic Plan to implement a citywide network of green spaces by 2030 [14]. The plan will connect the city's outer ring with its dynamic centre through a series of green axes (walking- and cycling-friendly regenerated or rewilded habitats along parks and rivers). The green network will enhance urban resilience (by providing flood mitigation for vulnerable spaces around Hamburg's vital ports, and by reducing urban heat island effects), improve air quality, and provide numerous health benefits for citizens.

A third project working at scale toward transformation is the Green City, Clean Waters programme [15] (launched 2011) led by the Philadelphia Water Department, on integrating Philadelphia's water systems to safeguard the ecological and economic future of the city. By implementing a range of measures, such as building green storm water infrastructure on public and private land, creating recreational spaces, and restoring biodiversity along water bodies, the programme aims to enhance the region's waterways by managing stormwater runoff and reducing dependence on additional underground grey infrastructure. The city is investing $2.4 billion over 25 years to manage more than one-third of the impervious cover within areas of the city served by combined sewers, a decision supported by a triple bottom-line analysis, comparing the green infrastructure approach with traditional alternatives. Through the Green City, Clean Waters programme, the aim is to generate new employment opportunities; increase $390 million in property value of homes over 45 years; avoid 140 fatalities due to heat reduction over the next 45 years; avoid 250 missed days of work or school per year; and eliminate 1.5 billion lbs of carbon dioxide emissions.

Each of these external case studies can be used to frame, in the context of NBS, the learnings from EIT Climate-KIC in real-world settings. All three view the need for increased application of NBS as not only a climate adaptation measure (a point solution), but rather adopt systems approaches when considering the magnitude of benefits from their interventions (as seen in the indicators of the Green City, Clean Waters programme). Each programme has the ambition to work at scale with a number of different intervention types (as seen by the Wild West End portfolio of solutions), with the support of the demand side (ranging from private land owners to local councils to municipalities to water departments), providing legitimacy and long-term direction for the programmes. Both the Wild West End and "My tree—My city" purposefully seek to engage with a range of stakeholders (from urban planners to local citizens to conservations to children and students). To differing degrees, these cases present practical examples of neighbourhood and city-scale transformation ambitions in the context of urban adaptation. They are not, however, presented as exemplar urban transformations but are instead given as examples through which the learnings from EIT Climate-KIC can be contextualised within an urban context.

5. Conclusions

Decision makers working toward urban resilience should consider these lessons learned from EIT Climate-KIC and the external case studies when seeking to address the risks they are exposed to. First, do not approach adaptation as a purely technical issue that can be solved through techno-centric innovation. Second, rather than focusing on the project-by-project or hazard-by-hazard scale, hold space at the aggregate to look across interventions or developments (taking a portfolio approach at scale). Third, to inspire and legitimise climate action work with the demand side and problem-owners that want to take on the mandate for change. Finally, to ensure the transition is socially just diverse communities need to be meaningfully engaged in decisions made about the future direction of resilience in their towns and cities.

Author Contributions: Conceptualization, S.L.and E.T.; methodology, S.L.and E.T.; investigation, E.T.; resources, E.T.; writing—original draft preparation, S.L.and E.T.; writing—review and editing, E.T..; project administration, E.T. All authors have read and agreed to the published version of the manuscript.

Funding: This research received no external funding.

Conflicts of Interest: The authors declare no conflict of interest.

References

1. NatCatSERVICE Analysis Tool, Download Center for Statistics on Natural Catastrophes. Available online: http://natcatservice.munichre.com/ (accessed on 16 August 2019).
2. Carbon Disclosure Project. Cities at Risk: Dealing with the Pressures of Climate Change. Available online: https://www.cdp.net/en/research/global-reports/cities-at-risk (accessed on 7 December 2019).
3. C40, 2018. The Future We Don't Want. Available online: https://c-production-images.s.amazonaws.com/other_uploads/images/_Future_We_Don\T\textquoterightt_Want_Report_1.4_hi-res_1.original.pdf (accessed on 7 December 2019).

4. EURACTIV, Junker Insists 2030 Climate Target Is EU's Top Priority, Not 2050 Plan. Available online: https://www.euractiv.com/section/climate-strategy-/news/juncker-insists--climate-target-is-eus-top-priority-not--plan/ (accessed on 15 August 2019).

5. Masson-Delmotte, V.; Zhai, P.; Pörtner, H.-O.; Roberts, D.; Skea, J.; Shukla, P.R.; Pirani, A.; Moufouma-Okia, W.; Péan, C.; Pidcock, R.; et al. (Eds.) 2018: Summary for Policymakers. Global Warming of 1.5 °C. An IPCC Special Report on the impacts of global warming of 1.5 °C above pre-industrial levels and related global greenhouse gas emission pathways, in the context of strengthening the global response to the threat of climate change. In *Sustainable Development, and Efforts to Eradicate Poverty*; Intergovernmental Panel on Climate Change: Geneva, Switzerland, 2018; In Press.

6. European Commission, Nature-Based Solutions. Available online: https://ec.europa.eu/research/environment/index.cfm?pg=nbs (accessed on 8 December 2019).

7. Innovating in Complexity: From Single-Point Solutions to Directional Systems Innovation. Available online: https://www.climate-kic.org/opinion/innovating-in-complexity/ (accessed on 16 August 2019).

8. AXILO. Available online: https://axilo.space (accessed on 4 August 2019).

9. EIT Climate-KIC. Transformation, in Time. 2019. Available online: https://www.climate-kic.org/wp-content/uploads///Transformation-in-time.pdf (accessed on 4 August 2019).

10. The New York Times, Climate Change Losing Earth. Available online: https://www.nytimes.com/interactive////magazine/climate-change-losing-earth.html (accessed on 16 August 2019).

11. EIT Climate-KIC, Deep Demonstrations of a Net-Zero Emissions, Resilient Future. Available online: https://www.climate-kic.org/wp-content/uploads///EIT-Climate-KIC-Deep-Demonstrations.pdf (accessed on 6 September 2019).

12. Wild West End. Available online: http://www.wildwestend.london/ (accessed on 15 August 2019).

13. Hamburg European Green Capital 2011, "My Tree—My City" Campaign. Available online: https://www.hamburg.de/projects//mytree-mycity/ (accessed on 6 September 2019).

14. Green Network Hamburg, on Green Paths through the City. Available online: https://www.hamburg.de/gruenes-netz//auf-gruenen-wegen-artikel/ (accessed on 6 September 2019).

15. Green City, Clean Waters. Available online: https://www.phila.gov/water/sustainability/greencitycleanwaters/Pages/default.aspx (accessed on 16 August 2019).

MDPI

St. Alban-Anlage 66

4052 Basel

Switzerland

Tel. +41 61 683 77 34

Fax +41 61 302 89 18

www.mdpi.com

Climate Editorial Office

E-mail: climate@mdpi.com

www.mdpi.com/journal/climate

Nucleic Acids Conjugates for Biotechnological Applications

Nucleic Acids Conjugates for Biotechnological Applications

Editors

Tamaki Endoh
Eriks Rozners
Takashi Ohtsuki

MDPI • Basel • Beijing • Wuhan • Barcelona • Belgrade • Manchester • Tokyo • Cluj • Tianjin

Editors

Tamaki Endoh
Frontier Institute for
Biomolecular Engineering
Research (FIBER)
Konan University
Kobe
Japan

Eriks Rozners
Department of Chemistry
Binghamton University
Binghamton
United States

Takashi Ohtsuki
Department of Interdisciplinary
Science and Engineering in
Health Systems
Okayama University
Okayama
Japan

Editorial Office
MDPI
St. Alban-Anlage 66
4052 Basel, Switzerland

This is a reprint of articles from the Special Issue published online in the open access journal *Applied Sciences* (ISSN 2076-3417) (available at: www.mdpi.com/journal/applsci/special_issues/ Heterogeneous_Conjugates_Nucleic_Acids).

For citation purposes, cite each article independently as indicated on the article page online and as indicated below:

LastName, A.A.; LastName, B.B.; LastName, C.C. Article Title. *Journal Name* **Year**, *Volume Number*, Page Range.

ISBN 978-3-0365-1514-4 (Hbk)
ISBN 978-3-0365-1513-7 (PDF)

Contents

Preface to "Nucleic Acids Conjugates for Biotechnological Applications"

Biological activities are maintained by the cooperative actions of biomolecules. The details of the functionalization mechanisms of these cooperative actions have been elucidated. With the development of biological and chemical technologies, artificial modifications of such systems consisting of various biomolecules has been attempted.

Nucleic acids not only store genetic information in their primary sequence but also exhibit biological functions through the formation of their unique structures. Based on such functions and specific and designable complementary recognition through base pairing, nucleic acids are considered attractive molecules as functional units for biotechnological applications.

This Special Issue focuses on the development of nucleic acid-based conjugates (including chemical conjugates with covalent bonds and complexed conjugate thorough intermolecular interaction) with different types of biomolecules such as proteins, lipids, and organic molecules for cellular engineering and biotechnological applications.

Tamaki Endoh, Eriks Rozners, Takashi Ohtsuki

Editors

Editorial

Nucleic Acids Chemistry and Engineering: Special Issue on Nucleic Acid Conjugates for Biotechnological Applications

Tamaki Endoh [1,*], Eriks Rozners [2] and Takashi Ohtsuki [3]

1 Frontier Institute for Biomolecular Engineering Research (FIBER), Konan University, Kobe 650-0047, Japan
2 Department of Chemistry, Binghamton University, New York, NY 13902, USA; erozners@binghamton.edu
3 Department of Interdisciplinary Science and Engineering in Health Systems, Okayama University, Okayama 700-8530, Japan; ohtsuk@okayama-u.ac.jp
* Correspondence: t-endoh@konan-u.ac.jp; Tel.: +81-78-303-1314

Citation: Endoh, T.; Rozners, E.; Ohtsuki, T. Nucleic Acids Chemistry and Engineering: Special Issue on Nucleic Acid Conjugates for Biotechnological Applications. *Appl. Sci.* **2021**, *11*, 3594. https://doi.org/10.3390/app11083594

Received: 13 April 2021
Accepted: 14 April 2021
Published: 16 April 2021

Publisher's Note: MDPI stays neutral with regard to jurisdictional claims in published maps and institutional affiliations.

1. Introduction

Nucleic acids not only store genetic information in their primary sequence but also exhibit biological functions through the formation of their unique structures. Based on such functions and specific and designable complementary recognition through base pairing, nucleic acids are considered attractive molecules as functional units for biotechnological applications. Life in biological systems is maintained by the cooperative actions of various biomolecules, such as nucleic acids, proteins, lipids, and small organic molecules. With the development of chemical and biological technologies related to nucleic acids, the details of the mechanisms of such cooperative actions between nucleic acids and other biomolecules have been elucidated. This Special Issue focuses on the biotechnological applications of nucleic acid conjugates that transcend the natural cooperative systems.

2. Addition of Superior Functions to Natural Nucleic Acid through Covalent Conjugations

Nucleic acids show complementary recognition by base pairing, which cannot be achieved by other biomolecules. A specific region on messenger RNA (mRNA) can be targeted by rationally designing oligonucleotides, such as antisense and small interfering RNA (siRNA), that enable the suppression of the protein expression from the target mRNA. An artificial nucleic acid mimic typified by peptide nucleic acid (PNA) further enables targeting a specific region on genomic double-stranded DNA through the formation of a double-duplex invasion complex. When these oligonucleotides are applied for gene regulation, the delivery of the oligonucleotides into cells and localization in appropriate cellular compartments are issues that need to be addressed. The conjugation of a functional unit for providing cell permeability to oligonucleotides is a promising strategy for the intracellular delivery. D. Stetsenko et al. synthesized a novel type of lipid-oligonucleotide conjugates (LONs) based on their original chemistry, performing the conjugation on solid support. They demonstrated that siRNAs built of LONs downregulated gene expression through self-internalization into cells [1]. Since the constructed LONs spontaneously formed particles, the additional functionality of resistance to degradation for in vivo application is expected. Localization in a specific cellular compartment can be achieved by the conjugation of a short peptide tag, such as nuclear localization signal (NLS). Y. Aiba and O. Shoji et al. have constructed an NLS-PNA conjugate, which was superior to natural PNA in its invasion efficiency into the target DNA duplex [2]. The demonstration motivates the applicability for in cell studies because the NLS-PNA could form the double-duplex invasion complex even in the presence of physiological salt concentrations, whereas the invasion efficiency of PNA decreased as the salt concentration increased.

In addition to the complementary sequence, nucleic acids form complex tertiary structures that recognize other biomolecules such as proteins and small molecule metabolites. Aptamers, nucleic acids with potential to specifically recognize another molecule, have been applied as functional units in biosensors. Biosensors require a unit for recognizing the target molecule and another unit for outputting signal, which can be easily recorded and quantified such as fluorescence and luminescence. Since aptamers only recognize the target molecule, the construction of aptamer-based biosensors has been demonstrated by conjugations with other functional units. For example, the conjugation of an aptamer with a protein is one of the useful strategies because some proteins exhibit unique photochemical output signals. M. Mie et al. have demonstrated the construction of nucleic acid–protein hybrid molecules using a protein involved in viral replication initiation that acts as a conjugation unit. They constructed an aptamer–protein hybrid that detects the target molecule by a bioluminescence resonance energy transfer (BRET) through designing a fusion protein of the conjugation unit between NanoLuc luciferase and Venus, which output luminescence and fluorescence as a donor and acceptor of BRET, respectively [3]. The construct of the aptamer–protein hybrid would be applicable for sensing various aptamer targets because change in the BRET signal occurs upon steric rearrangement of the protein units upon the interaction of the aptamer and its target. When the target of the aptamer is a small molecule, it may be difficult to cause a large steric rearrangement of the protein units. In such cases, aptamer-based biosensors can also be constructed by optimally locating a small fluorophore, in which fluorescence is sensitive to the microenvironment, close to the aptamer unit. T. Morii et al. have constructed fluorometric sensors based on a ribonucleopeptide (RNP), which acts as a scaffold to locate RNA aptamer and fluorophore units in proximity. Although the prototype RNP-based sensors were just formed by a noncovalent interaction between RNA and peptide, the present study covalently linked the complex to improve chemical and thermal stability through optimization of the conjugation conditions [4]. The covalently linked RNP sensor would be applicable for the simultaneous detection of multi targets by preventing the interexchange of the functional units between RNP sensors.

The formation of covalent linkage between nucleic acid and protein can also be used for entrapping proteins that interact with a nucleic acid of interest. Although crosslinking chemical reagents for biomolecules have been developed, in order to capture a protein that binds to a specific site of the target nucleic acid, it is necessary to conjugate the reactive unit at the specific site of nucleic acid. F. Nagatsugi et al. have constructed oligonucleotides bearing an unnatural nucleotide consisting of 4-amino-6-oxo-2-vinyltriazine as the reactive unit and acyclic linker in place of deoxyribose. The vinyl group in the reactive unit enables covalent linkage with amino acid residues with nucleophilic species at a proximity that captures the proteins interacting with the specific region on the duplex [5]. The formation of covalent linkage is useful for the basic study of nucleic acid binding proteins. In addition, it is also expected to be possible to use the reactive oligonucleotide as an irreversible inhibitor of the proteins like a suicide inhibitor of catalytic proteins.

3. Molecular Assemblies Including Nucleic Acids for Functional Materials and Systems

The original and fundamental function of nucleic acids is to store the genetic information required to synthesize proteins. Since mRNA encodes the amino acid sequence of the protein to be expressed, it is attracting attention as biomedical material exemplified as a vaccine against SARS-CoV-2. When mRNA is used as a nucleic acid-based drug, it should be delivered into cells to express the protein of interest. Due to the considerably large size of mRNA compared to oligonucleotide therapeutics, chemical conjugation through the technology introduced above becomes difficult. In this case, the encapsulation of the mRNA into an assembled complex, which shows cellular internalization function, would provide an efficient strategy to deliver the mRNAs. K. Matsuura et al. have constructed an artificial capsule by using a short peptide unit derived from tomato bushy stunt virus. They have chemically conjugated the peptide with a short poly-T oligonucleotide and succeeded to encapsulate mRNA, which hybridizes to the poly-T strand through its poly-A tail. The capsule was further modified to internalize into cells and express proteins by the modification of the outside of the capsule with a cell penetrating peptide [6]. The modification functions like a viral envelope and is expected to be applicable for targeting specific cells. In contrast to the use of nucleic acid as a drug, a strategy for internalizing a small drug molecule into an assembled complex via nucleic acid is also promising for the therapeutic application. M. Kuwahara et al. have constructed a bifunctional aptamer (bApt) consisting of two aptamer units for thrombin and a drug molecule, respectively. During the fibrin gel formation by the catalytic activity of thrombin, the thrombin aptamer unit enables the internalization of bApt into fibrin gel. Since another aptamer unit, which is interacting with the drug molecule, can be easily exchanged depending on the therapeutic target, the assembled fibrin gel can be a biocompatible material that stores and releases various drug molecules [7].

Assembly between nucleic acids and proteins is one of the major mechanisms that modulate gene expression in cells. For example, the assembly of Puf family proteins on the 3′ untranslated region of mRNA affects protein expression. Since the target RNA sequence of the Puf family proteins can be altered by rationally designing amino acids in their RNA binding domain, a strategy that assembles multiple Puf proteins on target mRNA would enable the efficient regulation of protein expression. M. Imanishi et al. have fused the engineered RNA binding domain of the Puf family protein with tristetraprolin (TTP), which is involved in mRNA decay and degradation. They demonstrated that a tandem assembly of four TTP-PUF fusion proteins effectively repressed protein expression from target mRNA [8]. Since the strong suppression was observed only when the four types of TTP-PUF fusion proteins are present as it is like a logic gate, the technology is expected to be applied as a gene regulation tool in the field of synthetic biology.

Molecular assembly is also applied for in vitro applications. The hybridization and self-organization of nucleic acids are designable based on the sequence complementarity. The reconstruction of assembly-based nucleic acid structures is applicable for controlling the subsequent output of functional units. As examples, Y. Ikawa et al. have demonstrated the regulation of catalytic activity of RNA through self-organization of split RNAs. They constructed an RNA triangle with six catalytic ribozyme units, in which the triangle complex was further supported by protein assembly on the corners. It is an attractive RNA-based nanoarchitecture and would be a unit for a system that exhibits more complex biological functions as the RNA triangle has succeeded to catalyze the trans-splicing of RNA with an arbitrary sequence [9]. S. Nishizawa et al. have synthesized a chemical probe for fluorometric detection of an orphan cytosine, which has no base paring partner due to an abasic nucleotide on the opposite strand. They also designed an oligonucleotide, which hybridizes to the target strand and locates abasic nucleotide opposite a specific cytosine, and demonstrated the detection of target nucleic acid by fluorescence signal through induction of the orphan cytosine and subsequent assembly of the chemical probe [10]. Since the assembly of the chemical probe is specific for the orphan cytosine sandwiched by matched base pairs, the technique is expected to be applicable for the detection of single nucleotide polymorphisms including viral mutations in relatively wide sequence variations.

4. Summary

In the published papers in this Special Issue, advanced research works involved in nucleic acid conjugates are reported in wide application fields, such as artificial gene regulation, biomolecular sensing, and therapeutics. In general, biotechnologies have been developed through discoveries and utilizations of functional mechanisms in life systems. The field of nucleic acid conjugates for technological applications will continue to develop since the full extent of natural nucleic acid functions has not been clarified.

Funding: This research received no external funding.

Acknowledgments: This Special Issue would not have been possible without the contributions of authors who responded to the invitation. We sincerely appreciate all authors for their efforts to summarize the research works and the editorial team in Applied Sciences for their support.

Conflicts of Interest: The authors declare no conflict of interest.

References

1. Derzhalova, A.; Markov, O.; Fokina, A.; Shiohama, Y.; Zatsepin, T.; Fujii, M.; Zenkova, M.; Stetsenko, D. Novel Lipid-Oligonucleotide Conjugates Containing Long-Chain Sulfonyl Phosphoramidate Groups: Synthesis and Biological Properties. *Appl. Sci.* **2021**, *11*, 1174. [CrossRef]
2. Aiba, Y.; Urbina, G.; Shibata, M.; Shoji, O. Investigation of the Characteristics of NLS-PNA: Influence of NLS Location on Invasion Efficiency. *Appl. Sci.* **2020**, *10*, 8663. [CrossRef]
3. Mie, M.; Hirashima, R.; Mashimo, Y.; Kobatake, E. Construction of an Enzymatically-Conjugated DNA Aptamer–Protein Hybrid Molecule for Use as a BRET-Based Biosensor. *Appl. Sci.* **2020**, *10*, 7646. [CrossRef]
4. Nakano, S.; Seko, T.; Zhang, Z.; Morii, T. RNA-Peptide Conjugation through an Efficient Covalent Bond Formation. *Appl. Sci.* **2020**, *10*, 8920. [CrossRef]
5. Odaira, K.; Yamada, K.; Ishiyama, S.; Okamura, H.; Nagatsugi, F. Design of the Crosslinking Reactions for Nucleic Acids-Binding Protein and Evaluation of the Reactivity. *Appl. Sci.* **2020**, *10*, 7709. [CrossRef]
6. Nakamura, Y.; Sato, Y.; Inaba, H.; Iwasaki, T.; Matsuura, K. Encapsulation of mRNA into Artificial Viral Capsids via Hybridization of a β-Annulus-dT20 Conjugate and the Poly(A) Tail of mRNA. *Appl. Sci.* **2020**, *10*, 8004. [CrossRef]
7. Fujita, H.; Kataoka, Y.; Kuwahara, M. Bifunctional Aptamer Drug Carrier Enabling Selective and Efficient Incorporation of an Approved Anticancer Drug Irinotecan to Fibrin Gels. *Appl. Sci.* **2020**, *10*, 8755. [CrossRef]
8. Sugimoto, M.; Suda, A.; Futaki, S.; Imanishi, M. Effective RNA Regulation by Combination of Multiple Programmable RNA-Binding Proteins. *Appl. Sci.* **2020**, *10*, 6803. [CrossRef]
9. Akagi, J.; Yamada, T.; Hidaka, K.; Fujita, Y.; Saito, H.; Sugiyama, H.; Endo, M.; Matsumura, S.; Ikawa, Y. An RNA Triangle with Six Ribozyme Units Can Promote a Trans-Splicing Reaction through Trimerization of Unit Ribozyme Dimers. *Appl. Sci.* **2021**, *11*, 2583. [CrossRef]
10. Wang, C.-X.; Sato, Y.; Sugimoto, T.; Teramae, N.; Nishizawa, S. Chloro-Substituted Naphthyridine Derivative and Its Conjugate with Thiazole Orange for Highly Selective Fluorescence Sensing of an Orphan Cytosine in the AP Site-Containing Duplexes. *Appl. Sci.* **2020**, *10*, 4133. [CrossRef]

Article

Novel Lipid-Oligonucleotide Conjugates Containing Long-Chain Sulfonyl Phosphoramidate Groups: Synthesis and Biological Properties

Alina Derzhalova [1,2,†], Oleg Markov [2,†], Alesya Fokina [1,3,†], Yasuo Shiohama [4], Timofei Zatsepin [5,6], Masayuki Fujii [7], Marina Zenkova [2] and Dmitry Stetsenko [1,3,*]

1   Department of Physics, Novosibirsk State University, 630090 Novosibirsk, Russia; alina.derzhalova@gmail.com (A.D.); alesya_fokina@mail.ru (A.F.)

2   Institute of Chemical Biology and Fundamental Medicine, Siberian Branch of the Russian Academy of Sciences, 630090 Novosibirsk, Russia; markov_oleg@list.ru (O.M.); marzen@niboch.nsc.ru (M.Z.)

3   Institute of Cytology and Genetics, Siberian Branch of the Russian Academy of Sciences, 63090 Novosibirsk, Russia

4   Tropical Biosphere Research Center, University of the Ryukyus, Okinawa 903-0213, Japan; h194005@comb.u-ryukyu.ac.jp

5   Skolkovo Institute of Science and Technology, Innovation Centre Skolkovo, 121205 Moscow, Russia; tsz@yandex.ru

6   Department of Chemistry, Lomonosov Moscow State University, 119991 Moscow, Russia

7   Department of Biological & Environmental Chemistry, Kindai University, Iizuka, Fukuoka 820-8555, Japan; mfujii@fuk.kindai.ac.jp

*   Correspondence: d.stetsenko@nsu.ru; Tel.: +7-383-363-4963

†   These authors contributed equally to the manuscript.

Citation: Derzhalova, A.; Markov, O.; Fokina, A.; Shiohama, Y.; Zatsepin, T.; Fujii, M.; Zenkova, M.; Stetsenko, D. Novel Lipid-Oligonucleotide Conjugates Containing Long-Chain Sulfonyl Phosphoramidate Groups: Synthesis and Biological Properties. *Appl. Sci.* **2021**, *11*, 1174. https://doi.org/10.3390/app11031174

Academic Editor: Tamaki Endoh, Eriks Rozners and Takashi Ohtsuki

Received: 8 December 2020

Accepted: 24 January 2021

Published: 27 January 2021

Publisher's Note: MDPI stays neutral with regard to jurisdictional claims in published maps and institutional affiliations.

Abstract: New lipid conjugates of DNA and RNA incorporating one to four [(4-dodecylphenyl)sulfonyl]phosphoramidate or (hexadecylsulfonyl)phosphoramidate groups at internucleotidic positions near the 3′ or 5′-end were synthesized and characterized. Low cytotoxicity of the conjugates and their ability to be taken up into cells without transfection agents were demonstrated. Lipid-conjugated siRNAs targeting repulsive guidance molecules a (RGMa) have shown a comparable gene silencing activity in PK-59 cells to unmodified control siRNA when delivered into the cells via Lipofectamine mediated transfection.

Keywords: therapeutic nucleic acid; drug delivery; nanoparticles; cytotoxicity; macrophages; cellular uptake; small interfering RNA; multiple sclerosis; repulsive guidance molecule a

1. Introduction

Nucleic acid derivatives such as antisense oligonucleotides or small interfering RNAs (siRNAs) are extensively studied as therapeutic agents that target biologically important RNAs; either mRNAs or noncoding RNAs such as micro-RNAs via either enzyme-mediated RNA cleavage [1–3] or steric blocking of RNA function [4,5]. While the former may be applicable to a wide range of diseases including some genetic disorders [6], cancer [7], viral [8] and bacterial [9] infections, where the target RNA can be eliminated, the latter is often employed where a fine tuning of the RNA activity is required such as in splice switching [10]. The progress in the development of oligonucleotide-based therapies in the past few years has resulted in the recent approval of several antisense and siRNA drugs by the FDA and EMA [11]. However, widespread application of therapeutic oligonucleotides still suffers from their relatively inefficient cellular uptake and often poor bioavailability [12,13]. The ways of promoting cellular delivery of oligonucleotides and their analogues include their structural modification [14,15], noncovalent binding [16] or covalent conjugation [17,18] to the molecules that could improve translocation through the outer membrane and direct intracellular trafficking of nucleic acids. Among these, lipids in a wider sense including biomolecules such as cholesterol, tocopherol (vitamin E) or fatty acids, and their synthetic analogues such as, e.g., DOTMA [19] have been shown to improve cellular uptake of oligonucleotides, most relevantly, in the form of lipid-oligonucleotide conjugates (LONs) [20–22].

In particular, lipid conjugates of antisense oligonucleotides and siRNAs have improved transfection ability and extended half-life in the bloodstream with beneficial impact on therapeutic activity in vivo [23–25]. We describe herein a series of novel DNA and RNA lipid conjugates, which have been chemically modified by one to four long-chain *N*-(sulfonyl)phosphoramidate groups, namely, [(4-dodecylphenyl)sulfonyl]-phosphoramidate group (δ) or (hexadecylsulfonyl)phosphoramidate group (η) at the internucleotidic position(s) near the 3′- or 5′-end of the oligonucleotide chain (Figure 1). The oligonucleotides were obtained by an automated phosphoramidite solid-phase synthesis protocol that employs the Staudinger reaction between the support-bound phosphite triester and the respective sulfonyl azide instead of usual aqueous iodine oxidation to introduce either of the lipophilic modifications following a previously developed procedure [26–28].

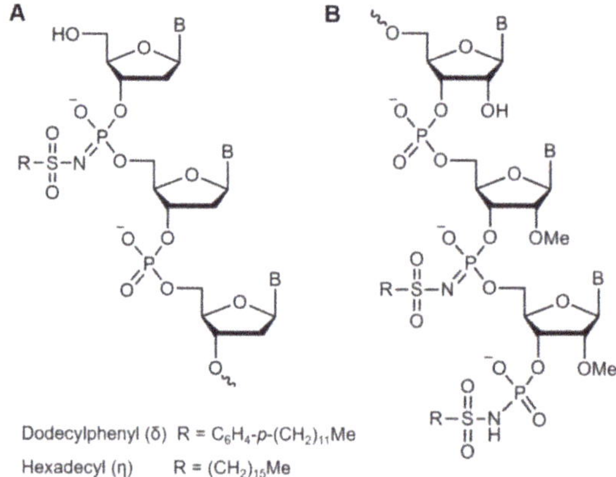

Figure 1. Structures and positions of hydrophobic groups in lipid-oligonucleotide conjugates used in this study: oligodeoxynucleotide (**A**); small interfering RNA (siRNA) (**B**).

The obtained lipid conjugates of oligodeoxyribonucleotides with one or two of the respective hydrophobic groups were nearly non-cytotoxic for either murine or human macrophages up to 20 μM concentration. Both types of lipid conjugates were able to enter the cells in a carrier free manner. Both features are beneficial for potential therapeutic application. The corresponding lipid conjugates of siRNAs with either two or four hydrophobic groups were designed to downregulate a validated target in multiple sclerosis, namely, repulsive guidance molecule a (RGMa) [29]. The conjugates with two groups were shown to possess silencing ability comparable to the unmodified siRNA when delivered by lipofectamine. This type of lipid conjugate may represent a useful addition to the inventory of therapeutic oligonucleotide derivatives.

2. Materials and Methods

2.1. Oligonucleotide Synthesis

Oligodeoxynucleotides and siRNAs containing ((4-dodecylphenyl)sulfonyl)phosphoramidate (δ) or (hexadecylsulfonyl)phosphoramidate (η) groups were assembled by automated solid-phase synthesis on a Mermaid MM-12 DNA/RNA synthesizer according to a modified β-cyanoethyl phosphoramidite protocol replacing aqueous iodine oxidation by a Staudinger reaction with either 4-dodecylbenzenesulfonyl azide or 1-hexadecanesulfonyl azide (0.5 M in acetonitrile for (δ) or in acetonitrile—THF (2:1 *v/v*) for (η), 120 min at ambient temperature) in an appropriate cycle of chain elongation [27,28]. Standard *N*-protected 5′-DMTr-3′-β-cyanoethyl-*N*,*N*′-phosphoramidites of deoxy, 2′-*O*-TBDMS-ribo and 2′-*O*-methylribonucleotides (Sigma-Aldrich Inc., St Louis, MO, USA) and 1000 Å CPG polymer supports (Glen Research Corp, Sterling, VA, USA) were used. After the completion of the synthesis, oligonucleotides were cleaved from support and deprotected by standard procedures followed by consecutive PAGE and RP-HPLC purification as described previously [30]. Molecular masses of the oligonucleotides were verified by ESI LC-MS (Table 1).

Table 1. Structures and molecular masses of oligonucleotides and their lipid conjugates (LONs).

Type	Designation	Oligonucleotide Sequence, 5′–3′ [1]	Molecular Mass, Da[2]	
			Calc. [M]	Exp. [M]
Model antisense oligonucleotide	ODN	agtctcgacttgctacc [3]	5121.25	5120.98
	ODN-*F* [4]	agtctcgacttgctacc-Flu	5688.74	5687.93
	1δ	a$^{\delta}$gtctcgacttgctacc	5248.86	5426.08
	2δ	a$^{\delta}$g$^{\delta}$tctcgacttgctacc	5736.36	5733.28
	1δ-*F*	a$^{\delta}$gtctcgacttgctacc-Flu	5996.35	5993.21
	2δ-*F*	a$^{\delta}$g$^{\delta}$tctcgacttgctacc-Flu	6303.85	6300.41
	1η	a$^{\eta}$gtctcgacttgctacc	5408.87	5406.11
	2η	a$^{\eta}$g$^{\eta}$tctcgacttgctacc	5696.38	5693.34
	1η-*F*	a$^{\eta}$gtctcgacttgctacc-Flu	5976.36	5973.24
	2η-*F*	a$^{\eta}$g$^{\eta}$tctcgacttgctacc-Flu	6263.87	6260.47
siRNA	EGFP	*Sense* CCUACGGCAAGCUGACCCU<u>GA</u>	6698.10	6695.84
		Antisense AGGGUCAGCUUGCCGUAGG<u>UG</u>	6812.10	6811.57
	RGMa	*Sense* CCAUCAUCUUCAAGAACUU<u>CC</u>	6563.99	6565.38
		Antisense AAGUUCUUGAAGAUGAUGG<u>UG</u>	6806.20	6811.30
	2Δ-RGMa	*Sense* CCAUCAUCUUCAAGAACUU<u>C</u>$^{\delta}$<u>C</u>$^{\delta}$-p[6]	7219.40	7220.40
		Antisense AAGUUCUUGAAGAUGAUGG<u>UG</u>	6806.20	6811.30
	4Δ-RGMa	*Sense* CCAUCAUCUUCAAGAACUU<u>C</u>$^{\delta}$<u>C</u>$^{\delta}$-p	7219.40	7220.40
		Antisense AAGUUCUUGAAGAUGAUGG<u>C</u>$^{\delta}$<u>C</u>$^{\delta}$-p	7460.60	7466.90
	2H-RGMa	*Sense* CCAUCAUCUUCAAGAACUU<u>C</u>$^{\eta}$<u>C</u>$^{\eta}$-p	7260.00	7262.90
		Antisense AAGUUCUUGAAGAUGAUGG<u>UG</u>	6806.20	6811.30
	4H-RGMa	*Sense* CCAUCAUCUUCAAGAACUU<u>C</u>$^{\eta}$<u>C</u>$^{\eta}$-p	7260.00	7262.90
		Antisense AAGUUCUUGAAGAUGAUGG<u>U</u>$^{\eta}$<u>G</u>$^{\eta}$-p	7501.20	7502.00

[1] Lowercase letters indicate deoxyribonucleotides, uppercase letters—ribonucleotides, underlined uppercase—2′-*O*-methylribonucleotides. [2] According to ESI LC-MS in the negative ions mode. [3] Control oligonucleotide with random sequence specially constructed not to affect gene expression in cells. [4] Flu—3′-terminal fluorescein label. [5] The symbols ($^{\delta}$), ($^{\eta}$) mark the positions of dodecylphenyl or hexadecyl groups, respectively. [6] p—3′-terminal phosphate group.

4-Dodecylbenzenesulfonyl azide was obtained from a commercial supplier (Sigma-Aldrich Inc., St Louis, MO, USA). 1-Hexadecanesulfonyl azide was prepared from the corresponding sulfonyl chloride and a slight excess of sodium azide in acetonitrile—THF (1:1 *v*/*v*) as a white solid as described in [31]. IR, ^{1}H and ^{13}C NMR spectra of the compound were consistent with the structure.

2.2. Dynamic Light Scattering (DLS) Characterization

Lipid-oligonucleotide conjugates (LONs) were analyzed at room temperature by dynamic light scattering (DLS) using a Zetasizer Nano ZS90 (Malvern Instruments Ltd., Malvern, UK). Size was measured in a DTS0012 cuvette (Malvern Instruments Ltd., Malvern, UK). All the scattered photons were collected at a 173° scattering angle. Measurement conditions were: material protein (RI: 1.450; Absorption: 0.001), dispersant water (Viscosity: 0,8872 cP; RI: 1.330) temperature at 25 °C and equilibration time was 3 s. Each test was triplicated. Nanoparticle size was determined in deionized water at 5 µM oligonucleotide concentration. The scattering intensity data was processed using the instrumental software to obtain the hydrodynamic diameter (Dh).

2.3. Cell Lines

RAW 264.7 cells were cultivated in Dulbecco's modified Eagle's medium (DMEM) (Sigma-Aldrich Inc., St Louis, MO, USA) supplemented with 10% heat-inactivated fetal bovine serum (FBS) (GE Healthcare, Chicago, IL, USA) and 1% antibiotic/antimycotic mix (100 unit/mL penicillin, 0.1 mg/mL streptomycin and 0.25 μg/mL amphotericin) at 37 °C in a humidified atmosphere with 5% CO_2 (standard conditions, SC) and passaged regularly to maintain exponential growth.

THP-1 cells were cultivated in Roswell Park Memorial Institute (RPMI) medium (Thermo Fisher Scientific, Waltham, MA, USA) supplemented with 10% FBS, 1% antibiotic/antimycotic mix (100 unit/mL penicillin, 0.1 mg/mL streptomycin and 0.25 μg/mL amphotericin), 1% GlutaMAX (Thermo Fisher Scientific, Waltham, MA, USA), 1% HEPES (Thermo Fisher Scientific, Waltham, MA, USA) and 1% glucose (Sigma-Aldrich Inc., St Louis, MO, USA) under SC and passaged regularly to maintain exponential growth.

To generate human macrophages (Mφ) THP-1 cells were cultivated in complete RPMI medium in the presence of 200 nM PMA (Sigma-Aldrich Inc., St Louis, MO, USA) under SC for 48 h. Then, the culture medium was replaced with complete RPMI medium in the absence of PMA and cells were incubated for an additional 24 h under SC.

PK-59 human pancreatic cancer cells were cultivated in complete RPMI-1640 medium (Thermo Fisher Scientific, Waltham, MA, USA) under SC and passaged regularly to maintain exponential growth.

2.4. MTT Assay of Cell Viability

RAW264.7 cells were seeded in 96-well plates at a density of 5×10^3 cells/well in complete DMEM and incubated for 24 h under SC to adhere. THP-1 cells were seeded in 96-well plates at a density of 50×10^3 cells/well in complete DMEM medium supplemented with 200 nM PMA and incubated for 72 h under SC to generate human Mφ as described above. Then, RAW264.7 and Mφ cells were incubated in the corresponding complete culture media supplemented with 5, 10 and 20 μM of the respective lipid-oligonucleotide conjugate (LON) for an additional 24 h under SC. Cytotoxicity of LONs was assessed by MTT assay. Briefly, MTT solution (Sigma-Aldrich, Darmstadt, Germany) was added to the cells up to 0.5 mg/mL and the incubation was continued under SC for an additional 2 h. Then, the MTT-containing medium was aspirated and the formazan crystals formed in living cells were solubilized with 100 μL/well of DMSO. The absorbance of each well was read at the test and reference wavelengths (λ) of 570 and 620 nm, respectively, on a Multiscan RC plate reader (Thermo LabSystems, Helsinki, Finland). The percentage of living cells was calculated as follows:

$$\text{Living cells (\%)} = (\text{OD}_{570\text{exp}} - \text{OD}_{620\text{exp}})/(\text{OD}_{570\text{cont}} - \text{OD}_{620\text{cont}}) \times 100, \qquad (1)$$

where $\text{OD}_{570\text{exp}}$ and $\text{OD}_{570\text{cont}}$ correspond to optical density in experimental and control wells, respectively, at λ 570 nm, and $\text{OD}_{620\text{exp}}$ and $\text{OD}_{620\text{cont}}$ correspond to optical density in experimental and control wells, respectively, at λ 620 nm. All experimental points were run in triplicate for statistical analysis.

2.5. Flow Cytometry

Intracellular accumulation of the lipid-oligonucleotide conjugates labeled with fluorophores, namely δ-Flu, 2δ-Flu, η-Flu and 2η-Flu, was assessed in RAW264.7 and human Mφ by flow cytometry. RAW264.7 cells were seeded into 24-well plates at a density of 80×10^3 cells/well in complete DMEM and cultivated overnight to adhere. THP-1 cells were seeded into 24-well plates at a density of 100×10^3 cells/well in complete RPMI medium supplemented with 200 nM PMA and incubated for 72 h under SC to generate human Mφ as described above. Subsequently, RAW 264.7 and Mφ were incubated in the corresponding complete media supplemented with 5 µM of the respective lipid-oligonucleotide conjugate under SC for 24 h. After incubation, cells were detached from the plate with 2% trypsin (MP Biomedicals, Irvine, CA, USA), resuspended in a complete culture medium, centrifuged at $200 \times g$ for 5 min, washed with PBS and fixed with 2% formaldehyde in a PBS buffer (10 min, room temperature). Cells were analyzed on a NovoCyte flow cytometer (ACEA Biosciences, Santa Clara, CA, USA) and the data were processed with a NovoExpress (ACEA Biosciences, Santa Clara, CA, USA). All experiments were run in triplicate for statistical analysis. The transfection efficiency (TE) was characterized by two values: percentage of fluorescence-positive cells in a population and mean fluorescent intensity (MFI) of cells in a sample.

2.6. Confocal Microscopy

RAW264.7 cells were seeded on coverslips (Marienfeld, Lauda-Königshofen, Germany) in 24-well plates at a density of 20×10^3 cells/well in complete DMEM medium and incubated for 24 h under SC to adhere. THP-1 cells were seeded on coverslips in 24-well plates at a density of 100×10^3 cells/well in complete RPMI medium supplemented with 200 nM PMA to obtain human Mφ as described above. Afterwards, RAW 264.7 cells and human Mφ were incubated in the corresponding complete media in the presence of 5 µM of the respective lipid-oligonucleotide conjugates under SC for 24 h. After incubation, coverslips with cells were washed twice with PBS, fixed with 3.7% formaldehyde in PBS buffer (15 min, 37 °C), and washed twice with PBS. The cells were stained with the Phalloidin-iFluor532 (Abcam, Cambridge, UK) or Phalloidin-TRITC (ECM Biosciences, Versailles, KY, USA) according to the manufacturer's protocols. After staining, the cells were washed twice with PBS and placed on slides in a drop of ProLongTM Glass Antifade Mountant with NucBlueTM (Thermo Fisher Scientific, Waltham, MA, USA). Mounted samples were allowed to cure on a flat, dry surface for 18–24 h at room temperature in the dark. Intracellular localization of lipid-oligonucleotide conjugates was assessed by confocal fluorescent microscopy on an LSM710 (Zeiss, Oberkochen, Germany) using an αPlan-Apochromat 100x/1.46 Oil DIC M27 objective. Analysis of intracellular accumulation of LONs and Z-stacks was performed using ZEN software (Zeiss, Oberkochen, Germany). Confocal microscopic analysis was performed in three channels (blue, green, red). Fluorescence in the blue channel corresponded to DAPI (nuclei staining); the green channel corresponded to fluorescence of LONs labeled with fluorescein (Flu), and the red channel corresponded to Phalloidin-iFluor532 or Phalloidin-TRITC (cytoskeleton staining).

2.7. Gene Silencing

Experiments on siRNA-mediated silencing of repulsive guidance molecule a (RGMa) were carried out in the RGMa-expressing PK-59 human pancreatic cancer cell line. On the day of transfection, PK-59 cells were separately seeded in 24-well flat-bottomed plates at a density of 4×10^5 cells/well in 0.5 mL of growth medium without antibiotics. The cells were transfected with lipid-conjugated RGMa siRNAs carrying either two (passenger strand only) or four (both passenger and antisense strands) dodecylphenyl (Δ) or hexadecyl (H) groups at their 3'-end, or unmodified control RGMa siRNA (Table 1) at concentrations 1, 10, or 100 nM in a complex with Lipofectamine® 2000 (2 µL/well) according to the manufacturer's instructions (Invitrogen Life Technologies Corporation, Frederick, MD, USA). The cells were then incubated at 37 °C in the atmosphere of 5% CO_2 in a humidified incubator. Cells were harvested and assayed 24 h after transfection.

The level of RGMa expression was quantified by real-time, two-step reverse transcriptase PCR (qRT-PCR) assay. Briefly, total RNA was extracted from the collected cells using ISOGNE (Nippon Gene, Ltd., Toyama, Japan), treated with DNase I® (ThermoFisher Scientific, Waltham, MA, USA) and retro-transcribed using Oligo $(dT)_{20}$ primer and ReverTra Ace® (Toyobo Co., Ltd., Japan). The resulting cDNA was quantified using the AriaMx Real-Time PCR System® (Agilent Technologies, Ltd., Santa Clara, CA, USA) using the following primers: 5'–TGGCCGCTCATCGACAATAA–3', forward; and 5'–

GCTTGTCCCCACCGTTCTTA–3′, reverse. Quantifications of RGMa mRNA expressions were normalized by the Comparative Ct Method (Ct) using 18S rRNA as an endogenous control to ensure accuracy; primers: 5′–GTAACCCGTTGAACCCCATT–3′, forward; and 5′–CCATCCAATCGGTAGTAGCG–3′, reverse [32]. All the data are the means of triplicate determinations. Each determination was calculated from triplicate measurements. Each error bar indicates a standard deviation of the triplicate determinations. Non-silencing siRNA targeting EGFP was used as a negative control, which was designed to have a minimum of 4 mismatches to all human, mouse and rat genes, and confirmed to have minimal targeting by genome-wide microarray analysis [33]. Significant differences in the results were determined by the two-tailed Student's t-test using GraphPad Prism. A probability value of $p < 0.05$ was considered to be significant in the present study.

3. Results

3.1. Synthesis of Lipid-Oligonucleotide Conjugates (LONs) Incorporating Long-Chain (Sulfonyl)Phosphoramidate Groups

The synthesis of lipid conjugates of oligodeoxynucleotides having one or two ((4-dodecylphenyl)sulfonyl)phosphoramidate (δ) or (hexadecylsulfonyl)phosphoramidate (η) modifications of the phosphate linkages adjacent to the 5′-end (Figure 1A) was accomplished according to our previously reported protocols of automated solid-phase assembly via the phosphoramidite method employing the Staudinger reaction of the support-bound phosphite triester formed during the phosphoramidite coupling with 0.5 M solution of either 4-dodecylbenzenesulfonyl azide or 1-hexadecanesulfonyl azide in acetonitrile for 120 min at ambient temperature instead of the usual aqueous iodine oxidation [23,24]. Oligonucleotides containing 5′-dimethoxytrityl group (DMTr) were isolated by reverse-phased (RP) HPLC or, if necessary, purified to homogeneity by denaturing PAGE under the same conditions as normal phosphodiester oligodeoxynucleotides, followed by RP-HPLC to remove phosphodiester impurities. The molecular masses of the δ- and η-oligodeoxynucleotides obtained have been verified by ESI LC-MS (Table 1).

Sequences of siRNAs targeting RGMa mRNA were obtained from literature [25]. Lipid-siRNA conjugates containing 2–4 dodecylphenyl (δ) or hexadecyl (η) groups in the antisense and/or passenger chains were prepared according to the method above (Table 1). The 3′-overhangs of the siRNAs were modified by 2′-O-methyl diribonucleotides to increase enzymatic stability and avoid possible side reactions associated with the neighboring 2′-OH group when sulfonyl phosphoramidate groups were introduced into the overhang and at the 3′-terminal phosphate group (Figure 1B).

3.2. Spontaneous Nanoparticle Formation by LONs in Aqueous Media

As amphiphiles, lipid-oligonucleotide conjugates (LONs) can be expected to spontaneously form micelles in aqueous solutions. Indeed, dynamic light scattering (DLS) analysis revealed that both 2δ and 2η LONs formed nanoparticles in water characterized by mean diameters of 21.6 ± 8.8 and 79.7 ± 16.4 nm for 2η and 2δ, respectively (Figure 2). Interestingly, longer aliphatic chain (16-carbon vs. 12-carbon) at the internucleotidic position favored the formation of smaller nanoparticles, which may be responsible for a better cellular uptake of a more hydrophobic 2η LON (Figure 2).

3.3. LONs were Nontoxic to Murine or Human Macrophages at Therapeutic Concentrations

Introduction of lipophilic moieties into oligonucleotide chains can alter such properties of LONs as cyototoxicity. Cytotoxicity of the most hydrophobic LONs, namely, 2δ and 2η, was analyzed with respect to murine RAW264.7 and human THP-1-derived macrophages. Cells were incubated in complete media supplemented with different concentrations of one of LONs for 24 h followed by estimation of cell viability using colorimetric MTT assay. As depicted in Figure 3, both 2δ and 2η oligonucleotides were nontoxic to murine and human macrophages. IC_{50} obtained by extrapolation of dose-response curves ranged from 74 to 120 μM for 2δ and from 77 to 85 μM for 2η oligonucleotides, whereas the working concentration of LONs in subsequent experiments was chosen to be 5 µM. It is worth mentioning that murine macrophages were more sensitive to LONs in comparison with human THP-1-derived macrophages (IC_{50} were 74.2 ± 29.6 µM and 120.3 ± 31.6 µM for 2δ oligonucleotides, respectively; 77.0 ± 9.2 µM and 84.6 ± 17.2 µM for 2η oligonucleotides, respectively) (Figure 3).

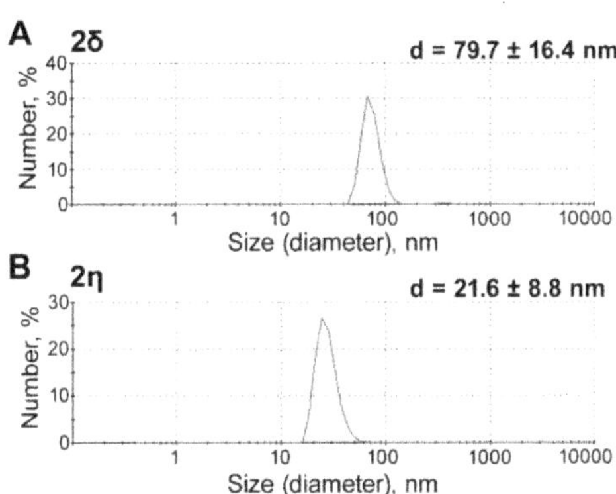

Figure 2. Hydrodynamic diameters of nanoparticles formed spontaneously in water by 2δ (**A**) and 2η (**B**) lipid-oligonucleotide conjugates. Dynamic light scattering (DLS) data.

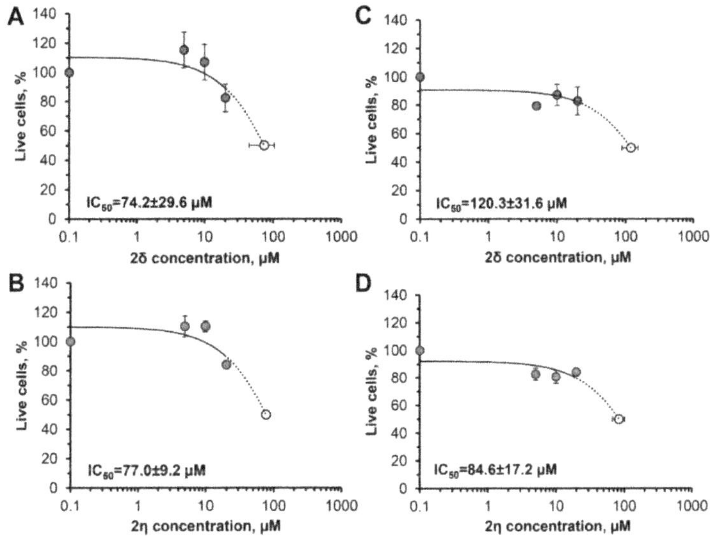

Figure 3. Dose-response curves of 2δ and 2η LONs for RAW264.7 (**A**,**B**) and human THP-1 induced macrophages (**C**,**D**). Cells were incubated with LONs in a complete medium for 24 h. The results are shown as a percentage of viable cells observed after cell incubation with LON relative to untreated cells (100%). All experimental points were run in triplicate for statistical analysis. Data are presented as MEAN ± SD. The dotted curve shows extrapolation to 50% of viable cells.

3.4. Conjugates with Longer Lipidic Chains Accumulated in Cells with Higher Efficiency

The ability of LONs to accumulate in eukaryotic cells in the absence of a transfection agent (gymnosis) was studied by flow cytometry using fluorescein-labeled LONs and murine macrophages RAW264.7 or THP-1-induced human macrophages. The efficiency of intracellular accumulation of LONs was assessed by two parameters: the number of fluorescein-positive cells in the population and average level of the conjugate content in the cells (mean fluorescence intensity, MFI).

Accumulation of LONs in RAW264.7 and THP-1-induced macrophages differed considerably. LONs accumulated in RAW264.7 cells with moderate efficiency: 20–50% of cells were transfected with a MFI of 6–11 × 10^3 RFU (Figure 4A,B). As expected, the cells incubated and control oligonucleotide ODN-*F* displayed the lowest efficiency: 10% of fluorescent cells with MFI of 4 × 10^3 RFU, which corresponded to the baseline level of oligonucleotide taken up by the macrophages and/or adhered to the cell surface. We observed that more hydrophobic LONs, that is, the ones containing either two hydrocarbon chains instead of only one, or those having a longer aliphatic group: hexadecyl instead of dodecylphenyl, were showing more efficient accumulation in RAW264.7 cells (Figure 4A,B: compare conjugates having two vs. one group, and η- vs. δ-modified LONs) than their less hydrophobic counterparts and unmodified control ODN-*F*.

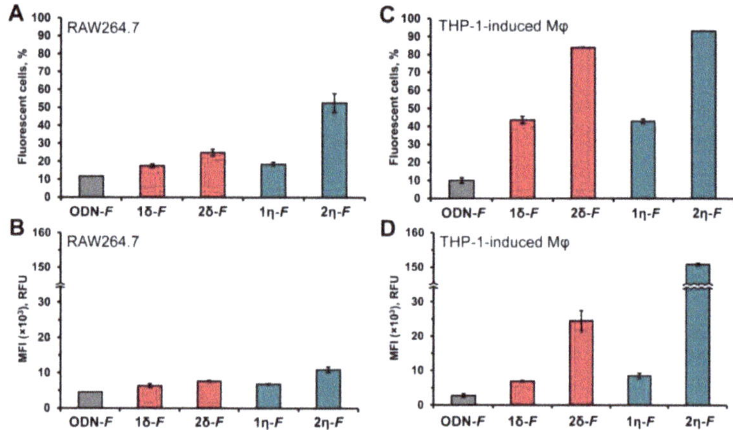

Figure 4. Accumulation of fluorescein-labeled LONs in murine macrophages RAW264.7 (**A**,**B**) and THP-1-induced human macrophages (**C**,**D**). The percentage of fluorescent cells (**A**,**C**) and mean fluorescence intensity (**B**,**D**) were measured by flow cytometry 24 h post transfection. Data are presented as MEAN ± SD.

A similar but much more prominent accumulation of lipid-conjugated oligodeoxynucleotides was observed in human THP-1-induced macrophages. LONs accumulated in these cells with higher efficiency both in terms of the number of transfected cells and mean fluorescence intensity (Figure 4C,D). Conjugates 1δ-*F* and 1η-*F* containing one lipophilic group displayed similar transfection efficiency regardless of the type of lipid chains, either dodecylphenyl (δ) or hexadecyl (η): 43% cells were transfected with MFI of 8 × 10^3 RFU, which was fourfold higher as compared to the control ODN-*F* (Figure 4C,D). Increasing the number of lipophilic groups from one to two resulted in a significant enhancement of transfection efficiency. 2δ LON yielded 85% cells with fluorescence intensity 25 × 10^3 RFU whereas 2η conjugate was taken up with the highest efficiency, accumulating in nearly 100% cells with fluorescence intensity 150 × 10^3 RFU (Figure 4C,D), which is on par with the level achieved by liposome-mediated oligonucleotide delivery [34].

3.5. Lipid-Oligonucleotide Conjugates are Efficiently Taken up by Cells in the Absence of Transfection Agents and Localize Mostly in the Cytoplasm within Endosomes

Confocal microscopic analysis was carried out to verify and characterize intracellular localization of LONs. RAW 264.7 and human THP-1-induced macrophages were incubated with 5.0 μM fluorescein-labeled LONs in complete medium for 24 h. After incubation, cells were stained with Phalloidin-iFluor532 or Phalloidin-TRITC (actin microfilaments) and NucBlue (nuclei) (Figures 5–7).

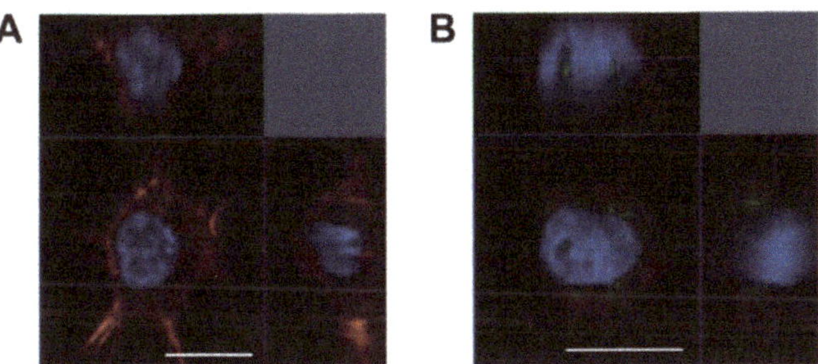

Figure 5. Intracellular localization of 2η LON in murine macrophages RAW264.7 (**A**) and THP-1-induced human macrophages (**B**). Analysis was performed by confocal fluorescent microscopy (αPlan-Apochromat 100x/1.46 Oil DIC M27 objective) 24 h post transfection. Data are presented as orthogonal projections. Blue signal is DAPI (nuclear staining), green signal is fluorescein-labeled 2η LON, red signal is Phalloidin-iFluor532 (**A**) or Phalloidin-TRITC (**B**). Scale bar 10 μm.

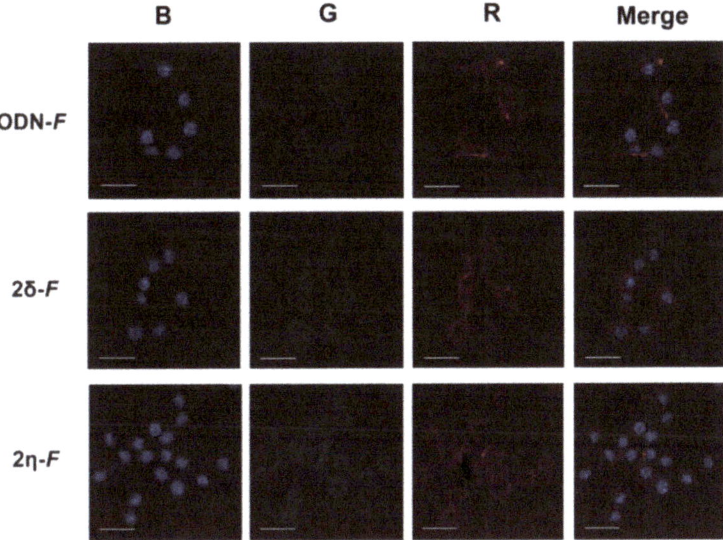

Figure 6. Intracellular accumulation of LONs in RAW264.7. Analysis of samples was performed 24 h after addition of fluorescein-labeled conjugates (5 μM) to the cells. Analysis was performed using αPlan-Apochromat 100x/1.46 Oil DIC M27 objective. Three-channel (**BGR**) pictures were obtained using staining by DAPI (nuclei staining) (**B**); fluorescein-labeled LONs (**G**); Phalloidin-iFluor532 (cytoskeleton staining) (**R**). Scale bars 20 μm. ODN-*F*—control unmodified oligonucleotide labeled with fluorescein.

Firstly, to prove that LONs do not expose on the cell membranes but do penetrate into the cells, confocal microscopic analysis was performed and Z-stack images were built. It was demonstrated that LON 2η, the most efficient one according to flow cytometry data, was indeed localized in the cytosol of RAW264.7 cells and THP-1-induced macrophages (Figure 5). In addition, LON 1η was shown to efficiently accumulate in KB-8-5 cells (see Supplementary Materials, Figure S4).

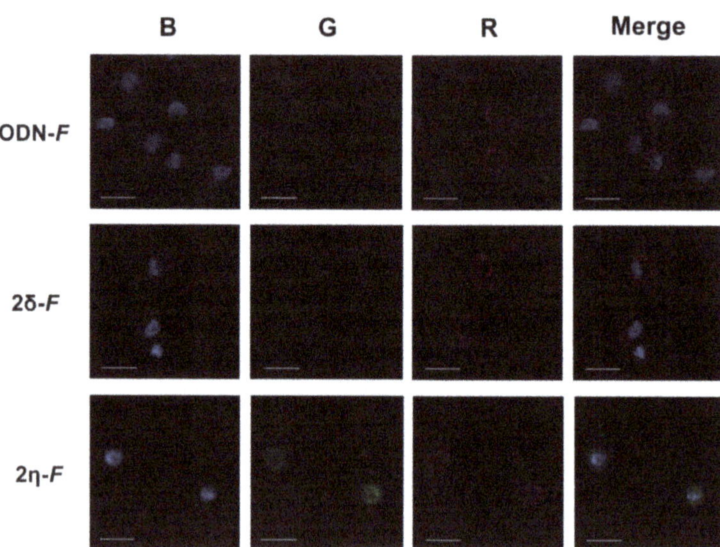

Figure 7. Intracellular accumulation of LONs in human THP-1-induced macrophages. Analysis of samples was performed 24 h after addition of fluorescein-labeled conjugates (5 μM) to the cells. Analysis was performed using αPlan-Apochromat 100x/1.46 Oil DIC M27 objective. Three-channel (**BGR**) pictures were obtained using staining by DAPI (nuclear staining) (**B**); fluorescein-labeled LONs (**G**); Phalloidin-TRITC (cytoskeleton staining) (**R**). Scale bars 20 μm. ODN-*F*—control unmodified oligonucleotide labeled with fluorescein.

It can be seen that the efficiency of intracellular accumulation of LONs according to confocal microscopy (Figures 6 and 7) correlate well with the obtained flow cytometry data (Figure 4). Oligonucleotides containing two lipophilic groups accumulated in cells with much higher efficiency in comparison with LONs containing only one group, LON 2η with two hexadecyl groups being the best penetrator. LONs 1δ-*F* and 1η-*F* containing only one lipophilic modification accumulated in cells with efficiency not significantly higher than that of the control unmodified oligonucleotide ODN-*F*. However, these LONs penetrated into the cells whereas ODN-*F* mainly adhered to the cell membrane (Figure S5, Supplementary Materials).

In detail, LONs were evenly distributed in the cytosol, localizing mainly in endosomes; this was conjectured from the presence of numerous bright green fluorescent dots containing LONs in the cytosol of both cell lines tested (Figures 6 and 7).

3.6. Lipid-Conjugated RGMa siRNAs Demonstrated Gene-Silencing Ability Similar to Unmodified RGMa siRNA in PK-59 Human Pancreatic Carcinoma Cell Culture

The gene silencing ability of lipid-conjugated siRNAs in comparison with unconjugated siRNA was elucidated in human pancreatic carcinoma PK-59 cells expressing repulsive guidance molecule a (RGMa). RGMa is a glycoprotein, which is a potent inhibitor of axonal re-myelination and, thus, is regarded as a potential therapeutic target for multiple sclerosis (MS) treatment [35]. Inhibition of RGMa by human monoclonal antibodies promotes neuronal regeneration and repair [36]. RGMa siRNA sequence was obtained from literature [29]. Lipid-conjugated RGMa siRNAs were modified at the 3'-end by either two (passenger strand only) or four (both passenger and antisense strands) dodecylphenyl (Δ) or hexadecyl (H) groups. To compare silencing activity under similar conditions, transfection was carried out at concentrations of siRNA 1, 10 and 100 nM in the presence of Lipofectamine® 2000. The levels of RGMa expression as assessed by RT-PCR are shown in Figure 8.

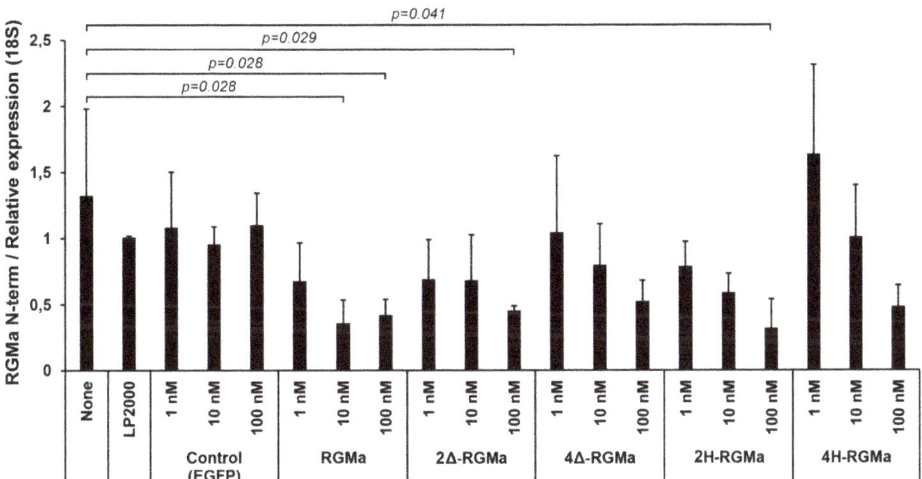

Figure 8. Inhibition of RGMa expression in PK-59 human pancreatic carcinoma cell culture by lipid conjugates of RGMa siRNAs vs. unmodified siRNA (RGMa). The conjugates were modified near the 3′-end by two (passenger chain only) or four (both chains) lipophilic groups (Δ—dodecylphenyl, H—hexadecyl).

The gene silencing effect of siRNA conjugates 2Δ-RGMa and 2H-RGMa that carried two lipophilic groups at the 3′-end of the passenger strand was comparable to that of the unconjugated RGMa siRNA at a concentration of 100 nM. At the same time, the activity of the conjugates 4Δ-RGMa and 4H-RGMa containing four lipophilic groups at both 3′-ends of passenger and antisense strands was reduced as compared to the unmodified control siRNA.

4. Discussion

A previously developed methodology of Staudinger reaction-mediated in-line oligonucleotide modification [26–28] was employed herein to obtain lipophilic conjugates of oligodeoxynucleotides and siRNAs carrying one or more extended hydrocarbon chains, 4-dodecylphenyl or hexadecyl, attached to the internucleotidic phosphate groups via the (sulfonyl)phosphoramidate linkage, which is hydrolytically stable and carries negative charge under physiological conditions (Figure 1A,B). The scheme allows one to modify any internucleotidic or terminal phosphate group within a DNA or 2′-OMe RNA chain from the 5′- to 3′-end by a facile chemical reaction with a stable sulfonyl azide precursor during automated DNA/RNA assembly without recourse to a special phosphoramidite monomer, which is usually obtained by a multi-step synthesis [18,20,21]. The obtained lipid-oligonucleotide conjugates (LONs) can be conveniently isolated by the usual C18 reverse-phased HPLC, due to significant difference in retention times with unmodified oligonucleotides (see Supplementary Materials) for HPLC profiles and ESI-MS spectra). Our preliminary data have shown that such lipophilic groups are fully compatible with other phosphate group modifications such as phosphorothioate, mesyl phosphoramidate [27,37,38] and phosphoryl guanidine [39].

Notably, we have observed spontaneous nanoparticle formation in aqueous solution by LONs containing two adjacent hydrophobic chains, with nanoparticle size dependent on the type of the side-chain (Figure 2). Nanoparticle formation by oligonucleotides is thought to be essential for good cellular uptake [40], structured RNA studies [41] and design of stimuli-responsive drug delivery systems [42].

Low cytotoxicity is another beneficial characteristic of such LONs, which allows one to vary the concentration of an oligonucleotide over a wide range to optimize its therapeutic dosage for further in vitro and in vivo studies. We have confirmed that LONs used in this study evinced low cytotoxicity. Calculated IC_{50} of studied LONs fell within the 75 to 120 μM range, which was significantly higher than concentrations of antisense oligonucleotides targeting therapeutically relevant RNAs commonly used in vitro and in vivo: nanomolar concentrations [43–46] or <5 μM [47] or therapeutic doses of siRNAs and antisense oligonucleotides approved by FDA and EMA for clinical practice: for siRNAs 21 nmol/kg for Patisiran [48] and 153 nmol/kg for Givosiran [49]; for antisense oligonucleotides <5 μmoles/kg for Golodirsen [50], Volanesorsen [51] and Inotersen [52]. It is worth noting that the increase in hydrophobicity of the conjugates have no influence on their cytotoxicity (Figure 3). This result prompts us to expect that the increase of the number of side-chains or introduction of more hydrophobic groups, for example, hydrocarbon chains longer than 16 carbon atoms into LONs will not increase significantly the overall cytotoxicity of conjugates.

We demonstrated that LONs accumulated in the cells with intracellular distribution analogous to that of oligonucleotides delivered by liposomal formulations (Figures 4 and 5) [53–55]. It appears that LONs penetrated through the outer membrane and localized in cytosol predominantly in endosomes 24 h after addition (Figures 6 and 7). It is known that endosomal escape of therapeutic oligonucleotides is one of the rate-limiting steps of their delivery and the key to successful intracellular performance of nucleic acid therapeutics [56]. Although longer term kinetics of LON accumulation (>24 h) was not studied, it can be assumed that such LONs will require longer times for their endosomal escape, similarly to cholesterol-conjugated siRNAs [57].

Cytoplasmic distribution of the developed LONs may be advantageous as far as the targets of therapeutic oligonucleotides (antisense oligonucleotides, siRNAs and anti-miRs) are localized in the cytosol such as mRNAs and many miRNAs. Application of the developed LONs to modulate splicing of pre-mRNA requires additional investigation including optimization of the number and length of the side-chains, selection of optimal delivery conditions etc. to achieve nuclear delivery of the conjugates. We demonstrated that dodecylphenyl and hexadecyl modifications did not significantly impede gene-silencing activity of lipophilic siRNAs delivered into cells by Lipofectamine (Figure 8). However, further studies on elucidating a wider spectrum of biological activities of siRNAs and antisense oligonucleotides carrying such (sulfonyl)phosphoramidate-based lipophilic groups are required.

One potential application for lipid-oligonucleotide conjugates as described herein could be the design of therapeutic oligonucleotides for the treatment of multiple sclerosis (MS), a severe neurodegenerative disease characterized by the destruction of the myelin sheaths of neurons [58]. Kitayama et al. [29] uncovered a link between the destruction of axonal myelin sheaths during MS and repulsive guidance molecule a (RGMa): a glycoprotein, which is a powerful inhibitor of neuronal regeneration. Inhibition of RGMa by human monoclonal antibodies stimulate re-myelination and regeneration of damaged neurons in the laboratory in vivo model of MS [36]. Thus, reducing the level of RGMa by downregulation of the corresponding gene by antisense oligonucleotide or siRNA could be a promising strategy for therapeutic intervention in MS.

We studied RGMa inhibition in the RGMa-expressing PK-59 pancreatic cancer cell line using lipid-siRNA conjugates containing either two or four dodecylphenyl or hexadecyl chains at the 3′-end of one or both of the complementary RNA strands, respectively (Figure 1B; Table 1). It was evident that siRNA conjugates containing two adjacent lipid groups in the passenger strand were able to downregulate RGMa mRNA with a similar efficiency to that of the unmodified control RGMa siRNA at 100 nM concentration while those carrying four of either dodecylphenyl or hexadecyl groups were significantly inferior (Figure 8). We foresee that such therapeutic lipid-oligonucleotide conjugates (LONs) displaying low toxicity and improved cellular uptake may demonstrate high efficiency and minimal side effects in future experiments in vivo.

5. Conclusions

To conclude, we have obtained lipid conjugates of oligodeoxynucleotides and siRNAs carrying one, two or four hydrophobic ((4-dodecylphenyl)sulfonyl)phosphoramidate or (hexadecylsulfonyl)phosphoramidate groups at either 5′- or 3′-internucleotidic positions by a modified protocol of the phosphoramidite synthesis replacing usual aqueous iodine oxidation by the Staudinger reaction between a solid-supported phosphite triester and the corresponding long-chain sulfonyl azide.

The oligodeoxynucleotide conjugates were characterized by low cytotoxicity in either murine or human macrophages and improved cell uptake into the same in the absence of a transfection agent. Spontaneous formation of nanoparticles was observed in aqueous solutions of doubly-conjugated oligonucleotides having either of the two lipophilic modifications at the 5′-end, which may be responsible for the improved cellular uptake of the respective conjugates. Notably, the size of the nanoparticles decreased with the increase in the length of the hydrocarbon chain from dodecylphenyl to hexadecyl. Flow cytometry revealed the superior uptake of hexadecyl conjugates over the dodecyl ones into human macrophages. Intracellular localization of the conjugates according to confocal microscopy was predominantly cytosolic with the majority of the fluorescence within endosomes.

Finally, two adjacent lipophilic dodecylphenyl or hexadecyl groups placed within the 3′-terminal overhang and at the 3′-terminal phosphate of the passenger strand did not significantly impair the gene-silencing ability of siRNAs targeting repulsive guidance molecule (RGMa) mRNA in PK-59 cell culture, whereas the siRNAs carrying four of the same at the 3′-ends of both passenger and antisense strands were considerably less active. Thus, lipid-siRNA conjugates with optimized number and position of the lipid chains could be a prospective candidate for future experiments in an in vivo model of multiple sclerosis and, potentially, other diseases when RNA targeting is applicable.

Supplementary Materials: The following are available online at https://www.mdpi.com/2076-3417/11/3/1174/s1, Figures S1–S3: IR, ^{1}H and ^{13}C spectra of 1-hexadecanesulfonyl azide; reverse-phased HPLC profiles and ESI mass spectra of modified oligonucleotides, Figure S4: Intracellular accumulation of η-*F* conjugate in KB-8-5 cells, Figure S5: Intracellular accumulation of δ-*F* and η-*F* conjugates in RAW264.7 cells and THP-1-induced macrophages.

Author Contributions: The first three authors (A.D., O.M. and A.F.) made equal contributions to the work. Conceptualization, D.S.; methodology, M.Z., M.F., D.S. and T.Z.; software, O.M.; validation, D.S., M.F. and M.Z.; formal analysis, O.M. and Y.S.; investigation, A.D. and Y.S.; resources, M.Z. and T.Z.; data curation, O.M., A.F. and Y.S.; writing—original draft preparation, O.M. and D.S.; writing—review and editing, O.M., M.Z. and D.S.; visualization, O.M. and A.D.; supervision, O.M. and A.F.; project administration, M.Z., M.F. and D.S.; funding acquisition, M.Z., D.S. and M.F. All authors have read and agreed to the published version of the manuscript.

Funding: This research was funded by Russian Science Foundation, grant No. 19-74-30011 (cytotoxicity, confocal microscopy, flow cytometry, dynamic light scattering), Russian Foundation for Basic Research, grant No. 18-515-57006 (siRNA synthesis), the Ministry of Science and Higher Education of the Russian Federation, project No. FSUS 2020-0035 to Novosibirsk State University (oligodeoxynucleotide synthesis), and Japan Medical Research Foundation, grant No. 2018JP002 (biological activity).

Data Availability Statement: Data sharing is not applicable to this article.

Conflicts of Interest: The authors declare no conflict of interest.

References

1. Crooke, S.T. Molecular mechanisms of antisense oligonucleotides. *Nucleic Acid Ther.* **2017**, *27*, 70–77. [CrossRef]
2. Shen, X.; Corey, D.R. Chemistry, mechanism and clinical status of antisense oligonucleotides and duplex RNAs. *Nucleic Acids Res.* **2018**, *46*, 1584–1600. [CrossRef]
3. Smith, C.I.E.; Zain, R. Therapeutic Oligonucleotides: State of the Art. *Annu. Rev. Pharmacol. Toxicol.* **2019**, *59*, 605–630. [CrossRef] [PubMed]

4. Evers, M.M.; Toonen, L.J.A.; van Roon-Mom, W.M.C. Antisense oligonucleotides in therapy for neurodegenerative disorders. *Adv. Drug Deliv. Rev.* **2015**, *87*, 90–103. [CrossRef] [PubMed]
5. Aung-Htut, M.T.; McIntosh, C.S.; Ham, K.A.; Pitout, I.L.; Flynn, L.L.; Greer, K.; Fletcher, S.; Wilton, S.D. Systematic Approach to Developing Splice Modulating Antisense Oligonucleotides. *Int. J. Mol. Sci.* **2019**, *20*, 5030. [CrossRef] [PubMed]
6. Bennett, C.F.; Baker, B.F.; Pham, N.; Swayze, E.; Geary, R.S. Pharmacology of Antisense Drugs. *Annu. Rev. Pharmacol. Toxicol.* **2017**, *57*, 81–105. [CrossRef]
7. Castanotto, D.; Stein, C.A. Antisense oligonucleotides in cancer. *Curr. Opin. Oncol.* **2014**, *26*, 584–589. [CrossRef]
8. Wagner, A.; Bock, C.T.; Fechner, H.; Kurreck, J. Application of modified antisense oligonucleotides and siRNAs as antiviral drugs. *Future Med. Chem.* **2015**, *7*, 1637–1642. [CrossRef]
9. Hegarty, J.P.; Stewart, D.B. Advances in therapeutic bacterial antisense biotechnology. *Appl. Microbiol. Biotechnol.* **2018**, *102*, 1055–1065. [CrossRef]
10. Shimo, T.; Maruyama, R.; Yokota, T. Designing Effective Antisense Oligonucleotides for Exon Skipping. In *Duchenne Muscular Dystrophy. Methods and Protocols*; Bernardini, C., Ed.; Humana Press: New York, NY, USA, 2018.
11. Ma, C.-C.; Wang, Z.-L.; Xu, T.; He, Z.-Y.; Wei, Y.-Q. The approved gene therapy drugs worldwide: From 1998 to 2019. *Biotechnol. Adv.* **2020**, *40*, 107502. [CrossRef]
12. Juliano, R.L. The delivery of therapeutic oligonucleotides. *Nucleic Acids Res.* **2016**, *44*, 6518–6548. [CrossRef] [PubMed]
13. Roberts, T.C.; Langer, R.; Wood, M.J.A. Advances in oligonucleotide drug delivery. *Nat. Rev. Drug. Discov.* **2020**, *19*, 673–694. [CrossRef] [PubMed]
14. Pattanayak, S.; Khatra, H.; Saha, S.; Sinha, S. A cationic morpholino antisense oligomer conjugate: Synthesis, cellular uptake and inhibition of Gli1 in the hedgehog signalling pathway. *RSC Adv.* **2014**, *4*, 1951–1954. [CrossRef]
15. Hnedzko, D.; McGee, D.W.; Karamitas, Y.A.; Rozners, E. Sequence-selective recognition of double-stranded RNA and enhanced cellular uptake of cationic nucleobase and backbone-modified peptide nucleic acids. *RNA* **2017**, *23*, 58–69. [CrossRef] [PubMed]
16. Wang, M.; Wu, B.; Shah, S.N.; Lu, P.; Lu, Q. Polyquaternium-mediated delivery of morpholino oligonucleotides for exon-skipping in vitro and in mdx mice. *Drug Deliv.* **2017**, *24*, 952–961. [CrossRef] [PubMed]
17. Morcos, P.A.; Li, Y.; Jiang, S. Vivo-Morpholinos: A non-peptide transporter delivers Morpholinos into a wide array of mouse tissues. *Biotechniques* **2018**, *45*, 613–623. [CrossRef]
18. Craig, K.; Abrams, M.; Amiji, M. Recent preclinical and clinical advances in oligonucleotide conjugates. *Expert Opin. Drug Deliv.* **2018**, *15*, 629–640. [CrossRef]
19. Ren, T.; Song, Y.K.; Zhang, G.; Liu, D. Structural basis of DOTMA for its high intravenous transfection activity in mouse. *Gene Ther.* **2000**, *7*, 764–768. [CrossRef]
20. Patwa, A.; Gissot, A.; Bestel, I.; Barthélémy, P. Hybrid lipid oligonucleotide conjugates: Synthesis, self-assemblies and biomedical applications. *Chem. Soc. Rev.* **2011**, *40*, 5844–5854. [CrossRef]
21. Zhao, B.; Tian, Q.; Bagheri, Y.; You, M. Lipid–oligonucleotide conjugates for simple and efficient cell membrane engineering and bioanalysis. *Curr. Opin. Biomed. Eng.* **2020**, *13*, 76–83. [CrossRef]
22. Li, X.; Feng, K.; Li, L.; Yang, L.; Pan, X.; Yazd, H.S.; Cui, C.; Li, J.; Moroz, L.; Sun, Y.; et al. Lipid-oligonucleotide conjugates for bioapplications. *Nat. Sci. Rev.* **2020**, *7*, 1933–1953. [CrossRef]
23. Osborn, M.F.; Khvorova, A. Improving siRNA Delivery In Vivo Through Lipid Conjugation. *Nucleic Acid Ther.* **2018**, *28*, 128–136. [CrossRef] [PubMed]
24. Østergaard, M.E.; Jackson, M.; Low, A.; Chappell, A.E.; Lee, R.G.; Peralta, R.Q.; Yu, J.; Kinberger, G.A.; Dan, A.; Carty, R.; et al. Conjugation of hydrophobic moieties enhances potency of antisense oligonucleotides in the muscle of rodents and non-human primates. *Nucleic Acids Res.* **2019**, *47*, 6045–6058. [CrossRef] [PubMed]
25. Chernikov, I.V.; Meschaninova, M.I.; Chernolovskaya, E.L. Preparation, Determination of Activity, and Biodistribution of Cholesterol-Containing Nuclease-Resistant siRNAs In Vivo. In *RNA Interference and CRISPR Technologies. Technical Advances and New Therapeutic Opportunities*; Sioud, M., Ed.; Humana Press: New York, NY, USA, 2020.
26. Prokhorova, D.V.; Chelobanov, B.P.; Burakova, E.A.; Fokina, A.A.; Stetsenko, D.A. New oligodeoxyribonucleotide derivatives bearing internucleotide *N*-tosyl phosphoramidate groups: Synthesis and complementary binding to DNA and RNA. *Russ. J. Bioorg. Chem.* **2017**, *43*, 38–42. [CrossRef]
27. Chelobanov, B.P.; Burakova, E.A.; Prokhorova, D.V.; Fokina, A.A.; Stetsenko, D.A. New oligodeoxynucleotide derivatives containing *N*-(methanesulfonyl)-phosphoramidate (mesyl phosphoramidate) internucleotide group. *Russ. J. Bioorg. Chem.* **2017**, *43*, 664–668. [CrossRef]
28. Burakova, E.A.; Derzhalova, A.S.; Chelobanov, B.P.; Fokina, A.A.; Stetsenko, D.A. New Oligodeoxynucleotide Derivatives Containing *N*-(Sulfonyl)-Phosphoramide Groups. *Russ. J. Bioorg. Chem.* **2017**, *45*, 662–668. [CrossRef]
29. Kitayama, M.; Ueno, M.; Itakura, T.; Yamashita, T. Activated Microglia Inhibit Axonal Growth through RGMa. *PLoS ONE* **2011**, *6*, e25234. [CrossRef]
30. Farzan, V.M.; Ulashchik, E.A.; Martynenko-Makaev, Y.V.; Kvach, M.V.; Aparin, I.O.; Brylev, V.A.; Prikazchikova, T.A.; Maklakova, S.Y.; Majouga, A.G.; Ustinov, A.V.; et al. Automated Solid-Phase Click Synthesis of Oligonucleotide Conjugates: From Small Molecules to Diverse N-Acetylgalactosamine Clusters. *Bioconjug. Chem.* **2017**, *28*, 2599–2607. [CrossRef]
31. Matano, Y.; Ohkubo, H.; Honsho, Y.; Saito, A.; Seki, S.; Imahori, H. Synthesis and Charge-Carrier Transport Properties of Poly(phosphole P-alkanesulfonylimide)s. *Org. Lett.* **2013**, *15*, 932–935. [CrossRef]

32. Kamerkar, S.; LeBleu, V.S.; Sugimoto, H.; Yang, S.; Ruivo, C.F.; Melo, S.A.; Lee, J.J.; Kalluri, R. Exosomes facilitate therapeutic targeting of oncogenic KRAS in pancreatic cancer. *Nature* **2017**, *546*, 498–503. [CrossRef]

33. Song, E.; Lee, S.K.; Wang, J.; Ince, N.; Ouyang, N.; Min, J.; Chen, J.; Lieberman, S.P. RNA interference targeting Fas protects mice from fulminant hepatitis. *J. Nat. Med.* **2003**, *9*, 347–351. [CrossRef] [PubMed]

34. Markov, O.V.; Filatov, A.V.; Kupryushkin, M.S.; Chernikov, I.V.; Patutina, O.A.; Strunov, A.A.; Chernolovskaya, E.L.; Vlassov, V.V.; Pyshnyi, D.V.; Zenkova, M.A. Transport Oligonucleotides—A Novel System for Intracellular Delivery of Antisense Therapeutics. *Molecules* **2020**, *25*, 3663. [CrossRef] [PubMed]

35. Kubo, T.; Tokita, S.; Yamashita, T. Repulsive guidance molecule-a and demyelination: Implications for multiple sclerosis. *J. Neuroimmune Pharmacol.* **2012**, *7*, 524–528. [CrossRef] [PubMed]

36. Mothe, A.J.; Tassew, N.G.; Shabanzadeh, A.P.; Penheiro, R.; Vigouroux, R.J.; Huang, L.; Grinnell, C.; Cui, Y.F.; Fung, E.; Monnier, P.P.; et al. RGMa inhibition with human monoclonal antibodies promotes regeneration, plasticity and repair, and attenuates neuropathic pain after spinal cord injury. *Sci. Rep.* **2017**, *7*, 10529. [CrossRef] [PubMed]

37. Miroshnichenko, S.K.; Patutina, O.A.; Burakova, E.A.; Chelobanov, B.P.; Fokina, A.A.; Vlassov, V.V.; Altman, S.; Zenkova, M.A.; Stetsenko, D.A. Mesyl phosphoramidate antisense oligonucleotides as an alternative to phosphorothioates: Improved biochemical and biological properties. *Proc. Natl. Acad. Sci. USA* **2019**, *116*, 1229–1234. [CrossRef] [PubMed]

38. Patutina, O.A.; Gaponova (Miroshnichenko), S.K.; Sen'kova, A.V.; Savin, I.A.; Gladkikh, D.V.; Burakova, E.A.; Fokina, A.A.; Maslov, M.A.; Shmendel', E.V.; Wood, M.J.A.; et al. Mesyl phosphoramidate backbone modified antisense oligonucleotides targeting miR-21 with enhanced in vivo therapeutic potency. *Proc. Natl. Acad. Sci. USA* **2020**, *117*. in press. [CrossRef]

39. Kupryushkin, M.S.; Pyshnyi, D.V.; Stetsenko, D.A. Phosphoryl Guanidines: A New Type of Nucleic Acid Analogues. *Acta Nat.* **2014**, *6*, 116–118. [CrossRef]

40. Ezzat, K.; Aoki, Y.; Koo, T.; McClorey, G.; Benner, L.; Coenen-Stass, A.; O'Donovan, L.; Lehto, T.; Garcia-Guerra, A.; Nordin, J.; et al. Self-Assembly into Nanoparticles Is Essential for Receptor Mediated Uptake of Therapeutic Antisense Oligonucleotides. *Nano Lett.* **2015**, *15*, 4364–4373. [CrossRef]

41. Deo, S.; Patel, T.R.; Dzananovic, E.; Booy, E.P.; Zeid, K.; McEleney, K.; Harding, S.E.; McKenna, S.A. Activation of 2'5'-Oligoadenylate Synthetase by Stem Loops at the 5'-End of the West Nile Virus Genome. *PLoS ONE* **2014**, *9*, e92545. [CrossRef]

42. Awino, J.K.; Gudipati, S.; Hartmann, A.K.; Santiana, J.J.; Cairns-Gibson, D.F.; Gomez, N.; Rouge, J.L. Nucleic Acid Nanocapsules for Enzyme-Triggered Drug Release. *J. Am. Chem. Soc.* **2017**, *139*, 6278–6281. [CrossRef]

43. Soutschek, J.; Akinc, A.; Bramlage, B.; Charisse, K.; Constien, R.; Donoghue, M.; Elbashir, S.; Geick, A.; Hadwiger, P.; Harborth, J.; et al. Therapeutic silencing of an endogenous gene by systemic administration of modified siRNAs. *Nature* **2004**, *432*, 173–178. [CrossRef] [PubMed]

44. Karaki, S.; Benizri, S.; Mejías, R.; Baylot, V.; Branger, N.; Nguyen, T.; Vialet, B.; Oumzil, K.; Barthélémy, P.; Rocchi, P. Lipid-oligonucleotide conjugates improve cellular uptake and efficiency of TCTP-antisense in castration-resistant prostate cancer. *J. Control. Release* **2017**, *258*, 1–9. [CrossRef] [PubMed]

45. Zhou, Y.; Zhu, F.; Liu, Y.; Zheng, M.; Wang, Y.; Zhang, D.; Anraku, Y.; Zou, Y.; Li, J.; Wu, H.; et al. Blood-brain barrier-penetrating siRNA nanomedicine for Alzheimer's disease therapy. *Sci. Adv.* **2020**, *6*, eabc7031. [CrossRef] [PubMed]

46. Sathyaprakash, C.; Manzano, R.; Varela, M.A.; Hashimoto, Y.; Wood, M.J.A.; Talbot, K.; Aoki, Y. Development of LNA Gapmer Oligonucleotide-Based Therapy for ALS/FTD Caused by the C9orf72 Repeat Expansion. In *Gapmers: Methods and Protocols*; Yokota, T., Maruyama, R., Eds.; Springer: New York, NY, USA, 2020; pp. 185–208.

47. Scoles, D.R.; Meera, P.; Schneider, M.D.; Paul, S.; Dansithong, W.; Figueroa, K.P.; Hung, G.; Rigo, F.; Bennett, C.F.; Otis, T.S.; et al. Antisense oligonucleotide therapy for spinocerebellar ataxia type 2. *Nature* **2017**, *544*, 362–366. [CrossRef] [PubMed]

48. Adams, D.; Gonzalez-Duarte, A.; O'Riordan, W.D.; Yang, C.C.; Ueda, M.; Kristen, A.V.; Tournev, I.; Schmidt, H.H.; Coelho, T.; Berk, J.L.; et al. Patisiran, an RNAi therapeutic, for hereditary transthyretin amyloidosis. *N. Engl. J. Med.* **2018**, *379*, 11–21. [CrossRef] [PubMed]

49. Scott, L.J. Givosiran: First Approval. *Drugs* **2020**, *80*, 335–339. [CrossRef]

50. Heo, Y.-A. Golodirsen: First Approval. *Drugs* **2020**, *80*, 329–333. [CrossRef]

51. Paik, J.; Duggan, S. Volanesorsen: First Global Approval. *Drugs* **2019**, *79*, 1349–1354. [CrossRef]

52. Keam, S.J. Inotersen: First Global Approval. *Drugs* **2018**, *78*, 1371–1376. [CrossRef]

53. Remaut, K.; Lucas, B.; Braeckmans, K.; Demeester, J.; De Smedt, S.C. Pegylation of liposomes favours the endosomal degradation of the delivered phosphodiester oligonucleotides. *J. Control. Release* **2007**, *117*, 256–266. [CrossRef]

54. Chen, J.; Yu, Z.; Chen, H.; Gao, J.; Liang, W. Transfection efficiency and intracellular fate of polycation liposomes combined with protamine. *Biomaterials* **2011**, *32*, 1412–1418. [CrossRef] [PubMed]

55. Ara, M.N.; Matsuda, T.; Hyodo, M.; Sakurai, Y.; Hatakeyama, H.; Ohga, N.; Hida, K.; Harashima, H. An aptamer ligand based liposomal nanocarrier system that targets tumor endothelial cells. *Biomaterials* **2014**, *35*, 7110–7120. [CrossRef] [PubMed]

56. Varkouhi, A.K.; Scholte, M.; Storm, G.; Haisma, H.J. Endosomal escape pathways for delivery of biologicals. *J. Control. Release* **2011**, *151*, 220–228. [CrossRef] [PubMed]

57. Petrova, N.S.; Chernikov, I.V.; Meschaninova, M.I.; Dovydenko, I.S.; Venyaminova, A.G.; Zenkova, M.A.; Vlassov, V.V.; Chernolovskaya, E.L. Carrier-free cellular uptake and the gene-silencing activity of the lipophilic siRNAs is strongly affected by the length of the linker between siRNA and lipophilic group. *Nucleic Acids Res.* **2012**, *40*, 2330–2344. [CrossRef] [PubMed]

58. Dobson, R.; Giovannoni, G. Multiple sclerosis—A review. *Eur. J. Neurol.* **2019**, *26*, 27–40. [CrossRef]

applied sciences

Communication

Investigation of the Characteristics of NLS-PNA: Influence of NLS Location on Invasion Efficiency

Yuichiro Aiba *,†[image_ref id="3" /], Gerardo Urbina †, Masanari Shibata and Osami Shoji *

Department of Chemistry, Graduate School of Science, Nagoya University, Furo-Cho, Chikusa-Ku, Nagoya, Aichi 464-8602, Japan; gerardoaus@gmail.com (G.U.); shibata.masanari@b.mbox.nagoya-u.ac.jp (M.S.)
* Correspondence: aiba.yuichiro@i.mbox.nagoya-u.ac.jp (Y.A.); shoji.osami@a.mbox.nagoya-u.ac.jp (O.S.)
† These authors equally contributed to this work.

Received: 1 November 2020; Accepted: 30 November 2020; Published: 3 December 2020

Abstract: Peptide nucleic acid can recognise sequences in double-stranded DNA (dsDNA) through the formation of a double-duplex invasion complex. This double-duplex invasion is a promising method for the recognition of dsDNA in cellula because peptide nucleic acid (PNA) invasion does not require the prior denaturation of dsDNA. To increase its applicability, we developed PNAs modified with a nuclear localisation signal (NLS) peptide. In this study, the characteristics of NLS-modified PNAs were investigated for the future design of novel peptide-modified PNAs.

Keywords: PNA; invasion; DNA; NLS

1. Introduction

The sequence-specific recognition of double-stranded DNA (dsDNA) has been an important research subject owing to its wide range of potential applications [1–7]. For this purpose, various methods have been developed, including those that make use of DNA-binding proteins [1,3–5,8–12], small molecules such as minor groove binders [13–16], and artificial DNA [17–24]. In addition, the recognition of dsDNA by peptide nucleic acid (PNA), an artificial nucleic acid mimic, has been reported by Nielsen et al. [7,21,24–31]. In 1991, PNA was first designed and synthesised as an analogue of natural nucleic acids, where the negatively charged sugar-phosphate backbone of DNA was substituted with an electrostatically neutral artificial N-(2-aminoethyl)glycine backbone (Figure 1a) [25]. Consequently, electrostatic repulsion between PNA and DNA is absent, enabling PNA to form more stable duplexes with complementary DNA via Watson–Crick base pairing than those between complementary DNA strands [31]. Furthermore, when pseudo-complementary PNAs (pcPNAs: see Figure S1)—where conventional adenine (A) and thymine (T) have been replaced by 2,6-diaminopurine (D) and 2-thiouracil (U)—are used, an effective invasion into dsDNA becomes possible, with two strands of pcPNA forming a double-duplex invasion complex (or, simply, invasion complex, Figure 1b) with dsDNA [26,32]. The introduction of D and U into PNA destabilises the formation of PNA/PNA duplexes through steric repulsions between the amino group of D and the thioketone group of U. The thermal stability of the PNA/DNA duplex is enhanced by the formation of stable base pairs between D/U and the natural nucleobases. PNA's unique DNA-recognition mode has been used in many applications, including site-directed mutagenesis [2,7], inhibition of enzymatic activity [26,33,34], and the development of artificial DNA cutters [35], with all of these depending on the sequence specificity of the invasion complex. The pseudopeptide backbone of PNA facilitates easy oligomer synthesis and modification for further improvements and imparts enhanced resistance to enzymatic degradation. Thus, double-duplex invasion is a promising method for the targeting of dsDNA in cellula.

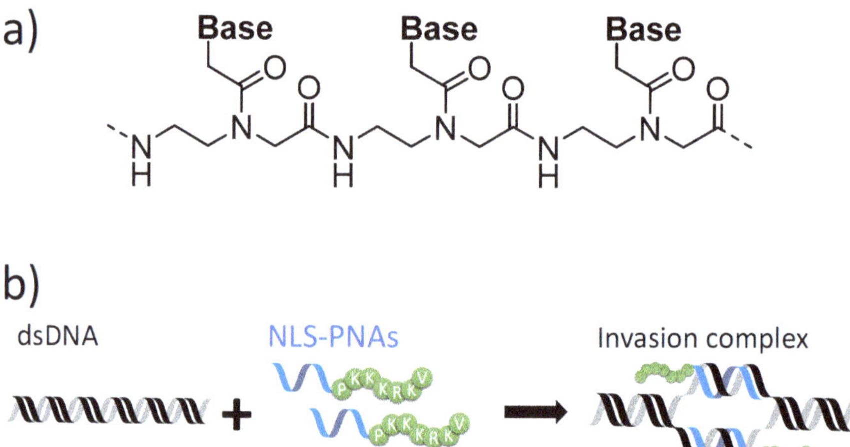

Figure 1. (**a**) Chemical structure of peptide nucleic acid (PNA). (**b**) Stable invasion complex formation using nuclear-localisation-signal-modified PNAs (NLS-PNAs).

However, to realise such applications, an important issue must be overcome—namely, the high salt concentrations encountered in the cellular environment. High salt concentrations increase the stability of the DNA/DNA duplexes whilst slightly destabilising the PNA/DNA duplexes [36], which results in an overall decrease in invasion efficiency [37]. Several approaches have been attempted to enhance invasion efficiency in physiological salt concentrations, such as chemical modification of PNA and conjugation with functional molecules [37–41]. Our group developed conjugates of PNA with Ru-complexes and PNAs modified with a nuclear localisation signal (NLS) peptide to increase their applicability for DNA recognition at high salt concentrations (Figure 1b). These two modified PNAs displayed higher DNA affinity than corresponding unmodified pcPNAs and effectively formed an invasion complex even at physiological salt concentrations, whereunder unmodified PNAs stop functioning efficiently. It is important to understand the properties and limitations of these modified PNAs in order to further improve PNAs. Of these two methods, NLS modification is the simpler PNA modification method as the resulting NLS-modified PNAs (NLS-PNAs) can be procured by simply conjugating PNA with the functional peptide, and can be readily introduced during the synthesis of PNAs. Thus, in this study, we focused on NLS-PNAs and investigated their characteristics to obtain insight into how they can be improved.

2. Materials and Methods

2.1. Materials

Solvents and reagents were purchased from FUJIFILM Wako Pure Chemical Co. (Tokyo, Japan); Tokyo Chemical Industry Co., Ltd. (Tokyo, Japan); Merck (Darmstadt, Germany); Sigma-Aldrich Co., LLC. (St. Louis, MO, USA); nacalai tesque Inc. (Kyoto, Japan); and Kanto Chemical Co., Inc. (Tokyo, Japan); and were used without further purification. PNA monomers were purchased from ASM Research Chemicals (Hannover, Germany). Oligonucleotides were purchased from Fasmac (Kanagawa, Japan).

PNAs with or without an NLS peptide were synthesised by standard Boc-chemistry-based solid-phase peptide synthesis according to a literature procedure [42], purified by reversed-phase high-performance liquid chromatography (HPLC) (Jasco Co.; Tokyo, Japan), and characterised by matrix-assisted laser desorption/ionization–time of flight mass spectrometer (MALDI-TOF MS) (ultraflex III, Bruker Daltonics; Billerica, MA, USA) (Figures S2 and S3). The concentration of PNAs and DNAs was determined based on their absorbance at 260 nm by using a UV-visible spectrophotometer (molar extinction coefficients for PNA monomers (in M^{-1} cm^{-1}): $\varepsilon(D)$ = 7600; $\varepsilon(C)$ = 6600; $\varepsilon(G)$ = 11,700; $\varepsilon(U)$ = 10,200).

2.2. Invasion Experiments

Target 130-bp DNA from pBR322 was incubated with a pair of pcPNAs in a 5 mM HEPES buffer (pH 7.0) at 50 °C for 1 h. After incubation, the solutions were subjected to a microchip electrophoresis system (MultiNA, Shimadzu; Kyoto, Japan) and the invasion efficiency was evaluated.

3. Results and Discussion

3.1. A Positive Effect on Invasion Efficiency by the Introduction of an NLS into Short PNAs

In a previous report, we successfully demonstrated that conjugation of an NLS to pcPNAs results in enhanced invasion efficiency [37]. Specifically, an NLS peptide with the sequence PKKKRKV found in the SV40 virus, which has been widely studied, was conjugated at the C-terminus of a pentadecamer PNA. If the introduction of an NLS, even in short PNAs, exerts a positive effect upon invasion efficiency, then the properties of NLS-PNAs should be more easily elucidated by examining them with short PNAs, whose synthesis requires less effort than longer PNAs. Thus, we checked the effectiveness of the NLS in decamer PNAs and employed two types of PNA: (1) unmodified PNAs (U-PNAs), i.e., a pair of decamer pcPNAs without an NLS (with only a few lysine residues for solubility) and (2) C-PNAs, which are identical in sequence to U-PNAs but with an NLS at their C-termini.

U-PNAs and C-PNAs were synthesised by standard *tert*-butoxycarbonyl (Boc)-based solid-phase peptide synthesis. The PNA sequences are listed in Table 1. PNA1 and PNA2 represent each strand of pcPNA that is complementary to 10 bp in the pBR322 plasmid (PNA1: GUUDCUGDUG and PNA2: CDUCDGUDDC). All PNAs have a free N-terminal amino group, with the C-terminal carboxylic acid being converted to an amide. The efficiency of invasion complex formation was evaluated by an electrophoresis mobility shift assay (EMSA). The formation of the invasion complex results in a lower electrophoretic mobility, mainly due to changes in the local structure of the dsDNA at the invasion site [43]. This allows the separation of the invasion complex from the free dsDNA. An EMSA of the invasion complex formation in the presence of 75 mM NaCl is shown in Figure 2.

Table 1. PNAs synthesised and used in this research.

Name	Sequences [a] (N to C)
U-PNA1	H- *K* **GUUDCUGDUG** *KK* -NH$_2$
U-PNA2	H- *K* **CDUCDGUDDC** *KK* -NH$_2$
N-PNA1	H- *PKKKRKV* **GUUDCUGDUG** *KK* -NH$_2$
N-PNA2	H- *PKKKRKV* **CDUCDGUDDC** *KK* -NH$_2$
C-PNA1	H- *K* **GUUDCUGDUG** *PKKKRKV* -NH$_2$
C-PNA2	H- *K* **CDUCDGUDDC** *PKKKRKV* -NH$_2$
N-Pro-PNA1	H- *PKKKRKVP* **GUUDCUGDUG** *KK* -NH$_2$
N-Pro-PNA2	H- *PKKKRKVP* **CDUCDGUDDC** *KK* -NH$_2$

[a] Bold: PNAs; Italic: amino acids; *K* = lysine; *P* = proline; *R* = arginine; *V* = valine; D = 2,6-diaminopurine; U = 2-thiouracil.

In lane 1, a single band corresponding to free dsDNA was observed between 100 and 150 bp. In the presence of U-PNA, the invasion complex was barely formed under the high-salt conditions employed, giving only a faint band with lower mobility between 180 and 200 bp (lane 2), which was assignable to the invasion complex. However, the C-PNAs with NLS at their C-termini gave an intense band corresponding to the invasion complex (lane 3). It is clear from these results that the invasion efficiency was improved by the conjugation of an NLS to the short PNAs, leading to the formation of a stable invasion complex even at elevated salt concentrations. As a result, it was confirmed that the NLS modifications in short PNAs showed a beneficial effect on invasion. Thus, we decided to investigate the properties of NLS-PNAs by using these short decamer PNAs. The effect of NLS modification appears to be mainly attributable to electrostatic interactions. However, the conjugation of PNA with repeating cationic peptides (e.g., polylysine) also increases non-specific interactions with the DNA backbone, resulting in a negative effect on the formation of the invasion complex [44]. Presumably, in biological motifs, unfavourable sequences (e.g., those provoking non-specific binding to DNA) have been evolutionarily excluded. Thus, we can expect that NLS peptides possess an ideal amino acid sequence that shows a moderate affinity to DNA as well as a positive effect on invasion efficiency.

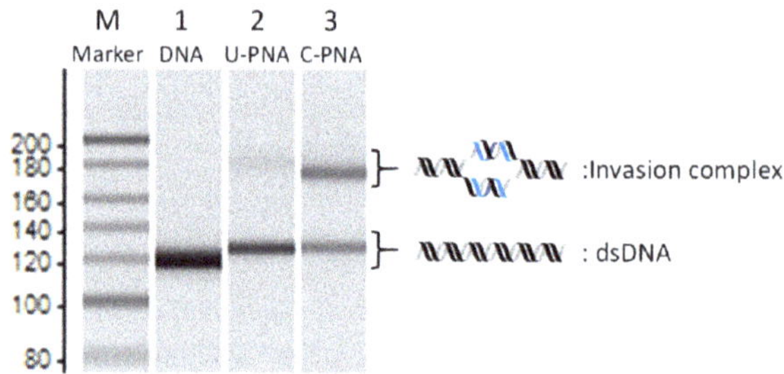

Figure 2. Comparison of U-PNA and C-PNA invasion efficiency in the presence of 75 mM NaCl. M: 20-bp DNA Ladder (TaKaRa); lane 1: 130-bp DNA only; lane2 U-PNA; lane 3: C-PNA. Invasion conditions: [DNA] = [each PNA] = 50 nM, [HEPES (pH 7.0)] = 5 mM, and [NaCl] = 75 mM at 50 °C for 1 h.

3.2. Changing the Location of the NLS from the C-Terminus to the N-Terminus of the PNAs

As mentioned above, in the previous report on NLS-PNAs, the NLS peptide was conjugated to the C-termini of the PNAs (C-PNAs). The synthesis of NLS-PNA was carried out from the C-terminus to the N-terminus by solid-phase peptide synthesis, and the NLS peptide was introduced to the resin first, followed by the PNA. When examining various functional peptides as alternatives to NLS for future research, N-terminally modified PNA conjugates, rather than C-terminally modified PNA conjugates, would significantly reduce the time and effort required for investigation. This is because it is possible to synthesise a large number of peptide-modified PNAs at once by elongating the PNAs first and then splitting the resin followed by peptide conjugation. Therefore, we evaluated whether changing the location of the NLS from the C-terminus to the N-terminus of the PNA (N-PNA) would have a detrimental effect upon invasion efficiency. PNAs modified with NLS peptide at the N-termini (N-PNAs) were synthesised using the same method as C-PNAs, and the invasion experiments with a series of PNAs were carried out under a broad range of salt concentrations, ranging from 0 to 125 mM NaCl.

As a control, the invasion efficiency of U-PNAs was also examined at increasing salt concentrations. Figure 3 shows that the band intensity of the invasion complex with U-PNA decreased with increasing salt concentration and almost disappeared when the NaCl concentration exceeded 50 mM. This result reaffirms the aforementioned phenomenon, wherein elevated salt concentrations stabilise dsDNA whilst weakening PNA/DNA interaction, resulting in a lower invasion efficiency. When the same experiment was conducted using C-PNAs, however, a high invasion efficiency beyond 50 mM NaCl was observed. Unlike in our previous report with pentadecamer NLS-PNAs, in the present study, a decrease in invasion efficiency was observed above 75 mM NaCl. This is due to the lower concentration of PNA, which we selected herein to amplify any effect. The introduction of NLS, even in shorter PNAs, was once more confirmed to be beneficial for invasion at high salt concentrations. Next, the effect of NLS position on invasion efficiency was investigated using N-PNAs. At first glance, both C-PNAs and N-PNAs exhibited similar behaviour in terms of invasion efficiency. The invasion efficiencies of both NLS-PNAs remained high, up to about 75 mM NaCl, although this value gradually decreased with increasing salt concentration. This comparison confirms that, regardless of the location of the NLS, PNA-NLS conjugates display similar performance. However, N-PNAs were slightly better in terms of resistance to higher salt concentrations (over 75 m>M) than C-PNAs, with a more pronounced effect at very high salt concentrations of 125 mM.

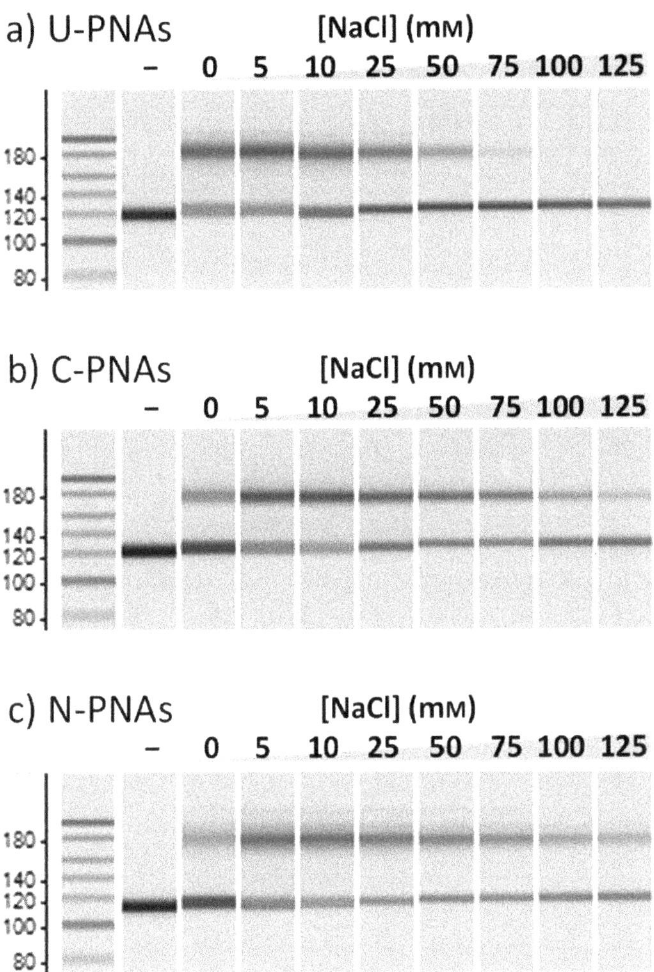

Figure 3. Comparison of (**a**) U-PNAs', (**b**) C PNAs', and (**c**) N-PNAs' invasion efficiency. Invasion conditions: [DNA] = [each PNA] = 50 nM, [HEPES (pH 7.0)] = 5 mM, and [NaCl] = 0–125 mM at 50 °C for 1 h.

3.3. Insertion of Proline between the PNAs and Valine of the NLS

Besides the location of the NLS itself, there is a further difference between C-PNAs and N-PNAs. In order to maintain the directionality of the NLS, in C-PNAs, proline (P) of the NLS connects to the PNA, whereas in N-PNAs, the NLS is connected via valine (V). Unlike valine, proline possesses a unique cyclic structure bestowing it with a much higher backbone rigidity than any of the other canonical amino acids. In order to verify its significance, we synthesised N-Pro-PNAs, which are identical to N-PNAs except that a proline residue has been inserted between the PNA and valine. Interestingly, the insertion of proline proved unfavourable for invasion efficiency, and N-Pro-PNAs displayed invasion properties between those of N-PNAs and C-PNAs (Figure 4). These results strongly indicate that the rigid amino acid connecting the NLS to PNA may unfavourably affect invasion.

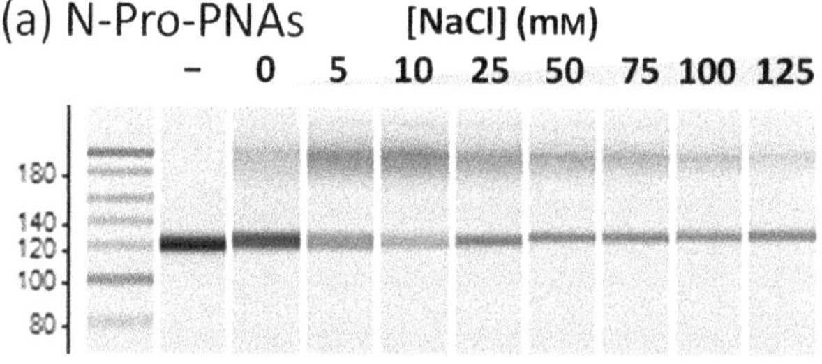

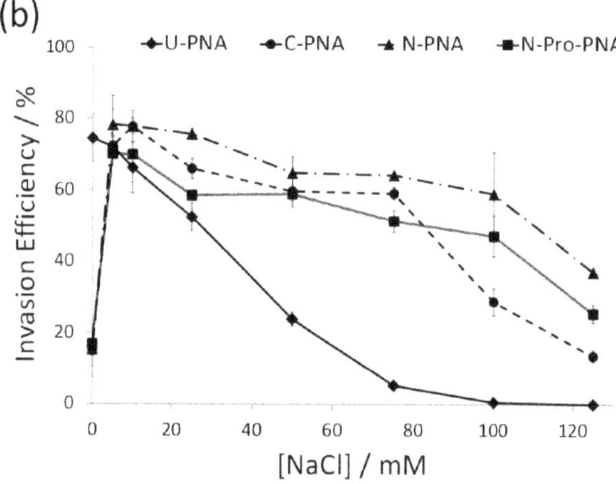

Figure 4. (**a**) Invasion experiments for N-Pro-PNAs to evaluate the effect of an amino acid between NLS and PNA. Invasion conditions: [DNA] = [each PNA] = 50 nM, [HEPES (pH 7.0)] = 5 mM, and [NaCl] = 0–125 mM at 50 °C for 1 h. (**b**) Effect of salt concentration on invasion efficiency with various PNAs (U-, C-, N-, and N-Pro-PNAs). The value for invasion efficiency was obtained by comparing the band intensities in a microchip electrophotogram.

4. Conclusions

In summary, we investigated the characteristics of NLS-PNA by employing a series of NLS-modified PNAs. These NLS-PNA conjugates were found to be superior to unmodified PNAs at elevated salt concentrations. The difference between C-terminally modified and N-terminally modified NLS-PNAs was not very significant, signifying that N-terminally modified PNA-NLS conjugates can be readily used for research purposes using the same method as previously reported C-terminally modified conjugates. This facilitates the development of new functional peptide-modified PNA conjugates, since solid-phase peptide synthesis requires synthesis from the C-terminus to the N-terminus. As an example, for C-terminally modified PNAs, when multiple functional peptides bound to the same PNA need to be screened, each functional peptide must be synthesised separately. In the case of N-terminal modification, the synthesis of peptide-modified PNAs can be simplified by synthesising the individual PNAs only once, separating the resin beads into as many samples as needed and continuing the synthesis of each NLS. In addition to the finding that the position of the NLS has some influence on invasion, we demonstrated that the amino acids between the NLS and PNA also play an important role. These findings also provide new perspectives for the future design of novel peptide-modified PNAs.

Supplementary Materials: The following are available online at http://www.mdpi.com/2076-3417/10/23/8663/s1, Figure S1: chemical structures of 2,6-diaminopurine (D) and 2-thouracil (U), Figure S2: HPLC charts of purified PNAs, Figure S3: MALDI-TOF MS spectra of purified PNAs.

Author Contributions: Investigation, G.U. and M.S.; writing—original draft preparation, Y.A. and G.U.; writing—review and editing, Y.A., G.U., M.S. and O.S.; supervision, Y.A. and O.S.; project administration, Y.A.; funding acquisition, Y.A. and O.S. All authors have read and agreed to the published version of the manuscript.

Funding: This work was supported by JSPS KAKENHI grant no. 19K05730 to Y.A. from the Ministry of Education, Culture, Sports, Science, and Technology (Japan). This work was also supported in part by the Izumi Science and Technology Foundation to Y.A., Kato Memorial Bioscience Foundation to Y.A., and Integrated Research Consortium on Chemical Sciences to Y.A.

Acknowledgments: We thank Joshua Kyle Stanfield for checking the English of this manuscript.

Conflicts of Interest: The authors declare no conflict of interest.

References

1. Kim, Y.G.; Cha, J.; Chandrasegaran, S. Hybrid restriction enzymes: Zinc finger fusions to Fok I cleavage domain. *Proc. Natl. Acad. Sci. USA* **1996**, *93*, 1156–1160. [CrossRef] [PubMed]

2. Kim, K.H.; Nielsen, P.E.; Glazer, P.M. Site-directed gene mutation at mixed sequence targets by psoralen-conjugated pseudo-complementary peptide nucleic acids. *Nucleic Acids Res.* **2007**, *35*, 7604–7613. [CrossRef] [PubMed]

3. Urnov, F.D.; Rebar, E.J.; Holmes, M.C.; Zhang, H.S.; Gregory, P.D. Genome editing with engineered zinc finger nucleases. *Nat. Rev. Genet.* **2010**, *11*, 636–646. [CrossRef] [PubMed]

4. Christian, M.; Cermak, T.; Doyle, E.L.; Schmidt, C.; Zhang, F.; Hummel, A.; Bogdanove, A.J.; Voytas, D.F. Targeting DNA Double-Strand Breaks with TAL Effector Nucleases. *Genetics* **2010**, *186*, 757–761. [CrossRef] [PubMed]

5. Qi, L.S.; Larson, M.H.; Gilbert, L.A.; Doudna, J.A.; Weissman, J.S.; Arkin, A.P.; Lim, W.A. Repurposing CRISPR as an RNA-guided platform for sequence-specific control of gene expression. *Cell* **2013**, *152*, 1173–1183. [CrossRef]

6. Komiyama, M.; Yoshimoto, K.; Sisido, M.; Ariga, K. Chemistry Can Make Strict and Fuzzy Controls for Bio-Systems: DNA Nanoarchitectonics and Cell-Macromolecular Nanoarchitectonics. *Bull. Chem. Soc. Jpn.* **2017**, *90*, 967–1004. [CrossRef]

7. Economos, N.G.; Oyaghire, S.; Quijano, E.; Ricciardi, A.S.; Saltzman, W.M.; Glazer, P.M. Peptide Nucleic Acids and Gene Editing: Perspectives on Structure and Repair. *Molecules* **2020**, *25*, 735. [CrossRef]

8. Wolfe, S.A.; Nekludova, L.; Pabo, C.O. DNA recognition by Cys(2)His(2) zinc finger proteins. *Annu. Rev. Biophys. Biomol. Struct.* **2000**, *29*, 183–212. [CrossRef]

9. Jantz, D.; Amann, B.T.; Gatto, G.J.; Berg, J.M. The design of functional DNA-binding proteins based on zinc finger domains. *Chem. Rev.* **2004**, *104*, 789–799. [CrossRef]

10. Dhanasekaran, M.; Negi, S.; Sugiura, Y. Designer zinc finger proteins: Tools for creating artificial DNA-binding functional proteins. *Acc. Chem. Res.* **2006**, *39*, 45–52. [CrossRef]

11. Boch, J.; Scholze, H.; Schornack, S.; Landgraf, A.; Hahn, S.; Kay, S.; Lahaye, T.; Nickstadt, A.; Bonas, U. Breaking the Code of DNA Binding Specificity of TAL-Type III Effectors. *Science* **2009**, *326*, 1509–1512. [CrossRef] [PubMed]

12. Xu, X.S.; Qi, L.S. A CRISPR-dCas Toolbox for Genetic Engineering and Synthetic Biology. *J. Mol. Biol.* **2019**, *431*, 34–47. [CrossRef] [PubMed]

13. Dervan, P.B.; Burli, R.W. Sequence-specific DNA recognition by polyamides. *Curr. Opin. Chem. Biol.* **1999**, *3*, 688–693. [CrossRef]

14. Dervan, P.B. Molecular recognition of DNA by small molecules. *Biorg. Med. Chem.* **2001**, *9*, 2215–2235. [CrossRef]

15. Dervan, P.B.; Edelson, B.S. Recognition of the DNA minor groove by pyrrole-imidazole polyamides. *Curr. Opin. Struct. Biol.* **2003**, *13*, 284–299. [CrossRef]

16. Bando, T.; Sugiyama, H. Synthesis and biological properties of sequence-specific DNA-alkylating pyrrole-imidazole polyamides. *Acc. Chem. Res.* **2006**, *39*, 935–944. [CrossRef]

17. Thuong, N.T.; Hélène, C. Sequence-Specific Recognition and Modification of Double-Helical DNA by Oligonucleotides. *Angew. Chem. Int. Ed. Engl.* **1993**, *32*, 666–690. [CrossRef]

18. Praseuth, D.; Guieysse, A.L.; Hélène, C. Triple helix formation and the antigene strategy for sequence-specific control of gene expression. *Biochim. Biophys. Acta Gene Struct. Expr.* **1999**, *1489*, 181–206. [CrossRef]

19. Fox, K.R. Targeting DNA with triplexes. *Curr. Med. Chem.* **2000**, *7*, 17–37. [CrossRef]

20. Rapireddy, S.; He, G.; Roy, S.; Armitage, B.A.; Ly, D.H. Strand invasion of mixed-sequence B-DNA by acridine-linked, gamma-peptide nucleic acid (gamma-PNA). *J. Am. Chem. Soc.* **2007**, *129*, 15596–15600. [CrossRef]

21. Vilaivan, T. Pyrrolidinyl PNA with alpha/beta-Dipeptide Backbone: From Development to Applications. *Acc. Chem Res.* **2015**, *48*, 1645–1656. [CrossRef] [PubMed]

22. Hrdlicka, P.J.; Karmakar, S. 25 years and still going strong: 2′-O-(pyren-1-yl) methylribonucleotides-versatile building blocks for applications in molecular biology, diagnostics and materials science. *Org. Biomol. Chem.* **2017**, *15*, 9760–9774. [CrossRef] [PubMed]

23. Nakamura, S.; Kawabata, H.; Fujimoto, K. Double duplex invasion of DNA induced by ultrafast photo-cross-linking using 3-cyanovinylcarbazole for antigene methods. *Chem. Commun.* **2017**, *53*, 7616–7619. [CrossRef] [PubMed]

24. Muangkaew, P.; Vilaivan, T. Modulation of DNA and RNA by PNA. *Bioorg. Med. Chem. Lett.* **2020**, *30*, 127064. [CrossRef] [PubMed]

25. Nielsen, P.E.; Egholm, M.; Berg, R.H.; Buchardt, O. Sequence-Selective Recognition of DNA by Strand Displacement with a Thymine-Substituted Polyamide. *Science* **1991**, *254*, 1497–1500. [CrossRef] [PubMed]

26. Lohse, J.; Dahl, O.; Nielsen, P.E. Double duplex invasion by peptide nucleic acid: A general principle for sequence-specific targeting of double-stranded DNA. *Proc. Natl. Acad. Sci. USA* **1999**, *96*, 11804–11808. [CrossRef] [PubMed]

27. Demidov, V.V.; Frank-Kamenetskii, M.D. Two sides of the coin: Affinity and specificity of nucleic acid interactions. *Trends Biochem. Sci.* **2004**, *29*, 62–71. [CrossRef]

28. Nielsen, P.E. Peptide Nucleic Acids (PNA) in Chemical Biology and Drug Discovery. *Chem. Biodivers.* **2010**, *7*, 786–804. [CrossRef]

29. Sharma, C.; Awasthi, S.K. Versatility of peptide nucleic acids (PNAs): Role in chemical biology, drug discovery, and origins of life. *Chem. Biol. Drug Des.* **2017**, *89*, 16–37. [CrossRef]

30. Manicardi, A.; Rozzi, A.; Korom, S.; Corradini, R. Building on the peptide nucleic acid (PNA) scaffold: A biomolecular engineering approach. *Supramol. Chem.* **2017**, *29*, 784–795. [CrossRef]

31. Egholm, M.; Buchardt, O.; Christensen, L.; Behrens, C.; Freier, S.M.; Driver, D.A.; Berg, R.H.; Kim, S.K.; Norden, B.; Nielsen, P.E. PNA hybridizes to complementary oligonucleotides obeying the Watson–Crick hydrogen-bonding rules. *Nature* **1993**, *365*, 566–568. [CrossRef] [PubMed]

32. Haaima, G.; Lohse, A.; Buchardt, O.; Nielsen, P.E. Peptide Nucleic Acids (PNAs) Containing Thymine Monomers Derived from Chiral Amino Acids: Hybridization and Solubility Properties of D-Lysine PNA. *Angew. Chem. Int. Ed.* **1996**, *35*, 1939–1942.

33. Izvolsky, K.I.; Demidov, V.V.; Nielsen, P.E.; Frank-Kamenetskii, M.D. Sequence-specific protection of duplex DNA against restriction and methylation enzymes by pseudocomplementary PNAs. *Biochemistry* **2000**, *39*, 10908–10913. [CrossRef] [PubMed]

34. Protozanova, E.; Demidov, V.V.; Nielsen, P.E.; Frank-Kamenetskii, M.D. Pseudocomplementary PNAs as selective modifiers of protein activity on duplex DNA: The case of type IIs restriction enzymes. *Nucleic Acids Res.* **2003**, *31*, 3929–3935. [CrossRef] [PubMed]

35. Komiyama, M.; Aiba, Y.; Yamamoto, Y.; Sumaoka, J. Artificial restriction DNA cutter for site-selective scission of double-stranded DNA with tunable scission site and specificity. *Nat. Protoc.* **2008**, *3*, 655–662. [CrossRef]

36. Tomac, S.; Sarkar, M.; Ratilainen, T.; Wittung, P.; Nielsen, P.E.; Norden, B.; Graslund, A. Ionic effects on the stability and conformation of peptide nucleic acid complexes. *J. Am. Chem. Soc.* **1996**, *118*, 5544–5552. [CrossRef]

37. Aiba, Y.; Honda, Y.; Komiyama, M. Promotion of Double-Duplex Invasion of Peptide Nucleic Acids through Conjugation with Nuclear Localization Signal Peptide. *Chem. Eur. J.* **2015**, *21*, 4021–4026. [CrossRef]

38. Aiba, Y.; Hamano, Y.; Kameshima, W.; Araki, Y.; Wada, T.; Accetta, A.; Sforza, S.; Corradini, R.; Marchelli, R.; Komiyama, M. PNA-NLS conjugates as single-molecular activators of target sites in double-stranded DNA for site-selective scission. *Org. Biomol. Chem.* **2013**, *11*, 5233–5238. [CrossRef]

39. Aiba, Y.; Ohyama, J.; Komiyama, M. Transfection of PNA-NLS Conjugates into Human Cells Using Partially Complementary Oligonucleotides. *Chem. Lett.* **2015**, *44*, 1547–1549. [CrossRef]

40. Hibino, M.; Aiba, Y.; Watanabe, Y.; Shoji, O. Peptide Nucleic Acid Conjugated with Ruthenium-Complex Stabilizing Double-Duplex Invasion Complex Even under Physiological Conditions. *ChemBioChem* **2018**, *19*, 1601–1604. [CrossRef]

41. Hibino, M.; Aiba, Y.; Shoji, O. Cationic guanine: Positively charged nucleobase with improved DNA affinity inhibits self-duplex formation. *Chem. Commun.* **2020**, *56*, 2546–2549. [CrossRef] [PubMed]

42. Komiyama, M.; Aiba, Y.; Ishizuka, T.; Sumaoka, J. Solid-phase synthesis of pseudo-complementary peptide nucleic acids. *Nat. Protoc.* **2008**, *3*, 646–654. [CrossRef] [PubMed]

43. Kuhn, H.; Cherny, D.I.; Demidov, V.V.; Frank-Kamenetskii, M.D. Inducing and modulating anisotropic DNA bends by pseudocomplementary peptide nucleic acids. *Proc. Natl. Acad. Sci. USA* **2004**, *101*, 7548–7553. [CrossRef] [PubMed]

44. Ishizuka, T.; Yoshida, J.; Yamamoto, Y.; Sumaoka, J.; Tedeschi, T.; Corradini, R.; Sforza, S.; Komiyama, M. Chiral introduction of positive charges to PNA for double-duplex invasion to versatile sequences. *Nucleic Acids Res.* **2008**, *36*, 1464–1471. [CrossRef] [PubMed]

Publisher's Note: MDPI stays neutral with regard to jurisdictional claims in published maps and institutional affiliations.

Article

Construction of an Enzymatically-Conjugated DNA Aptamer–Protein Hybrid Molecule for Use as a BRET-Based Biosensor

Masayasu Mie ***, Rena Hirashima, Yasumasa Mashimo and Eiry Kobatake**

Department of Life Science and Technology, School of Life Science and Technology, Tokyo Institute of Technology, 4259 Nagatsuta-cho, Midori-ku, Yokohama 226-8502, Japan; nrmsrh410@gmail.com (R.H.); mashimo.y.ab@m.titech.ac.jp (Y.M.); kobatake.e.aa@m.titech.ac.jp (E.K.)
* Correspondence: mie.m.aa@m.titech.ac.jp

Received: 28 September 2020; Accepted: 27 October 2020; Published: 29 October 2020

Abstract: DNA-protein conjugates are useful molecules for construction of biosensors. Herein, we report the development of an enzymatically-conjugated DNA aptamer–protein hybrid molecule for use as a bioluminescence resonance energy transfer (BRET)-based biosensor. DNA aptamers were enzymatically conjugated to a fusion protein via the catalytic domain of porcine circovirus type 2 replication initiation protein (PCV2 Rep) comprising residues 14–109 (tpRep), which was truncated from the full catalytic domain of PCV2 Rep comprising residues 1–116 by removing the flexible regions at the N- and C-terminals. For development of a BRET-based biosensor, we constructed a fusion protein in which tpRep was positioned between NanoLuc luciferase and a fluorescent protein and conjugated to single-stranded DNA aptamers that specifically bind to either thrombin or lysozyme. We demonstrated that the BRET ratios depended on the concentration of the target molecules.

Keywords: DNA-protein conjugate; replication initiation protein; DNA aptamer; BRET-based biosensor

1. Introduction

DNA-protein conjugates are useful materials for biosensing applications [1]. Use of DNA in these conjugate molecules allows for both signal amplification as well as molecular recognition (e.g., using DNA aptamers) [2–5]. DNA aptamers bind to specific molecules with high affinity and specificity. Therefore, DNA aptamer-based biosensing systems have been constructed as conjugates of DNA aptamers and reporter molecules [6–8]. Single-stranded DNA (ssDNA) containing aptamer sequences are modified at the 5′ and/or 3′ ends with a fluorophore for use in construction of DNA aptamer-based biosensors [9–12]. In addition to modification of ssDNA with a fluorophore, a reporter protein can also be conjugated to ssDNA [3–5]. However, conventional methods for conjugation of ssDNA to a reporter protein require cumbersome procedures. To overcome these limitations, we developed a method for conjugation of ssDNA to a protein of interest fused with a replication initiation protein (Rep) [5,13]. In this method, a protein fused with Rep can be covalently linked to ssDNA via enzymatic reaction, without the need for any chemical modification of the ssDNA. Recently, the catalytic domain of porcine circovirus type 2 Rep comprising residues 1-116 (pRep) was employed to construct DNA-NanoLuc luciferase (NanoLuc) conjugates [5]. DNA-NanoLuc conjugates were applied for use in a DNA aptamer-sandwich assay system. Moreover, pRep retains its DNA binding activity regardless of whether the protein of interest is fused to the N-terminus or C-terminus.

Bioluminescence resonance energy transfer (BRET)-based biosensors are frequently used in biosensing methods [14–16]. For the design of intramolecular BRET sensors, sensory domains that lead to conformational changes are located between donor and acceptor molecules, such as luciferase and a fluorescent protein. Therefore, we decided to design a fusion protein for application as a BRET biosensor, in which pRep was positioned between luciferase and a fluorescent protein, and then a DNA aptamer was conjugated via pRep. It was anticipated that our designed fusion protein conjugates containing DNA aptamers might undergo conformational changes upon binding of a target molecule to the DNA aptamer (Figure 1). BRET efficiency is dependent on not only the distance between the donor and acceptor molecules, but also their relative orientations. In our design, it was speculated that larger change in the distance between donor and acceptor molecules would not be occurred even when a target molecule binds to the DNA aptamer. However, the BRET efficiency was expected to change through alteration of their orientation upon binding of a target molecule to the DNA aptamer.

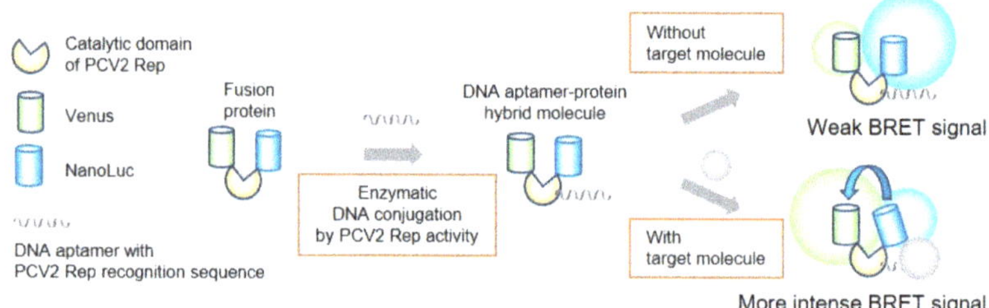

Figure 1. BRET-based biosensor with DNA-protein conjugates. The catalytic domain PCV2 Rep was fused with Venus and NanoLuc. A DNA aptamer with the PCV2 Rep recognition sequence was enzymatically conjugated to the fusion protein via Rep.

Herein, we tested this idea by constructing a fusion protein in which pRep was positioned between Venus as an acceptor fluorophore and NanoLuc as an energy donor. Overlap of the bioluminescent emission spectrum with the excitation spectrum of the acceptor fluorophore is important for selection of a BRET molecule pair. The combination of Venus and NanoLuc has been previously applied in several BRET-based sensors [15,16]. Energy transfer occurs when the donor and acceptor molecules are within 10 nm of each other in proximity [14,17]. Based on the known structure of pRep, we expected that BRET between NanoLuc and Venus would occur in the designed fusion protein. As noted above, the BRET efficiency depends not only on the distance between the donor and acceptor molecules, but also their relative orientations. To overcome limitations associated with molecular orientation, linkers are generally inserted between the sensory domain and the BRET molecules to allow for a certain degree of movement of the BRET molecules [14]. pRep exhibits flexible regions at the N- and C-terminals [18]. In the present study, we expected the orientation change upon binding of target molecules to sensor. To allow movement of the BRET molecules only when the target molecules bind to the DNA aptamer conjugated to the fusion protein, we constructed a truncated pRep variant from the catalytic domain of PCV2 Rep comprising residues 14–109 (tpRep). By using tpRep, the BRET efficiency is expected to change through alteration of the orientation upon binding of the target molecule to the DNA aptamer. Herein, truncated pReps were expressed in *Escherichia coli* and subsequently purified, and the DNA binding abilities were evaluated. Finally, tpRep was used to construct DNA-protein conjugates with thrombin- and lysozyme-binding DNA aptamers, and the resultant DNA-protein conjugates were evaluated for use as a BRET-based biosensor.

2. Materials and Methods

2.1. Construction of Plasmids

For expression of Rep mutants, the plasmids pET-His-pRep$_{1\text{-}109}$, pET-His-pRep$_{14\text{-}116}$ and pET-His-tpRep, were constructed as follows. pReps with the flexible N- and/or C-terminus regions truncated were constructed by site-directed mutagenesis. First, truncated pRep without the flexible 7-amino acid C-terminus, namely pRep$_{1\text{-}109}$, was constructed by site-directed mutagenesis using the primer set 5′-TGGAGCTGTCGACTAAGGTACCCTC-3′ and 5′-TAGTCGACAGCTCCACACTCCATTA-3′. pET-His-pRep, which was previously constructed in our laboratory, was used as a template [5]. The resultant plasmid was named pET-His-pRep$_{1\text{-}109}$. Truncated pRep without the flexible 13-amino acid N-terminus, namely pRep$_{14\text{-}116}$, and tpRep were constructed by the same procedure using the primer set 5′-CGAATTCCACAAACGTTGGGTCTTC-3′ and 5′-ACGTTTGTGGAATTCGCCCATGGCATG-3′. pET-His-pRep and pET-His-pRep$_{1\text{-}109}$ were used as templates, respectively. The resultant plasmids were named pET-His-pRep$_{14\text{-}116}$ and pET-His-tpRep, respectively.

The pET28-VenusΔ10-pRep-NanoLuc-His plasmid for expression of the fusion protein VenusΔ10-pRep-NanoLuc-His (VpRNH), in which the pRep protein was fused to the N-terminus of NanoLuc and the C-terminus of VenusΔ10, was constructed as follows. The VenusΔ10 fragment was amplified by PCR using the primer set 5′-tacgaattcagtaaaggagaagaactttc-3′ and 5′-ggcgaattcggtacccccagcagctgttac-3′, and Nano-lantern(cAMP-1.6)/pRSETB (Addgene #53591) was used as a template. The amplified fragment was cloned into pUC18 and its sequence was confirmed. The resultant plasmid, pUC18-VenusΔ10, was digested with *Eco*R I. The obtained fragment was inserted into the pET28-pRep-NanoLuc-His plasmid, which was previously constructed in our laboratory, and digested with the same restriction enzyme.

The pET28-VenusΔ10-tpRep-NanoLuc-His plasmid for expression of the fusion protein VenusΔ10-tpRep-NanoLuc-His (VtpRNH), in which the tpRep protein was fused to the N-terminus of NanoLuc and the C-terminus of VenusΔ10, was constructed as follows. To destroy an *Eco*R I restriction site located at the beginning of the VenusΔ10 sequence, a single nucleotide base of pET28-VenusΔ10-pRep-NanoLuc-His was changed by site-directed mutagenesis using the primer set 5′-GGCGATTTATGAGTAAAGGAGAAGAA-3′ and 5′-ACTCATAAATTCGCCCATGGTATATCT-3′. The resultant plasmid, pET28-delEcoRI-VenusΔ10-pRep-NanoLuc-His, was digested with *Eco*R I and *Sal* I to remove the fragment encoding pRep and the inserted tpRep fragment derived from pET-His-tpRep was digested with the same restriction enzymes. The resultant plasmid was named pET28-VenusΔ10-tpRep-NanoLuc-His.

2.2. Protein Expression and Purification

For protein expression, the respective plasmids were introduced into *E. coli* BL21(DE3)-competent cells. Transformed cells were inoculated in LB medium with 50 µg/mL ampicillin for expression of Rep mutants and with 20 µg/mL kanamycin for expression of the fusion proteins VpRNH and VtpRNH. Cells were then cultured at 37 °C until the OD$_{660}$ reached 0.6~0.8, followed by addition of 1 mM isopropyl-β-D(-)-thiogalactopyranoside (IPTG) for induction of protein expression. Cells were cultured overnight at 16 °C for expression of fusion proteins and for 6 h at 25 °C for expression of Rep mutants. Cells were then harvested by centrifugation. The collected cells were suspended in phosphate-buffered saline (PBS: 150 mM NaCl, 16 mM Na$_2$HPO$_4$, 4 mM NaH$_2$PO$_4$, pH 7.4) and disrupted by sonication, followed by centrifugation to obtain the soluble fraction. The supernatant was added to Profinity^{TM} IMAC Ni-Charged Resin (Bio-Rad) equilibrated with PBS followed by rotation at 4 °C for 30 min. After rotation, the samples were washed with Ni-NTA buffer (0.5 M NaCl, 20 mM phosphate buffer, pH 8.0) 5 times, and then washed twice with Ni-NTA buffer containing 10 mM imidazole. Proteins were eluted with Ni-NTA buffer containing 10, 100, 150 mM imidazole. The eluted samples were dialyzed against PBS 3 times using dialysis tubing. The concentrations of purified proteins were evaluated using a BCA assay kit (Pierce).

2.3. Evaluation of DNA Binding Ability of pRep

The ssDNA oligonucleotide Rep sub31 (5′-<u>AAGTATTAC</u>AAAAACCAGCGCAGTTGGGCAG-3′) was used for evaluation of the DNA binding ability of pRep. The underlined sequence denotes the PCV2 Rep recognition sequence. Purified proteins (5 µM) were mixed with Rep sub31 (5 µM) in 10 µL of reaction buffer, PBS with 2.5 mM MgCl$_2$. After incubation for 30 min at 37 °C, the samples were run on sodium dodecyl sulfate-polyacrylamide gel electrophoresis (SDS-PAGE), and the gels were stained with Coomassie Brilliant Blue (CBB).

2.4. Evaluation of Emission Spectra

BRET of purified fusion proteins were evaluated using the Nano-Glo Luciferase Assay System (Promega). Fusion proteins (1 μM) in 50 μL of PBS were mixed with the same volume of Nano-Glo Luciferase Assay Reagent. Emission spectra (350–650 nm) was acquired 10 s after reagent addition using a FP6500 spectrofluorophotometer (JASCO, Tokyo, Japan). Emission spectra were normalized to the peak emission of NanoLuc at 450 nm, which was set to an intensity of 1.00.

2.5. Homogeneous Assay with DNA Aptamer

The single-stranded DNA aptamers for lysozyme, namely Lysoapt 42 (5′-AAGTATTACATCTACGAATTCATCAGGGCTAAAGAGTGCAGAGTTACTTAG-3′), and for thrombin, namely TBA 29 (5′-AAGTATTACAGTCCGTGGTAGGGCAGGTTGGGGTGACT-3′), containing the Rep recognition sequence, were used in the homogeneous assay [19,20]. Human α-thrombin was purchased from Haematologic Technologies. Equal amounts (1 μM each) of VtpRNH and DNA aptamer were mixed in 50 μL of PBS containing 0.5 mM of $MgCl_2$ followed by incubation at 37 °C. After incubation for 30 min, 10 μL of solutions containing varying concentrations of target molecules was added to the mixture of VtpRNH and DNA aptamer followed by incubation at room temperature for 1 h. For the thrombin-binding DNA aptamer-VtpRNH conjugate, samples containing bovine serum albumin (BSA) (instead of thrombin) were evaluated as a negative control. Samples were mixed with the same volume of Nano-Glo Luciferase Assay Reagent. Emission spectra (350-650 nm) were acquired 10 sec after addition of the reagent using a FP6500 spectrofluorophotometer. Emission spectra were normalized to the peak emission of NanoLuc at 450 nm, which was set to an intensity of 1.00. The change in BRET ratio (ΔBRET) was calculated as follows:

$$\Delta BRET = \{(\text{emission at 528 nm})/(\text{emission at 450 nm})\} - BRET\ basal\ ratio$$

The BRET basal ratio is defined as (emission at 528 nm)/(emission at 450 nm) at 0 μM of the target molecule.

3. Results and Discussion

3.1. Truncation of the Catalytic Domain of PCV2 Rep

We previously developed a method for site-specific conjugation of ssDNA to a protein of interest via the fused replication initiator protein (Rep), such as conjugation of Gene A* from bacteriophage phiX 174 and the catalytic domain of porcine circovirus type 2 Rep comprising residues 1-116 (pRep) [5,13]. Recently, Gordon's group reported a similar strategy for construction of DNA-protein conjugates using several Reps, including pRep, as HUH-tags [21]. pRep is much smaller than Gene A* and well-expressed in E. coli. The structure of pRep (Protein Data Bank (PDB) ID:2HW0) exhibits flexible regions at the N- and C-terminals [18]. In our designed fusion protein, Rep was positioned between Venus and NanoLuc. For construction of BRET-based biosensors, flexible linkers are generally inserted between the sensory domain and the BRET molecule to allow for a certain degree of movement of the BRET molecule [14]. In the present study, to allow for movement of BRET molecules only when the target molecule binds to the DNA aptamer conjugated to the fusion protein, use of truncated variants of pRep was required. Truncated variants of pReps were constructed and their DNA binding abilities were evaluated.

Truncated variants of pRep, namely $pRep_{1-109}$ lacking the last 7 amino acids in the sequence, $pRep_{14-116}$ lacking the first 13 amino acids in the sequence, and tpRep ($pRep_{14-109}$) lacking the flexible regions at both the N- and C-terminals of pRep, were expressed in E. coli and subsequently purified. Truncated pReps were purified from the soluble fraction, similar to pRep. Proteins were mixed with ssDNA containing a Rep recognition sequence for evaluation of the DNA binding abilities of the truncated pReps. After incubation, samples were analyzed by SDS-PAGE (Figure 2). In addition to the bands appearing at the expected sizes of the protein, all samples also showed other bands at higher molecular weights than pReps without conjugation. Those bands appearing at higher molecular weights were attributed to pReps conjugated with ssDNA. Upon cleavage of specific sequence of ssDNA, PCV2 Rep is conjugate to DNA covalently. In addition of this reaction, PCV2 Rep also catalyze the reaction in the opposite direction [18]. PCV2 Rep has an ability to catalyze joining two ssDNA fragments, a free 3′-OH and the 5′-phosphate covalently linked to Rep, for regeneration of PCV2 Rep recognition sequence. Therefore, there is a possibility to exist unreacted protein. However, this joining activity is not efficient compared to cleavage activity followed by conjugation of DNA with Rep. In this experiment, the ratio of protein and ssDNA is 1:1. By increasing ssDNA concentration, the conjugation efficiencies should increase. These results demonstrate that tpRep retains the ability to covalently bind to ssDNA even after truncation of the flexible regions located at N- and C-terminals of pRep. Therefore, tpRep was used for construction of a BRET-based biosensor with a DNA-protein conjugate.

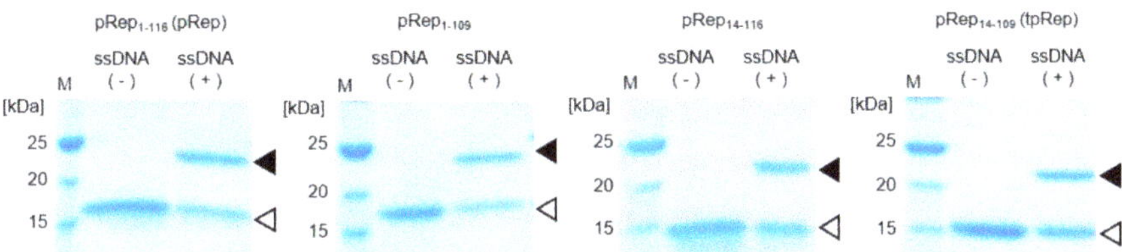

Figure 2. DNA binding abilities of truncated pRep proteins. The flexible region of the catalytic domain of PCV2 Rep comprising amino acids 1-116 was truncated from the full protein (pRep$_{1-116}$, abbreviated pRep; MW 15,100); pRep$_{1-109}$: pRep lacking the last 7 amino acids in the sequence (MW 14,300); pRep$_{14-116}$: pRep lacking the first 13 amino acids in the sequence (MW 13,700); and pRep$_{14-109}$, abbreviated tpRep: pRep lacking the flexible regions at both the N- and C-terminals of pRep (MW 12,900); white triangles indicate pRep without DNA and black triangles indicate pRep conjugated with DNA.

3.2. Construction of DNA-Protein Conjugates with NanoLuc and Venus for BRET-Based Biosensor

As shown in Figure 1, we designed a BRET-based sensor with DNA-protein conjugates. For construction of the DNA-protein conjugates, the catalytic domain of PCV2 Rep comprising residues 14–109 (tpRep) was fused to the N-terminus of NanoLuc and the C-terminus of VenusΔ10 with a His-tag (Figure 3A). The resultant fusion protein, VtpRNH, was expressed in E. coli and subsequently purified from the soluble fraction with a His-tag located at the C-terminus. After purification, the activities of each domain of VtpRNH were evaluated. First, the DNA binding ability of tpRep in the fusion protein was evaluated (Figure 3B). Even after fusion with NanoLuc and VenusΔ10, tpRe retained its DNA binding ability. The conjugation efficiencies were found increase with increasing DNA concentration. The maximum efficiency was around 50% (Figure 3C). We also evaluated the functions of both Venus and NanoLuc. Venus was shown to fluorescence in the fusion protein (Figure 3D). The emission peak was ~530 nm at an excitation wavelength of 500 nm [22].

Bioluminescence of NanoLuc in the fusion protein was also evaluated. The emission spectrum of the fusion protein exhibited a peak at ~450 nm without external excitation, which corresponds to the emission of NanoLuc [23]. These results suggest that VtpRNH retained its NanoLuc activity. Moreover, in addition to the emission peak around 450 nm, there was a small emission peak at ~530 nm. To identify the origin of this peak, the emission spectra were normalized to the peak emission of NanoLuc at 450 nm, which was set to an intensity of 1.00. As shown in Figure 3E, the emission peak of normalized intensity was observed at ~530 nm. The emission peak of Venus occurs at ~530 nm. These results suggest that BRET occurred between NanoLuc and Venus. We measured the emission spectrum after ssDNA binding via Rep to the fusion protein. Even after ssDNA binding to the fusion proteins via Rep, the emission peak at ~530 nm was observed (data not shown).

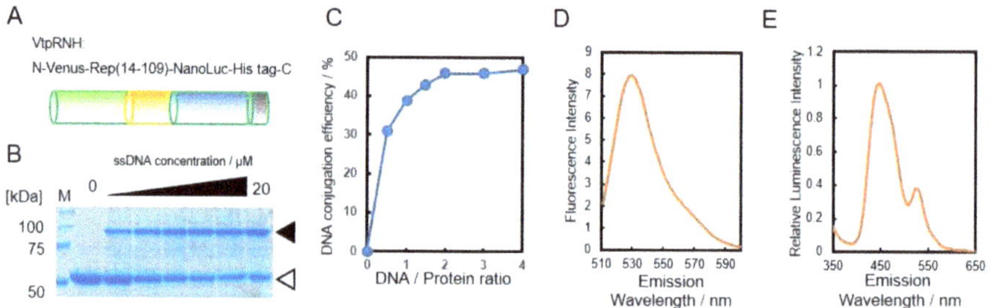

Figure 3. Evaluation of DNA binding ability of the fusion protein. Design of the constructed fusion protein, VtpRNH (MW 57,900) (**A**). DNA binding of tpRep in the fusion protein; the concentration of VtpRNH was 5 µM; white triangle indicates pRep without DNA conjugation and black triangle indicates pRep with DNA conjugation (**B**). DNA-conjugation efficiency of protein was evaluated by ImageJ (**C**). Emission spectrum of Venus at an excitation wavelength of 500 nm; the concentration of VtpRNH was 3 µM (**D**). NanoLuc activities of fusion proteins. The emission spectrum was normalized to the peak emission of NanoLuc at 450 nm; final concentration of VtpRNH was 0.5 µM (**E**).

3.3. Construction and Evaluation of BRET-Based Biosensor with DNA-Protein Conjugates

DNA aptamers are well known molecular binders with specificity to a specific molecule. In the present study, thrombin- and lysozyme-binding DNA aptamers were applied for the construction of a BRET-based biosensor with DNA-protein conjugates. For conjugation of DNA aptamer to the fusion protein via Rep, ssDNA comprised of the DNA aptamer and a Rep recognition sequence were reacted with the fusion protein.

Lysozyme-binding DNA aptamer was first conjugated to VtpRNH. After conjugation, the emission spectra of lysozyme binding DNA aptamer–protein conjugates with or without lysozyme were evaluated. As shown in insets of Figure 4A,B, the emission spectra were normalized to the peak emission of NanoLuc at 450 nm, which was set to an intensity of 1.00. In the presence of BSA (100 µM), the normalized emission spectra of both DNA-protein conjugates with or without BSA were similar (Figure 4A). On the other hand, presence of lysozyme, the normalized emission peaks at 528 nm were found to increase compared to the control without lysozyme (Figure 4B). These results suggest that BRET efficiencies change upon binding of lysozyme to the DNA aptamers. To confirm this speculation, we evaluated the emission spectra in the presence of different concentrations of lysozyme. As shown in Figure 4C, the changes in BRET ratio (ΔBRET) increased with increasing lysozyme concentration. These results confirm that the lysozyme-binding DNA aptamer-VtpRNH conjugate can be used as a BRET-based sensor. The K_d of the lysozyme-binding DNA aptamer is reported to be 30 nM [19]. Compared to this value, the sensitivity of our BRET-based biosensor with the lysozyme-binding DNA aptamer-VtpRNH conjugate is very low. As shown in Figure 3B, there were unreacted DNA. The sensitivity of sensor should be reduced by those unreacted DNA because unreacted DNA bind to target molecule competitively. However, in this case, it is not main reason for the low sensitivity. It was supposed that lysozyme formed aggregate in the presence of unreacted DNA caused by their net charge. Though monomeric lysozyme could not induce conformational change of fusion protein, BRET ratios would be changed by the aggregated lysozyme. The lysozyme binding DNA aptamer-VpRNH conjugate also showed similar results (Figure S1). We expected ΔBRET to be larger for the lysozyme-binding DNA aptamer-VtpRNH conjugate compared with the DNA-VpRNH conjugate because tpRep was truncated to remove the flexible amino acid sequences from the N- and C-terminals. In contrast, the change in BRET ratio for the DNA-VtpRNH conjugate was smaller than that for the DNA-VpRNH conjugate. In case of DNA-VpRNH with high concentration of lysozyme, it was speculated that BRET ratio was changed not only by changes of the orientation but also the distance.

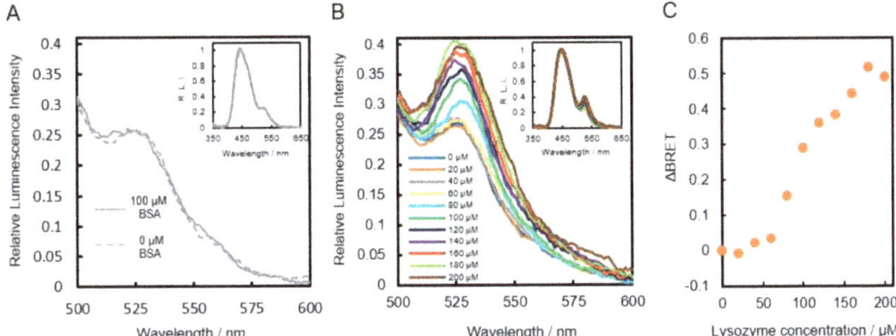

Figure 4. Evaluation of lysozyme-binding DNA aptamer–protein conjugates with and without target molecules. Normalized emission spectra of lysozyme-binding DNA aptamer-VtpRNH conjugate in the presence of BSA (**A**) and lysozyme (**B**). Change in BRET ratio (ΔBRET) as a function of lysozyme concentration (**C**).

We also evaluated the thrombin-biding DNA aptamer conjugated to VtpRNH for use as a BRET-based biosensor. As shown in insets of Figure 5A,B, the normalized emission spectra of thrombin-binding DNA aptamer- and lysozyme-binding DNA aptamer-VtpRNH conjugates were evaluated in the presence of different thrombin concentrations. As shown in Figure 5A, the emission spectra of the lysozyme-binding DNA aptamer-VtpRNH conjugate were unchanged in the presence of different thrombin concentrations. On the other hand, the emission spectra of thrombin-binding DNA aptamer-VtpRNH conjugate increased with increasing thrombin concentration (Figure 5B). These results were also confirmed by calculation of ΔBRET for each of the conjugates (Figure 5C). These results suggest that the BRET ratio changes only occurred with the appropriate combination of DNA aptamer and target molecules.

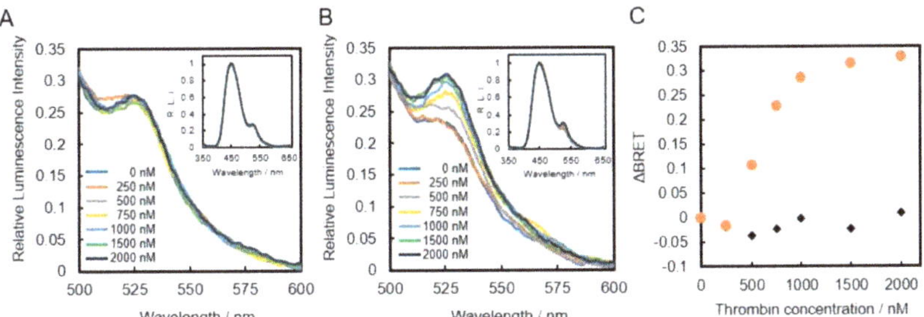

Figure 5. Evaluation of DNA aptamer-VtpRNH conjugate in the presence of different concentrations of thrombin. Conjugation with lysozyme-binding DNA aptamer (**A**) or thrombin-binding DNA aptamer (**B**). Change in BRET ratio (ΔBRET) as a function of thrombin concentration (**C**). Conjugation with lysozyme-binding DNA aptamer (black diamonds) or thrombin-binding DNA aptamer (orange circles).

Finally, to demonstrate specificity of the conjugates, we evaluated the BRET ratio of the thrombin-binding DNA aptamer-VtpRNH conjugate in the presence of more different concentrations of thrombin or BSA. As shown in Figure 6, ΔBRET of the thrombin-binding DNA aptamer-VtpRNH conjugate increased with increasing thrombin concentration. On the other hand, ΔBRET did not increase in the presence of different BSA concentrations. It is suggested that the ΔBRET depends on the specific binding between DNA aptamers and target molecules. These results further suggest that the thrombin-binding DNA aptamer-VtpRNH conjugate can be used as a BRET-based thrombin biosensor. The K_d of the thrombin-binding DNA aptamer is reported to be ~0.5 nM, which is different from the value obtained from our BRET-based biosensor with the thrombin-binding DNA aptamer-VtpRNH conjugate [20]. One of the reasons of that is existence of unconjugated DNA. In this experiment, same amount of DNA aptamer and VtpRNH were mixed. From the result of Figure 3C, around half of DNA aptamer would remain as unreacted DNA. It should decrease the sensitivity of this sensor. To evaluate the property of BRET sensors, separations of DNA aptamer-VtpRNH conjugates are required. In addition, steric effects would reduce the sensitivity of the sensor. Steric effects are required to induce conformational changes. However, steric effects may also reduce the binding ability of the DNA aptamer. Therefore, the K_d of DNA aptamers should increase in our BRET-based biosensor. The square of the correlation coefficient (R^2) of the linear fit equation between concentrations of 200 and 1500 nM is close to 1 ($R^2 = 0.97$). The square of the correlation coefficient was much closer to 1 compared to the results for the thrombin-binding DNA aptamer-VpRNH conjugate ($R^2 = 0.78$) (Figure S2). These differences may be caused by the existence of flexible regions with the pRep protein, which could allow for movement of the BRET molecules even after binding of the target molecule to the DNA aptamer. These results suggest that DNA aptamer-VtpRNH conjugates exhibit the potential for use as BRET-based biosensors. While the sensitivity of these biosensors was too low compared with other DNA aptamer sensors, the sensitivity might be improved by changing the design of the fusion protein and DNA aptamers.

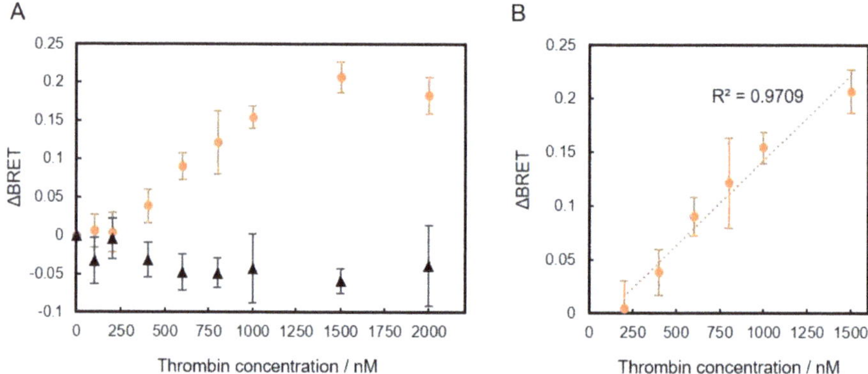

Figure 6. Change in BRET ratio (ΔBRET) of thrombin-binding DNA aptamer-VtpRNH conjugate as a function of thrombin concentration in the presence of thrombin (orange circles) and BSA (black triangles) (**A**). Correlation of ΔBRET and thrombin concentration between 200 and 1500 nM (**B**); each value represents the mean of 4 replicates ($n = 4$).

4. Conclusions

We constructed DNA aptamer–protein hybrid molecules for use as BRET-based biosensors. Fusion proteins were enzymatically conjugated to DNA aptamers via the catalytic domain of porcine circovirus type 2 replication initiation protein, which was positioned between NanoLuc luciferase and Venus. The catalytic domain of PCV2 Rep was shown to retain its DNA binding activity, even after elimination of the flexible regions at the N- and C-terminals. The resultant DNA aptamer–protein hybrid molecule exhibited a weak BRET signal, even in the absence of the target molecule of the DNA aptamer. However, the BRET signal was found to depend on the concentration of the target molecule. These results demonstrate the potential for use of the designed fusion protein as a DNA aptamer-based platform for construction of BRET-based biosensors.

Supplementary Materials: The following are available online at http://www.mdpi.com/2076-3417/10/21/7646/s1, Figure S1. Evaluation of lysozyme-binding DNA aptamer-VpRNH conjugates with and without target molecules. Normalized emission spectra of lysozyme-binding DNA aptamer-VpRNH conjugate in the presence of BSA (A) and lysozyme (B). Change in BRET ratio (ΔBRET) as a function of lysozyme concentration (C), Figure S2. ΔBRET of thrombin of thrombin-binding DNA aptamer-VpRNH conjugate as a function of thrombin concentration (A).

Correlation of ΔBRET and thrombin concentration between 200 and 1500 nM (B); each value represents the mean of 4 replicates ($n = 4$).

Author Contributions: Conceptualization, M.M.; methodology, M.M., R.H. and Y.M.; investigation, R.H.; writing—original draft preparation, M.M.; writing—review and editing, Y.M. and E.K.; supervision, M.M. and E.K. All authors have read and agreed to the published version of the manuscript.

Funding: This work was supported in part by JSPS KAKENHI Grant Numbers 16K01388 (M.M.) and 26289310 (E.K.).

Conflicts of Interest: The authors declare no conflict of interest.

References

1. Zhao, D.; Kong, Y.; Zhao, S.; Xing, H. Engineering Functional DNA–Protein Conjugates for Biosensing, Biomedical, and Nanoassembly Applications. *Top. Curr. Chem.* **2020**, *378*, 41. [CrossRef]

2. Sano, T.; Smith, C.; Cantor, C. Immuno-PCR: Very sensitive antigen detection by means of specific antibody-DNA conjugates. *Science* **1992**, *258*, 120–122. [CrossRef] [PubMed]

3. Shimada, J.; Maruyama, T.; Kitaoka, M.; Kamiya, N.; Goto, M. DNA–enzyme conjugate with a weak inhibitor that can specifically detect thrombin in a homogeneous medium. *Anal. Biochem.* **2011**, *414*, 103–108. [CrossRef] [PubMed]

4. Takahara, M.; Wakabayashi, R.; Minamihata, K.; Goto, M.; Kamiya, N. Primary Amine-Clustered DNA Aptamer for DNA–Protein Conjugation Catalyzed by Microbial Transglutaminase. *Bioconjug. Chem.* **2017**, *28*, 2954–2961. [CrossRef]

5. Mie, M.; Niimi, T.; Mashimo, Y.; Kobatake, E. Construction of DNA-NanoLuc luciferase conjugates for DNA aptamer-based sandwich assay using Rep protein. *Biotechnol. Lett.* **2019**, *41*, 357–362. [CrossRef]

6. Wang, G.; Wang, Y.; Chen, L.; Choo, J. Nanomaterial-assisted aptamers for optical sensing. *Biosens. Bioelectron.* **2010**, *25*, 1859–1868. [CrossRef]

7. Sassolas, A.; Blum, L.J.; Leca-Bouvier, B.D. Homogeneous assays using aptamers. *Analyst* **2011**, *136*, 257–274. [CrossRef]

8. Citartan, M.; Gopinath, S.C.B.; Tominaga, J.; Tan, S.-C.; Tang, T.-H. Assays for aptamer-based platforms. *Biosens. Bioelectron.* **2012**, *34*, 1–11. [CrossRef]

9. Gokulrangan, G.; Unruh, J.R.; Holub, D.F.; Ingram, B.; Johnson, C.K.; Wilson, G.S. DNA Aptamer-Based Bioanalysis of IgE by Fluorescence Anisotropy. *Anal. Chem.* **2005**, *77*, 1963–1970. [CrossRef]

10. Rupcich, N.; Chiuman, W.; Nutiu, R.; Mei, S.; Flora, K.K.; Li, Y.; Brennan, J.D. Quenching of Fluorophore-Labeled DNA Oligonucleotides by Divalent Metal Ions: Implications for Selection, Design, and Applications of Signaling Aptamers and Signaling Deoxyribozymes. *J. Am. Chem. Soc.* **2006**, *128*, 780–790. [CrossRef] [PubMed]

11. Urata, H.; Nomura, K.; Wada, S.; Akagi, M. Fluorescent-labeled single-strand ATP aptamer DNA: Chemo- and enantio-selectivity in sensing adenosine. *Biochem. Biophys. Res. Commun.* **2007**, *360*, 459–463. [CrossRef]

12. Hao, L.; Zhao, Q. A fluorescein labeled aptamer switch for thrombin with fluorescence decrease response. *Anal. Methods* **2015**, *7*, 3888–3892. [CrossRef]

13. Mashimo, Y.; Maeda, H.; Mie, M.; Kobatake, E. Construction of Semisynthetic DNA-Protein Conjugates with Phi X174 Gene-A* Protein. *Bioconjug. Chem.* **2012**, *23*, 1349–1355. [CrossRef] [PubMed]

14. Brown, N.E.; Blumer, J.B.; Hepler, J.R. Bioluminescence Resonance Energy Transfer to Detect Protein-Protein Interactions in Live Cells. In *Protein-Protein Interactions*; Meyerkord, C.L., Fu, H., Eds.; Methods in Molecular Biology; Springer: New York, NY, USA, 2015; Volume 1278, pp. 457–465. ISBN 978-1-4939-2424-0.

15. Yang, J.; Cumberbatch, D.; Centanni, S.; Shi, S.; Winder, D.; Webb, D.; Johnson, C.H. Coupling optogenetic stimulation with NanoLuc-based luminescence (BRET) Ca++ sensing. *Nat. Commun.* **2016**, *7*, 13268. [CrossRef]

16. Inagaki, S.; Tsutsui, H.; Suzuki, K.; Agetsuma, M.; Arai, Y.; Jinno, Y.; Bai, G.; Daniels, M.J.; Okamura, Y.; Matsuda, T.; et al. Genetically encoded bioluminescent voltage indicator for multi-purpose use in wide range of bioimaging. *Sci. Rep.* **2017**, *7*, 42398. [CrossRef]

17. Dale, N.C.; Johnstone, E.K.M.; White, C.W.; Pfleger, K.D.G. NanoBRET: The Bright Future of Proximity-Based Assays. *Front. Bioeng. Biotechnol.* **2019**, *7*, 56. [CrossRef]

18. Vega-Rocha, S.; Byeon, I.-J.L.; Gronenborn, B.; Gronenborn, A.M.; Campos-Olivas, R. Solution Structure, Divalent Metal and DNA Binding of the Endonuclease Domain from the Replication Initiation Protein from Porcine Circovirus 2. *J. Mol. Biol.* **2007**, *367*, 473–487. [CrossRef] [PubMed]

19. Kirby, R.; Cho, E.J.; Gehrke, B.; Bayer, T.; Park, Y.S.; Neikirk, D.P.; McDevitt, J.T.; Ellington, A.D. Aptamer-Based Sensor Arrays for the Detection and Quantitation of Proteins. *Anal. Chem.* **2004**, *76*, 4066–4075. [CrossRef] [PubMed]

20. Tasset, D.M.; Kubik, M.F.; Steiner, W. Oligonucleotide inhibitors of human thrombin that bind distinct epitopes. *J. Mol. Biol.* **1997**, *272*, 688–698. [CrossRef] [PubMed]

21. Lovendahl, K.N.; Hayward, A.N.; Gordon, W.R. Sequence-Directed Covalent Protein–DNA Linkages in a Single Step Using HUH-Tags. *J. Am. Chem. Soc.* **2017**, *139*, 7030–7035. [CrossRef] [PubMed]

22. Nagai, T.; Ibata, K.; Park, E.S.; Kubota, M.; Mikoshiba, K.; Miyawaki, A. A variant of yellow fluorescent protein with fast and efficient maturation for cell-biological applications. *Nat. Biotechnol.* **2002**, *20*, 87–90. [CrossRef] [PubMed]

23. Hall, M.P.; Unch, J.; Binkowski, B.F.; Valley, M.P.; Butler, B.L.; Wood, M.G.; Otto, P.; Zimmerman, K.; Vidugiris, G.; Machleidt, T.; et al. Engineered Luciferase Reporter from a Deep Sea Shrimp Utilizing a Novel Imidazopyrazinone Substrate. *ACS Chem. Biol.* **2012**, *7*, 1848–1857. [CrossRef] [PubMed]

Publisher's Note: MDPI stays neutral with regard to jurisdictional claims in published maps and institutional affiliations.

applied sciences

MDPI

Article

RNA-Peptide Conjugation through an Efficient Covalent Bond Formation

Shun Nakano, Taiki Seko, Zhengxiao Zhang and Takashi Morii *

Institute of Advanced Energy, Kyoto University, Gokasho, Uji, Kyoto 611-0011, Japan;
snaka@iae.kyoto-u.ac.jp (S.N.); seko.taiki.56a@st.kyoto-u.ac.jp (T.S.);
zhang.zhengxiao.38m@st.kyoto-u.ac.jp (Z.Z.)
* Correspondence: t-morii@iae.kyoto-u.ac.jp

Received: 12 October 2020; Accepted: 10 December 2020; Published: 14 December 2020

Abstract: Many methods for modification of an oligonucleotide with a peptide have been developed to apply for the therapeutic and diagnostic applications or for the assembly of nanostructure. We have developed a method for the construction of receptor-based fluorescent sensors and catalysts using the ribonucleopeptide (RNP) as a scaffold. Formation of a covalent linkage between the RNA and the peptide subunit of RNP improved its stability, thereby expanding the application of functional RNPs. A representative method was applied for the formation of Schiff base or dihydroxy-morpholino linkage between a dialdehyde group at the 3′-end of sugar-oxidized RNA and a hydrazide group introduced at the C-terminal of a peptide subunit through a flexible peptide linker. In this report, we investigated effects of the solution pH and contribution of the RNA and peptide subunits to the conjugation reaction by using RNA and peptide mutants. The reaction yield reached 90% at a wide range of solution pH with reaction within 3 h. The efficient reaction was mainly supported by the electrostatic interaction between the RNA subunit and the cationic peptide subunit of the RNP scaffold. Formation of the RNP complex was verified to efficiently promote the reaction for construction of the RNA-peptide conjugate.

Keywords: ribonucleopeptide (RNP); RNA-peptide conjugate; Schiff base; aptamer; fluorescent sensors

---

1. Introduction

RNA and peptide conjugates have been constructed for versatile therapeutic application, such as the delivery of siRNA [1–3] or the screening of a library of peptides in a mRNA display method [4,5]. Efficient reaction for the formation of covalent linkage between RNA and peptide is important for developing practical applications. Many methods have been reported to conjugate an oligonucleotide (RNA or DNA) with an oligopeptide by applying various chemistries [6–9]. Post-synthetic coupling of the respective oligonucleotide and peptide fragments is the common and reliable method to construct oligonucleotide-peptide conjugates because the stepwise solid-phase synthesis of the conjugates encounters difficulties in finding the compatible protecting groups for both nucleobases and amino-acid side chains. Several chemical reactions were applied for coupling the nucleotide and peptide, such as amidation [10,11], disulfide bond formation [12–14], the modification of the thiol group by haloacetyl [15,16], or the maleimide group [12,17,18], native chemical ligation [19–21], and click reaction [22,23]. These reactions are certainly useful for the selective coupling of oligonucleotides and oligopeptides. However, these reactions often associate with drawbacks, such as the instability of the linkage, the necessity for introducing additional chemical groups into both the oligonucleotide and the oligopeptide chains by multistep reactions, limitation for the sequence of oligopeptide due to the solubility or reactivity of the side chains, and/or the potential formation of the side products and the stereoisomers. In some cases, even applying the long reaction time, the yields of the product were rather low. One of the alternative methods for the modification of nucleotide with functional molecules was based on incorporating the aldehyde group into the nucleotide. Oxidation of the sugar moiety by perchloric acid or periodate is one of the common methods for introducing the aldehyde group on DNA or RNA [24–27]. By coupling the aldehyde group with alkoxyamine, cysteine, and hydrazine, formation of the covalent linkage through oxime, thiazolidine, and hydrazone (or a morpholine-like structure), respectively, has been reported [27–29]. These reactions proceed in mild conditions in a relatively short time to provide high coupling yields in aqueous solution. While this strategy also displays some of the limitations, as mentioned above, it has the advantage of having no need for the introduction of unnatural nucleic acids into oligonucleotide and facile preparation of the reactive peptide.

We have developed a stepwise method for the construction of receptor-based fluorescent sensors by using ribonucleopeptide (RNP) as a scaffold (Figure 1A). In this method, a complex of Rev peptide and Rev Responsive Element (RRE) RNA [30] was utilized to construct the RNP library with a randomized RNA sequence in the RNA subunit. RNP receptors [31] for a target molecule were selected from the RNP library by applying the in vitro selection method [32,33]. The peptide subunit of the selected RNP receptors was further modified with a fluorophore to construct a fluorophore-modified RNP receptor (F-RNP) library [34–42]. By screening the F-RNP library, fluorescent RNP sensors showing measurable fluorescent intensity changes upon binding the substrate were selected. Many kinds of RNP receptors and sensors were constructed for various target molecules, such as ATP [34–38,41,42], GTP [34,35,41], dopamine [39], and a tetra peptide containing a phosphorylated tyrosine residue [40]. Furthermore, the RNA subunit and the peptide subunit of fluorescent RNP sensors were covalently linked to improve the chemical and thermal stability [41]. The covalently linked RNP (c-RNP) sensors were applicable for simultaneous detection of multiple target molecules in the solution. Time-course monitoring of the concentration changes of the substrate and product in an enzymatic reaction was demonstrated by simultaneous application of sensors for the substrate and the product [41,42]. As mentioned above, for the covalent bond formation with the aldehyde group, the ribose moiety at the 3′-end of RNA was oxidized by sodium periodate to a dialdehyde group (Figure 1B). A hydrazide group was introduced at the C-terminal of peptide subunit through a ten-amino-acid flexible linker. Coupling of the RNA dialdehyde group and the C-terminal hydrazide group of peptide formed a covalent linkage between the RNA and peptide subunits of the Rev-RRE complex. This reaction proceeded rapidly in a quantitative yield. It is likely that a proximity effect between the reactive groups on the RNA and peptide subunit originated from the complex formation of RNP to assist the efficient coupling reaction. Here, we investigated several conditions for covalent linkage formation between the dialdehyde group of RNA and the hydrazide group of peptide within the ribonucleopeptide scaffold to elucidate the effect of solution pH for the reaction yield. Furthermore, the role of proximity effect in the coupling reaction was investigated by using mutants of RNA and peptide. Our results demonstrated that the efficient coupling reaction within the RNP was indeed the outcome of the proximity effect of reactive groups and provided an optimal condition for the construction of the covalently linked RNA-peptide complex.

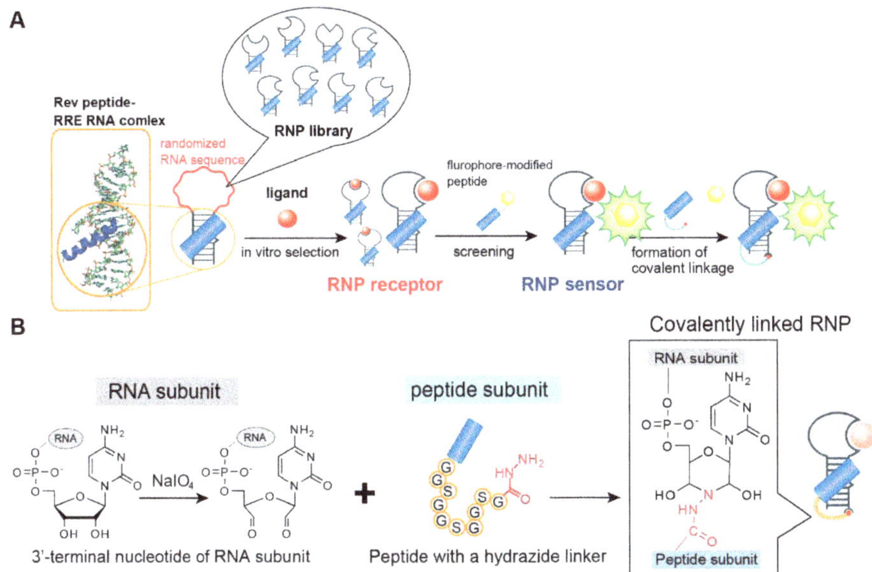

Figure 1. Schematic illustration for the construction of the covalently linked fluorescent ribonucleopeptide (RNP) sensor. RNA receptors selected from the RNA-derived RNP library were converted into a fluorescent RNP sensor library through modification of the peptide subunit by a fluorophore. (**A**) A fluorescent RNP sensor, selected from the fluorescent RNP sensor library, was converted to a covalently linked RNP (c-RNP) sensor by the crosslinking reaction between RNA and the peptide subunit [41]. (**B**) Scheme of covalent linkage formation between RNA and the peptide subunit of RNP.

2. Materials and Methods

PrimeSTAR HS DNA polymerase for PCR reactions was obtained from TaKaRa Bio Inc. (Shiga, Japan), and the T7-Scribe Standard RNA IVT Kit was obtained from CELLSCRIPT (Madison, WI, USA). N-α-Fmoc-protected amino acids, 2-(1H-benzotriazol-1-yl)-1,1,3,3-tetramethyluronium hexafluorophosphate (HBTU), 1-hydroxybenzotriazole (HOBt), distilled N,N-dimethylformamide (DMF), and 4-Hydroxymethylbenzoic acid-polyethylene glycol (PEG) resin (HMBA-PEG resin) were obtained from Watanabe Chemical Industries (Hiroshima, Japan). N,N-diisopropylethylamine (DIEA), diisopropylcarbodiimide (DIC), and 2′,4′,6′-trihydroxyacetophenone monohydrate (THAP) were obtained from Sigma-Aldrich (St. Louis, MO, USA). N,N-dimethyl-4-aminopyridine (DMAP), sodium periodate, hydrazine monohydrate, gel electrophoresis grade acrylamide, bisacrylamide, phenol, thioanisole, and 1,2-ethanedithiol were purchased from Wako Chemicals (Tokyo, Japan). Diammonium hydrogen citrate (DAHC) was obtained from Nacalai Tesque (Kyoto, Japan). A reversed-phase C18 column ULTRON VX-ODS (For analysis: 4.6 × 150 mm; For purification: 20 × 250 mm, particle size 5 μm) was purchased from Shinwa Chemical Industries (Kyoto, Japan).

2.1. Preparation of RNA

Forward and Reverse primers for the construction of the double-stranded template DNA for An16 RNA (5′-TCTAATACGACTCACTATAGGTCTGGGCGCA-3′ and 5′-GGCCTGTACCGTC-3′) and for scrAn16 RNA (5′-TCTAATACGACTCACTATAGGAGGCTTCAGCTTCG-3′ and 5′-CACAC AACCGCCCCG-3′) were purchased from Sigma–Aldrich, Japan (T7 RNA promoter is underlined). The double-stranded DNA templates and RNAs were prepared as previously described [42]. RNA subunits of RNP receptors were purified by means of denaturing polyacrylamide gel electrophoresis (8 M urea, 12%). The concentrations of purified RNA were quantified by measuring the absorption at 260 nm (An16 = 439,500; scrAn16 = 438,800 M^{-1} cm^{-1}).

2.2. Construction of a Covalently Linked RNP

The peptide subunit for the formation of a covalent linkage (Ac-TRQARRNRRRRWRERQR-GGSGGSGGSG-HZ) was synthesized as follows. A HMBA-PEG resin was placed in a dry flask, and a sufficient amount of DMF was added to soak the resin; this mixture was allowed to swell for 30 min. N-α-Fmoc-glycine (10 equivalent relative to resin loading) dissolved in DMF was mixed with a solution of DIC (5 equivalent relative to resin loading) in dry DMF on ice and then incubated for 20 min. The solution of an activated first amino acid was added to the resin prepared above. A DMF solution of DMAP (0.1 equivalent relative to resin loading) was added to the resin/amino acid mixture and incubated at room temperature for 1 h with occasional swirling. This procedure of coupling the first amino acid residue was repeated twice. The subsequent synthesis was performed on an automated peptide synthesizer (PSSM-8; Shimadzu, Kyoto, Japan) according to the Fmoc chemistry protocol using protected Fmoc-amino acids and HBTU. Acetylation of the N–terminal of peptide was performed by mixing with 1 M acetic acid anhydride and 1 M 1-methyl imidazole in DMF for 1 h at room temperature. The following procedures, cleavage of the protected peptide from the resin with hydrazine monohydrate, deprotection of the protected peptide, and purification of the peptide were performed as described previously [41]. The synthesized peptides were characterized by MALDI-TOF mass spectrometry (AXIMA-LNR, Shimadzu), as follows: Acetylated Rev peptide hydrazide (Ac-Rev-(GGS)$_3$G-HZ), m/z 3154.9 (calcd for [M+H]$^+$ 3155.5); Acetylated flexible peptide linker hydrazide (Ac-(GGS)$_3$G-HZ), m/z [M+H]$^+$ 735.3, [M+Na]$^+$ 757.8 (calculated for [M+H]$^+$ 734.3).

Crosslinking reaction of the RNA and the peptide subunits of RNP was carried out as described previously [41,42] with a slight modification of the conditions. Freshly prepared 0.01 M sodium periodate (5 μL; 50 nmol; 50 equiv) was added to 200 μM RNA (5 μL; 1 nmol) in 15 μL of 0.03 M sodium acetate (pH 5.2), and the reaction mixture (25 μL) was incubated for 1 h at 37 °C in the dark. After the reaction, 2.5 μL of 10 M glycerol was added to the reaction mixture to reduce an excess amount of sodium periodate. The resulting oxidized RNA was purified by ethanol precipitation. A coupling reaction between the 3′-modified RNA (40 μM) and Ac-Rev-(GGS)$_3$G-HZ (88 μM) was performed in 0.03 M sodium acetate (for pH 4 and 5) or 0.03 M sodium phosphate (for pH 6 and 7), containing 0.01 M NaCl (total 25 μL) at 37 °C in the dark. The reaction mixture was extracted by phenol/chloroform and purified by ethanol precipitation to remove unreacted peptide, then dissolved in TE (25 μL).

2.3. Evaluation of the Reaction Yields of Covalently-Linked RNPs

Denaturing polyacrylamide gel electrophoresis (PAGE) (8 M urea, 12%) was performed to separate the unreacted RNA and covalently linked RNP by loading the same volume of the purified samples. A total of 10 pmol of RNA was loaded on the gel as the control of RNA band mobility. The above stock solution of phenol/chloroform extracted RNP (25 μL) was further diluted with TE to 200 μL. From this RNP solution, 2 μL was analyzed by PAGE for each reaction. The acrylamide gel was stained with ethidium bromide to detect RNA and RNP. The yield of each reaction was evaluated from the ratio of intensity of bands corresponding to unreacted RNA and RNP. Actual isolation yields of RNP ranged from 30 to 40% after the PAGE purification and successive purification by ethanol precipitation.

2.4. MALDI-TOF Mass Analysis of the Reaction Solution

The reaction solution removed the unreacted peptide by phenol/chloroform extraction, and ethanol precipitation was characterized by MALDI-TOF mass spectrometry (AXIMA-Confidence, Shimadzu, Kyoto, Japan). A matrix solution was prepared as a mixture (*v/v*: 9/1) of 10 mg/mL of THAP in acetonitrile/H$_2$O (*v/v*: 1/1) and 50 mg/mL of DAHC in pure water. The sample solution was mixed with the same volume of matrix solution (1 μL each). DNA oligonucleotides (molecular weight: 5828.9,

11,397.5, 13,976.2, 23,164.3) were used as the external calibration standard sample. All measurements were performed in linear positive mode.

3. Results

3.1. Effect of Solution pH for the Formation of Covalent Linkage

The effect of solution pH for the formation of covalent linkage between the 3′-dialdehyde group of the oxidized RNA subunit and the hydrazide group at the C-terminal of peptide subunit of the Rev-RRE complex was investigated by carrying out the reaction at various pH levels. The RNA subunit of ATP-binding of the RNP receptor, An16, was oxidized by using sodium periodate in a sodium acetate buffer at pH 5. The reaction was performed at 37 °C for 1 h. After isolation of the oxidized RNA by ethanol precipitation, acetylated Rev peptide, modified with a C-terminal hydrazide group through a flexible linker (GGS)$_3$G (Ac-Rev-(GGS)$_3$G-HZ), was mixed with the oxidized RNA in sodium acetate buffer at pH 4 or 5 or in a sodium phosphate buffer at pH 6 or 7. The reaction mixtures were incubated for 3 h at 37 °C, then the reactions were stopped by the addition of phenol/chloroform and vigorously mixed. The water layer was condensed by ethanol precipitation. The reaction yield of c-RNP in each solution was evaluated by quantitation of the band intensity, corresponding to free RNA and conjugated c-RNP in a denaturing polyacrylamide gel electrophoresis (PAGE) (Figure 2). Over 80% of RNA was reacted with peptide in all conditions. The yields of reaction were 92 ± 4% at pH 4, 92 ± 5% at pH 5, 86 ± 3% at pH 6, and 82 ± 3% at pH 7 (Figure S1A). These results indicated that the solution pH ranging from 4 to 7 did not affect much to the yield of RNA-peptide conjugate. Prolonged incubation over 15 h at pH 5 did not show any difference in the ratio of the band intensity for c-RNP and unreacted RNA (data not shown).

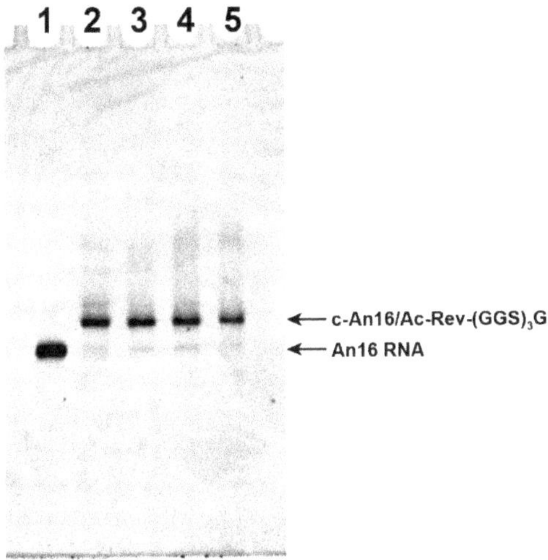

Figure 2. A denaturing polyacrylamide gel electrophoresis (PAGE) analysis (8 M Urea) of the conjugation reaction products of c-RNPs at different pH conditions. The yield was calculated from the ratio of intensities of the bands corresponding to c-RNP and RNA. Lane 1: An16 RNA only; Lane 2: An16 and Ac-Rev-(GGS)$_3$G-HZ reacted at pH 4; Lane 3: An16 and Ac-Rev-(GGS)$_3$G-HZ reacted at pH 5; Lane 4: An16 and Ac-Rev-(GGS)$_3$G-HZ reacted at pH 6; Lane 5: An16 and Ac-Rev-(GGS)$_3$G-HZ reacted at pH 7.

3.2. Contibution of the RNA-Peptide Interaction for the Reaction Yield

RNA sequence dependency of the reaction was next evaluated to address the proximity effect of reactive groups. The specific interaction between the Rev peptide and RRE RNA of the RNP scaffold was expected to have an influence on the efficiency of covalent linkage formation. A sequence-scrambled scrAn16 RNA that contained the same number of each nucleotide to An16 RNA was prepared (Figure 3A). The reaction of the oxidized scrAn16 with Ac-Rev-(GGS)$_3$G-HZ was performed at pH 5. The MALDI TOF MS analysis using the reaction solution after removing the unreacted peptide was performed to characterize the products. The production of covalently linked RNP was confirmed by observing respective MS peaks (Figures S2 and S3). In the five trials, scrAn16 showed a similar averaged reaction yield (92 ± 5%) to that of the parent An16 (92 ± 5%) when conjugated with the hydrazide group of Ac-Rev-(GGS)$_3$G-HZ (Figure 3B and Figure S1B,D). The result showed that the RNA sequence did not affect the efficiency of reaction. The non-specific electrostatic interaction between the RNA and the cationic Rev peptide would relate to the reaction efficiency. To verify this notion, the reaction for the formation of covalently linked RNP was carried out by using a truncated peptide of Ac-Rev-(GGS)$_3$G-HZ. A truncated peptide possessing only the GGS peptide linker moiety, Ac-GGSGGSGGSG-HZ, was designed by deleting the Rev peptide sequence from Ac-Rev-(GGS)$_3$G-HZ. The truncated peptide showed a significant decrease in the yield to 57 ± 1%, even after a prolonged reaction time for 14 h (Figure 4 and Figure S1C). This result supported the notion that the electrostatic interaction between RNA and the peptide mainly contributed to increasing the reaction yield. Thus, the RNP scaffold was useful for the efficient production of RNA-peptide conjugates.

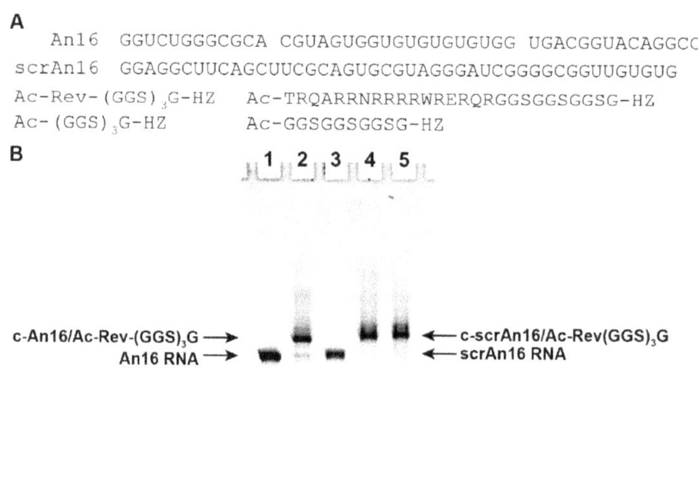

Figure 3. (**A**) Nucleotide and peptide sequence of An16 RNA, scrAn16 RNA, Ac-Rev-(GGS)$_3$G-HZ peptide, and Ac-(GGS)$_3$G-HZ peptide; (**B**) a denaturing PAGE (8 M Urea) analysis of the conjugation reaction products of c-RNPs constructed from An16 or scrAn16 RNA with Ac-Rev-(GGS)$_3$G-HZ. Lane1: An16 RNA; Lane 2: An16 and Ac-Rev-(GGS)$_3$G-HZ reacted for 3 h; Lane 3: scrAn16 RNA; Lane 4: scrAn16 and Ac-Rev-(GGS)$_3$G-HZ reacted for 3 h; Lane 5: scrAn16 and Ac-Rev-(GGS)$_3$G-HZ reacted for 14 h.

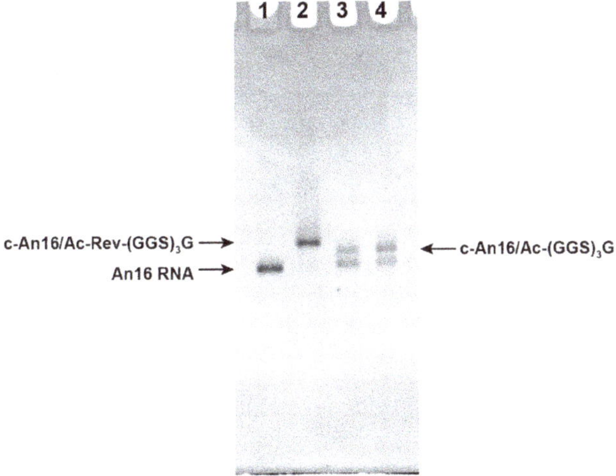

Figure 4. A denaturing PAGE (8 M Urea) analysis of the conjugation reaction products of c-RNP constructed from An16 and Ac-(GGS)$_3$G-HZ. Lane 1: An16 RNA; Lane 2: c-An16/Ac-Rev-(GGS)$_3$G-HZ reacted for 3 h; Lane 3: c-An16/Ac-(GGS)$_3$G-HZ reacted for 3 h; Lane 4: c-An16/Ac-(GGS)$_3$G-HZ reacted for 14 h.

4. Discussion

A Schiff base formation between the hydrazide and aldehyde group rapidly progressed at a mild acidic condition although the stability was less than that at a neutral pH. Because the reaction efficiency of this crosslinking method was based on the balance of such properties, the assays under the wide range of pH helped us to understand a limitation and the applicability of this reaction under the physiological condition for RNA-protein or RNA-peptide conjugation. Formation of the covalent linkage between RNA and the Rev peptide subunit in the Rev-RRE complex quantitatively proceeded in a mild acidic to a neutral pH condition within 3 h. The result indicated the versatility of this reaction in a physiological condition for the crosslink formation of peptide and RNA. Unlike the sequence-specific nature for the stable noncovalent RNP complex formation, formation of the covalent linkage between the 3′-dialdehyde group of RNA and the hydrazide group at the C-terminal of Rev peptide required no specific RNA sequence for the RNA binding of Rev peptide. Even the nonspecific RNA-peptide complex formation driven by the electrostatic interaction was sufficient for exerting the proximity effect of reactive groups, thereby resulting an efficient formation of covalent linkage between the Rev peptide and RNA. The significant decrease of the reaction efficiency by deletion of the cationic moiety in the peptide sequence also showed the contribution of the electrostatic interaction for the efficient proceeding of the crosslinking reaction.

We designed a DNA sequence-specific protein-tag that underwent a proximity driven intermolecular crosslinking between protein and DNA [43–49]. Selective DNA modification by a self-ligating protein tag conjugated with a DNA-binding domain, termed as a modular adaptor, was achieved by relying on the chemoselectivity of the protein tag [50–52]. By tuning the alkylation kinetics for the protein-tag and its substrate, the sequence-specific crosslinking reaction of the modular adaptor was exclusively driven by the DNA recognition when the dissociation rate of the DNA complex was much larger than the rate constant for the alkylation reaction [45–49]. In connection with these findings, a sequence-specific crosslinking reaction between the 3′-dialdehyde group of RNA and the hydrazide group of RNA-binding peptide would be realized by tuning the reaction conditions, such as the concentration of sodium salt. As we have reported previously, the crosslinking reaction between the 3′-dialdehyde group of RNA and the hydrazide group of peptide was applied for the facile construction of covalently crosslinked fluorescent RNP sensors [41,42]. The covalently crosslinked fluorescent RNP sensor not only realized an improved stability of RNP sensors but also expand its application, such as the simultaneous detection of multiple targets in the solution. The noncovalent RNP scaffold was effective for the library-based selection and the cooperative functionalization of RNP receptor, such as the fluorescent sensors and catalysts [53]. Once a functional noncovalent RNP was obtained, formation of a covalent linkage within the RNA-peptide complex provided stable RNP for a wide range of applications. A natural RNA oligonucleotide and facile preparation of the reactive peptide was only needed to perform this reaction. Our investigations regarding the effect of the pH, RNA, and a peptide sequence for the reaction efficiencies in this reaction would help us to understand the limitation and applicability of this method for construction of RNA-peptide or protein conjugate. It would also be helpful for the development of an emerging application of oligonucleotide-peptide conjugates, such as a self-assembled scaffold of nanostructure or an efficient delivery system of functional RNA into the cell.

Supplementary Materials: The following are available online at http://www.mdpi.com/2076-3417/10/24/8920/s1, Figure S1: The results of band intensity analysis of denaturing PAGE, Figure S2: MALDI TOF MS analysis of the reaction solution, Figure S3: Summary of the results of MALDI TOF MS analysis.

Author Contributions: S.N. and T.M. conceived and designed the experiments; S.N., T.S. and Z.Z. performed the experiments; T.M. supervised the project. All authors have read and agreed to the published version of the manuscript.

Funding: This work was supported by JSPS KAKENHI Grant Numbers 18K14335 (S.N.) and 17H01213 (T.M.), Japan and by JST CREST Grant Number JPMJCR18H5 (T.M.), Japan.

Conflicts of Interest: The authors declare no conflict of interest.

References

1. Roberts, T.C.; Langer, R.; Wood, M.J.A. Advances in oligonucleotide drug delivery. *Nat. Rev. Drug Discov.* **2020**, *19*, 673–694. [CrossRef] [PubMed]

2. Tai, W. Current aspects of siRNA bioconjugate for in vitro and in vivo delivery. *Molecules* **2019**, *24*, 2211. [CrossRef] [PubMed]

3. Tai, W.; Gao, X. Functional peptides for siRNA delivery. *Adv. Drug Deliv. Rev.* **2017**, *110–111*, 157–168. [CrossRef] [PubMed]

4. Roberts, R.W.; Szostak, J.W. RNA-peptide fusions for the in vitro selection of peptides and proteins. *Proc. Natl. Acad. Sci. USA* **1997**, *94*, 12297–12302. [CrossRef]

5. Nemoto, N.; Miyamoto-Sato, E.; Husimi, Y.; Yanagawa, H. In vitro virus: Bonding of mRNA bearing puromycin at the 3'-terminal end to the C-terminal end of its encoded protein on the ribosome in vitro. *FEBS Lett.* **1997**, *414*, 405–408.

6. Tung, C.H.; Stein, S. Preparation and applications of peptide-oligonucleotide conjugates. *Bioconjug. Chem.* **2000**, *11*, 605–618. [CrossRef]

7. Venkatesan, N.; Kim, B.H. Peptide conjugates of oligonucleotides: synthesis and applications. *Chem. Rev.* **2006**, *106*, 3712–3761. [CrossRef]

8. Lu, K.; Duan, Q.-P.; Ma, L.; Zhao, D.-X. Chemical strategies for the synthesis of peptide-oligonucleotide conjugates. *Bioconjug. Chem.* **2010**, *21*, 187–202. [CrossRef]

9. Singh, Y.; Murat, P.; Defrancq, E. Recent developments in oligonucleotide conjugation. *Chem. Soc. Rev.* **2010**, *39*, 2054–2070. [CrossRef]

10. Haralambidis, J.; Duncan, L.; Angus, K.; Tregear, G.W. The synthesis of polyamide-oligonucleotide conjugate molecules. *Nucleic Acids Res.* **1990**, *18*, 493–499. [CrossRef]

11. Kachalova, A.; Zubin, E.; Stetsenko, D.; Gait, M.; Oretskaya, T. Oligonucleotides with 2'-*O*-carboxymethyl group: Synthesis and 2'-conjugation via amide bond formation on solid phase. *Org. Biomol. Chem.* **2004**, *2*, 2793–2797. [CrossRef] [PubMed]

12. Eritja, R.; Pons, A.; Escarceller, M.; Giralt, E.; Albericio, F. Synthesis of defined peptide-oligonucleotide hybrids containing a nuclear transport signal sequence. *Tetrahedron* **1991**, *47*, 4113–4120. [CrossRef]

13. Vivès, E.; Lebleu, B. Selective coupling of a highly basic peptide to an oligonucleotide. *Tetrahedron Lett.* **1997**, *38*, 1183–1186. [CrossRef]

14. Turner, J.J.; Arzumanov, A.A.; Gait, M.J. Synthesis, cellular uptake and HIV-1 Tat-dependent trans-activation inhibition activity of oligonucleotide analogues disulphide-conjugated to cell-penetrating peptides. *Nucleic Acids Res.* **2005**, *33*, 27–42. [CrossRef] [PubMed]

15. Zhu, T.; Tung, C.-H.; Breslauer, K.J.; Dickerhof, W.A.; Stein, S. Preparation and physical properties of conjugates of oligodeoxynucleotides with poly(δ)ornithine peptides. *Antisense Res. Dev.* **1993**, *3*, 349–356. [CrossRef] [PubMed]

16. Arar, K.; Aubertin, A.-M.; Roche, A.-C.; Monsigny, M.; Mayer, R. Synthesis and antiviral activity of peptide oligonucleotide conjugates prepared by using NR-(Bromoacetyl) peptides. *Bioconjug. Chem.* **1995**, *6*, 573–577. [CrossRef]

17. Ghosh, S.S.; Kao, P.M.; McCue, A.W.; Chappelle, H.L. Use of maleimide-thiol coupling chemistry for efficient syntheses of oligonucleotide-enzyme conjugate hybridization probes. *Bioconjug. Chem.* **1990**, *1*, 71–76. [CrossRef]

18. Ede, N.J.; Tregear, G.W.; Haralambidis, J. Routine preparation of thiol oligonucleotides: Application to the synthesis of oligonucleotide-peptide hybrids. *Bioconjug. Chem.* **1994**, *5*, 373–378. [CrossRef]

19. McPherson, M.; Wright, M.C.; Lohse, P.A. Synthesis of an RNA-peptide conjugate by orthogonal ligation. *Synlett* **1999**, *S1*, 978–980. [CrossRef]

20. Stetsenko, D.A.; Gait, M.J. Efficient conjugation of peptides to oligonucleotides by "native ligation". *J. Org. Chem.* **2000**, *65*, 4900–4908. [CrossRef]

21. Takeda, S.; Tsukiji, S.; Nagamune, T. A cysteine-appended deoxyuridine for the postsynthetic DNA modification using native chemical ligation. *Tetrahedron Lett.* **2005**, *46*, 2235–2238. [CrossRef]

22. Gogoi, K.; Mane, M.V.; Kunte, S.S.; Kumar, V.A. A versatile method for the preparation of conjugates of peptides with DNA/PNA/analog by employing chemo-selective click reaction in water. *Nucleic Acids Res.* **2007**, *35*, e139. [CrossRef] [PubMed]

23. Brown, S.D.; Graham, D. Conjugation of an oligonucleotide to Tat, a cell-penetrating peptide, via click chemistry. *Tetrahedron Lett.* **2010**, *51*, 5032–5034. [CrossRef]

24. Proudnikov, D.; Timofeev, E.; Mirzabekov, A. Immobilization of DNA in polyacrylamide gel for the manufacture of DNA and DNA-oligonucleotide microchips. *Anal. Biochem.* **1998**, *259*, 34–41. [CrossRef] [PubMed]

25. Khym, J.X. The reaction of methylamine with periodate-oxidized adenosine 5'-phosphate. *Biochemistry* **1963**, *2*, 344–350. [CrossRef] [PubMed]

26. Brown, D.M.; Read, A.P. 932. Nucleotides. Part XLIX. The reduction of the adduct of periodate-oxidised adenosine-5' phosphate and methylamine. *J. Chem. Soc. Perkin 1.* **1965**, 5072–5074. [CrossRef] [PubMed]

27. Zatsepin, T.S.; Stetsenko, D.A.; Gait, M.J.; Oretskaya, T.S. Use of carbonyl group addition–elimination reactions for synthesis of nucleic acid conjugates. *Bioconjug. Chem.* **2005**, *16*, 471–489. [CrossRef] [PubMed]

28. Zatsepin, T.S.; Stetsenko, D.A.; Arzumanov, A.A.; Romanova, E.A.; Gait, M.J.; Oretskaya, T.S. Synthesis of peptide-oligonucleotide conjugates with single and multiple peptides attached to 2'-aldehydes through thiazolidine, oxime, and hydrazine linkages. *Bioconjug. Chem.* **2002**, *13*, 822–830. [CrossRef]

29. Forget, D.; Boturyn, D.; Defrancq, E.; Lhomme, J.; Dumy, P. Highly efficient synthesis of peptide-oligonucleotide conjugates: Chemoselective oxime and thiazolidine formation. *Chem. Eur. J.* **2001**, *7*, 3976–3984. [CrossRef]

30. Battiste, J.L.; Mao, H.; Rao, N.S.; Tan, R.; Muhandiram, D.R.; Kay, L.E.; Frankel, A.D.; Williamson, J.R. Alpha helix-RNA major groove recognition in an HIV-1 rev peptide-RRE RNA complex. *Science* **1996**, *273*, 1547–1551. [CrossRef]

31. Morii, T.; Hagihara, M.; Sato, S.; Makino, K. In vitro selection of ATP-binding receptors using a ribonucleopeptide complex. *J. Am. Chem. Soc.* **2002**, *124*, 4617–4622. [CrossRef] [PubMed]

32. Ellington, A.D.; Szostak, J.W. In vitro selection of RNA molecules that bind specific ligands. *Nature* **1990**, *346*, 818–822. [CrossRef] [PubMed]

33. Tuerk, C.; Gold, L. Systematic evolution of ligands by exponential enrichment: RNA ligands to bacteriophage T4 DNA polymerase. *Science* **1990**, *249*, 505–510. [CrossRef]

34. Hagihara, M.; Fukuda, M.; Hasegawa, T.; Morii, T. A modular strategy for tailoring fluorescent biosensors from ribonucleopeptide complexes. *J. Am. Chem. Soc.* **2006**, *128*, 12932–12940. [CrossRef] [PubMed]

35. Nakano, S.; Nakata, E.; Morii, T. Facile conversion of RNA aptamers to modular fluorescent sensors with tunable detection wavelengths. *Bioorg. Med. Chem. Lett.* **2011**, *21*, 4503–4506. [CrossRef] [PubMed]

36. Nakano, S.; Tamura, T.; Das, R.K.; Nakata, E.; Chang, Y.-T.; Morii, T. A diversity-oriented library of fluorophore-modified receptors constructed from a chemical library of synthetic fluorophores. *ChemBioChem* **2017**, *18*, 2212–2216. [CrossRef] [PubMed]

37. Hasegawa, T.; Hagihara, M.; Fukuda, M.; Morii, T. Stepwise functionalization of ribonucleopeptides: Optimization of the response of fluorescein ribonucleopeptide sensors for ATP. *Nucleosides Nucleotides Nucleic Acids* **2007**, *26*, 1277–1281. [CrossRef]

38. Nakano, S.; Mashima, T.; Matsugami, A.; Inoue, M.; Katahira, M.; Morii, T. Structural aspects for the recognition of ATP by ribonucleopeptide receptors. *J. Am. Chem. Soc.* **2011**, *133*, 4567–4579. [CrossRef]

39. Liew, F.F.; Hasegawa, T.; Fukuda, M.; Nakata, E.; Morii, T. Construction of dopamine sensors by using fluorescent ribonucleopeptide complexes. *Bioorg. Med. Chem.* **2011**, *19*, 4473–4481. [CrossRef]

40. Hasegawa, T.; Hagihara, M.; Fukuda, M.; Nakano, S.; Fujieda, N.; Morii, T. Context-dependent fluorescence detection of a phosphorylated tyrosine residue by a ribonucleopeptide. *J. Am. Chem. Soc.* **2008**, *130*, 8804–8812. [CrossRef]

41. Nakano, S.; Fukuda, M.; Tamura, T.; Sakaguchi, R.; Nakata, E.; Morii, T. Simultaneous detection of ATP and GTP by covalently linked fluorescent ribonucleopeptide sensors. *J. Am. Chem. Soc.* **2013**, *135*, 3465–3473. [CrossRef] [PubMed]

42. Nakano, S.; Shimizu, M.; Dinh, H.; Morii, T. Highly selective dual sensing of ATP and ADP using fluorescent ribonucleopeptide sensors. *Chem. Commun.* **2019**, *55*, 1611–1614. [CrossRef] [PubMed]

43. Nakata, E.; Liew, F.F.; Uwatoko, C.; Kiyonaka, S.; Mori, Y.; Katsuda, Y.; Endo, M.; Sugiyama, H.; Morii, T. Zinc-finger proteins for site-specific protein positioning on DNA-origami structures. *Angew. Chem. Int. Ed.* **2012**, *51*, 2421–2424. [CrossRef] [PubMed]

44. Ngo, T.A.; Nakata, E.; Saimura, M.; Kodaki, T.; Morii, T. A protein adaptor to locate a functional protein dimer on molecular switchboard. *Methods* **2014**, *67*, 142–150. [CrossRef]

45. Nakata, E.; Dinh, H.; Ngo, T.A.; Saimura, M.; Morii, T. A modular zinc finger adaptor accelerates the covalent linkage of proteins at specific locations on DNA nanoscaffolds. *Chem. Commun.* **2015**, *51*, 1016–1019. [CrossRef]

46. Ngo, T.A.; Nakata, E.; Saimura, M.; Morii, T. Spatially organized enzymes drive cofactor-coupled cascade reactions. *J. Am. Chem. Soc.* **2016**, *138*, 3012–3021. [CrossRef]

47. Nguyen, T.M.; Nakata, E.; Saimura, M.; Dinh, H.; Morii, T. Design of modular protein-tags for the orthogonal covalent bond formation at specific DNA sequences. *J. Am. Chem. Soc.* **2017**, *139*, 8487–8496. [CrossRef]

48. Nakata, E.; Nakano, S.; Rajendran, A.; Morii, T. Covalent bond formation by modular adaptors to locate multiple enzymes on a DNA scaffold. In *Kinetic Control in Synthesis and Self-Assembly*; Numata, M., Yagai, S., Hamura, T., Eds.; Academic Press: Cambridge, MA, USA, 2019; pp. 163–183.

49. Nguyen, T.M.; Nakata, E.; Zhang, Z.; Saimura, M.; Dinh, H.; Morii, T. Rational design of a DNA sequence-specific modular protein tag by tuning the alkylation kinetics. *Chem. Sci.* **2019**, *10*, 9315–9325. [CrossRef]

50. Keppler, A.; Gendreizig, S.; Gronemeyer, T.; Pick, H.; Vogel, H.; Johnsson, K. A general method for the covalent labeling of fusion proteins with small molecules in vivo. *Nat. Biotechnol.* **2003**, *21*, 86–89. [CrossRef]

51. Gautier, A.; Juillerat, A.; Heinis, C.; Correa, I.R.; Kindermann, M.; Beaufils, F.; Johnsson, K. An engineered protein tag for multiprotein labeling in living cells. *Chem. Biol.* **2008**, *15*, 128–136. [CrossRef]

52. Los, G.V.; Encell, L.P.; McDougall, M.G.; Hartzell, D.D.; Karassina, N.; Zimprich, C.; Wood, M.G.; Learish, R.; Ohana, R.F.; Urh, M.; et al. HaloTag: A novel protein labeling technology for cell imaging and protein analysis. *ACS Chem. Biol.* **2008**, *3*, 373–382. [CrossRef] [PubMed]

53. Tamura, T.; Nakano, S.; Nakata, E.; Morii, T. Construction of a library of structurally diverse ribonucleopeptides with catalytic groups. *Bioorg. Med. Chem.* **2017**, *25*, 1881–1888. [CrossRef] [PubMed]

Publisher's Note: MDPI stays neutral with regard to jurisdictional claims in published maps and institutional affiliations.

Article

Design of the Crosslinking Reactions for Nucleic Acids-Binding Protein and Evaluation of the Reactivity

Kenta Odaira, Ken Yamada, Shogo Ishiyama, Hidenori Okamura and Fumi Nagatsugi * 

Institute of Multidisciplinary Research for Advanced Materials, Tohoku University, Sendai 980-8577, Japan; k.odaira@kobayashi.co.jp (K.O.); Ken.Yamada@umassmed.edu (K.Y.); rare.earth.144@gmail.com (S.I.); hidenori.okamura.b8@tohoku.ac.jp (H.O.)
* Correspondence: nagatugi@tohoku.ac.jp; Tel.: +81-22-217-5633

Received: 2 October 2020; Accepted: 29 October 2020; Published: 30 October 2020

Featured Application: Alkylation for nucleic acids-binding protein.

Abstract: Selective chemical reactions of biomolecules are some of the important tools for investigations by biological studies. We have developed the selective crosslinking reactions to form covalent bonds to DNA or RNA using crosslinking oligonucleotides (CFO) bearing reactive bases. In this study, we designed the cross-linkable 4-amino-6-oxo-2-vinyltriazine derivative with an acyclic linker (*acy*AOVT) to react with the nucleic acids-binding protein based on our previous results. We hypothesized that the *acy*AOVT base would form a stable base pair with guanine by three hydrogen bonds at the positions of the vinyl group in the duplex DNA major groove, and the vinyl group can react with the nucleophilic species in the proximity, for example, the cysteine or lysine residue in the nucleic acids-binding protein. The synthesized oligonucleotides bearing the *acy*AOVT derivative showed a higher reactivity than that of the corresponding pyrimidine derivative without one nitrogen. The duplex containing *acy*AOVT-guanine (G) formed complexes with Hha1 DNMT even in the presence of 2-mercaptoethanol. We expect that our system will provide a useful tool for the molecular study of nucleic acids-binding proteins.

Keywords: oligonucleotide; crosslink; nucleic acid binding protein; cytosine methyltransferase

---

1. Introduction

The binding of proteins to nucleic acids (DNA and RNA) is central to all aspects of gene expression regulation. DNA-binding proteins include the transcription factor to bind to specific DNA sequences and control the transcription rate of genetic information from the DNA to RNA [1,2]. Many DNA modifying enzymes are also included in the DNA binding protein, for example, repair enzymes [3,4] and epigenetic modification enzymes. DNA methyl transferase is one of the epigenetic modification enzymes and catalyses the transfer of the methyl group to DNA modifying the function of genes and affecting gene expression [5,6]. RNA-binding proteins have important functions in the post-transcriptional process, such as splicing regulation [7] and modulation of the mRNA translation. Recent studies have revealed that post-transcriptional gene regulation by non-coding RNAs is involved in most biological activities. The association of the RNA binding protein with ncRNA plays a crucial role in these biological functions [8–10]. The chemical tools for the control of the interaction between the nucleic acids-binding protein and nucleic acids have the potential for the development as the new strategy for artificial control of the gene expression.

Crosslinking reactions between nucleic acids and binding proteins are powerful tools for analyzing these interactions. Photoreactive functional groups, such as diazirines and benzophenones, were utilized for DNA or RNA-protein crosslinking [11–14], but lack selectivity for the target amino acid. Furthermore, due to their intrinsic reactivity, these groups react with water and other reactive chemical species resulting in low yields. Oligodeoxynucleotides (ODN) with reactive groups are exploited for their effective DNA-protein crosslinking reactions by the proximity effect of the specific DNA recognition with the binding protein. The reactive groups activated by oxidation, such as a diol [15] and furan [16], react with a proximal lysine or arginine in the protein or peptides. Hocek et al. reported that the ODN bearing a vinylsulfone amide [17] or chloroacetamide group [18] efficiently cross-linked with the p53 protein through alkylation of the cysteine in the proximal position with the reactive groups. Recently, they demonstrated that ODN with 2-vinylhypoxantine reacted with a thiol-containing minor groove binding peptide by proximity effect [19].

In our previous study, we reported the synthesis of the ODN bearing 4-amino-6-oxo-2-vinyltriazine with an ethyl linker (Et-AOVT: (**1**)) as a cross linkable derivative and an evaluation of its properties [20]. ODN showed a relatively high reactivity with the pyrimidine nucleobases of the complementary DNAs and the lowest reactivity to guanine. We hypothesized that the AOVT base would form a stable base pair with guanine (G) by three hydrogen bonds to be positioned at the vinyl group in the DNA major groove. In this study, we have designed an AOVT derivative with an acyclic linker (*acy*AOVT: (**2**)) as a cross-linkable probe of the nucleic acids-binding protein. The distance between the reactive base and sugar moiety of **2** is shorter than that of **1**, and consequently, the *acy*AOVT (**2**) should form a base pair with guanine (G) similar to the natural cytosine (C)-G pair. This assumption is supported by the molecular modeling shown in Figure 1. The molecular modeling revealed that the structure of the base pair of *acy*AOVT (**2**)-G and the natural C-G significantly overlap each other (Figure 1C). On the other hand, the structure of Et-AOVT (**1**)-G is subtly shifted from the natural base pair one (Figure 1B). The duplex DNA containing the *acy*AOVT (**2**)-G is expected to bind to the nucleic acid binding proteins similar to the natural duplex and form covalent linkages with the nucleophilic residues of the protein in the proximal position of the vinyl group (Figure 2).

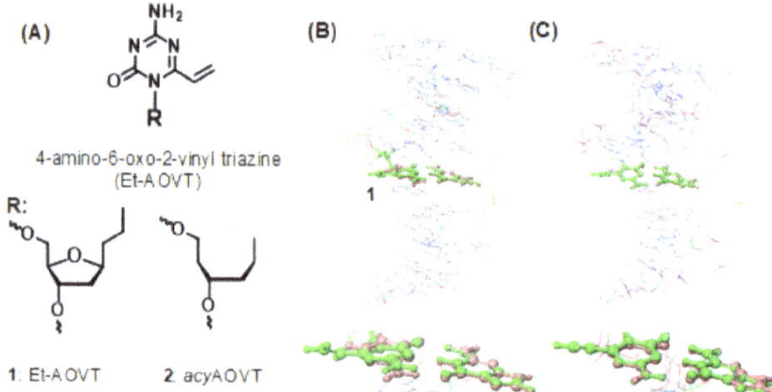

Figure 1. (**A**) Structures of cross-linkable derivatives; (**B,C**) Superimposed structures between the natural duplex (pink) and the duplex contained **1** (**B**) or **2** (**C**) (green).

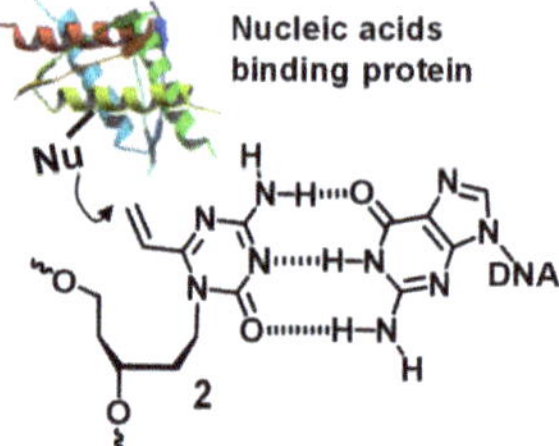

Figure 2. Design of the cross-linkable probe to nucleic acids-binding protein.

2. Materials and Methods

2.1. General

The ^{1}H-NMR spectra were obtained by 400 or 600 MHz spectrometers (Bruker, Billerica, MA, USA). The ^{1}H chemical shifts are described as δ values in ppm relative to acetone-d$_6$ (2.05 ppm), DMSO-d$_6$ (2.50 ppm), CDCl$_3$ (7.26 ppm), and tetramethylsilane (1H, 0.00 ppm). The ^{13}C NMR spectra were obtained by a Bruker 600 MHz spectrometer. The ^{13}C chemical shifts are described as the δ values in ppm relative to acetone-d$_6$ (29.84 ppm), DMSO-d$_6$ (39.52 ppm), and CDCl$_3$ (77.16 ppm). The ^{31}P-NMR spectra were recorded by a Bruker 500 MHz (202 MHz for ^{31}P). The multiplicity and qualifier abbreviations are as follows: s = singlet, d = doublet, t = triplet, q = quartet, quin. = quintet, sept. = septet, m = multiplet, br = broad. The electrospray ionization (ESI) mass spectra were recorded using a BioTOF II mass spectrometer or APEX III (Bruker Daltonics, Bruker, Billerica, MA, USA). The matrix-assisted laser desorption/ionization (MALDI-TOF) mass spectra were recorded Autoflex speed mass spectrometer and a laser at 337 nm in the negative mode using 3-hydroxypicolinic acid as the matrix or in the positive mode using 2,5-dihydroxybenzoic acid as the matrix. Thin-layer chromatography (TLC) was performed using silica gel 60 F$_{254}$ pre-coated plates. Column chromatography was performed using silica gel 60 N (spherical, neutral, 100–210 μm, Kanto Chemical, Tokyo, Japan). Flash chromatography was performed using Kanto Chemical silica gel 60 N (spherical, neutral, 40–50 μm). The ultraviolet-visible (UV-vis) absorption spectra were recorded using a DU800 spectrometer (Beckman Coulter, Brea, CA, USA). The ODN synthesis was carried out using an automated DNA synthesizer (392 DNA/RNA synthesizer, ABI, Foster City, CA, USA) following the standard phosphoramidite chemistry. High performance liquid chromatography (HPLC) was performed using a Cosmosil 5C18MSII column (4.6 or 10 × 250 mm, Nacalai Tesque, Kyoto, Japan), a PU-986 pump (JASCO, Tokyo, Japan), JASCO 2075 UV detector and a JASCO 2067 column oven. pH measurements were measured by a Seven Easy pH meter (Mettler Toledo, Columbus, OH) using an 8220BNWP electrode. The denaturing polyacrylamide gel plates were visualized and quantified using a FLA-5100 Fluor Imager (Fujifilm, Tokyo, Japan). Anhydrous methanol, DMF, THF, CH$_2$Cl$_2$, dioxane, pyridine, DMSO, CH$_3$CN, toluene and THF were purchased from Wako Pure Chemical Industries, Ltd. (Tokyo, Japan). Unless otherwise noted, all the synthetic reactions were carried out at ambient temperature. The reactions requiring anhydrous conditions were achieved under an argon atmosphere in flasks dried under 1–3 mmHg. Commercially available reagents were obtained from Wako Pure Chemical Industries, Ltd., TCI, Inc., (Tokyo, Japan), Sigma-Aldrich Co. LLC (St. Louis, MO) and Kanto Chemical Co., Inc (Tokyo, Japan) and used without further purification. Human DNA (cytosine-5) methyltransferase (DNMT1) was obtained from New England Bio Labs Japan, Inc. (Tokyo, Japan)

2.2. Synthesis of the Nucleoside Derivatives

2.2.1. Synthesis of 4-amino-6-oxo-1-(3′,5′-O-t-butyldimethylsilyl-2′,4′-dideoxy-D-ribityl) triazine (**4**)

The linker **3** (1.3 g, 3.0 mmol), which was prepared by a modified literature procedure [21], was added to 5-aza cytosine Na salt (1.0 g, 7.5 mmol) in DMSO (30 mL). The mixture was stirred at 60 °C for 18 h. The resulting mixture was diluted with EtOAc and quenched with NH$_4$Cl. The product was extracted with EtOAc and the combined organic layer was washed with brine, dried over Na$_2$SO$_4$, filtered, and evaporated under reduced pressure. The resulting oil was purified by silica gel column chromatography (CH$_3$OH/CHCl$_3$, 1:100, then CH$_3$OH/CHCl$_3$, 1:50) to afford the corresponding N1-alkylation (772 mg, 1.74 mmol, 58%) and N3-alkylation (126 mg, 0.29 mmol, 10%) products.

N1-alkylation product: ^{1}H-NMR (400 MHz, CDCl$_3$) δ 7.90 (s, 1H), 6.98 (brs, 1H), 5.65 (brs, 1H), 3.97 (ddd, 2H, J = 12.0, 12.0, 6.0 Hz), 3.90–3.77 (m, 2H), 3.64 (dd, 2H, J = 6.4, 6.4 Hz), 1.99–1.60 (m, 6H), 1.37–1.23 (m, 1 H), 0.89 (s, 9H), 0.87 (s, 9H), 0.07 (s, 3H), 0.06 (s, 3H), 0.03 (s, 6H); ^{13}C-NMR (125 MHz, CDCl3) δ 166.7, 158.7, 154.5, 67.4, 59.6, 45.1, 40.2, 36.2, 26.12, 26.09, 18.4, 18.3, −4.2, −4.3, −5.14, −5.16; HRMS (ESI) calcd. for C$_{20}$H$_{43}$N$_4$O$_3$Si$_2{}^+$ [M + H]$^+$ m/z 443.2868, found m/z 443.2879.

N3-alkylation product: ^{1}H-NMR (600 MHz, CDCl3) δ 8.34 (s, 1H), 5.90 (brs, 1H), 5.61 (brs, 1H), 4.42–4.35 (m, 1H), 4.05–4.01 (m, 1H), 3.70–3.64 (m, 2H), 1.99–1.95 (m, 1H), 1.89–1.84 (m, 2H), 1.7–1.67 (m, 2H), 0.871 (s, 9H), 0.867 (s, 9H), 0.046 (s, 3H), 0.027 (s, 3H), 0.025 (s, 3H), 0.018 (s, 3H); ^{13}C-NMR (125 MHz, CDCl3) δ 170.4, 167.93, 167.88, 66.3, 64.5, 59.6, 40.3, 35.9, 25.9, 25.8, 18.2, 18.0, −4.6, −4.8, −5.36, −5.38; HRMS (ESI) calcd. for $C_{20}H_{43}N_4O_3Si_2^+$ [M + H]$^+$ *m/z* 443.2868, found *m/z* 443.2879

2.2.2. Synthesis of 4-amino-2-(2-octylthioethyl)-6-oxo-1-(3′,5′-*O*-t-butyldimethylsilyl-2′, 4′-dideoxy-ᴅ-ribityl) triazine (**6**)

A 28% aq NH$_4$OH (10 mL) solution was added to a solution of the N1-alkylation product **4** (172 mg, 390 µmol) in dioxane-methanol (10 mL, 1/1, *v/v*), then sealed with a glass stopper. The mixture was stirred at 55 °C for 24 h. The resulting solution was evaporated under vacuum at 43 °C. The resulting moist solid was co-evaporated with ethanol × 3 to afford guanylurea **5**. This product was directly used for the next step without further purification. The orthoester (597 mg, 1.95 mmol, purity = 91%) was added to a solution of guanylurea (169 mg, 390 µmol) in DMF (0.2 M, 1.95 mL). The mixture was stirred at 120 °C for 2 h. The resulting yellow solution was diluted with Et$_2$O. The product was extracted with Et$_2$O and the combined organic layer was washed with NH$_4$Cl aq., NaHCO$_3$ aq., and brine, dried over Na$_2$SO$_4$, filtered and evaporated. The crude product was purified by silica gel column chromatography (CHCl$_3$ then, CHCl$_3$ with 1% Et$_3$N, then CH$_3$OH/CHCl$_3$, 1:300 with 1% Et$_3$N) to afford **6** (198.3 µmol, 51%) as a yellow oil: ^{1}H-NMR (600 MHz, CDCl$_3$) δ 5.70 (brs, 1H), 5.17 (brs, 1H), 4.02–3.89 (m, 3H), 3.65 (dd, 2H, *J* = 6.6, 6.6 Hz), 2.95–2.88 (m, 3H), 2.54 (t, 2H, *J* = 7.8 Hz), 1.93–1.88 (m, 1 H), 1.79–1.64 (m, 4H), 1.59 (quin, 2H, *J* = 7.2 Hz), 1.39–1.25 (m, 10H), 0.91 (s, 9H), 0.89 (t, 3H, *J* = 7.2 Hz), 0.88 (s, 9H), 0.089 (s, 3H), 0.086 (s, 3H), 0.04 (s, 6H); ^{13}C-NMR (150 MHz, CDCl$_3$) δ 168.8, 165.3, 155.8, 67.6, 59.6, 41.6, 40.4, 35.8, 34.3, 33.0, 32.1, 29.9, 29.5, 29.2, 28.2, 26.19, 26.18, 22.9, 18.5, 18.3, 14.4, 0.26, −4.18, −4.26, −4.30, −5.1; HRMS (ESI) calcd. for $C_{30}H_{63}N_4O_3SSi_2^+$ [M + H]$^+$ *m/z* 615.4154, found *m/z* 615.4158.

2.2.3. Synthesis of 2-(2-octylthioethyl)-6-oxo-4-acetylamino-1-(3′,5′-*O*-t-butyldimethylsilyl-2′, 4′-dideoxy-ᴅ-ribityl) triazine (**7**)

AcCl (5.3 µL, 74.7 µmol) was added to a solution of **6** (15 mg, 24.9 µmol) in pyridine (250 µL). The mixture was stirred at rt for 19 h. The resulting solution was diluted with Et$_2$O and quenched with NH$_4$Cl. The product was extracted with Et$_2$O and the combined organic layer was washed with brine, dried over Na$_2$SO$_4$, filtered and evaporated. The crude product was purified by silica gel column chromatography (CHCl$_3$, then CHCl$_3$ with 1% Et$_3$N) to afford **7** (13.4 µmol, 54%) as a colorless oil: ^{1}H-NMR (600 MHz, CDCl$_3$) δ 7.71 (br, 1H), 4.08–3.98 (m, 3H), 3.66 (t, 2H, *J* = 6.0 Hz), 3.65 (t, 2H, *J* = 6.0 Hz), 3.00 (t, 2H, *J* = 6.6 Hz), 2.91 (t, 2H, *J* = 6.6 Hz), 2.61 (s, 3H), 2.54 (t, 2H, *J* = 7.2 Hz), 1.95 (m, 1H), 1.81–1.64 (m, 3H), 1.58 (qt, 2H, *J* = 7.2, 7.2 Hz), 1.39–1.26 (m, 10 H), 0.92 (s, 9H), 0.91 (t, 3H, *J* = 7.2 Hz), 0.88 (s, 9H), 0.095 (s, 3H), 0.093 (s, 3H), 0.041 (s, 3H), 0.03 (s, 3H); ^{13}C-NMR (150 MHz, CDCl$_3$) δ 172.0, 170.2, 161.5, 154.8, 41.9, 40.0, 34.9, 34.1, 32.8, 29.7, 29.5, 29.2, 28.8, 27.7, 25.87, 25.86, 25.85, 22.6, 18.2, 18.0, 14.1, −4.52, −4.55, −5.38, −5.41; HRMS (ESI) calcd. for $C_{32}H_{64}N_4O_4SSi_2^+$ [M + H]$^+$ *m/z* 657.4246, found *m/z* 657.4246.

2.2.4. Synthesis of 2-(2-octylthioethyl)-6-oxo-4-acetylamino-1-(2′,4′-dideoxy-ᴅ-ribityl) triazine (**8**)

Boron trifluoride-diethylether complex (51 µL, 398 µmol) was slowly added to a solution of **7** (131 mg, 199 µmol) in CH$_3$CN (800 µL) at 0 °C. After the addition, the mixture was stirred at 0 °C for 2 h. The resulting solution was diluted with EtOAc, then quenched with phosphate buffer (pH = 6.0, 1 M). The product was extracted with EtOAc and the combined organic layer was washed with brine, dried over Na$_2$SO$_4$, filtered and evaporated. The crude product was purified by silica gel chromatography, eluting with CH$_3$OH/CHCl$_3$, 1:50, then 1:10 to give **8** (103 µmol, 52%) as a white solid: ^{1}H-NMR (600 MHz, CDCl$_3$) δ 7.80 (br, 1H), 4.38–4.33 (m, 1H), 4.00–3.96 (m, 1H), 3.91–3.88 (m, 1H), 3.85–3.81 (m, 2H), 3.15 (dt, *J* = 16.8, 1H, 7.2 Hz), 3.05 (dt, 1H, *J* = 16.8, 7.2 Hz), 2.93 (t, 2H, *J* = 7.2 Hz), 2.61 (s, 3H), 2.56 (t, 2H, *J* = 7.2 Hz), 1.95–1.90 (m, 1H), 1.82–1.71 (m, 3H), 1.59 (qt, 2H, *J* = 7.2, 7.2 Hz), 1.38–1.25 (m, 10H),

0.88 (t, 3H, J = 7.2 Hz); ^{13}C-NMR (150 MHz, CDCl$_3$) δ 171.9, 170.9, 161.7, 156.0, 68.1, 61.63, 61.59, 41.7, 38.0, 36.2, 34.1, 32.8, 31.8, 29.6, 29.2, 28.8, 27.9, 25.9, 22.6,14.1; HRMS (ESI) calcd. for C$_{20}$H$_{36}$N$_4$O$_4$S$^+$ [M + H]$^+$ *m/z* 429.2528, found *m/z* 429.2530.

2.2.5. Synthesis of 2-(2-octylthioethyl)-6-oxo-4-acetylamino-1-(5′-*O*-(4,4′-dimethoxytrityl)-2′, 4′-dideoxy-ᴅ-ribityl) triazine (9)

DMTrCl (58 mg, 170 µmol) was added to a solution of **8** (497 mg, 114 µmol) in pyridine (487 µL) at 0 °C. After the addition, the mixture was stirred at 0 °C for 1 h. The resulting mixture was diluted with CH$_2$Cl$_2$, then quenched with NaHCO$_3$. The product was extracted with CH$_2$Cl$_2$ and the combined organic layer was washed with brine, dried over Na$_2$SO$_4$, filtered and evaporated. The crude product was purified by chromatography on silica gel, eluting with CHCl$_3$ and 1% Et$_3$N, then CH$_3$OH/CHCl$_3$, 1:200 with 1% Et$_3$N to give **9** (80 µmol, 70%) as a colorless oil: ^{1}H-NMR (600 MHz, CDCl$_3$) δ 7.72 (br, 1H), 7.39–7.38 (m, 2H), 7.30–7.27 (m, 6H), 7.21–7.19 (m, 1H), 6.83–6.80 (m, 2H), 4.20 (dt, 1H, J = 13.2, 7.2 Hz), 4.04 (dt, 1H, J = 13.2, 7.2 Hz), 3.15 (dt, 1H, J = 16.8, 7.2 Hz), 3.06 (dt, 1H, J = 16.8, 7.2 Hz), 2.96 (t, 2H, J = 7.2 Hz), 2.60 (s, 3H), 2.54 (t, 2H, J = 7.2 Hz), 1.96–1.91 (m, 1H), 1.83–1.77 (m, 1H), 1.69–1.66 (m, 2H), 1.57 (qt, 2H, J = 7.2, 7.2 Hz), 1.37–1.26 (m, 10H), 0.87 (t, 3H, J = 7.2 Hz); ^{13}C-NMR (150 MHz, CDCl$_3$) δ 172.0, 170.8, 161.6, 158.5, 155.3, 144.5, 135.7 135.6, 129.88, 129.86, 128.0, 127.9, 126.9, 113.2, 86.9, 68.4, 62.4, 55.2, 42.1, 36.5, 35.5, 34.1, 32.6, 31.8, 29.5, 29.19, 29.17, 28.8, 25.9, 22.6, 14.1; HRMS (ESI) calcd. for C$_{41}$H$_{54}$N$_4$O$_6$S$^+$ [M + H]$^+$ *m/z* 731.3837, found *m/z* 731.3834.

2.2.6. Synthesis of 2-(2-octylthioethyl)-6-oxo-4-acetylamino-1-(3′-*N,N*-diisopropyl cyanoethyl-phosphoramidyl-5′-*O*-(4,4′-dimethoxytrityl)-2′,4′-dideoxy-ᴅ-ribityl) triazine (**10**)

DIPEA (83.2 µL, 14.3 µmol) was added to a solution of **9** (15.2 mg, 20.8 µmol) in CH$_2$Cl$_2$ (416 µL) and the mixture was cooled to 0 °C. NCCH$_2$CH$_2$OP[(N(*i*-Pr)$_2$]Cl (9.3 µL, 41.6 µmol) was then added to the mixture. After the addition, the mixture was stirred at 0 °C for 1 h, diluted with CH$_2$Cl$_2$, then quenched with NaHCO$_3$. The product was extracted with CH$_2$Cl$_2$ and the combined organic layer was washed with brine, dried over Na$_2$SO$_4$, filtered and evaporated. The crude product was co-evaporated with toluene, then purified by chromatography on silica gel, eluting with EtOAc/hexane, 1:1 with 1% Et$_3$N, then 2:3 with 1% Et$_3$N to give **10** (15.0 µmol, 72%) as a colorless oil mixture of two diastereomers: ^{1}H-NMR (600 MHz, CDCl$_3$) δ 7.65 (br, 1H), 7.41 (d, 2H, J = 7.2 Hz), 7.31–7.27 (m, 6H), 7.21–7.18 (m, 1H), 6.83–6.81 (m, 4H), 4.22–4.04 (m, 3H), 3.82–3.49 (m, 4H), 3.79 (s, 6H), 3.22–3.15 (m, 2H), 3.10–2.95 (m, 2H), 2.91–2.86 (m, 2H), 2.64–2.61 (m, 1H), 2.62–2.61 (m, 3H), 2.55–2.50 (m, 2H), 2.48–2.46 (m, 1H), 2.01–1.77 (m, 4H), 1.59–1.54 (m, 2H), 1.39–1.25 (m, 10H), 1.16–1.03 (m, 12H), 0.88 (t, 3H, J = 7.2 Hz); ^{13}C-NMR (150 MHz, CDCl$_3$) δ 172.04, 172.00, 170.4, 170.2, 161.44, 161.40, 158.3, 154.9, 154.7, 145.10, 145.05, 136.32, 136.30, 136.29, 136.27, 130.0, 128.1, 128.0, 127.7, 126.7, 126.6, 118.0, 117.6, 113.0, 86.0, 77.2, 77.0, 76.8, 70.4, 70.3, 70.0, 69.9, 60.4, 60.0, 58.0, 57.8, 57.4, 57.3, 55.18, 55.15, 43.08, 43.05, 43.00, 42.97, 42.0, 41.9, 36.82, 36.80, 36.25, 36.23, 34.4, 34.3, 34.2, 34.04, 33.95, 32.82, 32.79, 31.8, 29.53, 29.52, 29.18, 29.16, 28.84, 28.83, 27.7, 25.9, 24.80, 24.75, 24.60, 24.56, 24.52, 24.50, 22.6, 21.1, 20.50, 20.45; ^{31}P- NMR (202 MHz, CDCl$_3$) δ 147.70, 146.72; HRMS (ESI) calcd. for C$_{50}$H$_{71}$N$_6$O$_7$PS$^+$ [M + H]$^+$ *m/z* 931.4915, found *m/z* 931.4919.

2.3. Synthesis of the Oligodeoxynucleotides Containing acyAOVT Derivatives

ODN1 and **2** were synthesized on a 1 µmol scale by an ABI 392 DNA/RNA synthesizer with standard β-cyanoethyl chemistry. 5′-Terminal dimethoxytrityl-bearing ODN1 and **2** was removed from the solid support by treatment with 45 mM K$_2$CO$_3$-MeOH containing 10 mM 1-octanethiol (0.5 mL) and the residue was evaporated under reduced pressure. The crude product was purified by reverse phase HPLC with a C-18 column (Nacalai Tesque: Cosmosil 5C18-MS-II, 10 × 250 mm) by a linear gradient of 5–40%/25 min of acetonitrile in 0.1% TEAA buffer at the flow rate of 4 mL/min. The dimethoxytrityl group of the purified **ODN** was removed with 10% AcOH for 30 min and the mixture was additionally purified by reverse phase HPLC to afford **ODN1**; MALDI-TOF MS (*m/z*):

calcd. for [M-H]⁻ *m/z* 4045.8746; found 4045.780; **ODN2**; MALDI-TOF MS (*m/z*): calcd. for [M-H]⁻ *m/z* 6960.8115; found 6958.589.

To a solution of **ODN1** or **2** (70 µM) in ddH$_2$O was added a solution of MMPP (1 mM) in carbonate buffer (pH = 10) at room temperature. After 1 h, 5% AcOH was added to the mixture and the mixture was left for an additional 2 h to give **ODN3**. MALDI-TOF MS ODN (SOOct): calcd. for [M-H]⁻ *m/z* 4061.8736, found 4061.755, **ODN5** (vinyl): calcd. for [M-H]⁻ *m/z* 3899.5826, found 3899.304; **ODN6** (vinyl): calcd. for [M-H]⁻ *m/z* 6814.5195, found 6811.219

ODN7 was prepared by using 1.0 µM **ODN5** and 100 mM sodium thiomethoxide in 50 mM MES buffer (pH = 7.0) at 37 °C for 19 h. The reaction mixture was purified by reverse phase HPLC to afford **ODN7**: calcd. for [M-H]⁻ *m/z* 3947.6856, found 3947.309.

2.4. General Procedure for the Crosslinking Reactions

The reaction was performed using 5.0 µM **ODN5** and 2.0 µM of the target DNA or RNA labelled by fluorescein at the 5′-end in a buffer of 100 mM NaCl and 50 mM MES buffer (pH = 7.0). The reaction was incubated at 37 °C for 1–24 h. The reaction was quenched with the addition of loading dye (95% formamide, 20 mM EDTA, 0.05% xylene cyanol, and 0.05% bromophenol blue). The cross-linked products were analyzed by a denaturing 20% polyacrylamide gel electrophoresis containing urea (7 M) with TBE buffer at 200 V for 1 h. The labelled bands were visualized and quantified using a FLA-5100 Fluor Imager.

2.5. T_m Measurement

All samples for the T_m measurements consisted of 100 mM NaCl, 25 mM MES buffer (pH = 7.0) and 1 µM Duplex. The T_m measurements were performed using a temperature controller. Both the heating and cooling curves were measured three times over the temperature range of 25 °C to 80 °C at 0.5 °C/min. The absorbance at 260 nm was recorded every 0.5 °C.

2.6. Alkylation with ODN Probe to HhaI DNMT-1

The duplex oligo (final 1 µM) in pH 7.5 reaction buffer (50 mM Tris-HCl, 10 mM EDTA, 5 mM 2-mercaptoethanol) was pre-incubated at 37 °C for 5 min. HhaI Methyltransferase was then added to the reaction mixture and the mixture was pre-incubated at 37 °C (or 4 °C) for another 5 min. SAM was added to the mixture and the mixture was incubated at 37 °C (or 4 °C) for 0.5–24 h. A 4.5 µL aliquot of the reaction mixture (4.5 µL: for silver stain, 12 µL: for Flamingo gel stain) was used every time. Loading buffer (6×Tris.HCl, SDS, glycerol, 0.05% bromphenol blue, 200 mM DTT (final 20 mM)) was added and the mixture was heated at 65 °C for 15 min. The samples were separated by 10% SDS-PAGE (0.025 M Tris, 0.192 M glycine, 0.1% SDS) at room temperature (10 mA/1.5 h then 20 mA/1.0 h). Visualization of the 5′- FAM oligonucleotide was performed by fluorescence imaging using an FLA-5100 Fluor Imager. Visualization of the protein was performed by silver stain or Flamingo gel stain.

3. Results

3.1. Synthesis of the acyAOVT Nucleoside Phosphoramidite

We planned to synthesize the C6-substituted 5-azacytosine by ring opening of the 5-azacytosine moiety forming guanylurea and subsequent ring closing reaction that follows a previous report (Scheme 1) [20]. In our previous study, the acyclic linkers-protected MOM or Bn group produced the coupling products in low yields. The acyclic side chain 3-protected TBS group was synthesized from 2′-deoxy-ᴅ-ribose by a previous modified procedure [21]. The coupling reaction between the sodium salt of 5-azacytosine and the acyclic side chain compound **3** was done to form the N1- and N3- alkylated products. The reaction conditions were screened in order to optimize the formation of the N1-alkylated product **4** and a considerable amount of the N1-alkylation (58%) in DMSO at 60 °C was achieved with a N1:N3 ratio = 6:1. The formation of the glycosidic bond at the N1 position of 5-azacytosine was confirmed by HMQC and HMBC analyses.

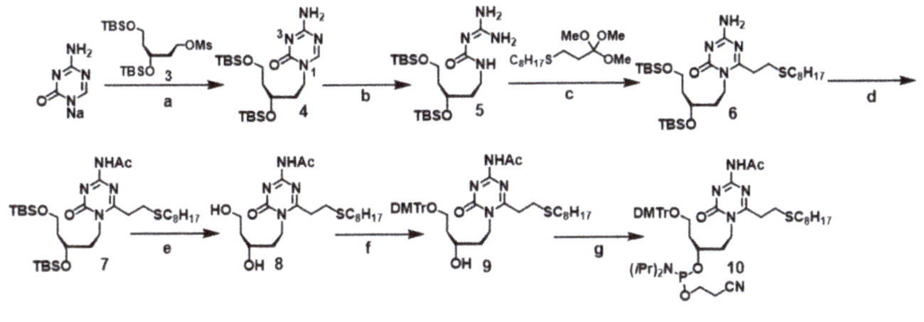

Scheme 1. *Reagents and conditions*: (**a**) 5-azacytosine sodium salt, DMSO, 60 °C, 18 h, 58%; (**b**) 7M NH$_4$OH, MeOH-1,4-dioxane (1:1), 55 °C, 24 h; (**c**) orthoester, DMF, 120 °C, 2 h, 51% (2 steps); (**d**) AcCl, pyridine, rt, 19 h, 54%; (**e**) BF$_3$·OEt, CH$_2$Cl$_2$, 0 °C, 2 h, 52%; (**f**) DMTrCl, pyridine, 0 °C, 1 h, 70%; (**g**) 2-Cyanoethyl-*N*,*N*-diisopropylchloropho sphoramidite, DIPEA, CH$_2$Cl$_2$, 0 °C, 1 h, 72%.

The 5-azacytidine derivative **4** was treated with NH$_3$ to give the guanylurea intermediate **5**, which was condensed with the orthoester to afford the desired C6-octylthioethyl-5-azacytidine derivative **6** in 45% yield (2 steps). After the 4-*N*-acetylation and the deprotection of the TBS groups by BF$_3$OEt$_2$, the 5′-hydroxyl group was selectively protected with the DMTr group, then the 3′-hydroxyl group was phosphitylated to yield the *acy*AOVT nucleoside phosphoramidite **10**.

3.2. Synthesis of the Oligonucleotides Containing acyAOVT and Evaluation of the Crosslinking Reactivity

We synthesized two kinds of **ODNs1, 2** by a DNA synthesizer using phosphoramidite **10** (Scheme 2). The synthesized ODNs were treated with 45 mM K$_2$CO$_3$/MeOH containing 10 mM 1-octanethiol to cleave them from the resin. After the DMTr-ON purification by reverse-phase (RP) HPLC, detritylation was carried out in an aqueous 10% AcOH solution at room temperature for 30 min. The obtained ODNs were further purified by RP-HPLC and were characterized by a MALDI-TOF MS analysis. The sulfide group of ODNs1, 2 was oxidized with magnesium monoperoxyphthalate (MMPP) and treated with aqueous 10% AcOH to give ODNs5 and 6 according to the reported procedure [20]. The characterization of the ODNs was performed by MALDI-TOF MS.

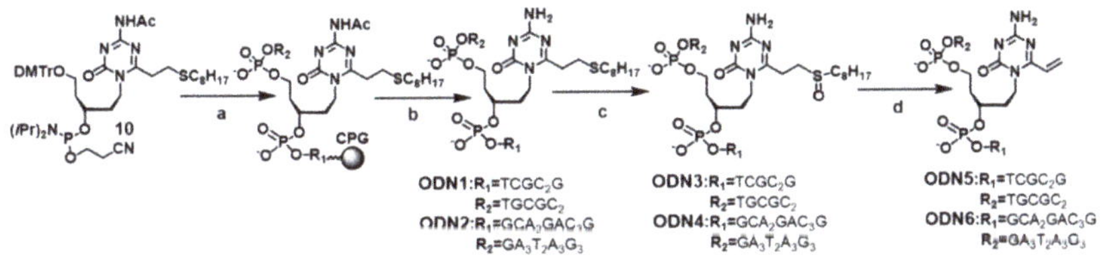

Scheme 2. *Reagents and conditions*: (**a**) DNA synthesizer; (**b**) (i) 45 mM K$_2$CO$_3$-MeOH, 10 mM 1-octanethiol, rt, 4 h; (ii) 10% AcOH, rt, 30 min.; (**c**) 1 mM MMPP (2 eq) in carbonate buffer (pH 10), rt, 1 h; (**d**) 5% aq. AcOH, rt, 2 h.

The crosslinking reactions using the reactive ODN5 to initially the complementary target DNA1 or RNA1 labelled with fluorescein at the 5′ end were investigated under neutral conditions. The reaction mixture was analyzed by 20% polyacrylamide gel electrophoresis (PAGE) containing 7 M urea and the yields of the cross-linked product were calculated based on the fluorescent intensity of each band observed on the gels. Figure 3 shows a comparison of the reactivity of ODN5 toward the different bases at the target site of the DNA1 or RNA1. ODN5 showed relatively high crosslinking reactivity to dC, dT, and dG, except for dA in the DNA. On the other hand, the crosslinked product was observed in high yields with rC and rU, and no significant products were observed with rG and rA in the RNA. The comparison of the reactivity to DNA or RNA with acyAOVT **2**, 4-amino-6-oxo-2-vinyl pyrimidine having an acyclic linker (**11:** acyAOVP) and Et-AOVT **1**, is shown in Figure 4. By comparison, acyAOVT **2** and acyAOVP **11** showed a significant difference in the reaction rate and the base selectivity to the target DNA and RNA (Figure 4A,B).

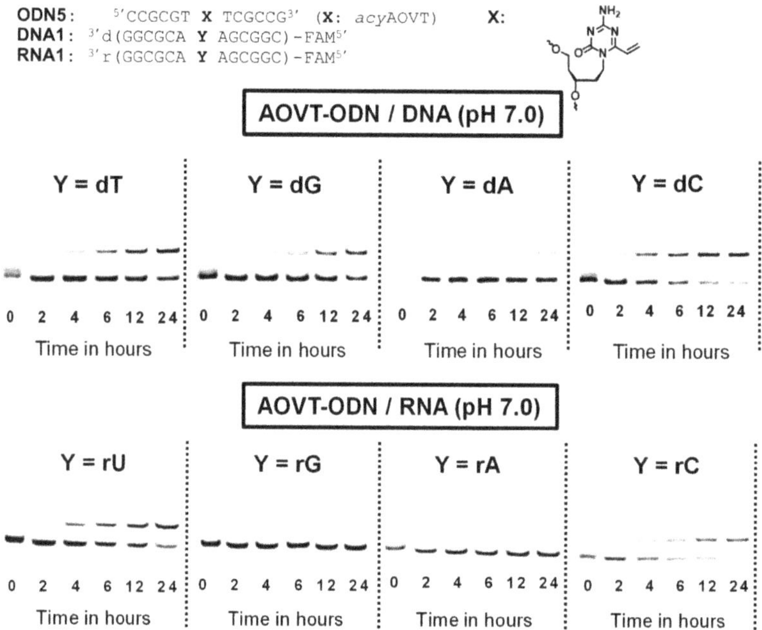

Figure 3. Denaturing gel electrophoresis of crosslink reaction to target DNA1 and RNA1 with **ODN5**. The reaction was performed in 50 mM MES-buffer (pH 7.0) containing 100 mM NaCl, **ODN5** (5 μM) and cDNAs or cRNAs (2 μM) at 37 °C.

The reaction rate with *acy*AOVT **2** was faster than that of *acy*AOVP **11**. The higher reaction rate with *acy*AOVT **2** might be attributed to the electron-withdrawing effect of the triazine increasing the reactivity of the 6-vinyl group of *acy*AOVT **2** than that of the pyrimidine-type *acy*AOVP **11**. Compared to the target base selectivity in **DNA1** with previously reported the *acy*AOVP **11** [21], AOVT **1** and **2** showed a drastically increased reactivity with the cytosine. On the other hand, the crosslinking yields to **RNA1** with AOVT **1** and **2** were significant higher to cytosine and uracil, and relatively lower to guanine than that of AOVP **11**. The reaction rate and selectivity to the pyrimidine bases (C, T and U) with *acy*AOVT **2** was comparable to that of Et-AOVT **1**. The reactivity to adenine (A) in DNA and RNA with **2** was significantly lower than that of Et-AOVT **1** (Figure 4A,C).

Next, we carried out a thermal denaturing study of the duplex containing the non-reactive *acy*AOVT. The **ODN7** bearing the non-reactive *acy*AOVT (SMe) was prepared by the addition of NaSMe to the vinyl derivative under weak acidic conditions for avoiding the decomposition of *acy*AOVT to guanylurea (Scheme 3).

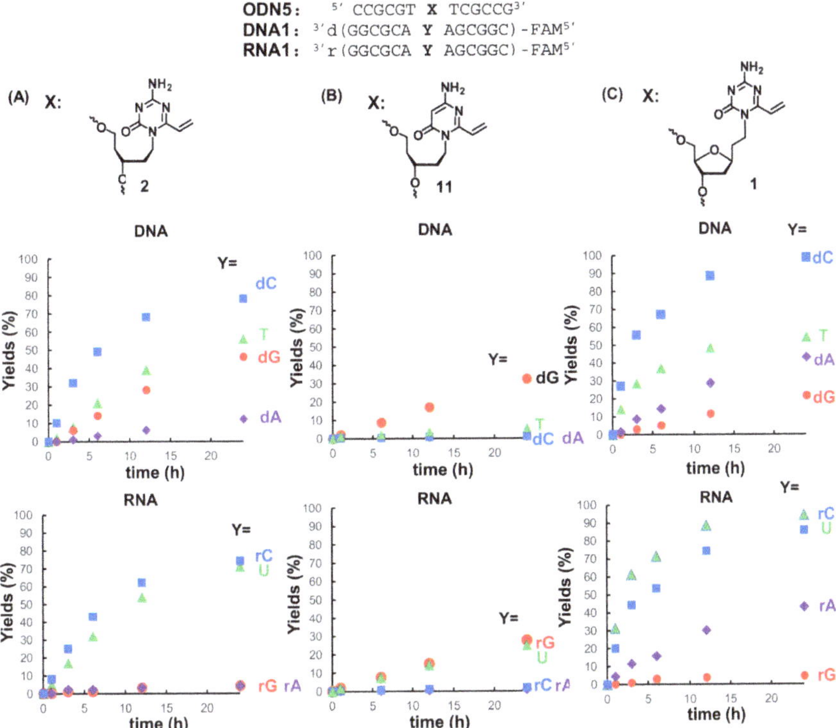

Figure 4. Comparison of the reaction yields calculated from the gel electrophoresis analysis of the cross-linking to target DNA1 (**Y** = dT, dG, dC, dA) or RNA1 (**N** = U, G, C, A) with *acy*AOVT (**2**) (**A**), *acy*AOVP (**11**) (**B**) and Et-AOVT (**1**) (**C**).

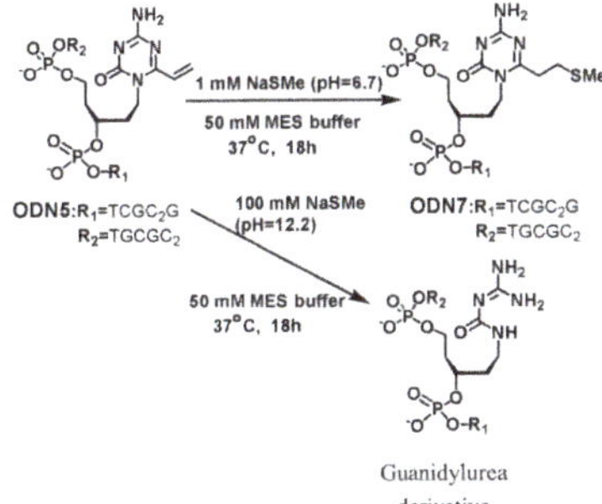

Scheme 3. Synthesis of the ODN bearing a stable precursor.

The melting temperature (*Tm*) of the **ODN7/DNA1** or **RNA1** is summarized in Figure 5. The *Tm* values of the **ODN7/DNA1** or **RNA1** were observed to be lower than that of the unmodified natural duplex (T_m = 67 °C). It should be noted that the *Tm* value for the duplex **ODN7/DNA1** (**Y** = dG) or **RNA1** (**Y** = rG) was 54 °C and 57 °C respectively, higher than the other duplexes. Taken together, our results of the T_m measurement suggest that AOVT might form stable triple hydrogen bonds with guanine to orient the C6-vinyl group of AOVT toward the opposite side of the complementary bases. Next, we attempted the crosslinking reactions to the nucleic acids-binding protein using the duplex DNA bearing AOVT derivative **2**.

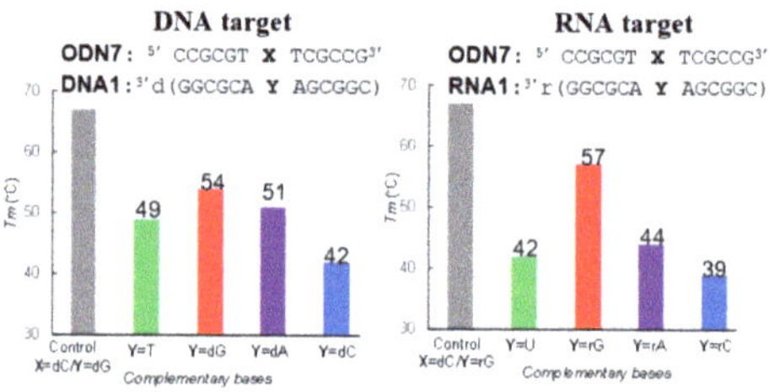

Figure 5. Comparative thermal denaturing analysis of duplexes **ODN7/DNA1** or **RNA1**. T_m values were measured in 50 mM MES buffer (pH 7.0) containing 100 mM NaCl.

3.3. Crosslinking Reactions to DNA Methyl Transferase

We chose cytosine-5-DNA methyl transferase (DNMT) as a model DNA binding protein. DNMT performs the methylation of cytosines for epigenetic modification of the genome that is involved in regulating many cellular processes [6]. In addition, aberrant methylation can cause many diseases, such as cancer [22–24]. The efficient inhibition of DNMT offers the possibility of interfering with the methylation process, which may allow control of the epigenetic regulation of cells and treat various cancers [25].

The mechanism of the cytosine-5 methylation with DNMT is illustrated in Figure 6A. A thiol of the Cys residue in the active site of DNMT acts as a nucleophile and attacks the C6 position of the cytosine to form an enzyme-linked intermediate. The resulting nucleophilic C5 position of the cytosine is then methylated by S-adenosyl-l-methionine (SAM). Subsequent abstraction of the proton at the 5 position and b-elimination provides the 5-substituted pyrimidine and active enzyme (Figure 6A) [26]. Based on this mechanism, the modified bases forming a covalent bond to DNMT were reported as irreversible inhibitors [27–31]. We expected that *acy*AOVT (**2**) replacement of the cytosine methylated by DNMT can react with the Cys residue in this enzyme (Figure 6B). In our investigation, we used the bacterial DNMT, i.e., the commercially-available HhaI DNMT-1 from *Haemophilus haemolyticus*. HhaI DNMT-1 methylated the internal 2′-deoxycytidine in the target sequence 5′-GCGC-3′, in which AOVT was substituted in place of the internal 2′-dC. We synthesized **ODN6**, which contained AOVT in the HhaI DNMT-1 recognition sequence and evaluated the crosslink reactivity to DNA. The base selectivity was similar, but the reactivity to the target DNA was significantly lower compared to that of the previous sequence (Figure S12).

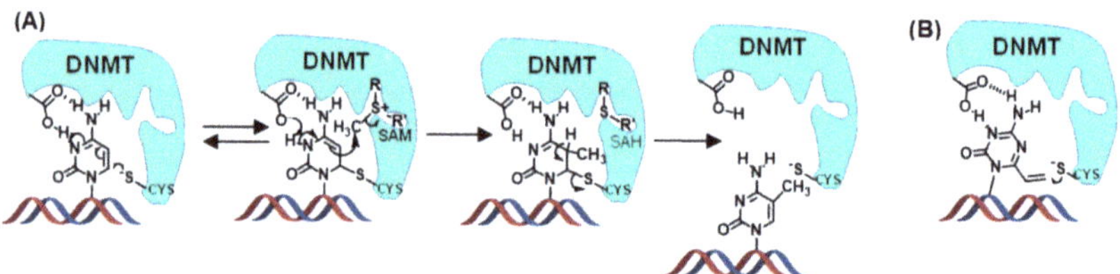

Figure 6. (**A**) Plausible mechanism for methylation by cytosine-5 methyltransferases (DNMTs) with SAM as a cofactor (**B**) Possible mechanism for covalent trapping with AOVT (**2**) derivative.

We selected DNA2 containing methyl cytosine as a target DNA because Dnmt1 methylates hemimethylated CpG sites on one strand of double-stranded DNA. The methylation conditions were determined using the duplex DNA between DNA2 and DNA3 contained cytosine instead of acyclic AOVT and the mono-methylated product in DNA3 was confirmed by MALDI-TOF MS. The duplex DNA (**ODN6** and **DNA2**) was incubated with HhaI DNMT-1 in the reaction buffer containing 5 mM 2-mercaptoethanol at 37 °C in the determined conditions. The reaction mixture was analyzed by sodium dodecyl sulfate −15% polyacrylamide gel electrophoresis (SDS–PAGE) and visualized with silver staining. The slower mobility band (*) was observed around 51kD and not observed with the non-reactive duplex DNA and Hha1 DNMT-1 (Figure 7A). We performed the reaction using the 5′ FAM-labelled **ODN6** and analyzed by SDS-PAGE. The slower mobility band stained silver matched the fluorescent band in the reaction using the 5′ FAM-labelled **ODN6** (Figure 7B) and this band was derived from the complex between Hha1 DNMT-1 and the duplex DNA. The quantification of the products was next performed by using the Flamingo™ Gel staining in-gel fluorescence analysis. We confirmed the linear correlation between the protein amount and the fluorescence intensity. Thus, the complex between the protein and ODN can be quantified by measuring the fluorescence intensity on the gel (Figure 7C). The reactions were performed at 37 °C and 4 °C. The yields were obtained by quantification of the fluorescence intensity for each band and reached 77% at 37 °C after 24 h.

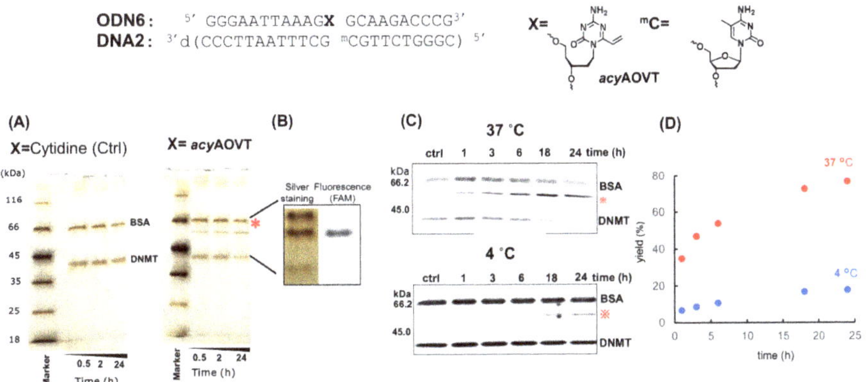

Figure 7. Evaluation of the crosslinking reaction between duplex DNA and Hha1 DNMT-1 (**A**) SDS-Page gel stained silver for analysis of reactions. Duplex DNA (**ODN6** (X = C or *acy*AOVT) and **DNA2**: 3 μM) and Hha1 DNMT-1 (1 μM) was incubated in the presence of SAM at 4 °C. (**B**) 5′-FAM labelled **ODN6** was used in the same reaction of (**A**) and the gel was visualized by silver staining and fluorescence. (**C**) SDS-Page gel stained Flamingo™ for analysis of reactions. Duplex DNA (**ODN6** (X = C for control or *acy*AOVT) and **DNA2**: 1 μM) and Hha1 DNMT-1 (1 μM) was incubated in the presence of SAM at 4 °C or 37 °C. (**D**) Time course of the yields at 4 °C or 37 °C.

The yields significantly decreased at 4 °C (Figure 7D), suggesting that the reaction rate depends on the reaction temperature. The reactions were performed using **ODN6** and **DNA4** containing 4 kinds of bases (G, A, C and T) at the complementary site for *acy*AOVT. The crosslinking yields to these targets were similar to that observed with G (Figure S13). These results suggested that DNMT can bind to the mismatch sequences and make a transition to flip out of the target *acy*AOVT to induce the crosslinking reactions. Taken together, the results of the evaluation for the reaction to Hha1 DNMT-1 indicated that the *acy*AOVT in the duplex would react with Hha1 DNMT-1 even in the presence of 2-mercaptoethanol.

4. Discussion

We have synthesized ODN containing the *acy*AOVT base **2** that is expected to react with the nucleophilic residue of the nucleic acid-binding protein in the proximity of the vinyl group. The reactivity with *acy*AOVT to DNA or RNA was significantly higher than that of *acy*AOVP. The interesting result that only one nitrogen substitution on AOVP drastically increased the reactivity could be attributed to the electron withdrawing effect of the nitrogen atom. The *acy*AOVT (**2**) showed relatively high yields to dC, dT and dG in DNA and rC, rU in RNA. The highest reactivity to cytosine with the AOVT derivatives **1** and **2** might be due to the high nucleophilicity of the amino group in cytosine and to the flexible linker, which makes it possible to access the vinyl group in the amino group of the cytosine. The cross-linking yields to adenine in the DNA or RNA with *acy*AOVT **2** were significantly lower than that of Et-AOVT **1**. The vinyl group in *acy*AOVT **2** might be located away from the reactive site of adenine because of being positioned at similar natural base-pair by the short linker with **2**, resulting in the low crosslinking yields to adenine (Figure S14). The melting temperature (T_m) values of the duplex for **ODN7/DNA1(G)** or **RNA1(G)** were the highest in all the duplexes containing the other AOVT-base pairing. These results suggested that the duplex containing *acy*AOVT-G would be stabilized by forming three hydrogen bonding, resulting in orientation of the vinyl group in *acy*AOVT to the opposite side of the complementary bases (Figure 8).

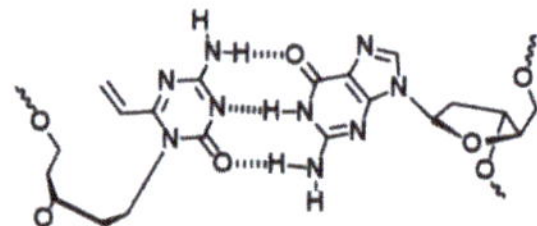

Figure 8. Hypothetical structure for *acy*AOVT and guanine in duplex DNA.

We attempted the reaction to the commercially-available DNMT using the duplex containing *acy*AOVT-G as a model reaction with the nucleic acids-binding protein. The results indicated that *acy*AOVT might react with the thiol group of the cysteine residue in DNMT in close proximity as shown in Figure 9.

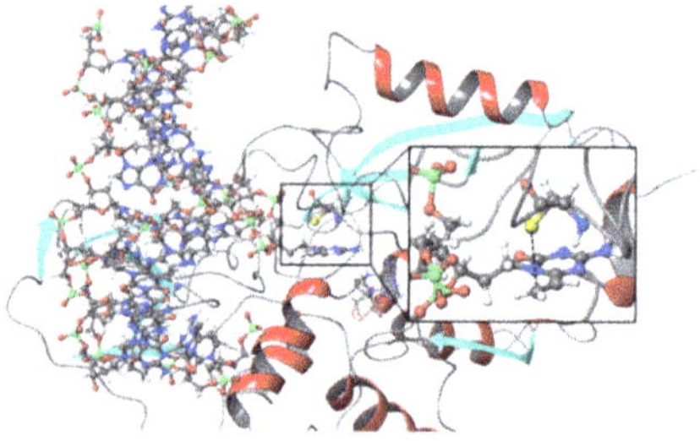

Figure 9. Predictive structure for reaction intermediate between *acy*AOVT and cysteine in *DNMT* (PDB-5MHT).

We expect that this *acy*AOVT-G base pairing system would provide a "cross-linkable duplex material" to react with the nucleic acid-binding proteins.

5. Conclusions

We have designed *acy*AOVT **2** as a reactive base to the nucleic acid-binding protein. The ODN containing **2** exhibited a higher reactivity to the pyridine bases (C, T and U) at the complementary site of **2** in duplexes DNA-DNA and DNA-RNA than that of *acy*AOVP. The results of the T_m measurements suggested that *acy*AOVT **2** can form a stable base pair with G, and consequently, the vinyl group of **2** might be located on the opposite side of the complementary bases. The preliminary results for the reaction to DNMT with the duplex DNA containing *acy*AOVT demonstrated the potential of *acy*AOVT-G base pairing system in the reaction to the nucleic acids-binding protein. This system can be used for the crosslinking reaction to the RNA binding protein. We expect that our system will provide a useful tool for the molecular study of the RNA binding protein in control of the RNA biological functions.

Supplementary Materials: The following are available online at http://www.mdpi.com/2076-3417/10/21/7709/s1.

Author Contributions: K.O. and K.Y. conceived and designed the experiments. K.O. and S.I. performed the experiments. F.N. and H.O. wrote the study. F.N. supervised the project. All authors have read and agreed to the published version of the manuscript.

Funding: This study was supported by a Grant-in-Aid for Scientific Research on Innovative Areas "Middle Molecular Strategy" (No. JP15H05838) from the Japan Society for the Promotion of Science (JSPS). This work was supported in part by "Dynamic Alliance for Open Innovation Bridging Human, Environment and Materials" from the Ministry of Education, Culture, Sports, Science and Technology of Japan (MEXT).

Conflicts of Interest: The authors declare no conflict of interest.

References

1. Bouhlel, M.A.; Lambert, M.; David-Cordonnier, M.H. Targeting transcription factor binding to DNA by competing with DNA binders as an approach for controlling gene expression. *Curr. Top. Med. Chem.* **2015**, *15*, 1323–1358. [CrossRef] [PubMed]

2. Todeschini, A.L.; Georges, A.; Veitia, R.A. Transcription factors: Specific DNA binding and specific gene regulation. *Trends Genet.* **2014**, *30*, 211–219. [CrossRef] [PubMed]

3. O'Brien, P.J. Catalytic promiscuity and the divergent evolution of DNA repair enzymes. *Chem. Rev.* **2006**, *106*, 720–752. [CrossRef] [PubMed]

4. Nevinsky, G.A. Structural, thermodynamic, and kinetic basis for the activities of some nucleic acid repair enzymes. *J. Mol. Recognit.* **2011**, *24*, 656–677. [CrossRef]

5. Gujar, H.; Weisenberger, D.J.; Liang, G.N. The roles of human DNA methyltransferases and their isoforms in shaping the epigenome. *Genes* **2019**, *10*, 171. [CrossRef]

6. Lyko, F. The DNA methyltransferase family: A versatile toolkit for epigenetic regulation. *Nat. Rev. Genet.* **2018**, *19*, 81–92. [CrossRef]

7. Taylor, K.; Sobczak, K. Intrinsic regulatory role of RNA structural arrangement in alternative splicing control. *Int. J. Mol. Sci.* **2020**, *21*, 5161. [CrossRef]

8. Janakiraman, H.; House, R.P.; Gangaraju, V.K.; Diehl, J.A.; Howe, P.H.; Palanisamy, V. The long (lncRNA) and short (miRNA) of it: TGF beta-mediated control of RNA-binding proteins and noncoding RNAs. *Mol. Cancer Res.* **2018**, *16*, 567–579. [CrossRef]

9. Ito, K.K.; Watanabe, K.; Kitagawa, D. The emerging role of ncRNAs and RNA-binding proteins in Mmitotic apparatus formation. *Non Coding RNA* **2020**, *6*, 13. [CrossRef]

10. Sternburg, E.L.; Karginov, F.V. Global approaches in studying RNA-binding protein interaction networks. *Trends Biochem. Sci.* **2020**, *45*, 593–603. [CrossRef]

11. Lercher, L.; McGouran, J.F.; Kessler, B.M.; Schofield, C.J.; Davis, B.G. DNA modification under mild conditions by Suzuki-Miyaura cross-coupling for the generation of functional probes. *Angew. Chem. Int. Ed.* **2013**, *52*, 10553–10558. [CrossRef] [PubMed]

12. Gerard-Hirne, T.; Thiebaut, F.; Sachon, E.; Desert, A.; Drujon, T.; Guerineau, V.; Michel, B.Y.; Benhida, R.; Coulon, S.; Saintome, C.; et al. Photoactivatable oligonucleotide probes to trap single-stranded DNA binding proteins: Updating the potential of 4-thiothymidine from a comparative study. *Biochimie* **2018**, *154*, 164–175. [CrossRef] [PubMed]

13. Smith, C.C.; Hollenstein, M.; Leumann, C.J. The synthesis and application of a diazirine-modified uridine analogue for investigating RNA-protein interactions. *RSC Adv.* **2014**, *4*, 48228–48235. [CrossRef]

14. Dziuba, D.; Hoffmann, J.E.; Hentze, M.W.; Schultz, C.A. Genetically Encoded Diazirine Analogue for RNA-Protein Photo-crosslinking. *Chembiochem* **2020**, *21*, 88–93. [CrossRef] [PubMed]

15. Wickramaratne, S.; Mukherjee, S.; Villalta, P.W.; Scharer, O.D.; Tretyakova, N.Y. Synthesis of sequence-specific DNA-protein conjugates via a reductive amination strategy. *Bioconj. Chem.* **2013**, *24*, 1496–1506. [CrossRef]

16. Carrette, L.L.G.; Morii, T.; Madder, A. Toxicity inspired cross-linking for probing DNA-peptide interactions. *Bioconj. Chem.* **2013**, *24*, 2008–2014. [CrossRef]

17. Dadova, J.; Orsag, P.; Pohl, R.; Brazdova, M.; Fojta, M.; Hocek, M. Vinylsulfonamide and acrylamide modification of DNA for cross-linking with proteins. *Angew. Chem. Int. Ed.* **2013**, *52*, 10515–10518. [CrossRef]

18. Olszewska, A.; Pohl, R.; Brazdova, M.; Fojta, M.; Hocek, M. Chloroacetamide-linked nucleotides and DNA for cross-linking with peptides and proteins. *Bioconj. Chem.* **2016**, *27*, 2089–2094. [CrossRef]

19. Matyasovsky, J.; Hocek, M. 2-Substituted 2'-deoxyinosine 5'-triphosphates as substrates for polymerase synthesis of minor-groove-modified DNA and effects on restriction endonuclease cleavage. *Org. Biomol. Chem.* **2020**, *18*, 255–262. [CrossRef]

20. Yamada, K.; Ishiyama, S.; Onizuka, K.; Nagatsugi, F. Synthesis and properties of cross-linkable DNA duplex using 4-amino-2-oxo-6-vinyl-1,3,5-triazine. *Tetrahedron* **2017**, *73*, 1424–1435. [CrossRef]

21. Kusano, S.; Ishiyama, S.; Lam, S.L.; Mashima, T.; Katahira, M.; Miyamoto, K.; Aida, M.; Nagatsugi, F. Crosslinking reactions of 4-amino-6-oxo-2-vinylpyrimidine with guanine derivatives and structural analysis of the adducts. *Nucl. Acids Res.* **2015**, *43*, 7717–7730. [CrossRef] [PubMed]

22. Robertson, K.D. DNA methylation and human disease. *Nat. Rev. Genet.* **2005**, *6*, 597–610. [CrossRef]

23. Jones, P.A.; Baylin, S.B. The epigenomics of cancer. *Cell* **2007**, *128*, 683–692. [CrossRef] [PubMed]

24. Pfeifer, G.P. Defining driver DNA methylation changes in human cancer. *Int. J. Mol. Sci.* **2018**, *19*, 1166. [CrossRef]

25. Zhou, Z.H.; Li, H.Q.; Liu, F. DNA Methyltransferase inhibitors and their therapeutic potential. *Curr. Top. Med. Chem.* **2018**, *18*, 2448–2457. [CrossRef] [PubMed]

26. Klimasauskas, S.; Kumar Roberts, R.J.; Cheng, X. HhaI methyltransferase flips its target base out of the DNA helix. *Cell* **1994**, *76*, 357–369. [CrossRef]

27. Shigdel, U.K.; He, C. A New 1'-methylenedisulfide deoxyribose that forms an efficient cross-link to DNA cytosine-5-methyltransferase (DNMT). *J. Am. Chem. Soc.* **2008**, *130*, 17634–17635. [CrossRef]

28. Sato, K.; Kawamoto, K.; Shimamura, S.; Ichikawa, S.; Matsuda, A. An oligodeoxyribonucleotide containing 5-formyl-2'-deoxycytidine (fC) at the CpG site forms a covalent complex with DNA cytosine-5 methyltransferases (DNMTs). *Bioorg. Med. Chem. Lett.* **2016**, *26*, 5395–5398. [CrossRef]

29. Sato, K.; Kunitomo, Y.; Kasai, Y.; Utsumi, S.; Suetake, I.; Tajima, S.; Ichikawa, S.; Matsuda, A. Mechanism-based inhibitor of DNA cytosine-5-methyltransferase by a SNAr reaction with an oligodeoxyribonucleotide containing a 2-amino-4-halopyridine-C-nucleoside. *Chembiochem* **2018**, *19*, 865–872. [CrossRef]

30. Utsumi, S.; Sato, K.; Ichikawa, S. Insight into the recognition mechanism of DNA cytosine-5 methyltransferases (DNMTs) by incorporation of acyclic 5-fluorocytosine (C-F) nucleosides into DNA. *Bioorg. Med. Chem. Lett.* **2018**, *28*, 2189–2194. [CrossRef]

31. Wildenhof, T.M.; Schiffers, S.; Traube, F.R.; Mayer, P.; Carell, T. Influencing epigenetic information with a hydrolytically stable carbocyclic 5-aza-2'-deoxycytidine. *Angew. Chem. Int. Ed.* **2019**, *58*, 12984–12987. [CrossRef]

 *applied sciences*

Article

Encapsulation of mRNA into Artificial Viral Capsids via Hybridization of a β-Annulus-dT$_{20}$ Conjugate and the Poly(A) Tail of mRNA

Yoko Nakamura [1], Yuki Sato [1], Hiroshi Inaba [1,2], Takashi Iwasaki [3] and Kazunori Matsuura [1,2,*] 

[1]   Department of Chemistry and Biotechnology, Graduate School of Engineering, Tottori University, Tottori 680-8552, Japan; yotasa0926@gmail.com (Y.N.); yuki.s09hr@gmail.com (Y.S.); hinaba@tottori-u.ac.jp (H.I.)

[2]   Centre for Research on Green Sustainable Chemistry, Tottori University, Tottori 680-8552, Japan

[3]   Department of Bioresources Science, Graduate School of Agricultural Sciences, Tottori University, Tottori 680-8553, Japan; itaka@tottori-u.ac.jp

*   Correspondence: ma2ra-k@tottori-u.ac.jp; Tel.: +81-857-31-5262

Received: 18 September 2020; Accepted: 9 November 2020; Published: 12 November 2020

check for updates

Abstract: Messenger RNA (mRNA) drugs have attracted considerable attention as promising tools with many therapeutic applications. The efficient delivery of mRNA drugs using non-viral materials is currently being explored. We demonstrate a novel concept where mCherry mRNA bearing a poly(A) tail is encapsulated into capsids co-assembled from viral β-annulus peptides bearing a 20-mer oligothymine (dT$_{20}$) at the N-terminus and unmodified peptides via hybridization of dT$_{20}$ and poly(A). Dynamic light scattering measurements and transmission electron microscopy images of the mRNA-encapsulated capsids show the formation of spherical assemblies of approximately 50 nm. The encapsulated mRNA shows remarkable ribonuclease resistance. Further, modification by a cell-penetrating peptide (His16) on the capsid enables the intracellular expression of mCherry of encapsulated mRNA.

Keywords: mRNA; poly(A) tail; artificial viral capsid; encapsulation; nanocapsule; self-assembly; β-annulus peptide; peptide-DNA conjugate

1. Introduction

Therapeutic mRNA have attracted attention in recent years as a new type of nucleic acid drugs. Such mRNAs show potential for use in protein replacement therapy and vaccination [1–6]. mRNA drugs have the advantage of inducing direct protein synthesis in the cytoplasm, thus differing from DNA therapy. Naked mRNA is; however, unstable outside of cells and is unable to effectively penetrate cell membranes due to electrostatic repulsion. Thus, various materials for mRNA delivery, such as liposomes, dendrimers, and polyion complex micelles, have been developed and are attractive for their flexibility in molecular design [2–11]. Conversely, RNA viruses package specific genome molecules inside their outer protein shell, the viral capsid [12,13]. Viral capsids are also attractive for precise mRNA packaging and delivery but the use of natural viruses for mRNA delivery has safety concerns. Virus-like artificial protein cages for RNA packaging are, therefore, attractive. Recently, Hilvert and coworkers demonstrated that a non-viral protein cage formed by *Aquifex aeolicus* lumazine synthase selectively packages mRNA via cationic peptide tags [14,15].

Eukaryotic mRNA possesses a long adenine nucleotide (poly(A)) tail of about 100~200 nt at the 3'-end to regulate translation [16–21]. We propose a novel strategy for the encapsulation of mRNA into virus-like peptide nanocapsules via hybridization between dT$_{20}$ and the poly(A) tails. We developed an "artificial viral capsid" with a size of 30–50 nm self-assembled from the β-annulus peptide (INHVGGTGGAIMAPVAVTRQLVGS) that participates in the formation of the dodecahedral internal skeleton of the tomato bushy stunt virus [22–35]. Artificial viral capsids encapsulate anionic guest molecules and His-tagged green fluorescence protein (GFP) into the cationic and Ni-NTA (Ni-nitrilotriacetic acid complex) modified interior [25–28]. Modification of the C-terminal, expected to be directed to the exterior of capsids, enabled the surface modification of artificial viral capsids with gold nanoparticles, coiled-coil peptides, single-stranded DNAs, and proteins [29–34]. Recently, we demonstrated that β-annulus peptide possessing ssDNA at the N-terminal self-assembled into ssDNA-encapsulated artificial viral capsids [35]. Thus, we designed a β-annulus peptide with dT_{20} at the N-terminal to direct mRNA to the interior of capsids via hybridization with poly(A) tails (Figure 1).

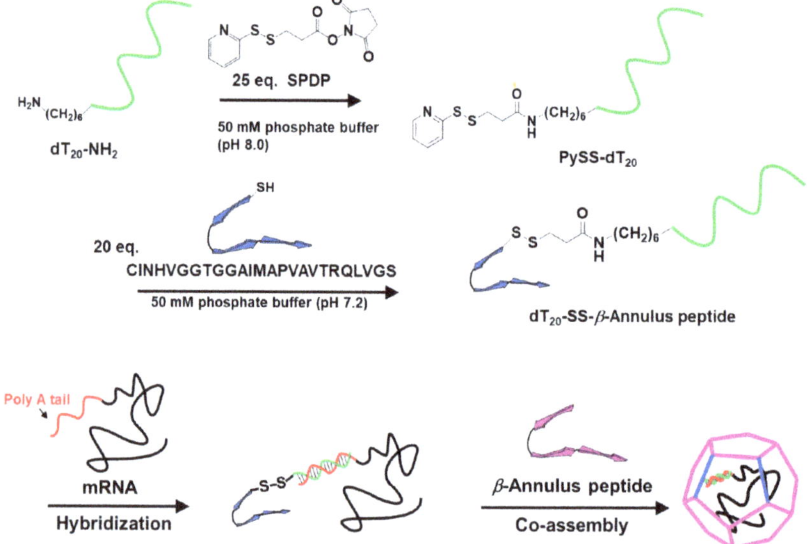

Figure 1. Schematic of synthesis of dT_{20} modified β-annulus peptide (dT_{20}-SS-β-annulus) and formation of an artificial viral capsid by co-assembly of β-annulus peptide and dT_{20}-SS-β-annulus hybridized with mRNA.

2. Materials and Methods

2.1. General

Reverse-phase HPLC was performed at ambient temperature with a Shimadzu LC-6AD liquid chromatograph equipped with a UV–Vis detector (220 nm and 260 nm, Shimadzu SPD-10AVvp, Kyoto, Japan) using an Inertsil WP300 C18 column (GL Science, 250 mm × 4.6 mm or 250 × 20 mm). MALDI-TOF mass spectra were obtained on an Autoflex T2 (Bruker Daltonics, Billerica, USA) in linear/positive mode with α-cyano-4-hydroxycinnamic acid (α-CHCA) or 3-hydroxypicolinic acid (3-HPA) with diammonium hydrogen citrate as the matrix. The UV–Vis spectra of DNA-conjugated peptides were measured at 260 nm using a Jasco V-630 (JASCO Corporation, Tokyo, Japan) with a quartz cell (S10-UV-1, GL Science). An mRNA coding mCherry fluorescent protein (mCherry mRNA, ~1 kb) with a 3′-polyadenylic acid (poly(A)) tail was purchased from OZ Biosciences (Marseille, France). RNase A was purchased from Nacalai Tesque (Kyoto, Japan). All other reagents were obtained from a commercial source and used without further purification. Deionized water of high resistivity (>18 MΩ cm) was purified using a Millipore Purification System (Milli-Q water, Merck Millipore, Burlington, USA) and used as a solvent.

2.2. Preparation of dT$_{20}$-SS-β-Annulus and β-Annulus Peptides

β-annulus (INHVGGTGGAIMAPVAVTRQLVGS) and Cys-β-annulus (CINHVGGTGGAIMA PVAVTRQLVGS) peptides were synthesized with a Biotage Initiator$^+$ (Biotage, Uppsala, Kingdom of Sweden) using standard Fmoc-based coupling chemistry as previously described [22,28]. MALDI-TOF-MS of the β-annulus peptide (matrix: α-CHCA): m/z = 2306 (exact mass: 2306) and of the Cys-β-annulus peptide (matrix: α-CHCA): m/z = 2409 (exact mass: 2409).

An amine-modified dT$_{20}$ (5′NH$_2$-(CH$_2$)$_6$-TTTTTTTTTTTTTTTTTTTT-3′, Gene Design Inc., Osaka, Japan) in 50 mM sodium phosphate buffer (pH 8.0) was added to 20-fold molar excess of *N*-succinimidyl 3-(2-pyridyldithio)propionate (SPDP) in acetonitrile and the mixture was incubated for 1 h at 25 °C. A Spectra/por7 membrane with a cutoff Mw 1000 (Spectrum Laboratories, Inc., Rancho Dominguez, USA) was used for dialysis against water for 24 h. The internal solution was lyophilized to afford a flocculent solid. This product was dissolved in water and the concentration was defined by UV–Vis spectroscopy. Product (PySS-dT$_{20}$) yield was 43.6 nmol (94.8%)—MALDI-TOF-MS (matrix: 3-HPA): m/z = 6409 (exact mass: 6398).

An aqueous solution of PySS-dT$_{20}$ (0.1 mM) in a 50 mM sodium phosphate buffer (pH 7.2) was mixed with the Cys-β-annulus peptide (2 mM) in a 50 mM sodium phosphate buffer (pH 7.2) and the mixture was incubated for 24 h at 25 °C. The solution was purified by reverse-phase HPLC eluted with a linear gradient of CH$_3$CN/0.1 M ammonium formate aqueous solution (10/90 to 100/0 over 95 min). The elution fraction was collected, concentrated in a centrifugal evaporator, and dialyzed as above against water for 20 h. The internal solution was lyophilized to afford a flocculent solid. This product was dissolved in water and the concentration was defined by UV–Vis spectroscopy. Product (dT$_{20}$-SS-β-annulus peptide) yield was 12.6 nmol (27.4%)—MALDI-TOF-MS (matrix: 3-HPA): m/z = 8710 (exact mass: 8698).

2.3. Preparation and Characterization of Artificial Viral Capsids

Lyophilized dT$_{20}$-SS-β-annulus and β-annulus peptides were dissolved in water, respectively. Peptide solutions were mixed at molar ratios of dT$_{20}$-SS-β-annulus:β-annulus = 1:9, 1:4.5, 1:2, and 1:1, sonicated for 5 min, and lyophilized. Stock solutions of peptides were prepared by dissolving peptide powders in 1× phosphate buffered saline (PBS, pH 7.4) and sonicating for 3 min.

Dynamic light scattering (DLS) was measured with a Zetasizer NanoZS (Malvern, Kobe, Japan) instrument at 25 °C using an incident He-Ne laser (633 nm). Correlation times of scattered light intensities G(τ) were measured several times and the means were fitted to Equation (1), where B is the baseline, A is amplitude, q is the scattering vector, $τ$ is the delay time, and D is the diffusion coefficient.

$$G(τ) = B + A \exp(-2q^2Dτ) \tag{1}$$

Hydrodynamic radii (R_H) of scattering particles were calculated using the Stokes-Einstein Equation (2), where $η$ is the solvent viscosity, k_B is Boltzmann's constant, and T denotes absolute temperature.

$$R_H = k_BT/6\pi ηD \tag{2}$$

Transmission electron microscopy (TEM) images were obtained with a JEOL JEM 1400 Plus (JEOL Ltd., Tokyo, Japan), using an accelerating voltage of 80 kV. Aliquots (5 µL) of DLS samples were applied to hydrophilized carbon-coated Cu-grids (C-SMART Hydrophilic TEM girds, Alliance Biosystems, Osaka, Japan) for 60 s and then removed. Subsequently, the TEM grid was instilled in the aqueous solution (5 µL) of 2% phosphotungstic acid, Na$_3$(PW$_{12}$O$_{40}$)(H$_2$O)$_n$. The staining solution was removed after 60 s and the sample-loaded grids were dried *in vacuo*.

2.4. Preparation of Complex of dT_{20}-SS-β-Annulus:β-Annulus Peptide Hybridized with mCherry mRNA

An aqueous solution of mCherry mRNA in 1× PBS (1 mg/mL, (nucleotide) ≈ 3 mM, and pH 7.4) was mixed with dT_{20}-SS-β-annulus:β-annulus peptide powders by gentle pipetting without sonication. The peptide solutions were diluted in 1× PBS (pH 7.4) so that the nucleotide concentration of mCherry mRNA was equal to dT_{20} ((mRNA nt) = (T) = 1 mM, (dT_{20}) = 50 μM, (RNA) = 1 μM). In a typical experiment, the dT_{20}-SS-β-annulus:β-annulus/mRNA solution was incubated for 30 min at 25 °C.

2.5. Electrophoretic Mobility Shift Assay and Nuclease Resistance Assay

Electrophoretic mobility shift assay (EMSA) was employed for the detection of peptide complexes with nucleic acids. The dT_{20}-SS-β-annulus:β-annulus/mRNA solutions were loaded onto 3% *w/v* agarose gels in a TAE buffer, pre-cast with GelRed (Wako Pure Chemical Industries, Osaka, Japan) for nucleic acid detection. Three microliters of the dT_{20}-SS-β-annulus:β-annulus/mRNA solution ((mRNA nt) = (T) = 1 mM) was loaded with 5 μL of Bluejuice Gel loading buffer (Thermo Ficher, Waltham, USA). Electrophoresis used 210 V for 30 min on an Atto AE-6100 with Atto mypower II300 AE-8130 (Atto, Tokyo, Japan). The bands were visualized with a UV illuminator (TP-15 MP, Atto, Tokyo, Japan) and the images were recorded with a digital camera.

Ribonuclease A from bovine pancreas (Nacalai Tesque, Kyoto, Japan) is an endoribonuclease which specifically hydrolyzes the 3′end of pyrimidine residues in single-stranded RNA. Each peptide/mRNA solution was incubated with a solution of RNase A (3 μL, 1 U/μL) in PBS for 10–60 min at 37 °C. Enzymatic degradation of mRNA was evaluated by EMSA.

2.6. Confocal Laser Scanning Microscopy (CLSM) Measurements of In-Cell Expression of mCherry mRNA

Confocal laser scanning microscopy (CLSM) used a Fluo View FV10i (Olympus, Tokyo, Japan). Human hepatoma HepG2 cells were cultured in DMEM (10 v/v% FBS, 100 μg/mL streptomycin, 100 units/mL penicillin, 1 mM sodium pyruvate, and 1 v/v% MEM nonessential amino acids) for 24 h at 37 °C in a 5% CO_2 atmosphere. The cells were seeded onto single-well bottom dishes with 2.0×10^4 cells/well in a final volume of 100 μL and incubated for 24 h at 37 °C and 5% CO_2. The medium was removed and 50 μM dT_{20}-SS-β-annulus, 450 μM β-annulus, and 1 μM mCherry mRNA in the fresh medium were added to the cells, then incubated for 48 h under 5% CO_2. The co-assembly mixture of 50 μM dT_{20}-SS-β-annulus, 450 μM β-annulus-His_{16}, and 1 μM mCherry mRNA was added to the cells for evaluation of expression of mCherry fluorescent protein. A complex of 2 μM TransIT-mRNA (Transfection kit, Mirus Bio LLC, Madison, USA) with 1 μM mCherry mRNA was added to cells as a positive control. CLSM images of mCherry fluorescence were obtained with excitation at 587 nm and an mCherry band-pass filter (Red). Fluorescence intensity was measured by cell from CLSM images by subtracting background intensity using ImageJ 1.51 software.

3. Results and Discussion

3.1. Synthesis and Self-Assembling Behavior of β-Annulus Peptide Bearing dT_{20} at the N-Terminus

We designed a β-annulus peptide bearing dT_{20} at the N-terminus that was directed to the interior of the capsid (Figure 1) to encapsulate mRNA via hybridization between dT_{20} and the poly(A) tail at the 3′end of mRNA. A 5′-terminal aminated 20-mer thymidine nucleotide (dT_{20}-NH_2) was reacted with *N*-succinimidyl 3-(2-pyridyldithio)propionate (SPDP) to obtain a dT_{20} bearing pyridyl disulfide group at the 5′-end (PySS-dT_{20}). A Cys-β-annulus peptide (CINHVGGTGGAIMAPVAVTRQLVGS) containing Cys at the N-terminus was synthesized by standard Fmoc-based solid-phase methods according to our reported procedure [28]. PySS-dT_{20} was attached to the Cys-β-annulus peptide with a disulfide-exchange reaction. A reversed-phase HPLC chart of the reaction mixture showed one peak at 26.7 min, which was different from the retention time of dT_{20}-NH_2 and PySS-dT_{20} (Figure 2A). The purified product was assigned by MALDI-TOF-MS as dT_{20}-modified β-annulus peptide (dT_{20}-SS-β-Annulus) (Figure 2B). DLS of a 50 μM solution of dT_{20}-SS-β-annulus in PBS (pH 7.4) exhibited formation of 77 ± 21 and 40 ± 8 nm assemblies (Figure 2C). TEM images of the aqueous solution stained with phosphotungstic acid showed the formation of spherical assemblies of approximately 40 nm in diameter (Figure 2D). The concentration dependence of the size distribution in DLS measurement indicated that dT_{20}-SS-β-annulus formed assemblies with sizes ranging from 30 to 100 nm in a concentration range of 5–125 µM (Figure S1). This tendency was similar to the findings in our previous report [35]. Notably, β-annulus peptide modified with highly charged dT_{20} formed relatively stable artificial viral capsids.

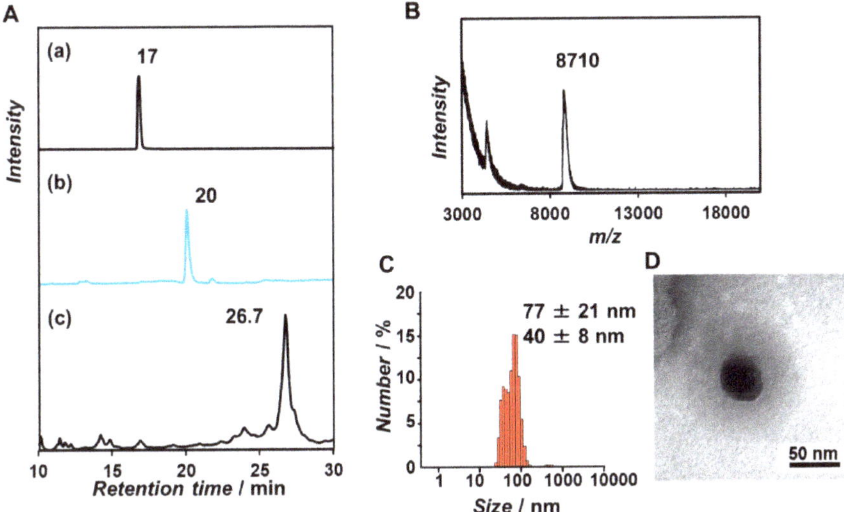

Figure 2. (**A**) Reverse-phase HPLC chart of (**a**) dT_{20}-NH_2, (**b**) PySS-dT_{20} detected at 260 nm, eluted with a linear gradient of CH_3CN/0.1 M NH_4HCO_2 aq (0/100 to 100/0 over 95 min), and (**c**) dT_{20}-SS-β-annulus detected at 260 nm, eluted with a linear gradient of CH_3CN/0.1 M NH_4HCO_2 aq (10/90 to 100/0 over 95 min). (**B**) MALDI-TOF-MS of the purified dT_{20}-SS-β-annulus (matrix: 3-HPA). (**C,D**) TEM image and size distribution obtained from dynamic light scattering (DLS) for an aqueous solution of 50 µM dT_{20}-SS-β-annulus in PBS (pH 7.4) at 25 °C.

3.2. Complexation of mCherry mRNA and Artificial Viral Capsid Bearing dT_{20}

We initially attempted encapsulation of mRNA within the artificial viral capsid consisting of dT_{20}-SS-β-annulus via hybridization between dT_{20} and mRNA. However, when poly(A) and dT_{20}-SS-β-annulus were mixed at equimolar base concentrations, capsid formation was insufficient (data are not shown). It is probably difficult to self-assemble among dT_{20}-SS-β-annulus peptides hybridized on a large poly(A) molecule due to the excluded volume effect. Therefore, we developed an alternate strategy to encapsulate mRNA with dT_{20}-SS-β-annulus and unmodified β-annulus peptide (INHVGGTGGAIMAPVAVTRQLVGS) (Figure 1). DLS and TEM images of the mixture of dT_{20}-SS-β-annulus/unmodified β-annulus peptide at 1:9 molar ratio in PBS showed the formation of spherical co-assemblies of size 45 ± 20 nm (Figure 3). The peptides mixture at other molar ratios (dT_{20}-SS-β-annulus:β-annulus = 1:4.5, 1:2, 1:1) also formed spherical co-assemblies of similar size (Figure S2). Next, we encapsulated mRNA coding in the mCherry fluorescent protein (mCherry mRNA) by co-assembly at different molar ratios of dT_{20}-modified and unmodified β-annulus peptide.

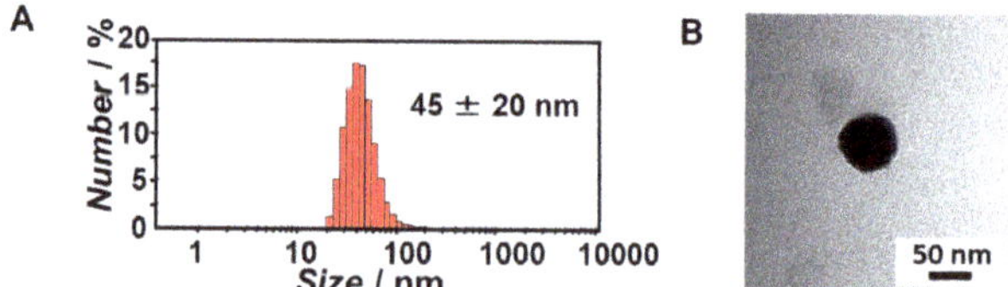

Figure 3. (**A**) Size-distributions obtained from DLS and (**B**) TEM images for co-assembly peptides (dT$_{20}$-SS-β-annulus:β-annulus = 1:9) in water at 25 °C.

An aqueous solution of mCherry mRNA in PBS was added to the lyophilized powder of dT$_{20}$-SS-β-annulus:β-annulus peptide followed by dilution with PBS and incubation for 30 min at 25 °C to prepare capsids containing mCherry mRNA. The electrophoretic mobility shift assay (EMSA) showed a band of mCherry mRNA significantly shifted by complexation with a co-assembly mixture of dT$_{20}$-SS-β-annulus:β-annulus peptide at a 1:9 molar ratio (Figure 4A, lane 4). In contrast, the other co-assembly mixtures showed only a slight retardation of mRNA (Figure 4A, lane 5–7), which reflect almost neutral net charge and relatively small size of the β-annulus peptide. The migration position of a mixed solution of unmodified β-annulus peptide and mCherry mRNA was not shifted from mCherry mRNA alone (Figure 4B). DLS and TEM images of the 1:9 co-assembly complexes with mCherry mRNA show the formation of spherical assemblies of size 52 ± 10 nm (Figure 5A,B). We previously reported that the N-terminus of the β-annulus peptide is directed toward the interior of the capsid and the surface ζ-potential of the capsid is almost zero at neutral pH [25]. Thus, when negatively charged RNA is encapsulated at neutral pH, the artificial viral capsid should minimally migrate from the applied position due to the capsid charge shield. Therefore, EMSA results indicate that the dT$_{20}$-SS-β-annulus:β-annulus peptide 1:9 co-assembly encapsulated mCherry mRNA via hybridization with dT$_{20}$ directed to the interior of capsids. EMSA shows that mCherry mRNA could be complexed with the 1:9 co-assembly at 4 °C, 25 °C, 37 °C, and 60 °C (Figure S3). However, complexation at 4 °C and 60 °C afforded large aggregates with bimodal size distribution (Figure S4). Thus, mCherry mRNA-encapsulated artificial viral capsid with unimodal size distribution can be obtained by incubation with the 1:9 co-assembly mixture at temperatures of 25 °C–37 °C.

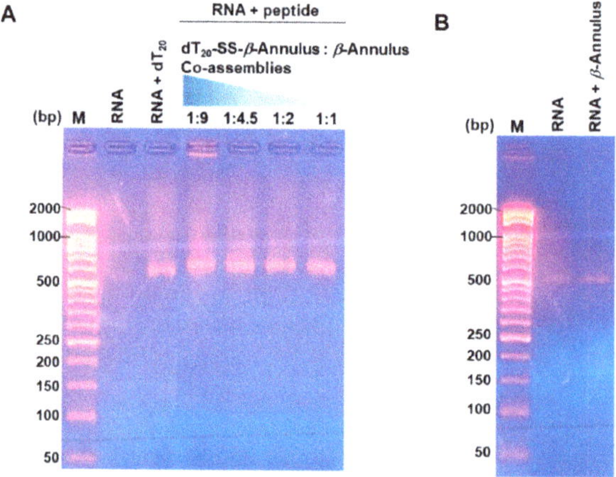

Figure 4. Gel shift assay of (**A**) co-assembly peptides hybridized with mCherry mRNA, (**B**) mixture of β-annulus peptide and mCherry mRNA in phosphate buffered saline (PBS) (pH 7.4) incubated at 25 °C for 30 min. M: DNA marker. All DNA markers used in the gel shift assay are the same. The gels were stained with GelRed.

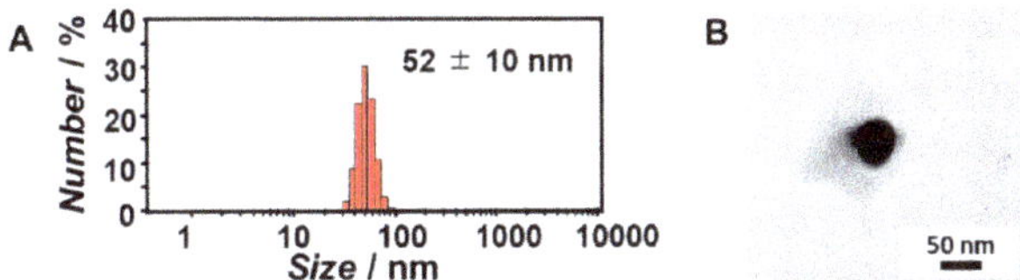

Figure 5. (**A**) Size-distributions obtained from DLS and (**B**) TEM images for co-assembly peptides (dT$_{20}$-SS-β-annulus:β-annulus = 1:9) and mCherry mRNA in PBS (pH 7.4) obtained after incubation at 25 °C.

3.3. Nuclease Resistance of mRNA-Encapsulated Artificial Viral Capsid

Resistance of mCherry mRNA-encapsulated artificial viral capsid to endoribonuclease (RNase A) was confirmed by EMSA. Naked mCherry mRNA was digested by RNase A within 10 min (Figure 6(Aa),B lane 3). Conversely, mRNA in artificial viral capsids (band near the loading well) was minimally digested after 60 min (Figure 6(Ac),B lane 5–8) due to protection by the capsid. For the artificial viral capsids modified with dT$_{20}$, the exteriors were constructed as a control by self-assembly of β-annulus peptide modified with dT$_{20}$ at the C-terminal that directed dT$_{20}$ to the exterior of the capsid. The mixture of mCherry mRNA and the dT$_{20}$-modified capsid at the exterior did not cause a significant mobility shift (Figure 6B lane 9), indicating that mRNA was not encapsulated. mRNA was rapidly digested by RNase A within 10 min (Figure 6(Ac),B lane 10). Thus, mRNA is protected from degradation by artificial viral capsids.

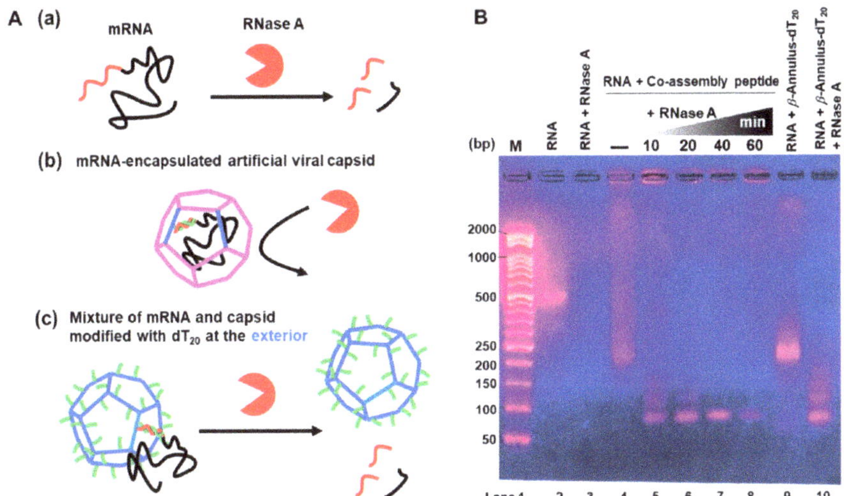

Figure 6. (**A**) Illustration of digestion of mRNA by RNase A: (**a**) mRNA alone, (**b**) mRNA-encapsulated artificial viral capsid, and (**c**) Mixture of mRNA and capsid modified with dT20 at the exterior. (**B**) Gel electrophoresis assay of digestions mRNA by RNase A at 37 °C for 10 min (mRNA alone and β-annulus-dT$_{20}$) and 10–60 min (dT$_{20}$-SS-β-annulus:β-annulus = 1:9 co-assembly). The gel was stained with GelRed.

3.4. In-Cell Expression of mCherry mRNA Encapsulated in Artificial Viral Capsid

Uptake of mCherry mRNA-encapsulated capsids into human hepatoma HepG2 cells were analyzed by CLSM to evaluate mCherry mRNA expression inside cells. mCherry mRNA and a strong transfection reagent (TransIT-mRNA) were incubated with the cells for 48 h at 37 °C as a positive control; significant red fluorescence of mCherry was observed (Figure 7(Aa)). Unfortunately, minimal expression of encapsulated mCherry mRNA was observed as similar to the naked mCherry mRNA(Figure 7A(b,c)). We employed a histidine 16-mer (His16), a cell-penetrating peptide uncharged under physiological conditions [36,37] to enhance cell permeability of capsids. We constructed artificial viral capsids modified with His16 on the exterior and using the 1:9 co-assembly of dT$_{20}$-SS-β-annulus peptide and β-annulus peptide modified with His16 at the C-terminal (β-annulus-His16). Significant red fluorescence of mCherry was observed in cells incubated with His16-modified capsids (Figure 7(Ad)). The expression level of mCherry was quantified by fluorescence intensity in the cell areas of the CLSM image using the software ImageJ. The expression level using His16-modified capsids was 26.5 times the expression observed with unmodified capsids (Figure 7B). This indicates that the intracellular expression of mCherry was significantly enhanced by the modification of His16 on the surface of the artificial viral capsid. We have demonstrated in our previous work that the reductive cleavage of the disulfide bond of the artificial viral capsid by dithiothreitol (DTT) caused the controlled-release of ssDNA with the destruction of the capsid [35]. mCherry mRNA was likely released from capsids in the reducing environment in cells.

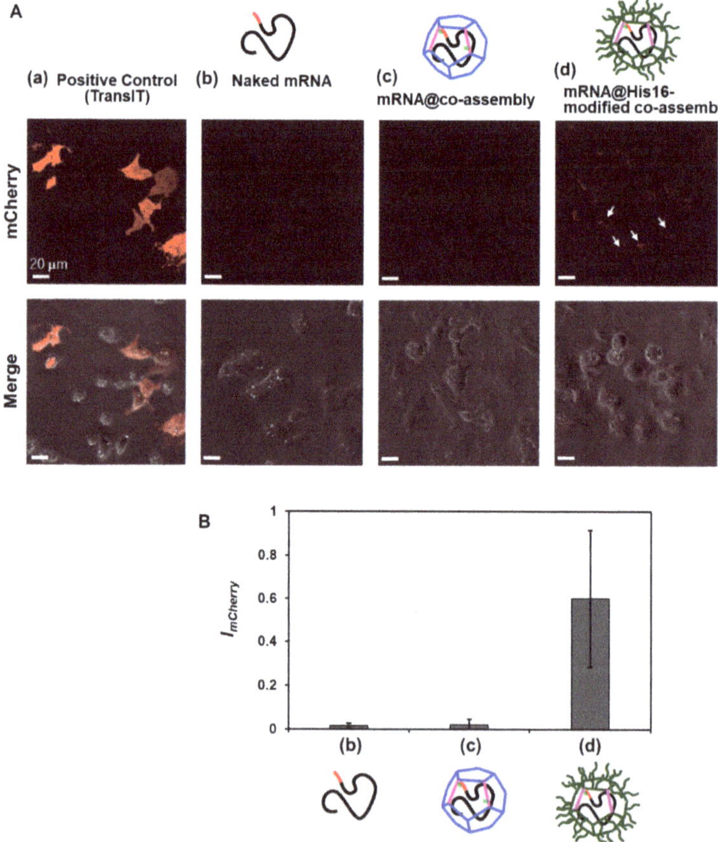

Figure 7. (**A**) Confocal laser scanning microscopy (CLSM) images of HepG2 Cells transfected with mCherry mRNA encapsulated in artificial viral capsids: (**a**) positive control (2 µM TransIT-mRNA with 1 µM mCherry mRNA), (**b**) 1 µM Naked mCherry mRNA, (**c**) mRNA-encapsulated artificial viral capsid (50 µM dT$_{20}$-SS-β-annulus, 450 µM β-annulus, and 1 µM mCherry mRNA), and (**d**) mRNA-encapsulated histidine 16-mer (His16)-artificial viral capsid (50 µM dT$_{20}$-SS-β-annulus, 450 µM β-annulus-His16, and 1 µM mCherry mRNA). (**B**) Relative fluorescence intensity of mCherry from HepG2 Cells.

4. Conclusions

We constructed artificial viral capsids co-assembled from viral β-annulus peptides bearing dT_{20} at the N-terminus and unmodified peptides. Capsids bearing interior dT_{20} enable encapsulation of mCherry mRNA bearing poly(A) tails via hybridization to form ribonuclease resistant spherical assemblies of approximately 50 nm in diameter. Further, we demonstrate novel material for mRNA delivery by His16-modification on the exterior of capsids. Further research will demonstrate potential applications for artificial viral capsids as delivery systems for nucleic acid drugs.

Supplementary Materials: The following are available online at http://www.mdpi.com/2076-3417/10/22/8004/s1, Figure S1: Concentration dependence of size distribution obtained from DLS for the aqueous solution of dT_{20}-SS-β-annulus in PBS (pH7.4) at 25 °C, Figure S2: Size-distributions obtained from DLS and TEM images for co-assembly peptides (dT_{20}-SS-β-annulus:β-annulus = 1:4.5 (a), 1:2 (b), and 1:1 (c) in water at 25 °C), Figure S3: Gel shift assay obtained after hybridization of co-assembly peptides and mCherry mRNA in PBS (pH 7.4) incubated at various temperatures for 10–30 min, and Figure S4: Size-distributions obtained from DLS and TEM images for co-assembly peptides (dT_{20}-SS-β-annulus:β-annulus = 1:9) and mCherry mRNA in PBS buffer (pH 7.4) obtained after incubation at 4 °C (a), 37 °C (b), and 60 °C (c) for 10–30 min.

Author Contributions: Conceptualization, K.M.; methodology, T.I. and H.I.; investigation, Y.N. and Y.S.; writing, Y.N. and K.M.; visualization, Y.S. and H.I.; supervision, K.M.; funding acquisition, K.M. All authors have read and agreed to the published version of the manuscript.

Funding: This research was funded by a Grant-in-Aid for Scientific Research on Innovative Areas "Chemistry for Multimolecular Crowding Biosystems" (grant number: JP18H04558) and a Grant-in-Aid for Scientific Research (B) (grant number: JP18H02089).

Conflicts of Interest: The authors declare no conflict of interest.

References

1. Sahin, U.; Karikó, K.; Türeci, Ö. mRNA-based therapeutics—Developing a new class of drugs. *Nat. Rev. Drug Discov.* **2014**, *13*, 759–780. [CrossRef] [PubMed]
2. Yin, H.; Kanasty, R.L.; Eltoukhy, A.A.; Vegas, A.J.; Dorkin, J.R.; Anderson, D.G. Non-viral vectors for gene-based therapy. *Nat. Rev. Genet.* **2014**, *15*, 541–555. [CrossRef] [PubMed]
3. Kauffman, K.J.; Webber, M.J.; Anderson, D.G. Materials for non-viral intracellular delivery of messenger RNA therapeutics. *J. Control. Release* **2016**, *240*, 227–234. [CrossRef] [PubMed]
4. Guan, S.; Rosenecker, J. Nanotechnologies in delivery of mRNA therapeutics using non-viral vector-based delivery systems. *Gene Ther.* **2017**, *24*, 133–143. [CrossRef] [PubMed]
5. Kowalski, P.S.; Rudra, A.; Miao, L.; Anderson, D.G. Delivering the Messenger: Advances in Technologies for Therapeutic mRNA Delivery. *Mol. Ther.* **2019**, *27*, 710–728. [CrossRef]
6. Meng, C.; Chen, Z.; Li, G.; Welte, T.; Shen, H. Nanoplatforms for mRNA Therapeutics. *Adv. Ther.* **2020**. [CrossRef]
7. Uchida, H.; Itaka, K.; Nomoto, T.; Ishii, T.; Suma, T.; Ikegami, M.; Miyata, K.; Oba, M.; Nishiyama, N.; Kataoka, K. Modulated Protonation of Side Chain Aminoethylene Repeats in N-Substituted Polyaspartamides Promotes mRNA Transfection. *J. Am. Chem. Soc.* **2014**, *136*, 12396–12405. [CrossRef]
8. Baba, M.; Itaka, K.; Kondo, K.; Yamasoba, T.; Kataoka, K. Treatment of neurological disorders by introducing mRNA In Vivo using polyplex nanomicelles. *J. Control. Release* **2015**, *201*, 41–48. [CrossRef]
9. Matsui, A.; Uchida, S.; Ishii, T.; Itaka, K.; Kataoka, K. Messenger RNA-based therapeutics for the treatment of apoptosis-associated diseases. *Sci. Rep.* **2015**, *5*, 15810. [CrossRef]
10. Li, J.; Wang, W.; He, Y.; Li, Y.; Yan, E.Z.; Zhang, K.; Irvine, D.J.; Hammond, P.T. Structurally Programmed Assembly of Translation Initiation Nanoplex for Superior mRNA Delivery. *ACS Nano* **2017**, *11*, 2531–2544. [CrossRef]
11. Benner, N.L.; McClellan, R.L.; Turlington, C.R.; Haabeth, O.A.W.; Waymouth, R.M.; Wender, P.A. Oligo(serine ester) Charge-Altering Releasable Transporters: Organocatalytic Ring-Opening Polymerization and their Use for In Vitro and In Vivo mRNA Delivery. *J. Am. Chem. Soc.* **2019**, *141*, 8416–8421. [CrossRef] [PubMed]
12. Rein, A. Retroviral RNA packaging: A review. In *Positive-Strand RNA Viruses*; Brinton, M.A., Calisher, C.H., Rueckert, R., Eds.; Springer: Vienna, Austria, 1994; Volume 9, pp. 513–522.
13. Perlmutter, J.D.; Hagan, M.F. Mechanisms of Virus Assembly. *Annu. Rev. Phys. Chem.* **2015**, *66*, 217–239. [CrossRef] [PubMed]
14. Azuma, Y.; Edwardson, T.G.W.; Terasaka, N.; Hilvert, D. Modular Protein Cages for Size-Selective RNA Packaging in Vivo. *J. Am. Chem. Soc.* **2018**, *140*, 566–569. [CrossRef] [PubMed]

15. Terasaka, N.; Azuma, Y.; Hilvert, D. Laboratory evolution of virus-like nucleocapsids from non-viral protein cages. *Proc. Natl. Acad. Sci. USA* **2018**, *115*, 5432–5437. [CrossRef]
16. Munroe, D.; Jacobson, A. Tales of poly(A): A review. *Gene* **1990**, *91*, 151–158. [CrossRef]
17. Sachs, A. The role of poly(A) in the translation and stability of mRNA. *Curr. Opin. Cell Biol.* **1990**, *2*, 1092–1098. [CrossRef]
18. Eckmann, C.R.; Rammelt, C.; Wahle, E. Control of poly(A) tail length. *Wiley Interdiscip. Rev. RNA* **2011**, *2*, 348–361. [CrossRef]
19. Weill, L.; Belloc, E.; Bava, F.-A.; Méndez, R. Translational control by changes in poly(A) tail length: Recycling mRNAs. *Nat. Struct. Mol. Biol.* **2012**, *19*, 577–585. [CrossRef]
20. Gruber, A.J.; Zavolan, M. Alternative cleavage and polyadenylation in health and disease. *Nat. Rev. Genet.* **2019**, *20*, 599–614. [CrossRef]
21. Nicholson, A.L.; Pasquinelli, A.E. Tales of Detailed Poly(A) Tails. *Trends Cell Biol.* **2019**, *29*, 191–200. [CrossRef]
22. Matsuura, K.; Watanabe, K.; Matsuzaki, T.; Sakurai, K.; Kimizuka, N. Self-Assembled Synthetic Viral Capsids from a 24-mer Viral Peptide Fragment. *Angew. Chem. Int. Ed.* **2010**, *49*, 9662–9665. [CrossRef]
23. Matsuurua, K. Rational design of self-assembled proteins and peptides for nano and micro-sized architectures. *RSC Adv.* **2014**, *4*, 2942–2953. [CrossRef]
24. Matsuura, K. Synthetic approaches to construct viral capsid-like spherical nanomaterials. *Chem. Commun.* **2018**, *54*, 8944–8959. [CrossRef]
25. Matsuura, K.; Watanabe, K.; Matsushita, Y.; Kimizuka, N. Guest-binding behavior of peptide nanocapsules self-assembled from viral peptide fragments. *Polym. J.* **2013**, *45*, 529–534. [CrossRef]
26. Fujita, S.; Matsuura, K. Inclusion of zinc oxide nanoparticles into virus-like peptide nanocapsules self-assembled from viral β-annulus peptide. *Nanomaterials* **2014**, *4*, 778–791. [CrossRef]
27. Fujita, S.; Matsuura, K. Encapsulation of CdTe Quantum Dots into Synthetic Viral Capsids. *Chem. Lett.* **2016**, *45*, 922–924. [CrossRef]
28. Matsuura, K.; Nakamura, T.; Watanabe, K.; Noguchi, T.; Minamihata, K.; Kamiya, N.; Kimizuka, N. Self-assembly of Ni-NTA-modified β-annulus peptides into artificial viral capsids and encapsulation of His-tagged proteins. *Org. Biomol. Chem.* **2016**, *14*, 7869–7874. [CrossRef]
29. Matsuura, K.; Ueno, G.; Fujita, S. Self-assembled artificial viral capsid decorated with gold nanoparticles. *Polym. J.* **2014**, *47*, 146–151. [CrossRef]
30. Nakamura, Y.; Yamada, S.; Nishikawa, S.; Matsuura, K. DNA-modified artificial viral capsids self-assembled from DNA-conjugated β-annulus peptide. *J. Pept. Sci.* **2017**, *23*, 636–643. [CrossRef]
31. Matsuura, K.; Matsuura, K. Self-assembled artificial viral capsids bearing coiled-coils at the surface. *Org. Biomol. Chem.* **2017**, *15*, 5070–5077. [CrossRef]
32. Matsuura, K.; Honjo, T. Artificial Viral Capsid Dressed Up with Human Serum Albumin. *Bioconjug. Chem.* **2019**, *30*, 1636–1641. [CrossRef] [PubMed]
33. Matsuura, K.; Ota, J.; Fujita, S.; Shiomi, Y.; Inaba, H. Construction of Ribonuclease-Decorated Artificial Virus-like Capsid by Peptide Self-assembly. *J. Org. Chem.* **2019**, *85*, 1668–1673. [CrossRef] [PubMed]
34. Matsuura, K. Dressing up artificial viral capsids self-assembled from C-terminal-modified β-annulus peptides. *Polym. J.* **2020**, *52*, 1035–1041. [CrossRef]
35. Nakamura, Y.; Inaba, H.; Matsuura, K. Construction of Artificial Viral Capsids Encapsulating Short DNAs via Disulfide Bonds and Controlled Release of DNAs by Reduction. *Chem. Lett.* **2019**, *48*, 544–546. [CrossRef]
36. Iwasaki, T.; Tokuda, Y.; Kotake, A.; Okada, H.; Takeda, S.; Kawano, T.; Nakayama, Y. Cellular uptake and In Vivo distribution of polyhistidine peptides. *J. Control. Release* **2015**, *210*, 115–124. [CrossRef] [PubMed]
37. Hayashi, T.; Shinagawa, M.; Kawano, T.; Iwasaki, T. Drug delivery using polyhistidine peptide-modified liposomes that target endogenous lysosome. *Biochem. Biophys. Res. Commun.* **2018**, *501*, 648–653. [CrossRef]

Publisher's Note: MDPI stays neutral with regard to jurisdictional claims in published maps and institutional affiliations.

Communication

Bifunctional Aptamer Drug Carrier Enabling Selective and Efficient Incorporation of an Approved Anticancer Drug Irinotecan to Fibrin Gels

Hiroto Fujita, Yuka Kataoka and Masayasu Kuwahara * 

Graduate School of Integrated Basic Sciences, Nihon University, 3-25-40 Sakurajosui, Setagaya-ku, Tokyo 156-8550, Japan; fujita.hiroto@nihon-u.ac.jp (H.F.); kataoka.yuka@nihon-u.ac.jp (Y.K.)
* Correspondence: mkuwa@chs.nihon-u.ac.jp; Tel.: +81-03-5317-9398

Received: 10 September 2020; Accepted: 4 December 2020; Published: 7 December 2020

Abstract: We have previously developed a bifunctional aptamer (bApt) binding to both human thrombin and camptothecin derivative (CPT1), and showed that bApt acts as a drug carrier under the phenomenon named selective oligonucleotide entrapment in fibrin polymers (SOEF), which enables efficient enrichment of CPT1 into fibrin gels, resulting in significant inhibition of tumor cell growth. However, although the derivative CPT1 exhibits anticancer activity, it is not an approved drug. In this study, we evaluated the binding properties of bApt to irinotecan, a camptothecin analog commonly used for anticancer drug therapy, in addition to unmodified camptothecin (CPT). Furthermore, we have revealed that irinotecan binds to bApt like CPT1 and is selectively concentrated on fibrin gels formed around the tumor cells under the SOEF phenomenon to suppress cell proliferation.

Keywords: drug delivery system; anticancer drug; camptothecin derivative; irinotecan

---

1. Introduction

Highly water-soluble derivatives of camptothecin, a type of quinoline alkaloid (irinotecan and topotecan), are currently known to have a broad spectrum of anticancer activity such as in lung cancer, colorectal cancer, ovarian cancer, and malignant lymphoma [1–7]. Its derivatives are the most clinically used because they bind to type I topoisomerase and inhibit recombination with DNA, inhibiting DNA synthesis, causing cancer cell apoptosis, and resulting in an extremely potent anticancer activity [4]. Irinotecan is biotransformed because of the action of carboxylesterase in SN-38 (7-ethyl-10-hydroxy-20(S)-camptothecin), which lacks a substituent at the 10-position on the piperidine ring. Additionally, the cytotoxicity of SN-38 is 100–1000 times greater than that of irinotecan (Figure S1) [8–10]. However, it causes serious side effects, and there is demand for developing a mechanism such as a drug delivery system (DDS) to deliver specific drugs and reduce dosages [11–16]. For example, a liposomal irinotecan called Onivyde has been reported and approved for medical use [17,18].

Previously, we have reported that thrombin-binding aptamer (TBA) is selectively and efficiently incorporated into the gel through thrombin during the fibrin gel formation known as the blood coagulation reaction [19–24]. In the fibrin gel formation process, first, thrombin cleaves FpA and FpB from the Aα and Bβ chains located at the N-terminus of fibrinogen to produce fibrin monomers [25]. The resulting fibrin monomer engulfs thrombin and polymerizes to form a gel. Therefore, it was suggested that by using this phenomenon named selective oligonucleotide entrapment in fibrin polymers (SOEF), the desired substance or functional group could be introduced into the fibrin gel with TBA [26].

Recently, it has been shown that fibrin plays an important role when cancer cells metastasize or invade other organs, and that fibrinogen is abundant in the cancer stroma [11]. When considering SOEF, the bifunctional aptamer (bApt), which is a combined 29-mer TBA and camptothecin binding aptamer, can deliver and condense camptothecin as a potent anticancer drug around cancer cells and efficiently suppress their cell growth (Figure 1). Previously, we developed a camptothecin (CPT)-binding modified DNA aptamer (CMA-70), which is a base-modified DNA aptamer obtained by the systematic evolution of ligands via the exponential enrichment (SELEX) method targeting the camptothecin derivative (CPT1) (Figure 1) [27–29]. Therefore, the ability to selectively introduce CPT1 into fibrin gel using bApt as a drug carrier under the SOEF phenomenon provides a new strategy for cancer chemotherapy [30]. However, as a shortcut to the practical application of the said DDS, it is desirable to use an approved anticancer drug such as irinotecan, and thus, this study was conducted.

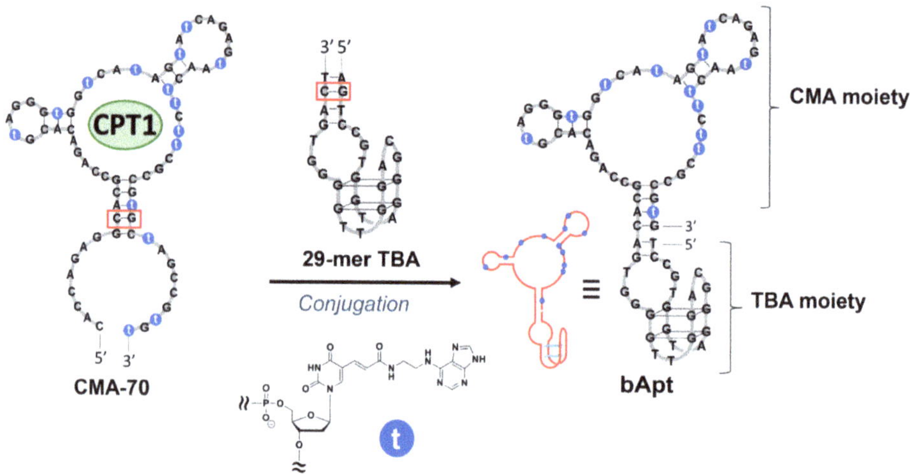

Figure 1. Design of bifunctional aptamer (bApt) conjugated with 29-mer thrombin-binding aptamer (TBA) and a camptothecin (CPT)-binding modified DNA aptamer (CMA-70) containing modified nucleic acids (2′-deoxyuridine-5′-phosphate (dU^{ad}TP)) represented with 't.' Red boxes indicate the conjugation site.

2. Materials and Methods

2.1. Materials

To synthesize CMA-70 and bApt, oligonucleotides, nucleoside triphosphates (dATP, dGTP, dCTP), and *KOD dash* DNA polymerase were, respectively, purchased from Japan Bio Services Co. Ltd. (Saitama, Japan), Roche Diagnostics K. K. (Tokyo, Japan), and Toyobo Co. Ltd. (Tokyo, Japan). For the fluorescence polarization assay, human fibrinogen and thrombin were purchased from Merck K. K. (Tokyo, Japan) and irinotecan and CPT were supplied by Tokyo Chemical Industry Co., Ltd. (Tokyo, Japan). For the cell cultivation and cell growth inhibition assay, Dulbecco's Modified Eagle's Medium (DMEM) low glucose were purchased from Wako Pure Chemical Industries, Ltd., fetal bovine serum (Gibco^{TM}) was purchased from Thermo Fischer Scientific K. K. (Tokyo, Japan), CellTracker^{TM} Green CMFDA (5-chloromethylfluorescein diacetate) Dye was purchased from Life Technologies Japan, Ltd. (Osaka, Japan), and HeLa cells (JCRB9004) were purchased from the National Institutes of Biomedical Innovation, Health and Nutrition (Osaka, Japan). All other reagents were of research grade.

2.2. Enzymatic Synthesis of CMA-70 and bApt

The CMA-70 aptamer was synthesized by one-primer PCR using a primer (CMA-70_P1), a template (CMA-70_Temp), three 2′-deoxyribonucleoside triphosphates (dATP, dGTP, and dCTP), and a modified 2′-deoxyuridine-5′-triphosphate (dU^{ad}TP) with *KOD Dash* DNA polymerase (Table S1) [27,29]. The antisense strand of 5′-monophosphate-labeled ODNs (CMA-70_Temp) was selectively degraded through λ-exonuclease treatment. Then, the resulting CMA-70 was purified via polyacrylamide gel electrophoresis.

The aptamer bApt was synthesized by one-primer PCR using a primer (TBA_P1), a template (T1), three 2′-deoxyribonucleoside triphosphates (dATP, dGTP, and dCTP), and a modified 2′-deoxyuridine-5′-triphosphate (dU^{ad}TP) with *KOD Dash* DNA polymerase (Table S1) [30]. The synthesized bApt was purified by polyacrylamide gel electrophoresis (Figure S2).

2.3. Fluorescence Polarization Assay for CMA-70 Versus CPT Derivatives

To analyze target binding specificity, a fluorescence polarization assay for a target (CPT1, irinotecan, and CPT; final concentration of 0.10 μM) with increasing concentrations of CMA-70 (final concentrations of 0, 0.010, 0.025, 0.050, 0.075, 0.10, 0.50, and 1.0 μM) were performed at 25 °C using an LS-55 fluorescence spectrometer.

First, CMA-70 was dissolved in 1 × phosphate-buffered saline (PBS) (11.8 mM HPO$_4^{2-}$, 140 mM Cl$^-$, 157 mM Na$^+$, 4.5 mM K$^+$; pH 7.4) at appropriate concentrations (2.0 μM), refolded by denaturing at 94 °C for 0.5 min to protect modified nucleic acids and subsequently cooled to 25 °C at a rate of 0.5 °C/min. Each CMA-70 solution (50 μL each) was mixed with 50 μL of a target solution (CPT1, irinotecan and CPT; 0.20 μM) in a PBS buffer, and incubated at 25 °C for 1 h. Fluorescence polarization for each mixture described above was recorded every 20–30 s for 20 min with excitation at 372 nm and monitoring at 456 nm of 25 °C. Thus, fluorescence polarization at 8 aptamer concentrations was generated as curve (Figure 2).

2.4. Fluorescence Polarization Assay for Time Course Analyses of bApt Complexes

Thrombin was dissolved in distilled water, and then, a 2.0 μM thrombin solution in 1 × PBS was prepared. Similarly, mother liquors of fibrinogen (20 μM) and irinotecan (1.0 μM) were prepared in 1 × PBS.

A solution containing bApt (1.0 μM) was prepared in 1 × PBS, and then, bApt was refolded by annealing (preheating at 94 °C for 0.5 min followed by cooling to 25 °C at a rate of 0.5 °C/min). First, the irinotecan solution (63 μL, 1.1 μM) was placed in a 1 cm cuvette and a started fluorescence polarization measurement. After 20 min, the bApt solution (7 μL, 1.0 μM) was put in the same cuvette. Subsequently, the thrombin solution (3.5 μL, 2.0 μM) was added to produce a ternary complex of thrombin, irinotecan, and bApt after 40 min of FP measurement. Finally, fibrinogen solution (3.5 μL, 20 μM) was added after 60 min of FP measurement and measured FP for another 20 min (Figure 3). The final concentration of thrombin/irinotecan/bApt was 0.10 μM and fibrinogen was 1.0 μM.

2.5. Cell Cultivation and Cell Growth Inhibition Assay

Trypsinized HeLa cells were diluted to a concentration of 1.0×10^4 cells/mL with a fresh medium containing DMEM-low glucose, a heat-inactivated serum (Gibco^{TM}, FBS, US origin), and an antibiotic solution (penicillin–streptomycin) at a percentage proportion of 90/9/1 (v/v/v%). The medium (300 μL) containing the cells was transferred to a 96-well tissue culture plate. Then, the cells were cultured in a humidified incubator containing 95% air and 5% CO$_2$ at 37 °C for 24 h. After 240 μL of the medium was removed, 240 μL of the new medium was added to each well. Then, 30 μL of the fibrinogen solution (10 μM) was added and mixed via gentle pipetting. Subsequently, the solution of the ternary thrombin/CPT1/bApt complex (30 μL, 1.0 μM) was added and mixed by gentle pipetting. Solutions of the binary thrombin/TBA complex and free CPT1 (30 μL, 1.0 μM each) were used as controls. Similarly, the abovementioned solutions for the other control experiments were examined. Furthermore, control experiments without fibrinogen (replaced with 1× PBS) were conducted. The cells were cultured in a humidified incubator containing 95% air and 5% CO$_2$ at 37 °C for 48 h.

Next, 300 µL of the medium was replaced with 100 µL of medium containing DMEM-low glucose and CellTracker™ Green CMFDA dye in dimethyl sulfoxide (10 mM) at a percentage proportion of 99.8/0.2 (v/v%). The sample was then incubated in a humidified incubator containing 95% air and 5% CO_2 at 37 °C for 30 min. After 300 µL of the medium was replaced with 300 µL of 1× PBS, the stained cell layer was examined under a BZ-X710 inverted fluorescence microscope to acquire fluorescence images (40×) using excitation light (480 nm) and a cut-off filter (<520 nm) (Figure 4). Three independent experiments were performed.

3. Results

3.1. Fluorescence Polarization Assay Using CMA-70

The fluorescence polarization measurement was conducted to examine the binding properties to CPT1, irinotecan, and camptothecin (CPT) (Figure 2A–C) using CMA-70, which is the mother aptamer of the camptothecin-derivative-binding site in bApt. From Figure 2D, we revealed that the fluorescence polarization degree (ΔFP) of irinotecan increases depending on the CMA-70 concentration, as in CPT1 and CPT. It was indicated that the affinity of CMA-70 for irinotecan exceeded that of CPT.

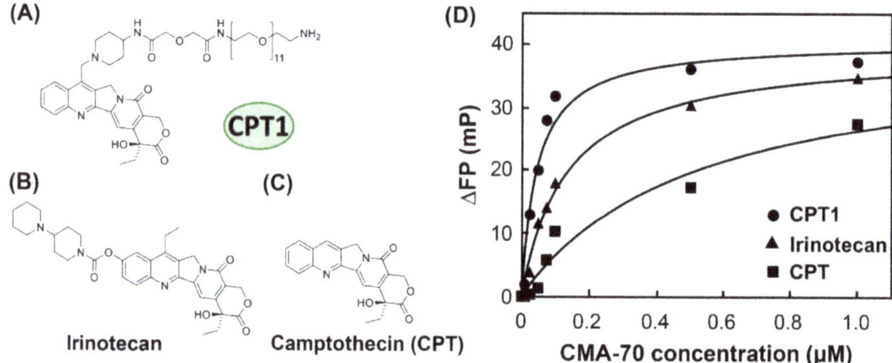

Figure 2. Chemical structures of (**A**) camptothecin derivative (CPT1), (**B**) irinotecan, and (**C**) camptothecin (CPT) and (**D**) titration curves for CPT1 (•), irinotecan (▲), and CPT (■) polarization versus CMA-70 concentration; polarization was monitored at 456 nm using the excitation wavelength of CPT1, irinotecan, and CPT (372 nm).

3.2. Fluorescence Polarization Assay Using bApt

Next, fluorescence polarization measurements of irinotecan were performed to observe the stepwise binding of bApt to its target and the uptake into fibrin gels (Figure 3). Figure 3A shows the time course in differential fluorescence polarization degrees by (i) addition of bApt to irinotecan, (ii) addition of thrombin, and (iii) addition of fibrinogen, followed by polymerization (gelation) of the fibrin monomer generated by thrombin cleavage activity. From Figure 3B, irinotecan bound to the CMA moiety of bApt, as the addition of bApt increased the fluorescence polarization by approximately 4 mP. Furthermore, by adding thrombin, thrombin bound to the TBA moiety of bApt, which increased the fluorescence polarization by approximately 18 mP. Finally, when fibrinogen was added, the fluorescence polarization degree increased by approximately 120 mP, indicating that irinotecan was incorporated into the gel along with bApt during the process of fibrin gel formation.

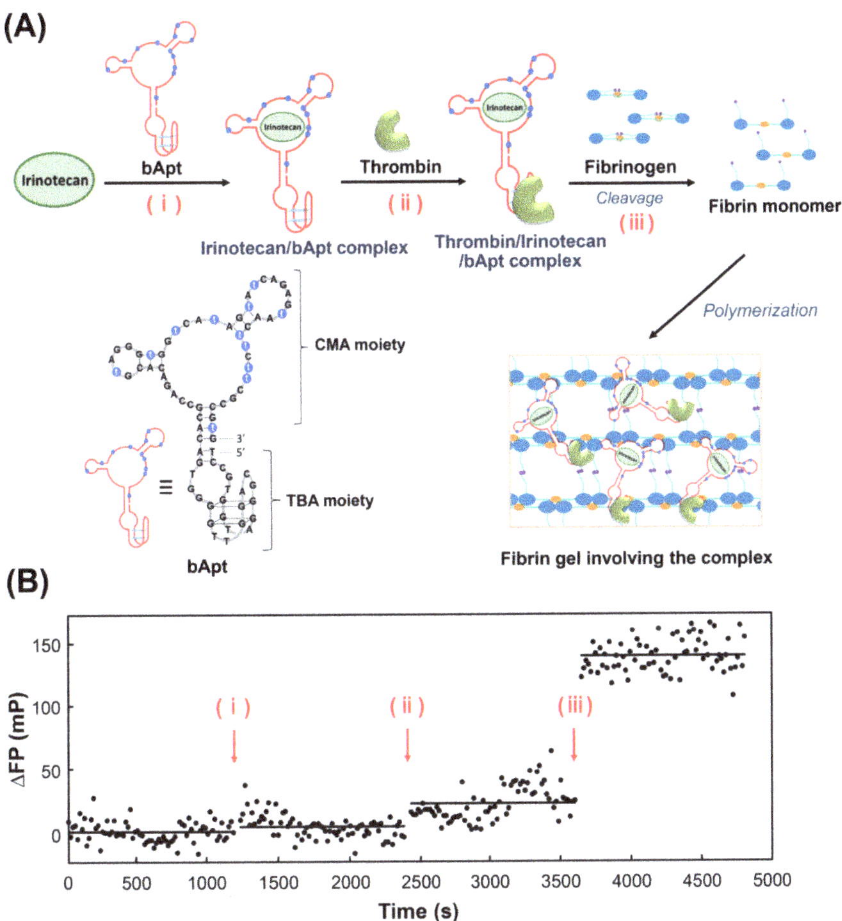

Figure 3. Mechanism of in situ condensation in fibrin gel based on the selective oligonucleotide entrapment in fibrin polymers (SOEF) phenomenon; entrapment of (**A**) the thrombin/irinotecan/bApt complex during the fibrin gel formation. (**B**) The titration curves for irinotecan polarization versus bApt concentration; polarization was monitored at 456 nm using an excitation wavelength of 372 nm.

3.3. Cell Growth Inhibition Assay

Finally, cell growth inhibition assays were performed using HeLa cells to verify whether bApt could be used as a drug carrier for irinotecan. Fluorescent staining uses CellTracker™, which can selectively stain only live cells. From Figure 4D,H, nonemissive images were obtained only in the presence of thrombin/CPT1/bApt complex or thrombin/irinotecan/bApt complex after 48 h incubation. Furthermore, the fibrin gels containing bApt are not cytotoxic [30]. Thus, it was demonstrated that a significant inhibition of cell proliferation occurs in irinotecan and CPT1 under the SOEF phenomenon.

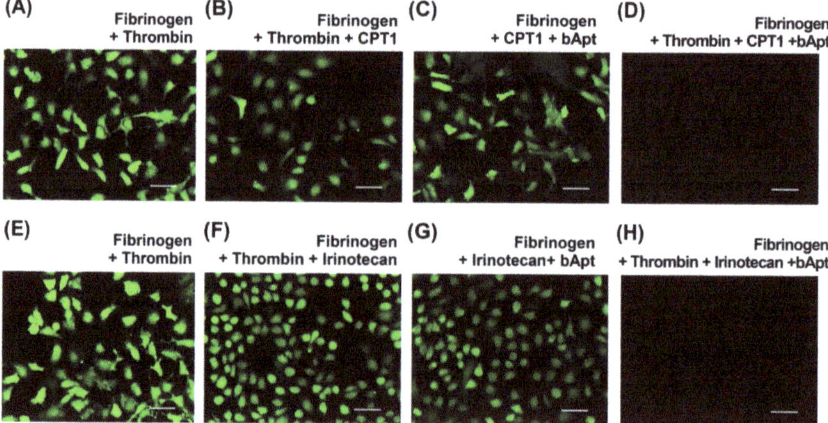

Figure 4. Fluorescence microscopy images of the HeLa cells incubated for 48 h at 37 °C after addition of (**A,E**) fibrinogen and thrombin, (**B,F**) fibrinogen, thrombin and CPT1 or irinotecan, (**C,G**) fibrinogen, CPT1 or irinotecan, and bApt, and (**D,H**) fibrinogen, thrombin, CPT1 or irinotecan, and bApt. The thrombin, CPT1 or irinotecan, and bApt concentrations are 100 nM each and fibrinogen concentration is 1000 nM. Scale bar = 50 μm.

4. Discussion

Fluorescence polarization measurements showed that the CMA part of bApt binds to irinotecan, and further, the irinotecan-binding bApt is efficiently incorporated into fibrin gels by binding the TBA part of bApt to thrombin. Furthermore, cell assays revealed that fibrin gels densely containing irinotecan can cause cancer cell death.

In this study, 29-mer TBA was used at the TBA part of bApt known to have a K_d value of 0.5 nM for thrombin. This binding to thrombin can delay the blood coagulation reaction by about 16 s compared to the normal condition without any inhibitors [24]. Meanwhile, in the presence of TBAs with different sequences of similar lengths (15–38 mer) with K_d values of 0.7–126 nM for thrombin, their blood coagulation reactions were delayed by about 2 s [24]. Therefore, the 29-mer TBA sequence used in this study has a relatively high affinity for thrombin and exhibits inhibitory activity although it does not completely inhibit the activity of thrombin.

Currently, phenomena in which thrombin bound to iron oxide nanoparticles is incorporated into fibrin gels were observed with electron microscopes [31]. Those reports agree with the efficient introduction of bApt into fibrin gels due to the high avidity of the TBA to thrombin and the efficient thrombin-mediated uptake of TBA into fibrin gels observed in this study [26,30].

Moreover, in this study, irinotecan was loaded at the CMA site of bApt instead of CPT1. Recently, we have demonstrated that using bApt carrying CPT1, CPT1 was concentrated in a thin layer of about 50 μm thickness formed by fibrin gels, and thereby the growth inhibitory effect on cancer cells appeared more significantly about 182 times higher than that of CPT1 only [30]. The inhibitory concentration (IC$_{50}$) values of CPT1 entrapped in fibrin gel (7.6 ± 0.36 nM) were approximately 170 times lower than that of CPT1 only in the culture medium (1300 ± 600 nM) [30]. Similarly, in this study, irinotecan was concentrated in a thin layer formed by fibrin gel and assimilated into the cell, resulting in a higher cancer cell growth inhibitory effect when compared with that obtained only using irinotecan. The IC$_{50}$ values of irinotecan for HeLa cell were 1320 ± 130 nM [2,32]. Therefore, the IC$_{50}$ values of irinotecan entrapped in fibrin gel were estimated to be approximately several nM to several tens of nM.

For the application of bifunctional molecular recognition biomolecules to pharmaceuticals, an antibody (emicizumab) with two target recognition sites is currently approved for treating hemophilia [33,34]. Conversely, for using a nucleic acid aptamer, it has been reported that aptamer–aptamer conjugates enable signal control of tyrosine kinase receptor [35] or delivery of doxorubicin to the tumor cell/brain lesions [36], and aptamer–antibody or aptamer–aptamer conjugate can attract T and cancer cells and destroy cancer cells [37–39]; however, they have not been practically used [35–40].

The bApt oligonucleotide has the activity of binding to both the anticancer drug and thrombin, and acts as a carrier for anticancer drugs, which enables the concentration of the anticancer drug in fibrin gels around growing cancer cells. In the future, in the design of bApt, the anticancer drug binding site is compatible with other functionalities. Therefore, the concept and methodology may also be applicable to diseases related to blood coagulation in the future.

5. Conclusions

We demonstrated that the approved drug irinotecan can be selectively and efficiently concentrated into fibrin gels. In the future, the bApt we have developed can be applied as a drug carrier to DDS of anticancer agents around cancer tissues where fibrin production is active. Furthermore, creating drug carriers using a combination of approved drugs and their aptamers and their application to DDS are expected to be applied to diseases related to fibrin, such as blood coagulation disorders.

Supplementary Materials: The following are available online at http://www.mdpi.com/2076-3417/10/23/8755/s1, Figure S1: Metabolism of irinotecan to SN-38, Figure S2: Enzymatic synthesis of bApt through a primer extension reaction, Table S1: Synthetic oligonucleotides used in this study.

Author Contributions: Conceptualization, M.K.; methodology, M.K.; validation, H.F. and Y.K.; formal analysis, H.F. and Y.K.; investigation, H.F. and Y.K.; data curation, H.F. and Y.K.; writing—original draft preparation, H.F. and Y.K.; writing—review and editing, H.F., Y.K. and M.K.; supervision, M.K.; project administration, M.K.; funding acquisition, M.K. All authors have read and agreed to the published version of the manuscript.

Funding: This paper was supported by a Grant for Translational Research Program (H355TS) from the Japan Agency for Medical Research and Development (AMED), and Nihon University Multidisciplinary Research Grant for 2020.

Conflicts of Interest: The authors declare no conflict of interest.

References

1. Shetty, D.; Skorjanc, T.; Olson, M.A.; Trabolsi, A. Self-assembly of stimuli-responsive imine-linked calix[4]arene nanocapsules for targeted camptothecin delivery. *Chem. Commun.* **2019**, *55*, 8876–8879. [CrossRef]

2. Zhou, M.; Liu, M.; He, X.; Yu, H.; Wu, D.; Yao, Y.; Fan, S.; Zhang, P.; Shi, W.; Zhong, B. Synthesis and biological evaluation of novel 10-substituted-7-ethyl-10-hydroxycamptothecin (SN-38) prodrugs. *Molecules* **2014**, *19*, 19718–19731. [CrossRef]

3. Li, X.Q.; Wen, H.Y.; Dong, H.Q.; Xue, W.M.; Pauletti, G.M.; Cai, X.J.; Xia, W.J.; Shi, D.; Li, Y.Y. Self-assembling nanomicelles of a novel camptothecin prodrug engineered with a redox-responsive release mechanism. *Chem. Commun.* **2011**, *47*, 8647–8649. [CrossRef] [PubMed]

4. Pommier, Y. Topoisomerase I inhibitors: Camptothecins and beyond. *Nat. Rev. Cancer* **2006**, *6*, 789–802. [CrossRef] [PubMed]

5. Cheng, K.; Rahier, N.J.; Eisenhauer, B.M.; Gao, R.; Thomas, S.J.; Hecht, S.M. 14-Azacamptothecin: A potent water-soluble topoisomerase I poison. *J. Am. Chem. Soc.* **2005**, *127*, 838–839. [CrossRef] [PubMed]

6. Pizzolato, J.F.; Saltz, L.B. The camptothecins. *Lancet* **2003**, *361*, 2235–2242. [CrossRef]

7. Arimondo, P.B.; Boutorine, A.; Baldeyrou, B.; Bailly, C.; Kuwahara, M.; Hecht, S.M.; Sun, J.S.; Garestier, T.; Helene, C. Design and optimization of camptothecin conjugates of triple helix-forming oligonucleotides for sequence-specific DNA cleavage by topoisomerase I. *J. Biol. Chem.* **2002**, *277*, 3132–3140. [CrossRef]

8. Fujita, K.; Kubota, Y.; Ishida, H.; Sasaki, Y. Irinotecan, a key chemotherapeutic drug for metastatic colorectal cancer. *World J. Gastroenterol.* **2015**, *21*, 12234–12248. [CrossRef]

9. Iyer, L.; King, C.D.; Whitington, P.F.; Green, M.D.; Roy, S.K.; Tephly, T.R.; Coffman, B.L.; Ratain, M.J. Genetic predisposition to the metabolism of irinotecan (CPT-11). Role of uridine diphosphate glucuronosyltransferase isoform 1A1 in the glucuronidation of its active metabolite (SN-38) in human liver microsomes. *J. Clin. Investig.* **1998**, *101*, 847–854. [CrossRef]

10. Kunimoto, T.; Nitta, K.; Tanaka, T.; Uehara, N.; Baba, H.; Takeuchi, M.; Yokokura, T.; Sawada, S.; Miyasaka, T.; Mutai, M. Antitumor activity of 7-ethyl-10-[4-(1-piperidino)-1-piperidino]carbonyloxy-camptothec in, a novel water-soluble derivative of camptothecin, against murine tumors. *Cancer Res.* **1987**, *47*, 5944–5947.

11. Gurav, D.; Varghese, O.P.; Hamad, O.A.; Nilsson, B.; Hilborn, J.; Oommen, O.P. Chondroitin sulfate coated gold nanoparticles: A new strategy to resolve multidrug resistance and thromboinflammation. *Chem. Commun.* **2016**, *52*, 966–969. [CrossRef]

12. Li, J.; Mo, L.; Lu, C.H.; Fu, T.; Yang, H.H.; Tan, W. Functional nucleic acid-based hydrogels for bioanalytical and biomedical applications. *Chem. Soc. Rev.* **2016**, *45*, 1410–1431. [CrossRef] [PubMed]

13. Li, H.; Arroyo-Curras, N.; Kang, D.; Ricci, F.; Plaxco, K.W. Dual-reporter drift correction to enhance the performance of electrochemical aptamer-based sensors in whole blood. *J. Am. Chem. Soc.* **2016**, *138*, 15809–15812. [CrossRef] [PubMed]

14. McKeague, M.; De Girolamo, A.; Valenzano, S.; Pascale, M.; Ruscito, A.; Velu, R.; Frost, N.R.; Hill, K.; Smith, M.; McConnell, E.M.; et al. Comprehensive analytical comparison of strategies used for small molecule aptamer evaluation. *Anal. Chem.* **2015**, *87*, 8608–8612. [CrossRef]

15. Coles, D.J.; Rolfe, B.E.; Boase, N.R.; Veedu, R.N.; Thurecht, K.J. Aptamer-targeted hyperbranched polymers: Towards greater specificity for tumours in vivo. *Chem. Commun.* **2013**, *49*, 3836–3838. [CrossRef]

16. Shoji, A.; Kuwahara, M.; Ozaki, H.; Sawai, H. Modified DNA aptamer that binds the (R)-isomer of a thalidomide derivative with high enantioselectivity. *J. Am. Chem. Soc.* **2007**, *129*, 1456–1464. [CrossRef]

17. Koeller, J.; Surinach, A.; Arikian, S.R.; Zivkovic, M.; Janeczko, P.; Cockrum, P.; Kim, G. Comparing real-world evidence among patients with metastatic pancreatic ductal adenocarcinoma treated with liposomal irinotecan. *Ther. Adv. Med. Oncol.* **2020**, *12*, 1–8. [CrossRef]

18. Kciuk, M.; Marciniak, B.; Kontek, R. Irinotecan-Still an Important Player in Cancer Chemotherapy: A Comprehensive Overview. *Int. J. Mol. Sci.* **2020**, *21*, 4919. [CrossRef] [PubMed]

19. Riccardi, C.; Napolitano, E.; Platella, C.; Musumeci, D.; Montesarchio, D. G-quadruplex-based aptamers targeting human thrombin: Discovery, chemical modifications and antithrombotic effects. *Pharmacol. Ther* **2020**, 107649, in press. [CrossRef] [PubMed]

20. Chabata, C.V.; Frederiksen, J.W.; Sullenger, B.A.; Gunaratne, R. Emerging applications of aptamers for anticoagulation and hemostasis. *Curr. Opin. Hematol.* **2018**, *25*, 382–388. [CrossRef] [PubMed]

21. Takahashi, S.; Sugimoto, N. Volumetric contributions of loop regions of G-quadruplex DNA to the formation of the tertiary structure. *Biophys. Chem.* **2017**, *231*, 146–154. [CrossRef] [PubMed]

22. Kasahara, Y.; Irisawa, Y.; Fujita, H.; Yahara, A.; Ozaki, H.; Obika, S.; Kuwahara, M. Capillary electrophoresis-systematic evolution of ligands by exponential enrichment selection of base- and sugar-modified DNA aptamers: Target binding dominated by 2'-O,4'-C-methylene-bridged/locked nucleic acid primer. *Anal. Chem.* **2013**, *85*, 4961–4967. [CrossRef] [PubMed]

23. Kasahara, Y.; Irisawa, Y.; Ozaki, H.; Obika, S.; Kuwahara, M. 2',4'-BNA/LNA aptamers: CE-SELEX using a DNA-based library of full-length 2'-O,4'-C-methylene-bridged/linked bicyclic ribonucleotides. *Bioorg. Med. Chem. Lett.* **2013**, *23*, 1288–1292. [CrossRef] [PubMed]

24. Tasset, D.M.; Kubik, M.F.; Steiner, W. Oligonucleotide inhibitors of human thrombin that bind distinct epitopes. *J. Mol. Biol.* **1997**, *272*, 688–698. [CrossRef]

25. Mosesson, M.W. Fibrinogen and fibrin structure and functions. *J. Thromb. Haemost.* **2005**, *3*, 1894–1904. [CrossRef]

26. Fujita, H.; Inoue, Y.; Kuwahara, M. Selective incorporation of foreign functionality into fibrin gels through a chemically modified DNA aptamer. *Bioorg. Med. Chem. Lett.* **2018**, *28*, 35–39. [CrossRef]

27. Fujita, H.; Kuwahara, M. Selection of Natural and Base-Modified DNA Aptamers for a Camptothecin Derivative. *Curr. Protoc. Nucleic Acid Chem.* **2016**, *65*, 9.10.1–9.10.19. [CrossRef]

28. Fujita, H.; Imaizumi, Y.; Kasahara, Y.; Kitadume, S.; Ozaki, H.; Kuwahara, M.; Sugimoto, N. Structural and affinity analyses of g-quadruplex DNA aptamers for camptothecin derivatives. *Pharmaceuticals* **2013**, *6*, 1082–1093. [CrossRef]

29. Imaizumi, Y.; Kasahara, Y.; Fujita, H.; Kitadume, S.; Ozaki, H.; Endoh, T.; Kuwahara, M.; Sugimoto, N. Efficacy of base-modification on target binding of small molecule DNA aptamers. *J. Am. Chem. Soc.* **2013**, *135*, 9412–9419. [CrossRef]

30. Kuwahara, M.; Fujita, H.; Kataoka, Y.; Nakajima, Y.; Yamada, M.; Sugimoto, N. In situ condensation of an anti-cancer drug into fibrin gel enabling effective inhibition of tumor cell growth. *Chem. Commun.* **2019**, *55*, 11679–11682. [CrossRef]

31. Ziv-Polat, O.; Skaat, H.; Shahar, A.; Margel, S. Novel magnetic fibrin hydrogel scaffolds containing thrombin and growth factors conjugated iron oxide nanoparticles for tissue engineering. *Int. J. Nanomed.* **2012**, *7*, 1259–1274. [CrossRef]

32. Takara, K.; Kitada, N.; Yoshikawa, E.; Yamamoto, K.; Horibe, S.; Sakaeda, T.; Nishiguchi, K.; Ohnishi, N.; Yokoyama, T. Molecular changes to HeLa cells on continuous exposure to SN-38, an active metabolite of irinotecan hydrochloride. *Cancer Lett.* **2009**, *278*, 88–96. [CrossRef] [PubMed]

33. Kitazawa, T.; Shima, M. Emicizumab, a humanized bispecific antibody to coagulation factors IXa and X with a factor VIIIa-cofactor activity. *Int. J. Hematol.* **2020**, *111*, 20–30. [CrossRef] [PubMed]

34. Sampei, Z.; Igawa, T.; Soeda, T.; Okuyama-Nishida, Y.; Moriyama, C.; Wakabayashi, T.; Tanaka, E.; Muto, A.; Kojima, T.; Kitazawa, T.; et al. Identification and multidimensional optimization of an asymmetric bispecific IgG antibody mimicking the function of factor VIII cofactor activity. *PLoS ONE* **2013**, *8*, e57479. [CrossRef] [PubMed]

35. Ueki, R.; Atsuta, S.; Ueki, A.; Sando, S. Nongenetic Reprogramming of the Ligand Specificity of Growth Factor Receptors by Bispecific DNA Aptamers. *J. Am. Chem. Soc.* **2017**, *139*, 6554–6557. [CrossRef] [PubMed]

36. Macdonald, J.; Denoyer, D.; Henri, J.; Jamieson, A.; Burvenich, I.J.G.; Pouliot, N.; Shigdar, S. Bifunctional Aptamer-Doxorubicin Conjugate Crosses the Blood-Brain Barrier and Selectively Delivers Its Payload to EpCAM-Positive Tumor Cells. *Nucleic Acid Ther.* **2020**, *30*, 117–128. [CrossRef]

37. Yang, Y.; Sun, X.; Xu, J.; Cui, C.; Safari Yazd, H.; Pan, X.; Zhu, Y.; Chen, X.; Li, X.; Li, J.; et al. Circular Bispecific Aptamer-Mediated Artificial Intercellular Recognition for Targeted T Cell Immunotherapy. *ACS Nano* **2020**, *14*, 9562–9571. [CrossRef]

38. Passariello, M.; Camorani, S.; Vetrei, C.; Cerchia, L.; De Lorenzo, C. Novel Human Bispecific Aptamer-Antibody Conjugates for Efficient Cancer Cell Killing. *Cancers* **2019**, *11*, 1268. [CrossRef]

39. Boltz, A.; Piater, B.; Toleikis, L.; Guenther, R.; Kolmar, H.; Hock, B. Bi-specific aptamers mediating tumor cell lysis. *J. Biol. Chem.* **2011**, *286*, 21896–21905. [CrossRef]

40. Riccardi, C.; Napolitano, E.; Musumeci, D.; Montesarchio, D. Dimeric and Multimeric DNA Aptamers for Highly Effective Protein Recognition. *Molecules* **2020**, *25*, 5227. [CrossRef]

Publisher's Note: MDPI stays neutral with regard to jurisdictional claims in published maps and institutional affiliations.

 *applied sciences*

Article

Effective RNA Regulation by Combination of Multiple Programmable RNA-Binding Proteins

Misaki Sugimoto, Akiyo Suda, Shiroh Futaki and **Miki Imanishi ***

Institute for Chemical Research, Kyoto University, Uji, Kyoto 611-0011, Japan;
sugimoto.misaki.25c@st.kyoto-u.ac.jp (M.S.); suda.akiyo.3v@kyoto-u.ac.jp (A.S.); futaki@scl.kyoto-u.ac.jp (S.F.)
* Correspondence: imiki@scl.kyoto-u.ac.jp; Tel.: +81-774-38-3212

Received: 7 September 2020; Accepted: 25 September 2020; Published: 28 September 2020

Featured Application: RNA targeting.

Abstract: RNAs play important roles in gene expression through translation and RNA splicing. Regulation of specific RNAs is useful to understand and manipulate specific transcripts. Pumilio and fem-3 mRNA-binding factor (PUF) proteins, programmable RNA-binding proteins, are promising tools for regulating specific RNAs by fusing them with various functional domains. The key question is: How can PUF-based molecular tools efficiently regulate RNA functions? Here, we show that the combination of multiple PUF proteins, compared to using a single PUF protein, targeting independent RNA sequences at the 3′ untranslated region (UTR) of a target transcript caused cooperative effects to regulate the function of the target RNA by luciferase reporter assays. It is worth noting that a higher efficacy was achieved with smaller amounts of each PUF expression vector introduced into the cells compared to using a single PUF protein. This strategy not only efficiently regulates target RNA functions but would also be effective in reducing off-target effects due to the low doses of each expression vector.

Keywords: RNA binding protein; PUF; RNA regulation

---

1. Introduction

Manipulation of specific RNAs is important for elucidating the roles of specific RNAs and for therapeutic purposes as represented by RNA interference and antisense strategies targeting specific mRNAs. Other than such complementary oligonucleotides, sequence-specific RNA-binding proteins, Pumilio and fem-3 mRNA-binding factor (PUF) proteins and pentatricopeptide repeat (PPR) proteins, and clustered regularly interspaced short palindromic repeats (CRISPR)-based systems, including dCas13 and dCas9-PAMer complexed with guide RNA, have been used for the customized sequence-specific regulation of mRNAs [1–6]. By combining them with various functional domains, such as translational regulation factors [7,8], splicing factors [9–11], RNA editing enzymes [12], and RNA (de)methylation enzymes [13–18], targeted control of RNAs can be achieved. The RNA-binding domains of PUF proteins consist of eight highly homologous structural repeats (repeats 1 to 8), flanked by N- and C-terminal repeats (repeats N and C) [19] (Figure 1a). Each repeat recognizes an RNA base through direct interaction between RNA bases and side chains of amino acids at the 12th and 16th positions [20] (Figure 1b). By changing the amino acid pairs of a PUF repeat, artificial RNA-binding domains corresponding to various 8-nucleotide (nt) RNA sequences can be designed [9,21,22]. As the length of RNA recognized by PUFs (8-nt) is not enough to specify the target RNAs within a huge transcriptome, efforts to increase the length of the RNA-binding sequences have been applied by increasing the number of PUF repeats [21,23–26]. Previous reports also showed the assembly of two PUF proteins in adjacent regions on the RNA to image specific transcripts [27–30]. By contrast, Cao et al. reported that increasing the number of the PUF-binding sequences in reporter mRNAs results in the effective regulation of reporter genes compared to reporter mRNAs having a single PUF-binding sequence [7]. This was also supported through the regulation of endogenous mRNA by PUFs fused to the RNA decay factor, tristetraprolin (TTP). Abil et al. designed artificial PUF proteins, targeting the 3′ untranslated region (UTR) of vascular endothelial growth factor A *(VEGFA)* mRNA. TTP-PUF fusion proteins effectively repressed VEGFA production when the particular 8-nt PUF-binding sequence existed at multiple places within the 3′ UTR of the target mRNA [31]. However, it is unusual that many repeated sequences are at the control region of a single mRNA. Instead of increasing the number of target RNA sequences, we intended to use four different PUF proteins, corresponding target RNA sequences that exist at the control region of a reporter RNA (Figure 1c). In this study, we expressed PUF RNA-binding proteins fused with TTP in mammalian cells and enabled the efficient repression of the reporter protein expression by using multiple TTP-PUF proteins.

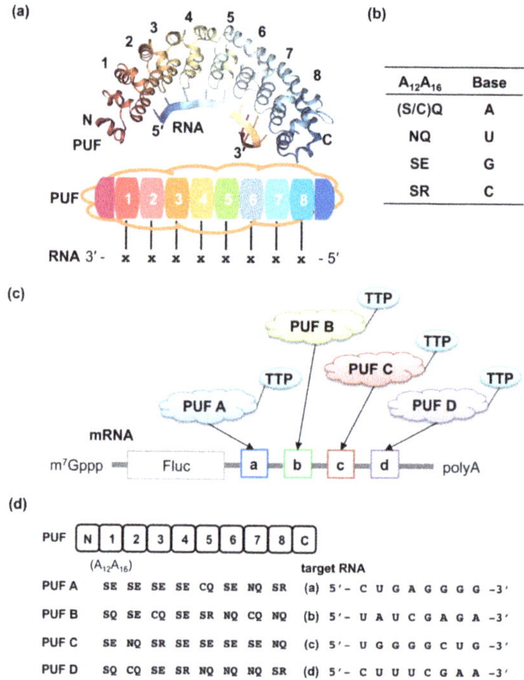

Figure 1. Programmable RNA binding protein, PUF, and regulation of RNA functions. (**a**) Structure of a PUF protein bound to RNA [PDB ID: 1M8Y] and schematic representation of RNA recognition by a PUF protein. (**b**) 12th and 16th amino acids (A$_{12}$ and A$_{16}$) in each PUF repeat interact with an RNA base. (**c**) Schematic representation of mRNA regulation by multiple artificial PUF proteins fused with TTP. (**d**) Design of PUF proteins, PUF (A)–(D), and their recognition RNA sequences, (a)–(d), used in this research.

2. Materials and Methods

2.1. Plasmid Construction

PUF expression vectors for *E. coli* expression were constructed by a Golden Gate Assembly using the PUF Assembly Kit (Addgene #1000000051) and pET28-GG-PUF as the receiving vector [31]. The amino acid sequences at the 12th and 16th positions of each repeat are shown in Figure 1d.

TTP-PUF expression vectors for mammalian cells were constructed by Golden Gate Assembly using the PUF Assembly Kit and pCMV-TTP(C147R)-GG-PUF as the receiving vector.

Luciferase reporter vectors were constructed, as described in the text that follows. The DNA oligonucleotides containing the corresponding sequences for 4 × RNA (a)–(d), or 1 × (a-b-c-d) and their complementary DNA oligonucleotides (Eurofins genomics) were phosphorylated by T4 polynucleotide kinase (NEB) and annealed; they were inserted into the 3′ UTR of the pmirGLO vector (Promega). The sequences of 4 × RNA (a)–(d), or 1 × (a-b-c-d) are shown in Supplementary Figure S1.

2.2. Protein Expression and Purification

PUF proteins were expressed in *E. coli* BL21 (DE3) cells (Nippon gene) and purified as described before [23]. Briefly, protein expression was induced by adding 0.1 mM isopropyl β-D-1-thiogalactopyranoside (IPTG), and *E. coli* cells were incubated at 18 °C overnight. PUF proteins were purified from the *E. coli*-soluble fraction using Ni-NTA HisTrap FF column (GE Healthcare) chromatography followed by ultrafiltration by an Amicon Ultra (10 kDa NMWL) (Merck Millipore).

2.3. Electrophoretic Mobility Shift Assay (EMSA)

RNA oligonucleotides, labeled with 6-FAM at the 5′-end, were obtained from FASMAC. Labeled-RNA (10 nM) and three-fold serially diluted concentrations of purified PUF proteins (0–300 nM) were mixed in the reaction buffer containing 10 mM HEPES, 50 mM KCl, 0.5 mM EDTA, 2 mM DTT, 0.01% Tween20, 0.1 mg/mL BSA, 0.02 U/μL RNasin Plus RNase Inhibitor (Promega), and 2.5% Ficoll. The mixture was incubated at 25 °C for 30 min. Free RNA and bound RNA were separated via gel electrophoresis using 8% nondenaturing polyacrylamide gels in 0.5× TBE at 4 °C. After electrophoresis, fluorescently labeled RNAs were detected using the Typhoon scanner RGB system (GE Healthcare). The peak intensity for bound and free labeled-RNA was measured using the ImageQuant TL software (GE Healthcare). The fractions of bound RNA were plotted against the protein concentrations. For calculating equilibrium dissociation constants (K_d), the plotted data were fit to the 1:1 binding equation using the Kaleida Graph software (Hulinks) as described before [23].

2.4. Cell Culture

NIH3T3 cells (RIKEN cell bank; RCB1862) were maintained in Dulbecco's modified Eagle's medium (DMEM) (Fujifilm Wako Pure Chemical) containing 10% FBS.

2.5. Western Blotting Assay

Cells were lysed using the RIPA buffer, and cell lysates were subjected to SDS-PAGE. The proteins were detected using mouse anti-FLAG antibody (Sigma Aldrich; St. Louis, MO, USA, F3165) or mouse anti-β-actin antibody (Sigma Aldrich; AC-74) as primary antibodies, and Goat anti-mouse IgG antibody (GeneTex) as the Horseradish peroxidase (HRP)-labeled secondary antibody. The bands were visualized using an Amersham ECL Prime (GE Healthcare, Chicago, IL, USA).

2.6. Luciferase Reporter Assay

Here, 1.5×10^4 of NIH3T3 cells were seeded onto 96-well plates the day before transfection. Transfection was performed using Lipofectamine LTX (Thermo Fisher Scientific, Waltham, MA, USA). Transfection mixtures contained 100 ng TTP-PUF expression vector and 1 ng reporter vector. For control, pCMV-TTP(C147R)-GG-PUF (Addgene #1000000051), in which the TTP gene but no PUF-encoding sequences were inserted, was used instead of TTP-PUF expression vectors. Then, 24 h after transfection, cells were lysed using Passive Lysis Buffer (Promega). Firefly luciferase and *Renilla* luciferase activities were measured using the Dual-Luciferase Reporter Assay System (Promega) with measurements taken on a GloMax-Multi Detection System (Promega). Firefly luciferase activity (Fluc) was normalized to *Renilla* luciferase activity (Rluc), and then relative luciferase activities (RLAs), compared to control samples, were calculated (Equation (1)). The degree of repression by a TTP-PUF relative to the control was indicated by the reciprocal of the relative luciferase activity (Equation (2)).

$$\text{RLA}_{\text{(TTP-PUF)}} = [\text{Fluc}_{\text{(TTP-PUF)}}/\text{Rluc}_{\text{(TTP-PUF)}}]/[\text{Fluc}_{\text{(control)}}/\text{Rluc}_{\text{(control)}}] \tag{1}$$

$$\text{Fold repression} = 1/\text{RLA} \tag{2}$$

3. Results and Discussion

3.1. RNA-Binding Properties of Programmable RNA-Binding Proteins

Four PUF proteins, PUF (A), (B), (C), and (D), targeting different RNA sequences, RNA (a), (b), (c) and (d), respectively, were designed by assembling PUF repeats corresponding to each RNA base in order (Figure 1d). EMSA revealed the high affinity and specificity of PUF proteins to target RNA sequences. In this assay, the binding of PUF proteins reduced the mobility of the target RNAs, allowing the evaluation of RNA-binding abilities of proteins. PUF (A)–(D) proteins were expressed in *E. coli* cells and purified via affinity chromatography using a Ni-NTA column. PUF (A), (B), (C), and (D) showed band-shifts for RNA (a), (b), (c), and (d), respectively (Figure 2, Figure S2a). The apparent dissociation constants between each PUF protein and corresponding RNA were 10.1 nM (PUF (A)/RNA (a)), 5.3 nM (PUF (B)/RNA (b)), <1 nM (PUF (C)/RNA (c)), and 9.1 nM (PUF (D)/RNA (d)) (Table 1, Figure S2b). Although the K_d values were somewhat different between PUFs, they were within the reported range of engineered PUF proteins (10^{-10}–10^{-8} M) [23,31]. Importantly, all PUF (A)–(D) proteins showed high specificity to their corresponding RNA sequences (Figure S2a). RNA (a) showed clear shift bands only when incubated with PUF (A) but not with the other three PUFs. Similarly, RNA (b)–(d) showed clear shift bands only when incubated with their corresponding PUFs.

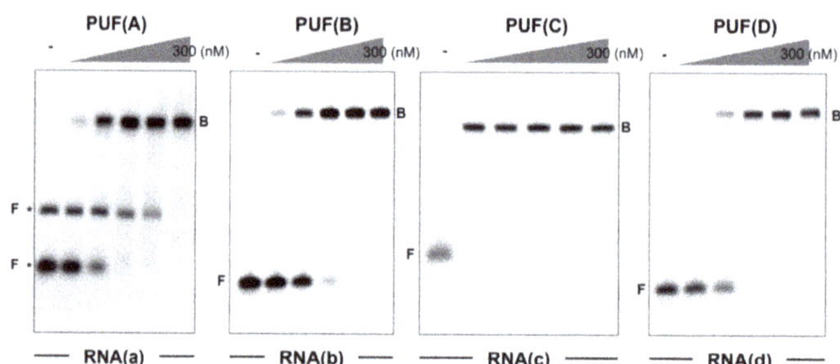

Figure 2. RNA binding of engineered PUF proteins to target RNAs. In the electrophoretic mobility shift assay (EMSA), RNA oligonucleotides, RNA (a)–(d), were incubated with corresponding PUF proteins, PUF (A)–(D), respectively, at the following concentrations (left to right): 0, 3.7, 11, 33, 100, 300 nM, and the mixtures were electrophoresed in native PAGE. "F" and "B" indicate protein-free RNA and protein-bound RNA, respectively. RNA (a) showed two protein-free bands (shown by *).

Table 1. RNA binding of engineered PUF proteins to specific RNA sequences.

	K_D (nM) [1]			
	RNA(a)	RNA(b)	RNA(c)	RNA(d)
PUF (A)	10.1 ± 1.4	n.d.	n.d.	n.d.
PUF (B)	n.d.	5.3 ± 2.7	n.d.	n.d.
PUF (C)	n.d.	n.d.	<1	n.d.
PUF (D)	n.d.	n.d.	n.d.	9.1 ± 1.4

[1] mean ± SD, n.d.; not detected.

3.2. Specific RNA Binding of TTP-PUF within Cells

The RNA binding specificities of PUF (A)–(D) for their corresponding target RNAs were also confirmed within living cells using luciferase reporter assays (Figure 3, Figure S1a,b). PUF (A)–(D) fused with tristetraprolin (TTP), an mRNA decay factor, were designed as described before [23,31]. As a control, a TTP expression vector that expresses only TTP without fusing with PUFs was used to estimate the effect of PUF [23,31]. Previous reports showed that a TTP-PUF fusion protein, which can bind to the 3′ UTR of the luciferase reporter gene transcript, represses luciferase activity [23,31]. The degree of luciferase repression reflects the RNA binding of TTP-PUF fusion proteins. Here, we prepared reporter vectors that express both firefly luciferase and *Renilla* luciferase genes driven by a PGK and an SV40 promoter, respectively (Figure S1a,b). A four-times repeat sequence of the PUF target sequences, 4 × (a), (b), (c) or (d), was inserted at the 3′ UTR of the firefly luciferase gene (Figure 3a). Each TTP-PUF expression or control vectors and a reporter vector were transfected into NIH3T3 cells. In this reporter system, we expected the firefly luciferase activity to be affected by TTP-PUFs, and *Renilla* luciferase activity to reflect the transfection efficiencies. Thus, the firefly luciferase activity, under the expression of each TTP-PUF, was normalized to the *Renilla* luciferase activity of the same sample. The normalized luciferase activity was compared to that of the control sample to evaluate the degree of repression. The expression of the TTP-PUF proteins was confirmed via Western blotting assay (Figure S1c). As shown in Figure 3b, TTP-PUF (A), (B), (C), and (D) showed a 4–6-fold repression in luciferase gene expression with corresponding RNA-binding sequences at the 3′ UTR. By contrast, they had little effect on luciferase gene expression without corresponding binding sites at the 3′ UTR. None of the TTP-PUFs repressed the luciferase activity in the no binding site (nbs) reporter. These results indicated that the PUF proteins specifically bound to the corresponding RNA sequences at the 3′ UTR of the luciferase gene even within cells.

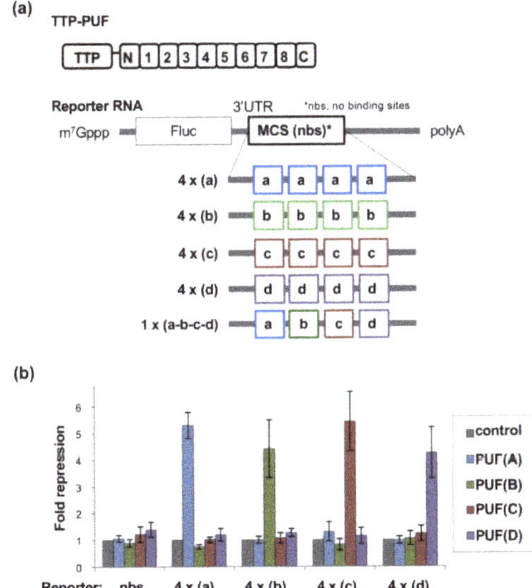

Figure 3. Specific binding of engineered PUF proteins to target RNAs in living cells. (**a**) Schematic representation of TTP-PUF proteins and reporter RNAs. TTP-PUF expression plasmids and luciferase reporter plasmids were co-transfected into NIH3T3 cells. The luciferase reporter genes were transcribed, and reporter RNAs with PUF-binding sequences at the 3′ untranslated region (UTR) region should have been produced in the cells. (**b**) Fold repression of luciferase reporter activities by TTP-PUF proteins. Data are presented as mean ± SD.

3.3. Effective Repression of Reporter Genes with Not-Repetitive Sequences by Multiple TTP-PUF Proteins

It is unusual that four or more times sequences exist at the regulatory region of single transcripts. Instead of targeting repeat sequences at the 3′ UTR, the synergistic effect of multiple TTP-PUFs was shown through the targeting of multiple RNA sequences at the 3′ UTR. As a model of transcripts, a reporter vector with a 1 × (a-b-c-d) sequence at the 3′ UTR of the firefly luciferase gene was designed, in which only one binding sequence for each PUF (A)–(D) existed at the 3′ UTR (Figure 3a). In contrast to the 4–6-fold repression of the luciferase genes with 4 × binding sequences, each TTP-PUF (A), (C), or (D) showed only about a 1.5-fold repression (Figure 4, lanes 3,5,6), and TTP-PUF (B) alone did not repress luciferase activity (Figure 4, lane 4), maybe because of the low TTP-PUF (B) expression compared to other TTP-PUF proteins (Figure S1c).

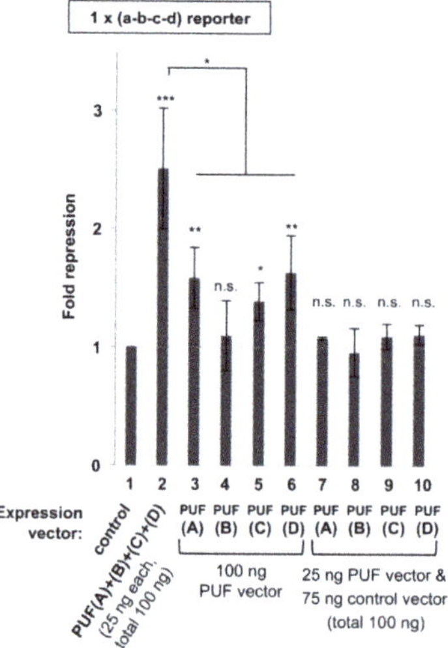

Figure 4. Combination of TTP-PUF (A)–(D) with 1 × (a-b-c-d) reporter efficiently repressed the luciferase reporter activities at the 3′ UTR of the luciferase mRNA. NIH3T3 cells were co-transfected with 1 × (a-b-c-d) reporter (1 ng) and 100 ng (total) of the following expression vectors. Lane 1, control vector; lane 2, mixture of TTP-PUF (A), (B), (C), and (D) (25 ng each); lanes 3–6, 100 ng of TTP-PUF (A) (lane 3), (B) (lane 4), (C) (lane 5), (D) (lane 6); lanes 7–10, 75 ng of control vector and 25 ng of TTP-PUF (A) (lane 7), (B) (lane 8), (C) (lane 9), (D) (lane 10). Data are presented as mean ± SD. (***, $p < 0.005$; **, $p < 0.01$; *, $p < 0.05$; n.s. = not significant).

When NIH3T3 cells were co-transfected with all of the expression vectors (TTP-PUF (A)–(D); 25 ng each) and the 1 × (a-b-c-d) reporter vector, significant repression (2.5-fold) of luciferase activity was observed, even though the amount of each expression vector (25 ng each) was one-fourth of that when used alone (100 ng) (Figure 4, lane 2). The mixture of four TTP-PUF expression vectors had little effect on the reporter lacking a PUF-binding sequence (Figure S3). Furthermore, when the cells were transfected with 25 ng of one of the four TTP-PUF vectors, the same amount used for the mixture, none of them showed significant repression of the 1 × (a-b-c-d) reporter (Figure 4, lanes 7–10). These results indicated that the luciferase activity of the 1 × (a-b-c-d) reporter vector was repressed by the sequence-specific binding of TTP-PUF proteins to their binding sites at the 3′ UTR. Additionally, the combination of four TTP-PUF proteins, targeting different RNA sequences, caused effective repression of luciferase activity compared to the use of a single TTP-PUF protein. Differences in repression ratios were observed between reporters with 4 × binding sequences by the corresponding single TTP-PUF (4–6-fold) (Figure 3b) and the 1 × (a-b-c-d) reporter by the mixture of TTP-PUF (A)–(D) (2.5-fold) (Figure 4). One plausible reason for the difference is avidity effects for the

4 × proximity binding sites. The overall RNA structure of the PUF binding sequences can affect the binding of TTP-PUFs. Further optimization will increase the synergistic effects of multiple functional PUF proteins.

So far, synergistic gene activation or repression was shown by a combination of multiple artificial transcription regulators based on transcription activator-like effectors (TALEs) and CRISPR-dCas9 [32,33]. The results using a combination of multiple RNA binding proteins are in accordance with these reports targeting DNA.

4. Conclusions

In this study, four independent PUF proteins were designed and shown to recognize the target RNA sequence with high specificity both in vitro and in cellulo. By combining these four PUF proteins, we showed the use of multiple TTP-PUF proteins in the effective repression of target gene expression, even though the amount of each transfected expression vector was smaller than that of a single PUF protein. Although further verification is needed to understand the off-target effects, we expect that this strategy will reduce off-target effects because of a lower dose of expression vectors. Sequence-specific RNA-binding molecules, like PUFs, can exert desired functions at specific RNA regions when fused with various functional domains. As the importance of RNAs in life science becomes clear, there is an increasing demand for methods to control specific RNAs. Combination of multiple PUF proteins would be a promising strategy to control endogenous RNAs effectively. Furthermore, this concept might be applied to other RNA-targeting molecular tools, such as PPR proteins and CRISPR-Cas13 based systems.

Supplementary Materials: The following are available online at http://www.mdpi.com/2076-3417/10/19/6803/s1, Figure S1: Plasmids used for luciferase reporter assays and expression of TTP-PUFs, Figure S2: Representative results of EMSA, Figure S3: Effect of the mixture of TTP-PUFs on the nbs reporter.

Author Contributions: Conceptualization, M.I. and M.S.; investigation, M.S. and A.S.; writing—M.I. and S.F. All authors have read and agreed to the published version of the manuscript.

Funding: This research was funded by JSPS/KAKENHI (19H02850 to M.I.).

Acknowledgments: We kindly thank Huimin Zhao for plasmids to construct PUFs (Addgene #1000000051).

Conflicts of Interest: The authors declare no conflict of interest.

References

1. Pei, Y.; Lu, M. Programmable RNA manipulation in living cells. *Cell. Mol. Life Sci.* **2019**, *76*, 4861–5867. [CrossRef] [PubMed]

2. Hall, T.M. De-coding and re-coding RNA recognition by PUF and PPR repeat proteins. *Curr. Opin. Struct. Biol.* **2016**, *36*, 116–121. [CrossRef] [PubMed]

3. Chen, Y.; Varani, G. Engineering RNA-binding proteins for biology. *FEBS J.* **2013**, *280*, 3734–3754. [CrossRef] [PubMed]

4. Terns, M.P. CRISPR-Based Technologies: Impact of RNA-Targeting Systems. *Mol. Cell* **2018**, *72*, 404–412. [CrossRef] [PubMed]

5. Filipovska, A.; Rackham, O. Designer RNA-binding proteins: New tools for manipulating the transcriptome. *RNA Biol.* **2011**, *8*, 978–983. [CrossRef]

6. Mackay, J.P.; Font, J.; Segal, D.J. The prospects for designer single-stranded RNA-binding proteins. *Nat. Struct. Mol. Biol.* **2011**, *18*, 256–261. [CrossRef]

7. Cao, J.; Arha, M.; Sudrik, C.; Schaffer, D.V.; Kane, R.S. Bidirectional regulation of mRNA translation in mammalian cells by using PUF domains. *Angew. Chem. Int. Ed.* **2014**, *53*, 4900–4904. [CrossRef]

8. Campbell, Z.T.; Valley, C.T.; Wickens, M. A protein-RNA specificity code enables targeted activation of an endogenous human transcript. *Nat. Struct. Mol. Biol.* **2014**, *21*, 732–738. [CrossRef]

9. Dong, S.; Wang, Y.; Cassidy-Amstutz, C.; Lu, G.; Bigler, R.; Jezyk, M.R.; Li, C.; Hall, T.M.; Wang, Z. Specific and modular binding code for cytosine recognition in Pumilio/FBF (PUF) RNA-binding domains. *J. Biol. Chem.* **2011**, *286*, 26732–26742. [CrossRef]

10. Wang, Y.; Cheong, C.G.; Hall, T.M.; Wang, Z. Engineering splicing factors with designed specificities. *Nat. Methods* **2009**, *6*, 825–830. [CrossRef]

11. Du, M.; Jillette, N.; Zhu, J.J.; Li, S.; Cheng, A.W. CRISPR artificial splicing factors. *Nat. Commun.* **2020**, *11*, 2973. [CrossRef] [PubMed]

12. Cox, D.B.T.; Gootenberg, J.S.; Abudayyeh, O.O.; Franklin, B.; Kellner, M.J.; Joung, J.; Zhang, F. RNA editing with CRISPR-Cas13. *Science* **2017**, *358*, 1019–1027. [CrossRef] [PubMed]

13. Liu, X.M.; Zhou, J.; Mao, Y.; Ji, Q.; Qian, S.B. Programmable RNA N^6-methyladenosine editing by CRISPR-Cas9 conjugates. *Nat. Chem. Biol.* **2019**, *15*, 865–871. [CrossRef] [PubMed]

14. Shinoda, K.; Suda, A.; Otonari, K.; Futaki, S.; Imanishi, M. Programmable RNA methylation and demethylation using PUF RNA binding proteins. *Chem. Commun.* **2020**, *56*, 1365–1368. [CrossRef]

15. Rau, K.; Roesner, L.; Rentmeister, A. Sequence-specific m^{6}A demethylation in RNA by FTO fused to RCas9. *RNA* **2019**, *25*, 1311–1323. [CrossRef]

16. Wilson, C.; Chen, P.J.; Miao, Z.; Liu, D.R. Programmable m^{6}A modification of cellular RNAs with a Cas13-directed methyltransferase. *Nat. Biotechnol.* **2020**. [CrossRef]

17. Li, J.; Chen, Z.; Chen, F.; Xie, G.; Ling, Y.; Peng, Y.; Lin, Y.; Luo, N.; Chiang, C.M.; Wang, H. Targeted mRNA demethylation using an engineered dCas13b-ALKBH5 fusion protein. *Nucleic Acids Res.* **2020**, *48*, 5684–5694. [CrossRef]

18. Zhao, J.; Li, B.; Ma, J.; Jin, W.; Ma, X. Photoactivatable RNA N^6-methyladenosine editing with CRISPR-Cas13. *Small* **2020**, *16*, e1907301. [CrossRef]

19. Wang, X.; Zamore, P.D.; Hall, T.M. Crystal structure of a Pumilio homology domain. *Mol. Cell* **2001**, *7*, 855–865. [CrossRef]

20. Wang, X.; McLachlan, J.; Zamore, P.D.; Hall, T.M. Modular recognition of RNA by a human pumilio-homology domain. *Cell* **2002**, *110*, 501–512. [CrossRef]

21. Filipovska, A.; Razif, M.F.; Nygard, K.K.; Rackham, O. A universal code for RNA recognition by PUF proteins. *Nat. Chem. Biol.* **2011**, *7*, 425–427. [CrossRef] [PubMed]

22. Cheong, C.G.; Hall, T.M. Engineering RNA sequence specificity of Pumilio repeats. *Proc. Natl. Acad. Sci. USA* **2006**, *103*, 13635–13639. [CrossRef] [PubMed]

23. Shinoda, K.; Tsuji, S.; Futaki, S.; Imanishi, M. Nested PUF proteins: Extending target RNA elements for gene regulation. *Chembiochem* **2018**, *19*, 171–176. [CrossRef] [PubMed]

24. Adamala, K.P.; Martin-Alarcon, D.A.; Boyden, E.S. Programmable RNA-binding protein composed of repeats of a single modular unit. *Proc. Natl. Acad. Sci. USA* **2016**, *113*, E2579–E2588. [CrossRef]

25. Zhao, Y.Y.; Mao, M.W.; Zhang, W.J.; Wang, J.; Li, H.T.; Yang, Y.; Wang, Z.; Wu, J.W. Expanding RNA binding specificity and affinity of engineered PUF domains. *Nucleic Acids Res.* **2018**, *46*, 4771–4782. [CrossRef]

26. Criscuolo, S.; Gatti Iou, M.; Merolla, A.; Maragliano, L.; Cesca, F.; Benfenati, F. Engineering REST-specific synthetic PUF proteins to control neuronal gene expression: A combined experimental and computational study. *ACS Synth. Biol.* **2020**, *9*, 2039–2054. [CrossRef]

27. Ozawa, T.; Natori, Y.; Sato, M.; Umezawa, Y. Imaging dynamics of endogenous mitochondrial RNA in single living cells. *Nat. Methods* **2007**, *4*, 413–419. [CrossRef]

28. Yoshimura, H.; Inaguma, A.; Yamada, T.; Ozawa, T. Fluorescent probes for imaging endogenous beta-actin mRNA in living cells using fluorescent protein-tagged pumilio. *ACS Chem. Biol.* **2012**, *7*, 999–1005. [CrossRef]

29. Yamada, T.; Yoshimura, H.; Inaguma, A.; Ozawa, T. Visualization of nonengineered single mRNAs in living cells using genetically encoded fluorescent probes. *Anal. Chem.* **2011**, *83*, 5708–5714. [CrossRef]

30. Ozawa, T.; Umezawa, Y. Genetically-encoded fluorescent probes for imaging endogenous mRNA in living cells. *Methods Mol. Biol.* **2011**, *714*, 175–188.

31. Abil, Z.; Denard, C.A.; Zhao, H. Modular assembly of designer PUF proteins for specific post-transcriptional regulation of endogenous RNA. *J. Biol. Eng.* **2014**, *8*, 7. [CrossRef] [PubMed]

32. Nomura, W.; Matsumoto, D.; Sugii, T.; Kobayakawa, T.; Tamamura, H. Efficient and orthogonal transcription regulation by chemically inducible artificial transcription factors. *Biochemistry* **2018**, *57*, 6452–6459. [CrossRef] [PubMed]
33. Perez-Pinera, P.; Ousterout, D.G.; Brunger, J.M.; Farin, A.M.; Glass, K.A.; Guilak, F.; Crawford, G.E.; Hartemink, A.J.; Gersbach, C.A. Synergistic and tunable human gene activation by combinations of synthetic transcription factors. *Nat. Methods* **2013**, *10*, 239–242. [CrossRef] [PubMed]

Article

An RNA Triangle with Six Ribozyme Units Can Promote a *Trans*-Splicing Reaction through Trimerization of Unit Ribozyme Dimers

Junya Akagi [1], Takahiro Yamada [1], Kumi Hidaka [2], Yoshihiko Fujita [3], Hirohide Saito [3], Hiroshi Sugiyama [2,4], Masayuki Endo [2,4], Shigeyoshi Matsumura [1] and Yoshiya Ikawa [1,*]

1   Department of Chemistry, Graduate School of Science and Engineering, University of Toyama, Gofuku 3190, Toyama 930-8555, Japan; clspinachaj0759@gmail.com (J.A.); m2041213@ems.u-toyama.ac.jp (T.Y.); smatsumu@sci.u-toyama.ac.jp (S.M.)
2   Department of Chemistry, Graduate School of Science, Kyoto University, Kyoto 606-8502, Japan; hidaka.kumi.8z@kyoto-u.ac.jp (K.H.); hs@kuchem.kyoto-u.ac.jp (H.S.); endo@kuchem.kyoto-u.ac.jp (M.E.)
3   Department of Life Science Frontiers, Center for iPS Cell Research and Application (CiRA), Kyoto University, Kyoto 606-8502, Japan; yoshihiko.fujita@cira.kyoto-u.ac.jp (Y.F.); hirohide.saito@cira.kyoto-u.ac.jp (H.S.)
4   Institute for Integrated Cell-Material Sciences, Kyoto University, Kyoto 606-8502, Japan
*   Correspondence: yikawa@sci.u-toyama.ac.jp; Tel.: +81-76-445-6599

Abstract: Ribozymes are catalytic RNAs that are attractive platforms for the construction of nanoscale objects with biological functions. We designed a dimeric form of the *Tetrahymena* group I ribozyme as a unit structure in which two ribozymes were connected in a tail-to-tail manner with a linker element. We introduced a kink-turn motif as a bent linker element of the ribozyme dimer to design a closed trimer with a triangular shape. The oligomeric states of the resulting ribozyme dimers (kUrds) were analyzed biochemically and observed directly by atomic force microscopy (AFM). Formation of kUrd oligomers also triggered *trans*-splicing reactions, which could be monitored with a reporter system to yield a fluorescent RNA aptamer as the *trans*-splicing product.

Keywords: catalytic RNA; group I ribozyme; RNA nanostructure; RNA nanotechnology; RNA-protein complex; *trans*-splicing

Citation: Akagi, J.; Yamada, T.; Hidaka, K.; Fujita, Y.; Saito, H.; Sugiyama, H.; Endo, M.; Matsumura, S.; Ikawa, Y. An RNA Triangle with Six Ribozyme Units Can Promote a *Trans*-Splicing Reaction through Trimerization of Unit Ribozyme Dimers. *Appl. Sci.* **2021**, *11*, 2583. https://doi.org/10.3390/app11062583

Academic Editors: Tamaki Endoh and Chih-Ching Huang

Received: 29 November 2020
Accepted: 11 March 2021
Published: 14 March 2021

Publisher's Note: MDPI stays neutral with regard to jurisdictional claims in published maps and institutional affiliations.

1. Introduction

Folded polypeptides work not only in monomeric states but also in assembled states, including homo- and heterooligomers [1–3]. In the course of enzyme evolution, symmetrical protein homooligomers with polygonal and polyhedral shapes emerged from their monomeric ancestors, presumably due to the advantages in their biological properties over the respective monomeric states, including enzymatic activities and structural stability [1–3]. Interfaces between monomer units play a key role in the assembly of protein oligomers. The artificial design of protein oligomers with symmetrical shapes has been explored with the use of protein–protein interaction interfaces extracted from naturally occurring protein oligomers, generating the field of protein nanostructure research [4–6]. Recent challenges in this field include de novo design of artificial protein–protein interaction interfaces that can serve as modular structural units for versatile protein nanostructure design [7,8]. Among the various forms of symmetrical protein homooligomers, symmetrical protein homodimers can be regarded as the simplest molecular parts for the rational design and assembly of protein structures [9–11].

RNA is a biopolymer that can fold into defined 3D structures with biological functions. Structured RNAs are known to perform many types of protein-like functions, including roles as catalysts and receptors [12–14]. Large and complex 3D RNA structures frequently formed through noncovalent assembly of structural modules, which can fold locally within a single polynucleotide chain [15]. Intramolecular assembly of RNA structural modules is supported and maintained by RNA–RNA interaction interfaces [15,16]. Their association is usually strong enough to reconstitute the catalytic activity in a multimolecular format, in which separately prepared modular RNAs are assembled noncovalently by RNA–RNA interaction interfaces [17–25].

We recently engineered an RNA–RNA interaction interface of the *Tetrahymena* group I ribozyme to design a prototype of ribozyme modules (abbreviated as RzM-0, Figure 1) with self-dimerizing capability [26,27]. The RNA–RNA interface in the RzM-0 homodimer is supported by two sets of tertiary interactions between the ΔP5 core ribozyme module and P5abc module, which are connected by flexible linker elements (P5-P5a linker) in the parent ribozyme and form strong intramolecular associations (WT group I ribozyme, Figure 1). We rationally replaced the flexible P5-P5a with a rigid duplex to prevent intramolecular ΔP5-P5abc assembly (Figure S1C), resulting in homodimerization of the resulting variant (RzM-0 in Figure 1) in a head-to-head manner by forming RzM:RzM interface **0:0** [26,27]. A pair of symmetrical ΔP5-P5abc interfaces yielding RzM:RzM homodimer was then engineered to asymmetrical recognition interfaces, such as RzM:RzM interface **1:1′**, yielding RzM:RzM heterodimers (Figure 1). We then connected a pair of RzM monomers in a tail-to-tail manner at their P6b region, enabling the resulting covalent dimer (termed unit ribozyme dimer and abbreviated as Urd) to homooligomerize in a directional manner (Figure 1) [28]. Unit ribozyme dimers (Urds) homooligomerized to form an open chain assembly (Figure 1). The extents of oligomerization were programmable by designing the specificity of the assembly interfaces between two ribozyme modules (RzMs) in Urds [28]. We also expected that the parent Urds could be modified to produce closed forms of symmetrical oligomers if we used a bent P6b-P6b linker. In this study, we constructed and examined a new series of Urds, which were designed to form closed trimers with a triangular shape.

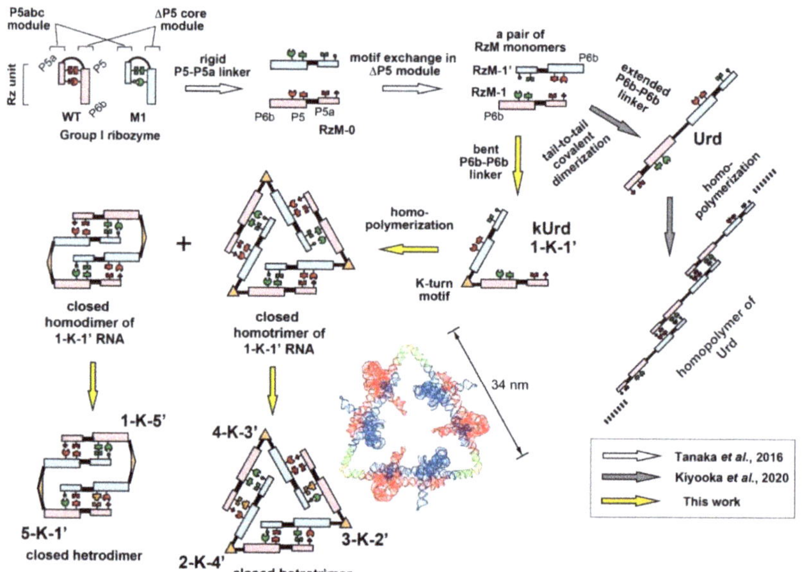

Figure 1. Design of closed oligomers based on engineered group I ribozyme dimers. Scheme of modular redesign to generate unit ribozyme dimers (kUrds) and their oligomerization. White arrows indicate structural redesign to prepare RzM RNAs and their noncovalent dimerization as reported by Tanaka et al. in 2016 [26]. Gray arrows indicate rational redesign for covalent dimerization of RzM RNAs reported by Kiyooka et al. in 2020 [28]. Yellow arrows indicate structural redesign to construct kUrd **1-K-1′** in this study. Sequences and secondary structures of wild-type ribozyme, M1 variant ribozyme and their structural elements are shown in Figure S1.

2. Materials and Methods

2.1. Molecular Design

Three-dimensional structural models of kink-turn unit ribozyme dimer (abbreviated as kUrd) **1-K-1′** and its closed homotrimer were constructed using the crystal structure of a shortened form of the *Tetrahymena* group I ribozyme (PDB ID: 1X8W) [29], a model 3D structure of its full-length form [27] and the KT-15 kink-turn motif (PDB ID: 1JJ2) [30]. Molecular modeling was performed using Discovery Studio (BIOVIA, San Diego, CA, USA), protocol of which has been described with an example to construct a 3D model structure of the RzM:RzM dimer [27]. In this study, the modeling was continued to connect three RzM:RzM dimers and three KT-15 motifs by pairs of A-form RNA duplexes (15 bp/10 bp).

2.2. Plasmid Construction and RNA Preparation

Plasmids encoding the sequences of α- and β-chains of kUrd RNAs were derived in two steps from plasmids encoding chimeric constructs of the wild-type *Tetrahymena* group I intron and its M1 variant-type intron [31]. PCR-based mutagenesis was used in each step of the stepwise plasmid construction. We first changed the sequences of P5-P5a linker elements of the starting plasmids to form the rigid P5-P5a duplex to yield RzM:RzM interfaces. The resulting plasmids were further modified by replacing the sequences of P6b elements to introduce the KT-15 kink-turn as the bent linker element. The resulting plasmids were used as templates for PCR. For each PCR, the sense primer contained the T7 promoter sequence. PCR-amplified DNA templates were used for synthesis of α- and β-chains of kUrd RNAs by in vitro transcription with T7 RNA polymerase. Transcription reactions were performed for 4.5 h at 37 °C in the presence of nucleotide triphosphates (1 mM each), 15 mM Mg^{2+}, 40 mM Tris-Cl (pH 7.5), 10 mM dithiothreitol and 2 mM spermidine. The DNA template in the reaction mixture was removed by DNase treatment for 30 min. The transcribed RNA was purified on 4% denaturing polyacrylamide gels. 3′-End labeling of RNAs with BODIPY fluorophore was performed according to the published protocol in which the diol moiety in the 3′ end ribose was oxidized by $NaIO_4$ to produce two aldehyde moieties, which were then connected with a primary amino group in the BODIPY derivative in the presence of $NaBH_3CN$ [32].

2.3. Preparation of L7Ae Protein and L7Ae–EGFP Fusion Protein

L7Ae protein and L7Ae–EGFP fusion protein were prepared according to the published protocol [33,34]. Briefly, the pET 28-b vector was used for cloning and construction of the recombinant protein L7Ae from *Archaeoglobus fulgidus* and its EGFP fusion protein. *Escherichia coli* BL-21 (DE3) (pLysS) cells were then used for protein production.

2.4. Electrophoretic Mobility Shift Assay (EMSA) of kUrd Oligomers

To analyze homooligomerization of kUrd **1-K-1′**, aqueous solution (18 μL) containing its α- and β-chain RNAs (0.14 μM) was heated at 80 °C for 2.5 min and then cooled to 37 °C. The 10× concentrated buffer (2 μL) containing 300 mM Tris-Cl (pH 7.5) and 100–200 mM $Mg(OAc)_2$ was added to the RNA solution. The resulting solution contained 30 mM Tris-Cl (pH 7.5), 10–20 mM $Mg(OAc)_2$ and 0.125 μM each α- and β-chain RNA to form 0.125 μM **1-K-1′**. To analyze heterotrimers, three kUrd solutions (6.67 μL each), each of which contained 0.375 μM each α- and β-chain RNA, were prepared separately and their mixed solution (20 μL) was then prepared. To analyze heterodimers, two kUrd solutions (10 μL each), each of which contained 0.25 μM each α- and β-chain RNA, were prepared separately and their mixed solution (20 μL) was then prepared. The resulting solution (20 μL) containing kUrd homo- or heterooligomers was incubated at 37 °C for 30 min and then 4 °C for 30 min and 6× loading buffer containing 50% glycerol and 0.1% xylene cyanol (4 μL) were added. The samples (24 μL) were loaded onto a 5% non-denaturing polyacrylamide gel (29:1 acrylamide:bisacrylamide) containing 50 mM Tris-acetate (pH 7.5) and 5–25 mM $Mg(OAc)_2$. Electrophoresis was performed at 4 °C, 200 V for the initial 5 min, followed by 75 V for 5 or 12 h. Gels were analyzed using a Pharos FX fluoroimager (BioRad,

Hercules, CA, USA). Each RzM:RzM interface consists of a pair of ΔP5/P5abc interfaces. The binding affinity between ΔP5 core module and P5abc module varied considerably depending on the identity of the ΔP5/P5abc interface [35,36]. Stability of the interface is improved by increasing Mg^{2+} concentration [35,36]. In each EMSA, we therefore tuned the concentration of Mg^{2+}.

2.5. Atomic Force Microscopy

Atomic force microscopy (AFM) was performed on a high-speed AFM (Nano Live Vision; RIBM, Tsukuba, Japan) according to the protocol reported previously [37]. The sample solutions containing kUrd RNAs were diluted to a final concentration of ~80 nM in folding buffer containing 17.5 mM Mg^{2+} and 2 μL of the sample was deposited onto the mica surface of the AFM.

2.6. Electrophoretic Mobility Shift Assay (EMSA) of the kUrd Trimer with L7Ae Protein

Three kUrd solutions (4.1 μL each), each of which contained 30 mM Tris-Cl (pH 7.5), 17.5 mM Mg(OAc)$_2$ and 0.61 μM each α- and β-chain RNA, were prepared separately according to the protocol for kUrd oligomer formation. To each kUrd solution (4.1 μL) was added 5× concentrated binding buffer (1.33 μL) containing 50 mM HEPES-KOH (pH 7.5), 750 mM KCl and 7.5 mM MgCl$_2$. To this solution, was added a solution (1.25 μL) of L7Ae (or L7Ae–EGFP) protein with appropriate concentration and the resulting solution (6.68 μL) was incubated at 37 °C for 30 min. Three kUrd+L7Ae solutions were then mixed. The resulting solution (20 μL) containing three kUrds (0.125 μM each) and L7Ae was incubated at 37 °C for 30 min and then 4 °C for 30 min and 6× loading buffer containing 50% glycerol and 0.1% xylene cyanol (4 μL) was added. The samples (24 μL) were analyzed in a same manner to EMSA of kUrd oligomers.

2.7. Substrate Cleavage Reaction

To prepare a set of kUrd monomer solutions, aqueous solutions containing appropriate pairs of α- and β-chain RNAs (final concentration of each kUrd: 0.25 μM) were heated at 80 °C for 5 min and then cooled to 37 °C. Then, 10× concentrated reaction buffer (final concentration: 30 mM Tris-Cl, pH 7.5 and 3 mM or 5 mM MgCl$_2$) and 2 mM guanosine triphosphate (final concentration: 0.2 mM) were added to the RNA solution and incubated at 37 °C for 30 min to form a given kUrd monomer. The resulting kUrd monomer solutions were then mixed to afford a kUrd oligomer solution, which was additionally incubated at 37 °C for 30 min. Ribozyme reactions were started by adding substrate-a (5'-FAM-GGCCCUCUAAAAA-3', final concentration: 0.25 μM) conjugated with carboxyfluorescein (FAM) at its 5' end [37]. The resulting solution containing 0.25 μM each kUrd and 0.25 μM substrate-a was kept at 37 °C. Aliquots were taken at given time points and the mixtures were electrophoresed on 15% denaturing polyacrylamide gels. The substrate and reaction products were analyzed using a Pharos FX fluoroimager. All of the ribozyme activity assays were repeated at least twice and representative gel images are shown.

2.8. Trans-Splicing Monitored by Fluorescence of Spinach RNA–DFHBI

To prepare three kUrd monomer solutions, aqueous solutions (17 μL) containing appropriate pairs of α- and β-chain RNAs (final concentration of each kUrd: 0.25 μM) were heated at 80 °C for 5 min and then cooled to 37 °C. Then, 10× concentrated reaction buffer (2 μL, final concentration: 40 mM Tris-Cl, pH 7.5, 125 mM KCl, 5 mM MgCl$_2$ and 10 μM DFHBI) and 2 mM guanosine triphosphate (1 μL, final concentration: 0.2 mM) were added to the RNA solution and incubated at 37 °C for 30 min to form a given kUrd monomer. The resulting kUrd monomer solutions were then mixed to afford a kUrd oligomer solution (60 μL) containing 0.25 μM each kUrd and initiate the *trans*-splicing reaction. The solution was transferred quickly to the wells of a microplate reader with a mineral oil overlay. The plate was then incubated at 37 °C in a plate reader (Infinite F200 Pro; Tecan, Männedorf,

Switzerland), which was pre-warmed at 37 °C and recorded fluorescence intensity of the sample solution.

3. Results

3.1. Design of Unit Ribozyme Dimers (Urds) with a Bent Linker

To design a new series of Urds to form closed trimers, we needed a suitable RNA motif consisting of two helices with fixed (~60°) angles as a bent linker unit. We selected the kink-turn RNA motif (Figure 1 and Figure S2) [30], which is one of the most extensively studied RNA motif families and provides a sharp (~60°) bend to the helical axis (Figure 2A). Kink-turn motifs have been identified in a diverse range of naturally occurring RNA structures with various functions [30]. Here, we chose the KT-15 kink-turn motif (Figure 2A), which not only serves as an RNA structural element but also acts as a recognition element for the ribosomal protein, L7Ae (Figure 2A) [30]. This RNA–protein complex has also been used as a module in synthetic RNA–protein nanostructures [33,34,38,39] and also synthetic RNA switches [40–43].

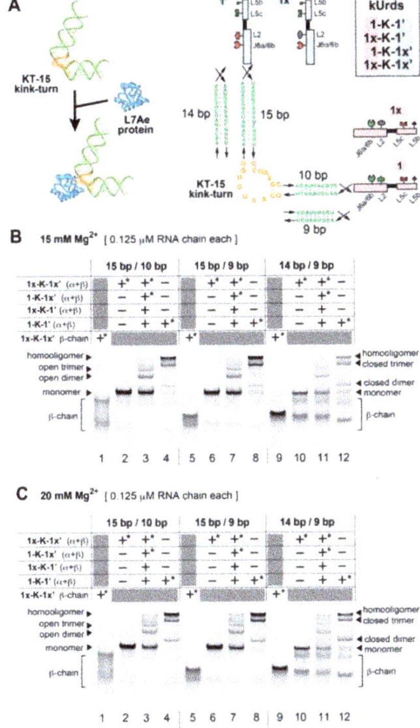

Figure 2. Structures of kUrds and their assembly analyzed by EMSA. (**A**) 3D structures of KT-15 kink-turn motif (orange) with a pair of duplexes (green) and its binding with L7Ae protein (blue). Sequence and secondary structures of the linker elements of kUrds with three distinct pairs of duplexes. Small red and green icons indicate RNA motifs. Mutations in L2 elements to disrupt RzM:RzM interaction are also shown as elliptic gray icons. Sequences and secondary structures of these RNA motifs are shown in Figure S1. (**B**) EMSA of **1-K-1′** homopolymers formed in the presence of 15 mM Mg^{2+}. Asterisks indicate RNA chains labeled with BODIPY fluorophore. (**C**) EMSA of **1-K-1′** homopolymers formed in the presence of 20 mM Mg^{2+}. Asterisks indicate RNA chains labeled with BODIPY fluorophore.

3.2. Homooligomerization of kUrds

A new class of Urd variants bearing the KT-15 motif was designated as kink-turn unit ribozyme dimer (abbreviated as kUrd, Figure 1). To optimize kUrd linkers, we compared three linker elements, each of which had the KT-15 motif flanked by a pair of duplexes with 15 bp/10 bp, 15 bp/9 bp, or 14 bp/9 bp (Figure 2A). The three distinct kUrd linkers were used to connect two RzM monomers (RzM-1 and RzM-1′) in a tail-to-tail manner, yielding three **1-K-1′** RNAs (Figure 2A and Figure S2). The three **1-K-1′** were designed to homotrimerize to form closed trimers (Figure 1). For each **1-K-1′**, we prepared **1x-K-1′** and **1-K-1x′** to eliminate assembly ability of the RzM on one side by introducing a UUCG tetraloop [44,45] into the L2 element (where x designates L2 UUCG, see Figures S1D and S3). We also prepared **1x-K-1x′** mutant for each **1-K-1′** to eliminate assembly ability of RzMs on both sides (Figure S3). We analyzed three **1-K-1′** kUrds with distinct pairs of helices by electrophoretic mobility shift assay (EMSA) to characterize their assembly properties. Each **1-K-1′** kUrd was composed of two RNA strands (α- and β-RNA chains) (Figures S2B and S3, Table S1). Preliminary experiment showed that the two RNA strands assembled efficiently to form the kUrd monomer (Figure S4). We first labeled the 3′-end of the β-chain RNA with the BODIPY fluorophore [32] to visualize kUrd and its assembly in native gels in the presence of 10 mM Mg^{2+} (Figure S5A), 15 mM Mg^{2+} (Figure 2B) and 20 mM Mg^{2+} (Figure 2C).

BODIPY-labeled β-chain RNA of **1x-K-1x′**, which can also be used as β-chain RNA of **1-K-1x′**, showed broad and multiple bands (lanes 1, 5 and 9 in Figure 2B,C). The β-chain RNA, however, showed a sharp and retarded band in the presence of an equimolar amount of the partner α-chain RNA (lanes 2, 6 and 10 in Figure 2B,C), indicating assembly of the α- and β-chains RNAs to form a structured kUrd **1x-K-1x′** monomer. Introduction of two L2-UUCG loops into both of the RzMs in kUrd afforded **1x-K-1x′** to selectively form a monomeric state (lanes 2, 6 and 10 in Figure 2B,C). We then confirmed the assembly state of **1-K-1′** and its mutants by preparing a solution containing equal amounts of four kUrds (**1x-K-1x′**, **1x-K-1′**, **1-K-1x′** and **1-K-1′**), prepared by mixing two α-chain RNAs and two β-chain RNAs in a single tube (lanes 3, 7 and 11 in Figure 2B,C). BODIPY-labeled β-chain RNAs for **1x-K-1x′** and **1-K-1x′** selectively visualized kUrd monomer (**1x-K-1x′**), kUrd dimer (**1x-K-1′:1-K-1x′**) and kUrd open trimer (**1x-K-1′:1-K-1′:1-K-1x′**). In addition, to kUrd monomers, two new bands with lower mobilities were observed (lanes 3, 7 and 11 in Figure 2B,C), which were expected to be a kUrd dimer consisting mainly of **1x-K-1′:1-K-1x′** and a kUrd open trimer consisting mainly of **1x-K-1′:1-K-1′:1-K-1x′**.

We then analyzed the solution containing only **1-K-1′** (lanes 4, 8 and 12 in Figure 2B,C), which was expected to form homooligomers, including a closed form homotrimer. The dominant band was retained in the gel slot, suggesting the formation of high-molecular weight oligomers. Two new bands, which were not seen in lanes 3, 7 and 11, were also observed. The lower mobility band would correspond to the desired closed trimer containing six ribozyme units (lanes 4, 8 and 12 in Figure 2B,C). The higher mobility band could be a closed dimer of Urds, which was not predicted in molecular modeling of **1-K-1′**. In the presence of 10 mM Mg^{2+}, the unit kUrd appeared to form efficiently (lanes 2, 6 and 10 in Figure S5A, see also Figure S5B,C) but oligomeric states of kUrds were formed less efficiently than those with 15 mM Mg^{2+} (Figure S5B,C). Formation of oligomeric states of kUrds seemed enhanced in the presence of 20 mM Mg^{2+} compared to 15 mM Mg^{2+} (Figure S5B). Among the three kUrds with distinct pairs of duplexes (15 bp/10 bp, 15 bp/9 bp and 14bp/9 bp) in the kink-turn linker elements, the 15 bp/10 bp and 15 bp/9 bp linkers worked better than the 14 bp/9 bp linker, which produced multiple bands in the presence of 20 mM Mg^{2+} (Figure 2C). In the following experiments, we used the 15 bp/10 bp pair as the kink-turn linker element.

3.3. Block Copolymerization of kUrds

Based on the homooligomerization of **1-K-1′** forming the desired closed trimer and unexpected closed dimer, we then examined selective formation of a closed trimer and a closed dimer through copolymerization of two or three different kUrds (Figure 1). We substituted two RzMs in **1-K-1′** to prepare a trio of Urds (**3-K-2′**, **2-K-4′** and **4-K-3′**) for triblock-type kUrd copolymer. For this purpose, we introduced three RzM:RzM interactions **2:2′**, **3:3′** and **4:4′** (Figure 3A and Figure S1D), which were designed to be orthogonal to one another in their recognition specificity [28]. In a similar manner, a pair of Urds (**1-K-5′** and **5-K-1′**) were also designed for diblock-type kUrd copolymer (Figure 3A and Figure S1D), which would selectively yield a closed dimer. In block copolymerization by two or three distinct kUrds, we first prepared each kUrd separately by assembling its α- and β-RNA chain RNAs (Figures S2B and S4). Solutions of preformed kUrds were then mixed to form block copolymers of kUrds.

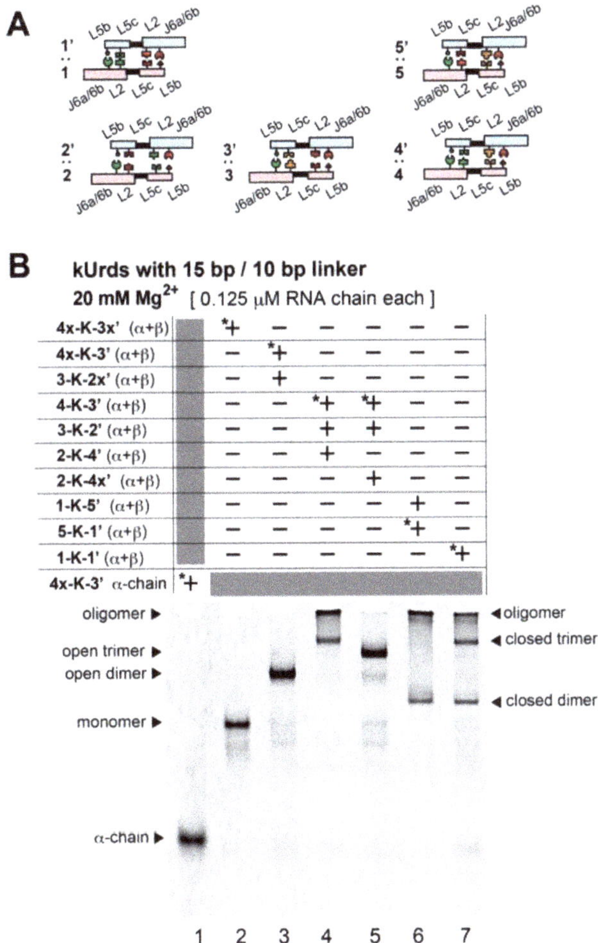

Figure 3. Assembly of a trio of kUrds to form a closed heterotrimer and a pair of kUrds to form a closed heterodimer. (**A**) Five distinct pairs of RzM:RzM interfaces, three of which were used in a closed heterotrimer configuration and the remaining two of which were used in a closed heterodimer configuration. Sequences and secondary structures of RNA motifs specifying RzM:RzM interfaces are shown in Figure S3. (**B**) EMSA of triblock and diblock kUrd copolymers formed in the presence of 20 mM Mg^{2+}. Asterisks indicate RNA chains labeled with BODIPY fluorophore.

We first examined copolymerization of **3-K-2′**, **2-K-4′** and **4-K-3′** to obtain the selective formation of a closed trimer (Figure 3B). To distinguish the closed trimer from the open trimer, we also examined **4-K-3x′**, with which ring closure would be blocked by disruption of the **3:3′** interface. As seen in lane 4 in Figure 3B, assembly of three kUrds provided a migrating band the mobility of which was close to that of the band corresponding to the possible closed trimer in homopolymerization of **1-K-1′** (lane 7 in Figure 3B). Higher oligomerization of kUrds seen in **1-K-1′** homopolymers (lane 7 in Figure 3B) was still seen in triblock copolymers (lane 4 in Figure 3B). In the presence of **4-K-3x′** in place of **4-K-3′**, no band was retained in the sample well (lane 5 in Figure 3B), consistent with the molecular design to form the open block trimer (**3-K-2′:2-K-4′:4-K-3x′**) lacking further oligomerization ability. Selective formation of the open form of kUrd trimer was suggested by the single migrating band (lane 5 in Figure 3B), which migrated slower than the open kUrd dimer (**4x-K-3:3-K-2′**, see lane 3 in Figure 3B) but faster than the possible closed trimers (lanes 5 and 7 in Figure 3B). Importantly, in the trio of kUrd designed for triblock copolymer (lane 4 in Figure 3B), no band corresponding to the possible closed dimer of **1-K-1′** (see lane 7 in Figure 3B) was observed. Bands corresponding to the open dimer and monomer were not seen in lane 4 (Figure 3B), suggesting that triblock oligomerization proceeded efficiently. Comparison between two trios of kUrds with/without ring closure ability (lanes 4 and 5 in Figure 3B) suggested that the closed trimer is thermodynamically stable and the dissociation of RzM interfaces in the closed trimer occurred poorly.

The pair of kUrds for diblock copolymer also formed oligomers with no migration (lane 6 in Figure 3B). This mixture also yielded a band the mobility of which corresponded to the possible closed dimer of **1-K-1′** (lanes 6 and 7 in Figure 3B). No band was seen in lane 6 at the position corresponding to that of the possible closed trimer. In the closed heterotrimer consisting of three distinct kUrds, each kUrd was formed by α- and β-chain RNAs. As the kUrd closed heterotrimer contained six distinct RNA chains, we labeled one of six RNA chains with BODIPY fluorophore and monitored the mobility of the closed trimer on native gels (Figure S6). Regardless of the identity of the BODIPY-labeled RNA chain, all EMSA showed similar behavior, supporting the formation of the closed heterotrimer by assembly of three kUrds.

3.4. Atomic Force Microscopy Analysis of kUrd Heterotrimer and Heterodimer

Based on the observation that three kUrds are likely to form closed trimers in both homopolymerization and triblock copolymerization, we visualized the molecular structures of the heterotrimer (**3-K-2′:2-K-4′:4-K-3′**) and heterodimer (**1-K-5′:5-K-1′**) by atomic force microscopy (AFM) [33,37–39]. In the presence of 17.5 mM Mg^{2+}, a sample containing three kUrds gave AFM images containing objects with triangular shapes (Figure 4A,B). The triangular objects were observed with limited frequency. We measured lengths of 21 sides of seven triangular objects observed in AFM images. While their lengths varied between 28.1 nm 39.6 nm, their average value was 34.5 nm. This value was closely similar to that predicted by molecular modeling of the closed trimer (approximately 34 nm, Figure 1). In some of the AFM images of kUrd triangles, we observed triangular shapes as three bright regions (Figure 4B), heights of which were 6~8 nm (Figure S7A).

We then analyzed AFM images of a solution containing **1-K-5′** and **5-K-1′** that formed a closed heterodimer, **1-K-5′:5-K-1′** (Figure 4C). Due to the limited efficacy of the closed dimer formation (Figure 3B), AFM provided the corresponding images inefficiently. A few RNA objects, however, were found to reflect closed dimers (Figure 4C), in which two bright spots were observed in each object. Heights of bright spots were 5~7 nm (Figure S7B). We also analyzed AFM images of two mutant kUrds with disrupted L2 interacting motifs (Figure S8). They retained some of RNA interacting motifs but were not able to form kUrd dimers (Figure S8A). The two mutant kUrds formed some aggregates on mica surfaces but gave no bright spots in the presence of 17.5 mM Mg^{2+} (Figure S8C,D).

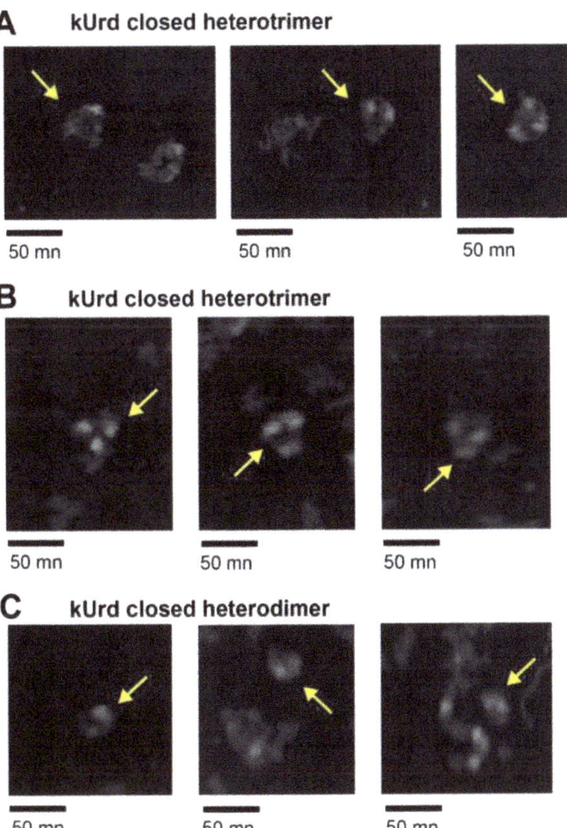

Figure 4. AFM imaging of closed kUrd heterotrimers and closed heterodimers. (**A**,**B**) AFM of closed kUrd trimers in the presence of 17.5 mM Mg^{2+}. Yellow arrows indicate closed kUrd trimers. (**C**) AFM of closed kUrd dimers in the presence of 17.5 mM Mg^{2+}. Yellow arrows indicate closed kUrd dimers.

3.5. Decoration of the kUrd Closed Trimer with Kink-Turn Binding Protein L7Ae

We then decorated the kUrd closed trimer with the RNA binding protein, L7Ae, for which the KT-15 kink-turn motif serves as a binding element (Figure 2A) [30]. We performed EMSA of the heterotrimer (**3-K-2′:2-K-4′:4-K-3′**) in the absence and presence of L7Ae protein. In the presence of each RNA chain at 0.125 μM (2.5 pmol each in 20 μL of solution), theoretical amounts of the KT-15 motif in the resulting three distinct kUrd units (7.5 pmol in 20 μL of solution) was also 0.375 μM (7.5 pmol in 20 μL of solution).

In the presence of 0.375 μM L7Ae protein, slight but distinct retardation of the closed trimer was observed (lane 6 in Figure 5A). A twofold molar excess amount of L7Ae protein (0.75 μM) over the theoretical amount of kink-turn motif (0.375 μM) did not cause further retardation of the band (lane 7 in Figure 5A). The mobility of the trimer band did not change in the presence of a greater excess of L7Ae protein (Figure S9). These observations suggested that the retarded band corresponded to the RNA–protein complex consisting of the kUrd closed trimer possessing three KT-15 kink-turns and three L7Ae protein molecules. In the kUrd trimer–L7Ae complex, fluorescent visualization of L7Ae proteins was also examined using an engineered L7Ae protein fused to enhanced green fluorescent protein (EGFP) (Figure 5B) [46].

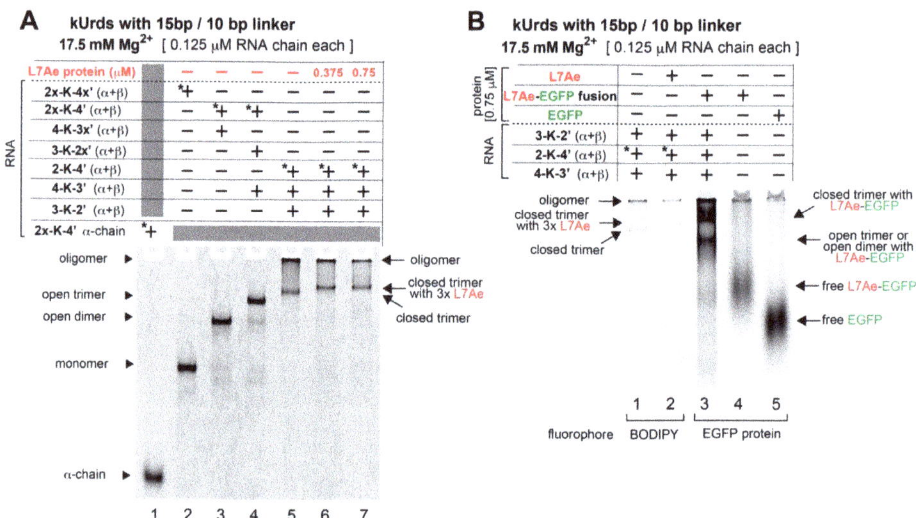

Figure 5. Complex formation of L7Ae protein with KT-15 kink-turn motifs in a closed kUrd heterotrimer. (**A**) EMSA of a closed kUrd heterotrimer in the absence and presence of L7Ae protein. (**B**) EMSA of a closed kUrd heterotrimer in the absence and presence of L7Ae protein and L7Ae-EGFP fusion protein.

L7Ae–EGFP fusion protein, which showed a broad band in free form (lane 4 in Figure 5B), shifted significantly in the presence of the kUrd heterotrimer (lane 3 in Figure 5B). In lane 3, the major band migrated more slowly than the free kUrd heterotrimer (lane 1) and also more slowly than the complex of closed trimer and L7Ae protein (lane 2). The major band in lane 3 corresponded to a complex of the closed trimer and L7Ae–EGFP fusion protein. The faster migrating band in lane 3 may correspond to an RNA–protein complex containing an open trimer or an open dimer because it was observed very weakly as a byproduct in assembly of the closed trimer (lane 4 in Figure S9B).

3.6. Catalytic Activities of kUrds and Their Oligomers

Although closed trimers and closed dimers were not formed exclusively without higher oligomers, it was interesting to see the catalytic abilities of ribozyme units in kUrds because their assembly would produce an active ribozyme unit consisting of the ΔP5 core ribozyme module and P5abc activator module (Figures 1 and 6A) [18].

As the ΔP5 core ribozyme can be activated by transient (weak) association of the P5abc module, activity assay required a lower Mg^{2+} concentration than EMSA [36]. In the substrate cleavage reactions by the wild-type ΔP5 module (Figure S1A) and M1 type ΔP5 module (Figure S1B), the efficiencies of P5abc-depedent activation were optimum in the presence of 3 mM Mg^{2+} and 5 mM Mg^{2+}, respectively. In the closed heterotrimer, six ΔP5-P5abc ribozyme units were formed, in which three had the wild-type ΔP5 modules (units 2, 3 and 4) with the remaining three having the M1 type ΔP5 module (units 2′, 3′ and 4′) (Figure 6A). We examined activation of the ΔP5 core ribozyme by the P5abc module upon formation of kUrd oligomers including the closed trimers.

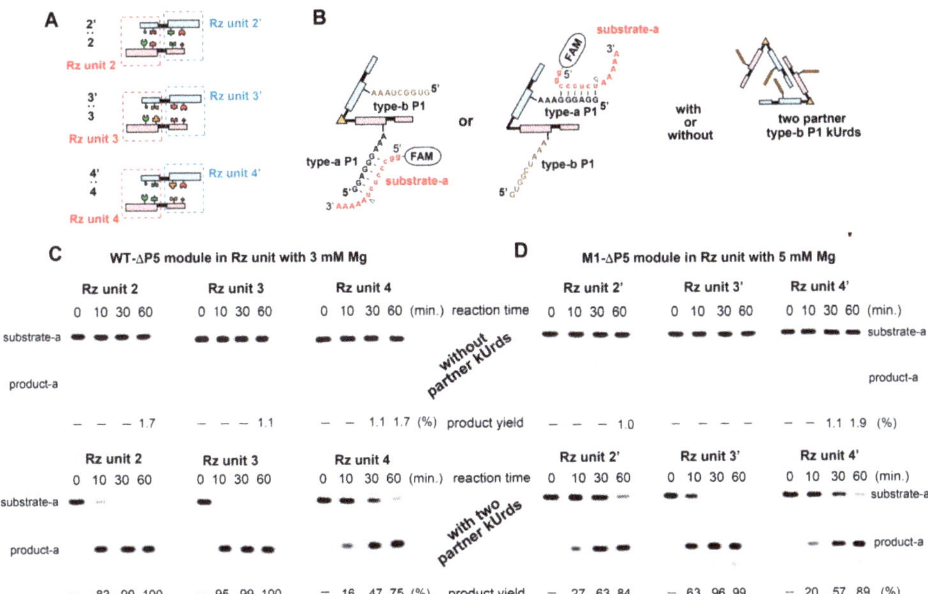

Figure 6. Activation of ribozyme unit upon assembly of kUrds. (**A**) Formation of two ribozyme units in each RzM:RzM interface. One ribozyme unit (2, 3, or 4) formed the wild-type ribozyme and the other ribozyme unit (2′, 3′, or 4′) formed the M1 type ribozyme. (**B**) Sequences of substrate-a and two types of P1 binding elements. Location of P1 element in the secondary structure of the *Tetrahymena* group I ribozyme are shown in Figure S1. (**C**) Cleavage of substrate-a by the wild-type ribozyme unit in a given kUrd without (top) or with (bottom) its partner kUrds in the presence of 3 mM Mg^{2+}. (**D**) Cleavage of substrate-a by the M1 type ribozyme unit in a given kUrd without (top) or with (bottom) its partner kUrds in the presence of 5 mM Mg^{2+}.

We developed an assay system with which each ΔP5-P5abc ribozyme unit in kUrd oligomers could be analyzed separately. In the ΔP5-P5abc ribozyme unit, the P1 substrate recognition element for the cleavage reaction could be introduced in a modular manner. In EMSA and AFM analysis, α- and β-RNA chain RNAs commonly possessed a P1 recognition element for type-a substrate. To analyze the single ribozyme units in kUrds and in their oligomers selectively, we employed the second P1 element recognizing type-b substrate (Figure 6B) [47]. The substrate recognition of type-a and type-b P1 elements are highly orthogonal to each other [37]. The type-a substrate-P1 pair was introduced selectively to one of six ribozyme units, while the remaining five units were designed to possess type-b P1 elements (Figure 6B). The cleavage reaction of type-a substrate thus, enabled us to selectively monitor the catalytic ability of a ribozyme unit with a type-a P1 element (Figure 6B).

In the isolated state of each kUrd, its wild-type ΔP5 ribozyme module possessing the type-a P1 element showed no activity in the presence of 3 mM Mg^{2+} (Figure 6C, top). Addition of the partner kUrds in which the ΔP5 ribozyme module possessed only type-b P1 elements, however, induced the catalytic ability to cleave the type-a substrate (Figure 6C, bottom). In a similar manner, the M1 type ribozyme unit in each kUrd was also inactive with 5 mM Mg^{2+} (Figure 6D, top) but activated by the addition of partner kUrds (Figure 6D, bottom). Such activation behavior was primarily conducted by the association of ΔP5 catalytic modules in each kUrd with P5abc activator modules with the partner kUrds. It should be noted that moderate differences were observed in the catalytic activities among the three ribozyme units sharing the wild-type (unit 2, 3 and 4) or M1 type (unit 2′, 3′ and 4′) ΔP5 module (Figure 6C,D). The catalytic activity of the ΔP5-P5abc ribozyme unit was governed by the identity of the ΔP5 module and also by the identity of the RNA–RNA interface between the ΔP5 module and the P5abc module (Figure 6A) [35,36].

3.7. Trans-Splicing Reaction Promoted by kUrd Oligomer Formation

The intron form of ΔP5 core ribozyme module can perform the self-splicing reaction to yield ligated exons in the presence of the P5abc activator module (Figure 7A) [18]. The pair of ΔP5 core ribozyme modules in each kUrd were composed of α- and β-chain RNAs. In the intron form of wild-type ΔP5 core ribozyme module on one side of kUrd, 5′ and 3′ exons were provided by α- and β-chain RNAs, respectively (Figure S10A). In the intron form of the M1 type ΔP5 core ribozyme module on the opposite side of kUrd, 5′ and 3′ exons were provided by β- and α-chain RNAs, respectively (Figure S6B). The organization of α- and β-chain RNAs in the intron form of the ΔP5 core ribozyme modules would conduct *trans*-splicing reaction joining 5′ and 3′ exons, which were provided by two separate RNA chains, to produce a mature exon sequence (Figure S10A,B) [31,48]. *Trans*-splicing promoted by group I ribozyme has been recognized as a promising strategy to repair mRNAs of disease-related genes bearing undesirable mutations [48–50]. Therefore, we examined *trans*-splicing reaction catalyzed by a kUrd in triblock copolymers (**3-K-2′:2-K-4′:4-K-3′**) including the closed trimer.

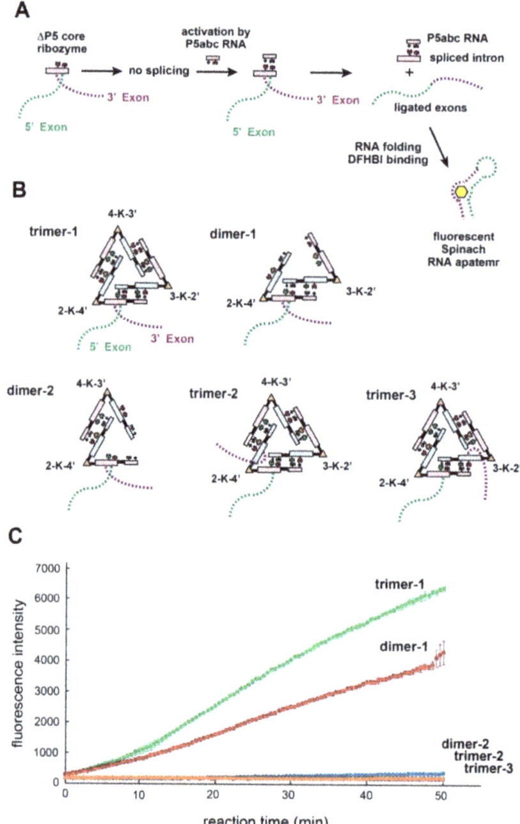

Figure 7. *Trans*-splicing reaction by the ribozyme unit 2 formed in the closed kUrd heterotrimer. (**A**) Parent *cis*-splicing reaction of the bimolecular group I ribozyme. 5′ and 3′ exons were both provided by the unimolecular ΔP5 ribozyme RNA, the activity of which is repressed in the absence of the P5abc RNA. Splicing reaction produced Spinach RNA as the ligated exons. Spinach RNA captured DFHBI and induced its fluorescence. (**B**) Closed kUrd trimers and their partial dimers used for analysis of the *trans*-splicing reaction. (**C**) Time-dependent increases in fluorescence of Spinach RNA–DFHBI complex. Splicing reactions were carried out at 37 °C.

We installed 5′ and 3′ exon sequences and their recognition elements into the α- and β- chain RNAs of **2-K-4′** kUrd, respectively (Figure S10C and Table S2). Exon sequences were appended to the ribozyme module consisting of the wild-type ΔP5 ribozyme module activated by the P5abc module provided by **3-K-2′** kUrd through the **2:2′** interface (Figure 7B and Figure S10C). We used a shortened derivative of the Spinach RNA aptamer as a ligated exon sequence dissected in precursor form (Figure 7A) [51]. Spinach RNA is an RNA receptor for DFHBI and its fluorescence is induced by restricting the molecular motion of DFHBI in the complex form [52]. The progression of *trans*-splicing reaction can be monitored in the presence of DFHBI [51,53]. To evaluate the effects of kUrd oligomer formation on the *trans*-splicing reaction, we prepared a set of kUrd variants some of which possessed 5′ and 3′ exon sequences (Figure 7B and Figure S10, see also Table S2). A trio of kUrds (**3-K-2′:2-K-4′:4-K-3′**) to form a triblock oligomer (**trimer-1**) exhibited an increase in emission (green plot in Figure 7C). We also examined two sets of triblock oligomers (**trimer-2** and **trimer-3**) in which the 5′ exon and 3′ exon were installed to incorrect RNA chains and the two exons were unable to meet in one ΔP5 ribozyme unit. The two sets of kUrds showed no increase in fluorescence from Spinach RNA complexed with DFHBI (purple plot and blue plot in Figure 7C). These data indicated that the correct assignments of 5′ and 3′ exons in six RNA chains are crucial in the oligomeric state of kUrds.

To evaluate the effects of triblock oligomerization (**3-K-2′:2-K-4′:4-K-3′**) in the splicing proficient trimer (**trimer-1**), we assayed a solution containing **2-K-4′** and **3-K-2′**, which form **dimer-1**, in which the intron form of the ribozyme unit in **2-K-4′** was formed correctly and **2-K-4′** was correctly activated by the P5abc module in **3-K-2′** (Figure 7B). The difference in activity between the presence and absence of the third kUrd unit (**4-K-3′**) mainly reflected the oligomerization ability. The solution of **dimer-1** exhibited emission (red plot in Figure 7C), which was, however, about two thirds that of the solution containing three kUrds. This observation suggested that **4-K-3′** improved the activity of the splicing ribozyme unit through the formation of oligomeric states. To examine the contribution of the P5abc module provided by **3-K-2′**, we depleted the **3-K-2′** unit from the three kUrd components. The sample solution containing **2-K-4′** and **4-K-3′** to form **dimer-2** showed no increase in emission (orange plot in Figure 7C), indicating the importance of activation by the P5abc module to conduct full splicing ability of the ΔP5 ribozyme module possessing 5′ and 3′ exons.

4. Discussion

In this study, we performed molecular modeling of kUrd to design a closed trimer, the formation of which was confirmed by EMSA and AFM analyses. On the other hand, significant formation of higher oligomers indicated that the current structural design was insufficient partly due to the 3D structure of the full-length *Tetrahymena* ribozyme used for the molecular design in this study. In molecular modeling of kUrds (a dimeric form of an engineered *Tetrahymena* ribozyme connected by the kink-turn motif), we used a model 3D structure of the full-length ribozyme and the X-ray crystallographic structure of a shortened form of the ribozyme lacking some peripheral elements [29]. There may have been some deviations from the full-length *Tetrahymena* ribozyme 3D structure, which has yet to be solved by X-ray crystallography. Improvement of 3D structural design, however, may be possible because a new 3D model structure of the full-length ribozyme was constructed recently by combining cryo-EM data, chemical probing data and molecular modeling [54]. Although the resolution of this new model is modest (6.8 Å), it will be useful to refine the design of ribozyme-based nanostructures.

The formation of the closed dimer is also an issue to be resolved for precise design of kUrd-based RNA nanostructures. Formation of the closed kUrd dimer, which was unexpected from molecular modeling, suggested the structural flexibility of kUrd RNAs and entropic effects in RNA assembly. The flexible nature of the kink-turn RNA motifs has been analyzed and reported previously [55–57]. Therefore, we hypothesized that the formation of the closed dimer may also be supported by the structural flexibility of the

KT-15 kink-turn motif. Distinct from our previous Urd with an extended linker, each kUrd can accept L7Ae protein at its linker element bearing the KT-15 kink-turn motif. EMSA of the heterotrimer in the presence of L7Ae suggested the formation of KT-15–L7Ae complex at each corner of the triangle-shaped closed kUrd trimer. We also performed AFM analysis of the heterotrimer in the presence of L7Ae but no significant differences were observed between images of kUrds with and without L7Ae. This was conceivable because L7Ae is much smaller than kUrd RNA. We also found that L7Ae protein strongly inhibited the catalytic ability of kUrds. Collaboration between L7Ae protein and the *Tetrahymena* ribozyme unit in catalysis is still an open issue.

In this study, we mainly analyzed the closed trimer and closed dimer of kUrds. The closed trimer contained six ribozyme units and three kink-turn motifs that interacted with L7Ae proteins. In reactions promoted by the ribozyme unit in kUrd oligomers, *trans*-splicing reaction proceeded more efficiently in the presence of three kUrds forming a heterotrimer than two kUrds providing the core ΔP5 ribozyme and P5abc activator. Although the extent of improvement was moderate, this was a promising result as an initial step toward RNA processing in vivo controlled by ribozyme-based nanostructures. The kUrd trimer containing six group I ribozyme units consists of nearly 2400 nucleotides, which is comparable to bacterial 16S and 23S ribosomal RNAs. The size of the triangle shape (with an edge of approximately 34 nm) was also comparable to the bacterial ribosome (approximately 20 nm in diameter). Although the structure and function of kUrd-based nanostructures are still primitive, this study represents an initial step toward artificial construction of complex RNA nanostructures capable of playing crucial roles in naturally occurring biological systems. *Trans*-splicing ribozymes have been investigated as a promising tool to repair mutated mRNAs causing genetic disorders [49,50]. Ribozyme dimers employed in this study may be used to repair two distinct mutated mRNA simultaneously and cooperatively. Two ribozyme units in a dimer also target the same mRNA sequence, to which two *trans*-splicing ribozyme units perform distinct splicing reactions to yield different products. For example, one ribozyme unit repairs the mutation of the target mRNA whereas the other ribozyme unit attaches a fluorescent RNA aptamer to the target mRNA for its fluorescent imaging. Fabrication of closed trimers using ribozyme dimers and KT-15 kink-turn would enable us to attach various protein components through L7Ae protein that binds to KT-15 [38,39]. Such ribozyme-based RNA-protein nanoparticle can be further decorated by RNA-based sensors and therapeutic RNA modules such as RNA aptamers and siRNAs, with which next generation of RNA-based nanomedicines may be developed.

Supplementary Materials: The following are available online at https://www.mdpi.com/2076-3417/11/6/2583/s1, Figure S1: Sequences and secondary structures of the *Tetrahymena* group I ribozymes and their structural elements used for modular engineering in this study, Figure S2: Scheme of rational redesign of unimolecular *Tetrahymena* ribozymes to generate kUrd **1-K-1'**, Figure S3: Kink-turn unit ribozyme dimers (kUrds) employed in this study, Figure S4: Formation of kUrds through the assembly of their α-chain and β-chain RNAs, Figure S5: Oligomerization of **1-K-1'** and its mutants, Figure S6: EMSA of triblock copolymers formed in the presence of 20 mM Mg^{2+}, Figure S7: AFM images of the closed heterotrimer and heterodimer and their cross sections, Figure S8: AFM images of the closed heterodimer and its mutant monomers, Figure S9: Complex formation of L7Ae proteins with KT-15 kink-turn motifs in a closed kUrd heterotrimer, Figure S10: Closed kUrd trimers and their partial dimers used for analysis of the *trans*-splicing reaction.

Author Contributions: Conceptualization, Y.I.; methodology, Y.I., H.S. (Hirohide Saito), H.S. (Hiroshi Sugiyama), M.E. and S.M.; investigation, J.A., K.H. and Y.F.; writing—original draft preparation, Y.I. and T.Y.; writing—review and editing, H.S. (Hirohide Saito), H.S. (Hiroshi Sugiyama), M.E. and S.M.; supervision, Y.I., H.S. (Hirohide Saito), H.S. (Hiroshi Sugiyama) and M.E. All authors have read and agreed to the published version of the manuscript.

Funding: This research was funded by University of Toyama Discretionary Funds of the President "Toyama RNA Collaborative Research" (to Y.I. and S.M.).

Institutional Review Board Statement: Not applicable.

Informed Consent Statement: Not applicable.

Data Availability Statement: Not applicable.

Conflicts of Interest: The authors declare no conflict of interest.

References

1. Goodsell, D.S.; Olson, A.J. Structural symmetry and protein function. *Annu. Rev. Biophys. Biomol. Struct.* **2000**, *29*, 105–153. [CrossRef]
2. Lesieur, C. The assembly of protein oligomers: Old stories and new perspectives with graph theory. In *Oligomerization of Chemical and Biological Compounds*; Lesieur, C., Ed.; IntecOpen: London, UK, 2014. [CrossRef]
3. Jones, S.; Thornton, J.M. Protein-protein interactions: A review of protein dimer structures. *Prog. Biophys. Mol. Biol.* **1995**, *63*, 31–65. [CrossRef]
4. Glover, D.J.; Clark, D.S. Protein calligraphy: A new concept begins to take shape. *ACS Cent. Sci.* **2016**, *2*, 438–444. [CrossRef]
5. Sun, H.; Luo, Q.; Hou, C.; Liu, J. Nanostructures based on protein self-assembly: From hierarchical construction to bioinspired materials. *Nano Today* **2017**, *14*, 16–41. [CrossRef]
6. Kuan, S.L.; Bergamini, F.R.G.; Weil, T. Functional protein nanostructures: A chemical toolbox. *Chem. Soc. Rev.* **2018**, *47*, 9069–9105. [CrossRef] [PubMed]
7. Boyken, S.E.; Chen, Z.; Groves, B.; Langan, R.A.; Oberdorfer, G.; Ford, A.; Gilmore, J.M.; Xu, C.; DiMaio, F.; Pereira, J.H.; et al. De novo design of protein homo-oligomers with modular hydrogen-bond network-mediated specificity. *Science* **2016**, *352*, 680–687. [CrossRef]
8. Fallas, J.A.; Ueda, G.; Sheffler, W.; Nguyen, V.; McNamara, D.E.; Sankaran, B.; Pereira, J.H.; Parmeggiani, F.; Brunette, T.J.; Cascio, D.; et al. Computational design of self-assembling cyclic protein homo-oligomers. *Nat. Chem.* **2017**, *9*, 353–360. [CrossRef]
9. Kuhlman, B.; O'Neill, J.W.; Kim, D.E.; Zhang, K.Y.; Baker, D. Conversion of monomeric protein L to an obligate dimer by computational protein design. *Proc. Natl. Acad. Sci. USA* **2001**, *98*, 10687–10691. [CrossRef]
10. Stranges, P.B.; Machius, M.; Miley, M.J.; Tripathy, A.; Kuhlman, B. Computational design of a symmetric homodimer using beta-strand assembly. *Proc. Natl. Acad. Sci. USA* **2011**, *108*, 20562–20567. [CrossRef] [PubMed]
11. Mou, Y.; Huang, P.S.; Hsu, F.C.; Huang, S.J.; Mayo, S.L. Computational design and experimental verification of a symmetric protein homodimer. *Proc. Natl. Acad. Sci. USA* **2015**, *112*, 10714–10719. [CrossRef]
12. Fedor, M.J.; Williamson, J.R. The catalytic diversity of RNAs. *Nat. Rev. Mol. Cell Biol.* **2005**, *6*, 399–412. [CrossRef]
13. Weinberg, C.E.; Weinberg, Z.; Hammann, C. Novel ribozymes: Discovery, catalytic mechanisms, and the quest to understand biological function. *Nucleic Acids Res.* **2019**, *47*, 9480–9494. [CrossRef] [PubMed]
14. Breaker, R.R. Prospects for riboswitch discovery and analysis. *Mol. Cell* **2011**, *4*, 867–879. [CrossRef]
15. Butcher, S.E.; Pyle, A.M. The molecular interactions that stabilize RNA tertiary structure: RNA motifs, patterns, and networks. *Acc. Chem. Res.* **2011**, *44*, 1302–1311. [CrossRef] [PubMed]
16. Ishikawa, J.; Fujita, Y.; Maeda, Y.; Furuta, H.; Ikawa, Y. GNRA/receptor interacting modules: Versatile modular units for natural and artificial RNA architectures. *Methods* **2011**, *54*, 226–238. [CrossRef]
17. Jarrell, K.A.; Dietrich, R.C.; Perlman, P.S. Group II intron domain 5 facilitates a *trans*-splicing reaction. *Mol. Cell. Biol.* **1988**, *8*, 2361–2366. [CrossRef]
18. van der Horst, G.; Christian, A.; Inoue, T. Reconstitution of a group I intron self-splicing reaction with an activator RNA. *Proc. Natl. Acad. Sci. USA* **1991**, *88*, 184–188. [CrossRef]
19. Engelhardt, M.A.; Doherty, E.A.; Knitt, D.S.; Doudna, J.A.; Herschlag, D. The P5abc peripheral element facilitates preorganization of the *Tetrahymena* group I ribozyme for catalysis. *Biochemistry* **2000**, *39*, 2639–2651. [CrossRef]
20. Doudna, J.A.; Cech, T.R. Self-assembly of a group I intron active site from its component tertiary structural domains. *RNA* **1995**, *1*, 36–45.
21. Ikawa, Y.; Shiraishi, H.; Inoue, T. *Trans*-activation of the *Tetrahymena* ribozyme by its P2-2.1 domains. *J. Biochem.* **1998**, *123*, 528–533. [CrossRef]
22. Rahman, M.M.; Matsumura, S.; Ikawa, Y. Effects of molecular crowding on a bimolecular group I ribozyme and its derivative that self-assembles to form ribozyme oligomers. *Biochem. Biophys. Res. Commun.* **2018**, *507*, 136–141. [CrossRef]
23. Pan, T. Higher order folding and domain analysis of the ribozyme from *Bacillus subtilis* ribonuclease P. *Biochemistry* **1995**, *34*, 902–909. [CrossRef]
24. Loria, A.; Pan, T. Domain structure of the ribozyme from eubacterial ribonuclease P. *RNA* **1996**, *2*, 551–563.
25. Kim, H.; Poelling, R.R.; Leeper, T.C.; Meyer, M.A.; Schmidt, F.J. In vitro transactivation of *Bacillus subtilis* RNase P RNA. *FEBS Lett.* **2001**, *506*, 235–238. [CrossRef]
26. Tanaka, T.; Matsumura, S.; Furuta, H.; Ikawa, Y. Tecto-GIRz: Engineered group I ribozyme the catalytic ability of which can be controlled by self-dimerization. *ChemBioChem* **2016**, *17*, 1448–1455. [CrossRef] [PubMed]
27. Tanaka, T.; Ikawa, Y.; Matsumura, S. Rational engineering of a modular group I ribozyme to control its activity by self-dimerization. *Methods Mol. Biol.* **2017**, *1632*, 325–340. [PubMed]

28. Kiyooka, R.; Akagi, J.; Hidaka, K.; Sugiyama, H.; Endo, M.; Matsumura, S.; Ikawa, Y. Catalytic RNA nano-objects formed by self-assembly of group I ribozyme dimers serving as unit structures. *J. Biosci. Bioeng.* **2020**, *130*, 253–259. [CrossRef] [PubMed]

29. Guo, F.; Gooding, A.R.; Cech, T.R. Structure of the Tetrahymena ribozyme: Base triple sandwich and metal ion at the active site. *Mol. Cell* **2004**, *16*, 351–362.

30. Klein, D.J.; Schmeing, T.M.; Moore, P.B.; Steitz, T.A. The kink-turn: A new RNA secondary structure motif. *EMBO J.* **2001**, *20*, 4214–4221. [CrossRef] [PubMed]

31. Tanaka, T.; Hirata, Y.; Tominaga, Y.; Furuta, H.; Matsumura, S.; Ikawa, Y. Heterodimerization of group I ribozymes enabling exon recombination through pairs of cooperative *trans*-splicing reactions. *ChemBioChem* **2017**, *18*, 1659–1667. [CrossRef]

32. Ikawa, Y.; Moriyama, S.; Furuta, H. Facile syntheses of BODIPY derivatives for fluorescent labeling of the 3′ and 5′ ends of RNAs. *Anal. Biochem.* **2008**, *378*, 166–170. [CrossRef]

33. Ohno, H.; Kobayashi, T.; Kabata, R.; Endo, K.; Iwasa, T.; Yoshimura, S.H.; Takeyasu, K.; Inoue, T.; Saito, H. Synthetic RNA-protein complex shaped like an equilateral triangle. *Nat. Nanotechnol.* **2011**, *6*, 116–120. [CrossRef] [PubMed]

34. Ohuchi, S.J.; Sagawa, F.; Sakamoto, T.; Inoue, T. A trifunctional, triangular RNA-protein complex. *FEBS Lett.* **2015**, *589*, 2424–2428. [CrossRef] [PubMed]

35. Tanaka, T.; Furuta, H.; Ikawa, Y. Installation of orthogonality to the interface that assembles two modular domains in the *Tetrahymena* group I ribozyme. *J. Biosci. Bioeng.* **2014**, *117*, 407–412. [CrossRef]

36. Rahman, M.M.; Matsumura, S.; Ikawa, Y. Artificial RNA motifs expand the programmable assembly between RNA modules of a bimolecular ribozyme leading to application to RNA nanostructure design. *Biology* **2017**, *6*, 37. [CrossRef] [PubMed]

37. Oi, H.; Fujita, D.; Suzuki, Y.; Sugiyama, H.; Endo, M.; Matsumura, S.; Ikawa, Y. Programmable formation of catalytic RNA triangles and squares by assembling modular RNA enzymes. *J. Biochem.* **2017**, *161*, 451–462. [CrossRef]

38. Fujita, Y.; Furushima, R.; Ohno, H.; Sagawa, F.; Inoue, T. Cell-surface receptor control that depends on the size of a synthetic equilateral-triangular RNA-protein complex. *Sci. Rep.* **2014**, *4*, 6422. [CrossRef]

39. Osada, E.; Suzuki, Y.; Hidaka, K.; Ohno, H.; Sugiyama, H.; Endo, M.; Saito, H. Engineering RNA-protein complexes with different shapes for imaging and therapeutic applications. *ACS Nano.* **2014**, *8*, 8130–8140. [CrossRef] [PubMed]

40. Saito, H.; Kobayashi, T.; Hara, T.; Fujita, Y.; Hayashi, K.; Furushima, R.; Inoue, T. Synthetic translational regulation by an L7Ae-kink-turn RNP switch. *Nat. Chem. Biol.* **2010**, *6*, 71–78. [CrossRef]

41. Wroblewska, L.; Kitada, T.; Endo, K.; Siciliano, V.; Stillo, B.; Saito, H.; Weiss, R. Mammalian synthetic circuits with RNA binding proteins for RNA-only delivery. *Nat. Biotechnol.* **2015**, *33*, 839–841. [CrossRef]

42. Matsuura, S.; Ono, H.; Kawasaki, S.; Kuang, Y.; Fujita, Y.; Saito, H. Synthetic RNA-based logic computation in mammalian cells. *Nat. Commun.* **2018**, *9*, 4847. [CrossRef]

43. Ohno, H.; Akamine, S.; Saito, H. Synthetic mRNA-based systems in mammalian cells. *Adv. Biosyst.* **2020**, *4*, e1900247. [CrossRef] [PubMed]

44. Ennifar, E.; Nikulin, A.; Tishchenko, S.; Serganov, A.; Nevskaya, N.; Garber, M.; Ehresmann, B.; Ehresmann, C.; Nikonov, S.; Dumas, P. The crystal structure of UUCG tetraloop. *J. Mol. Biol.* **2000**, *304*, 35–42. [CrossRef]

45. Hall, K.B. Mighty tiny. *RNA* **2015**, *21*, 630–631. [CrossRef] [PubMed]

46. Zhang, G.; Gurtu, V.; Kain, S.R. An enhanced green fluorescent protein allows sensitive detection of gene transfer in mammalian cells. *Biochem. Biophys. Res. Commun.* **1996**, *227*, 707–711. [CrossRef]

47. Campbell, T.B.; Cech, T.R. Mutations in the *Tetrahymena* ribozyme internal guide sequence: Effects on docking of the P1 helix into the catalytic core and correlation with catalytic activity. *Biochemistry* **1996**, *35*, 11493–11502. [CrossRef]

48. Ikawa, Y.; Matsumura, S. Engineered group I ribozymes as RNA-based modular tools to control gene expression. In *Applied RNA Bioscience*; Masuda, S., Izawa., S., Eds.; Springer: Berlin, German, 2018; pp. 203–220.

49. Müller, U.F. Design and experimental evolution of *trans*-splicing group I intron ribozymes. *Molecules* **2017**, *22*, 75. [CrossRef]

50. Lee, C.H.; Han, S.R.; Lee, S.W. Therapeutic applications of group I intron-based *trans*-splicing ribozymes. *Wiley Interdiscip. Rev. RNA* **2018**, *9*, e1466. [CrossRef]

51. Furukawa, A.; Maejima, T.; Matsumura, S.; Ikawa, Y. Characterization of an RNA receptor motif that recognizes a GCGA tetraloop. *Biosci. Biotechnol. Biochem.* **2016**, *80*, 1386–1389. [CrossRef] [PubMed]

52. Paige, J.S.; Wu, K.Y.; Jaffrey, S.R. RNA mimics of green fluorescent protein. *Science* **2011**, *333*, 642–646. [CrossRef]

53. Furukawa, A.; Tanaka, T.; Furuta, H.; Matsumura, S.; Ikawa, Y. Use of a fluorescent aptamer RNA as an exonic sequence to analyze self-splicing ability of a group I intron from structured RNAs. *Biology* **2016**, *5*, 43. [CrossRef] [PubMed]

54. Kappel, K.; Zhang, K.; Su, Z.; Watkins, A.M.; Kladwang, W.; Li, S.; Pintilie, G.; Topkar, V.V.; Rangan, R.; Zheludev, I.N.; et al. Accelerated cryo-EM-guided determination of three-dimensional RNA-only structures. *Nat. Methods* **2020**, *17*, 699–707. [CrossRef] [PubMed]

55. Matsumura, S.; Ikawa, Y.; Inoue, T. Biochemical characterization of the kink-turn RNA motif. *Nucleic Acids Res.* **2003**, *31*, 5544–5551. [CrossRef]

56. Goody, T.A.; Melcher, S.E.; Norman, D.G.; Lilley, D.M. The kink-turn motif in RNA is dimorphic, and metal ion-dependent. *RNA* **2004**, *10*, 254–264. [CrossRef] [PubMed]

57. Shi, X.; Huang, L.; Lilley, D.M.; Harbury, P.B.; Herschlag, D. The solution structural ensembles of RNA kink-turn motifs and their protein complexes. *Nat. Chem. Biol.* **2016**, *12*, 146–152. [CrossRef]

Article

Chloro-Substituted Naphthyridine Derivative and Its Conjugate with Thiazole Orange for Highly Selective Fluorescence Sensing of an Orphan Cytosine in the AP Site-Containing Duplexes

Chun-xia Wang †, **Yusuke Sato**, **Takashi Sugimoto, Norio Teramae and Seiichi Nishizawa** *

Department of Chemistry, Graduate School of Science, Tohoku University, Sendai 980-8578, Japan;
cxwang@iccas.ac.cn (C.-x.W.); yusuke.sato.a7@tohoku.ac.jp (Y.S.); sgmt6600@gmail.com (T.S.);
norio.teramae.d3@tohoku.ac.jp (N.T.)
* Correspondence: seiichi.nishizawa.c8@tohoku.ac.jp
† College of New Energy and Materials, China University of Petroleum Beijing, Beijing 102249, China.

Received: 21 May 2020; Accepted: 13 June 2020; Published: 16 June 2020

Abstract: Fluorescent probes with the binding selectivity to specific structures in DNAs or RNAs have gained much attention as useful tools for the study of nucleic acid functions. Here, chloro-substituted 2-amino-5,7-dimethyl-1,8-naphthyridine (ClNaph) was developed as a strong and highly selective binder for target orphan cytosine opposite an abasic (AP) site in the DNA duplexes. ClNaph was then conjugated with thiazole orange (TO) via an alkyl spacer (ClNaph–TO) to design a light-up probe for the detection of cytosine-related mutations in target DNA. In addition, we found the useful binding and fluorescence signaling of the ClNaph–TO conjugate to target C in AP site-containing DNA/RNA hybrid duplexes with a view toward sequence analysis of microRNAs.

Keywords: fluorescent probe; conjugate; abasic site; DNA; microRNA

---

1. Introduction

Much attention has been paid to the design of fluorescent probes capable of selective binding to specific structures in DNAs and RNAs [1,2]. This class of fluorescent probes has been designed to bind abasic (apyrimidinic or apurinic; AP) sites [3–8], bulges [9,10], mismatched sites [11,12] and overhanging structures [13–15], by which the biological functions of these noncanonical base-pairs have been examined. Moreover, these fluorescent probes have great potential to be applied for the analysis of target DNAs and RNAs based on their binding-induced fluorescence signaling. These probes also serve as the affinity-labeling agents in the label-free assays for detecting various analytes [16–18].

We developed AP site-binding ligands (APLs) that are conjugated with thiazole orange (TO) for the fluorescence sensing of orphan nucleobases in DNA/DNA duplexes (Figure 1A) [19,20]. APLs can form the pseudo-base pairing with target orphan nucleobases, which allows the selective binding to the target nucleobase opposite the AP site [3–7]. On the other hand, TO unit connected with the APL unit through an appropriate linker can function as a fluorescent intercalator [21,22]. The TO unit alone shows negligible fluorescence, but its fluorescence is greatly enhanced upon intercalation into the base pairs near the AP site. APL–TO conjugates thus enable the light-up sensing of target orphan nucleobase in the AP site-containing DNA (AP–DNA) duplexes. Such light-up probes are useful for a more sensitive analysis compared to fluorescence quenching probes. We succeeded in the detection of single base mutation in target DNA sequences based on the binding and light-up functions of APL–TO conjugates in combination with an AP–DNA probe (cf. Figure 1A). In addition, these conjugates were applicable to the sequence-selective analysis of microRNAs, a class of small non-coding RNAs associated with various diseases and cancers [23], by the selective binding of conjugates to target orphan nucleobases in the AP site-containing DNA/RNA (AP–DNA/RNA) hybrids. The properties of nucleobase-selectivity and fluorescence emission wavelength are tunable by adopting suitable APLs and cyanine dyes, respectively, which enables the development of multiplex analysis of target sequences by the simultaneous use of these conjugates in a single solution [19]. In this context, 2-amino-5,6,7-trimethyl-1,8-naphthyridine (ATMND, Figure 1B) was previously used as an APL unit in the conjugate with TO unit for the light-up sensing of target orphan C in the AP–DNA and AP–DNA/RNA hybrid duplexes [19]. The binding selectivity of ATMND toward target C can be rationalized by the complementary base pairing of the N1-protonated form of ATMND with target C through hydrogen bonding (Figure 1B) [3]. However, ATMND inherently has only moderate selectivity for C over T because of the possible protonation at N8 that allows the recognition of T (Figure 1B). This should limit the application of the conjugate for the sequence analysis of target DNAs and microRNAs such as the discrimination of C against T or U.

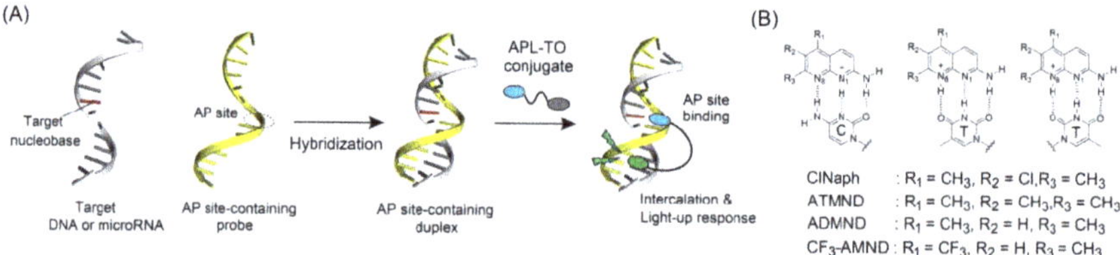

Figure 1. (**A**) Schematic illustration of the analysis of target DNA or microRNA sequences based on the binding of abasic (AP) site-binding ligands conjugated with thiazole orange conjugates (APL-TO) in combination with an AP site-containing probe for hybridization; (**B**) Chemical structures of chloro-substituted 2-amino-5,7-dimethyl-1,8-naphthyridine (ClNaph) and its related derivatives. Proposed binding of these ligands with cytosine and thymine (two possible patterns) are also shown.

This work describes the enhanced binding selectivity of a 2-amino-1,8-naphthyridine derivative for orphan C by the incorporation of a chloro group. We previously found that the introduction of the electron-withdrawing trifluoromethyl group into 2-amino-7-methyl-1,8-naphthrydine (AMND) led to remarkably enhanced C/T selectivity due to more favorable protonation at N1 compared to N8 [24]. However, the obtained CF$_3$-AMND (Figure 1B) showed much weaker affinity compared to ATMND. Herein, 2-amino-5,7-dimethyl-1,8-naphthyridine (ADMND), an AMND derivative with an additional methyl group, was explored as the scaffold for the introduction of an electron-withdrawing group considering the fact that ADMND can bind more strongly to target C than AMND [3]. The resulting compound carrying the chloro group at position 6 in the naphthyridine ring (ClNaph) was found to exhibit the strong binding affinity as well as the high selectivity for target C. The use of ClNaph as an APL unit in the conjugate with TO was shown to be useful for the design of a light-up probe toward target C over other nucleobases in the AP–DNA duplexes and AP–DNA/RNA hybrids with a view toward the detection of C-related mutations in DNA and microRNA sequences.

2. Materials and Methods

2.1. Materials

All of the DNAs and RNAs were purchased from Nihon Gene Research Laboratories, Inc. (Sendai, Japan) and Sigma-Genosys (Hokkaido, Japan), respectively. The other reagents were commercially available and used without further purification. The concentration of DNAs and RNAs were determined from the molar extinction coefficient at 260 nm (ε_{260}) according to the literature [25]. Water was deionized ($\geq$18.0 MΩ cm specific resistance) by an Elix 5 UV water purification system and a Milli-Q synthesis A10 system (Millipore Corp., Bedford, MA, USA). The other reagents were purchased from standard suppliers and used without further purification. ^{1}H NMR spectra were measured with a JEOL

ECA-600 spectrometer at 500 MHz. High-resolution ESI-MS spectra were measured with a Bruker APEX III mass spectrometer.

Unless otherwise mentioned, all measurements were performed in 10-mM sodium cacodylate buffer solutions (pH 7.0) containing 100-mM NaCl, 1.0-mM EDTA and ethanol (<2%). Before the measurements, target duplex-containing samples were annealed as follows: heated at 75 °C for 10 min and gradually cooled to 5 °C (3 °C/min), after which the solution temperature was raised again to 20 °C. The probe was then added to the samples.

2.2. Probe Synthesis

ClNaph: 2,6-diaminopyridine (3.17 g, 29.1 mmol) in phosphoric acid (50 mL) was added to 3-chloropentane-2,4-dione (3.99 g, 29.6 mmol) and the reaction mixture was stirred at 90 °C for 24 h. After neutralization with NaOH, the mixture was filtered. The brown residue was extracted with 300 mL of $CHCl_3$ three times and the organic phase was dried over Na_2SO_4. The solvent was evaporated *in vacuo*, affording ClNaph as a dark brown solid (3.57 g, 17.2 mmol, 59.2%). ^{1}H-NMR ($CDCl_3$): 8.03 (d, 1H, J = 8.4 Hz), 6.76 (d, 1H, J = 9.2 Hz), 2.75 (s, 3H), 2.64 (s, 3H). High resolution ESI-MS calcd ([M + H]$^+$) 207.0563; found, 207.0569.

ClNaph–TO and ClNaph–TO2 conjugates: The conjugates were synthesized according to our previous report [19]. Briefly, aminoethyl group was incorporated into ClNaph and the resulting derivative was attached with the carboxylate-terminated decanyl (C10) spacer-containing TO derivative [26,27]. ClNaph–TO conjugate, ^{1}H-NMR (DMSO-d_6) 8.80 (d, 1H, J = 8.8 Hz), 8.60 (d,1H, J = 7.2 Hz), 8.13 (d, 1H, J = 8.8 Hz), 8.05 (d, 1H, J = 7.6 Hz), 8.00 (t, 1H, J = 7.6 Hz), 7.80 (d, 1H, J = 8.4 Hz), 7.76 (t, 1H, J = 8.0 Hz), 7.63 (t, 1H, J = 8.4 Hz), 7.41 (t,1 H, J = 8.0 Hz), 7.36 (d, 1H, J = 7.6 Hz), 6.93 (s, 1H), 6.78 (d, 1H, J = 9.2 Hz), 4.58 (t, 2H, J = 7.2 Hz), 3.44–3.47 (q, 4H), 2.58 (s, 3H), 2.54 (s, 3H), 2.05 (t, 2H, J = 7.2 Hz), 1.84 (t, 2H), 1.42 (t, 2H), 1.34–1.16 (q, 14H). High resolution ESI-MS calcd ([M + H]$^+$) 707.3299; found, 707.3292. ClNaph–TO2 conjugate, ^{1}H-NMR (DMSO-d_6) 8.72 (d, 1H, J = 8.8 Hz), 8.61 (d, 1H, J = 6.8 Hz), 8.07–8.01 (m, 4H), 7.78–7.74 (m, 2H), 7.62–7.58 (t, 1H, J = 7.6 Hz), 7.43–7.37 (m, 2H), 6.92 (s, 1H), 6.77 (d, 2H, J = 4.0 Hz), 4.61 (t, 2H), 4.17 (s, 3H), 3.50–3.42 (m, 4H), 2.58 (s, 3H), 2.55 (s, 3H), 2.00 (t, 2H, J = 7.2 Hz), 1.59–1.12 (m, 18H). High resolution ESI-MS calcd ([M + H]$^+$) 707.3299; found, 707.3286.

2.3. Fluorescent Measurements

Fluorescence spectra were measured with a JASCO model FP-6500 spectrofluorophotometer equipped with a thermoelectrically temperature-controlled cell holder (Japan Spectroscopic Co., Ltd., Tokyo, Japan) using a 3 × 3 mm quartz cell.

The dissociation constant (K_d) of the probe was determined at 20 °C by fluorescent titration experiments. The changes in fluorescence intensity were analyzed by nonlinear least-squares regression based on a 1:1 binding isotherm [19]. Errors in the K_d values are the standard deviations obtained from three independent experiments (N = 3).

3. Results and Discussion

First, we examined the binding ability of ClNaph (500 nM) to target orphan nucleobases in model 21-meric AP–DNA duplexes (500 nM; 5′-d(GCA GCT CCC A**X**A GTC TCC TCG)-3′/3′-d(CGT CGA GGG T**N**T CAG AGG AGC)-5′, **X** = AP site (Spacer C3, a propyl residue), **N** = target nucleobase; G, C, A or T) by the fluorescence measurements. As shown in Figure 2, ClNaph showed the emission with the maximum at 393 nm in the absence of DNAs. The addition of AP–DNA duplexes caused a decrease in the fluorescence intensity of ClNaph. The largest fluorescence quenching response was observed for target C, which indicates that ClNaph shows the preferential binding to target C compared to target G, A and T. The binding affinity of ClNaph was assessed by the fluorescence titration experiments (Inset of Figure 2). ClNaph showed the quenching response for target DNA duplexes in a concentration-dependent manner and the resulting titration curves were analyzed by a

1:1 binding model for the determination of the dissociation constants (K_d). The K_d value for C was obtained as 70 ± 5.3 nM. Significantly, this affinity is much stronger compared to the parent ADMND and the previously developed CF$_3$-AMND and is almost comparable to ATMND (Table 1). In addition, ClNaph showed useful binding selectivity toward target C, in which the K_d value for target C was two orders of magnitude smaller than those for other target nucleobases (K_d/nM: T, 2020 ± 240, G and A > 5000). The observed selectivity for C over T is much superior to ADMND and ATMND [3] whereas it is slightly moderate in comparison with CF$_3$-AMND [24]. We reason that this is due to the more favorable protonation of ClNaph at the N1 position for the complementary base-pairing with orphan C compared to the N8 protonation (Figure 1B), where the electron-withdrawing effect of the chloro group would be essential as was observed for CF$_3$-AMND [24]. We note that the type of the AP site affects the binding of ClNaph. The use of Spacer C3 (propyl residue) allowed the stronger binding to target C compared to the tetrahydrofuranyl residue (dSpacer), presumably due to less steric hindrance (Figure S3). The observed strong and highly selective binding properties for target C should render ClNaph a useful APL unit in the conjugate with TO unit.

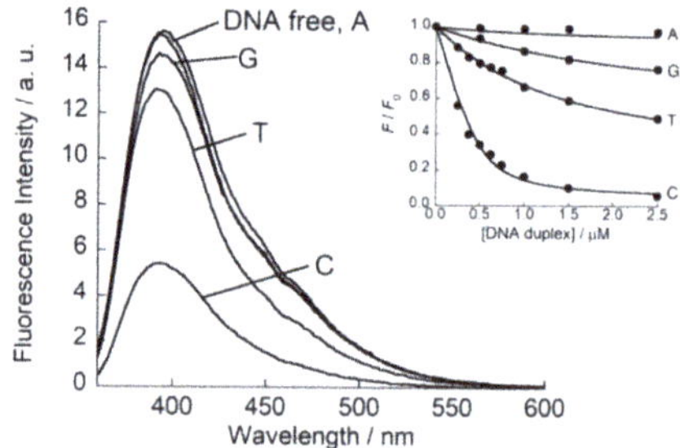

Figure 2. Fluorescence response of ClNaph (500 nM) to target AP–DNA duplexes (500 nM). Inset: Titration curves for the binding of ClNaph (500 nM) to target nucleobases. Measurements were done in solutions buffered to pH 7.0 (10-mM sodium cacodylate) containing 100-mM NaCl and 1.0-mM EDTA. F and F_0 denote the fluorescence intensities of ClNaph in the presence and absence of DNA duplexes. Excitation, 350 nm. Analysis, 393 nm. Temperature, 20 °C.

Table 1. Dissociation constants (K_d) of ClNaph and its related derivatives for target C in 21-meric AP–DNA duplexes [a].

	K_d/nM
ClNaph	70 ± 5.3
ADMND [b]	164
ATMND [b]	53
CF$_3$-AMND [c]	1400

[a] K_d values measured in solutions buffered to pH 7.0 (10-mM sodium cacodylate) containing 100-mM NaCl and 1.0-mM EDTA. Temperature, 20 °C. [b] Values taken from [3]. [c] Value taken from [24].

ClNaph was coupled with the quinolone ring of the TO unit through a long alkyl (C10)-linker according to our previous report [19], which affords the ClNaph–TO conjugate (Figure 3A). We measured the fluorescence response of the TO unit of the conjugate (500 nM) for the same AP–DNA duplexes (500 nM) as those used for the examination of ClNaph (cf. Figure 2). As shown in Figure 3A, TO unit shows negligible emission in the absence of DNAs due to the free rotation of the benzothiazole and quinolone rings [21]. In contrast, we observed a remarkable light-up response of the TO unit for target C-containing AP–DNA duplex, in which the fluorescence intensity at 526 nm increased by 207-fold. This can be explained by the restriction of the rotation of the TO unit by intercalation into the duplex region (cf. Figure 1A) [19]. In addition, the fluorescence titration experiments revealed that the K_d value of ClNaph–TO conjugate for target C reached 6.6 ± 0.3-nM (Figure S4). This affinity is one order of magnitude stronger than ClNaph itself (cf. Table 1). Apparently, the conjugation with the TO unit led to the enhanced binding affinity for target C in the AP–DNA duplexes. It should be noted that ClNaph–TO conjugate has high binding selectivity for target C over other three nucleobases (Figure S4: K_d values for T, G and A > 1500 nM). Considering that TO lacks the selectivity for the orphan nucleobases [19], the ClNaph unit is responsible for the observed C-selectivity of the conjugate. This was confirmed by the fluorescence quenching response of the ClNaph unit with high selectivity to target C (Figure S5). These results showed the useful binding and light-up functions of the ClNaph–TO conjugate for the detection of C-related mutations in target DNA sequences. It is also noteworthy that the strong emission of ClNaph–TO conjugate for target C can be seen even with the naked eyes under UV light irradiation (Figure 3B). Fluorescence response for target C is clearly distinguishable from those for other target nucleobases, which thus facilitates a simple and rapid analysis of target DNA sequences.

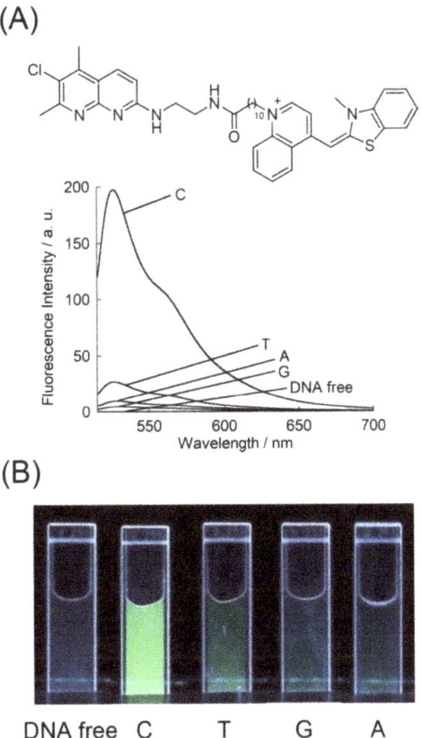

Figure 3. (**A**) Fluorescence response of ClNaph–TO conjugate (500 nM) to target AP–DNA duplexes (500 nM). Excitation, 504 nm. Temperature, 20 °C. (**B**) Images of fluorescence emission of ClNaph–TO conjugate (500 nM) in the absence and presence of target AP–DNA duplexes (500 nM) under excitation light with the wavelength of 365 nm, obtained by a digital camera at room temperature. Other solution conditions are the same as those given in Figure 2.

We found that the kind of heterocycles of the TO unit, benzothiazole or quinolone rings, that connect to ClNaph unit in the conjugate affected the binding and fluorescence signaling abilities for the binding to AP–DNA duplexes. When the C10-linker was appended to the benzothiazole ring, the resulting conjugate (ClNaph–TO2) showed the selective light-up response for target C in the AP–DNA duplexes (Figure S6); however, the degree of the response was smaller compared to ClNaph–TO (cf. Figure 3A). This is attributable to the reduced binding affinity of ClNaph–TO2 conjugate to target C

(K_d = 71 nM). This indicates that the conjugation of ClNaph unit into the benzothiazole ring hinders the effective intercalation of the TO unit for DNA duplexes as reported in the literature [27]. In addition, we observed moderate selectivity of ClNaph–TO2 conjugate for target C over other three nucleobases in AP–DNA duplexes (Figure S6) while the reason for this is unclear yet. Hence, the connection of ClNaph unit to the quinolone ring of the TO unit is effective for the strong binding and large light-up response of the conjugate for target C in the AP–DNA duplexes.

As described above, the ClNaph–TO conjugate can serve as a useful light-up probe for target C in the AP–DNA duplexes. Meanwhile, we noticed weak binding of the conjugate for AP–RNA duplexes (Figure 4A and Figure S7). The binding affinity was estimated as > 1300-nM for target C in the AP–RNA duplex (5′-r(GCA GCU CCC A**X**A GUC UCC UCG)-3′/3′-r(CGU CGA GGG U**C**U CAG AGG AGC)-5′, **X** = AP site (Spacer C3), **C** = target nucleobase), which was three orders of magnitude larger than that for the AP–DNA duplex (cf. Figure S4). We consider that this arises from the preferential binding of the ClNaph–TO conjugate to B-formed AP–DNA duplexes relative to A-formed AP–RNA duplexes [6]. However, when targeting C in single stranded RNA, the use of AP–DNA probe is highly effective. The resulting duplex is the hybrid between RNA and AP–DNA ((5′-d(GCA GCT CCC A**X**A GTC TCC TCG)-3′/3′-r(CGU CGA GGG U**C**U CAG AGG AGC)-5′, **X** = AP site (Spacer C3), **C** = target nucleobase) that can adopt A-form/B-form intermediate structure [28], under the same measurement conditions used to examine AP–RNA duplexes. As shown in Figure 4B, we observed the significant light-up response of the conjugate for the AP–DNA/RNA hybrid, in which the response was 22-fold larger compared to that for the AP–RNA duplex. The K_d for target C in the hybrid was obtained as 22 nM (Figure S7), where the affinity is much superior to that for AP–RNA duplexes. Importantly, the ClNaph–TO conjugate retains its selectivity for target C over U opposite the AP site in the DNA/RNA hybrid, as was observed in AP–DNA duplexes (cf. Figure 3B). These results suggest the potential use of ClNaph–TO conjugate for the detection of microRNAs in combination with AP–DNA hybridization probe (cf. Figure 1A).

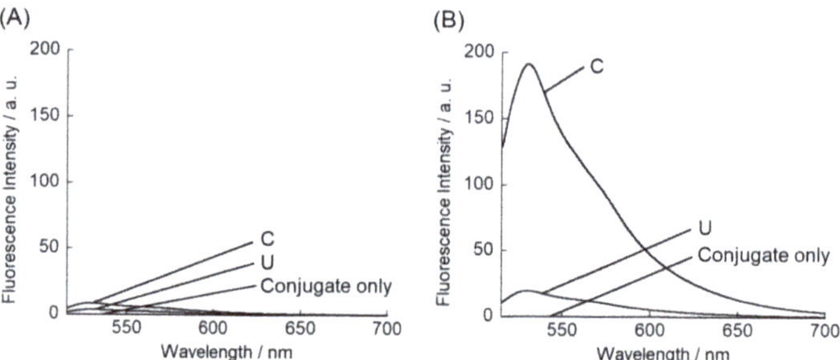

Figure 4. Fluorescence response of ClNaph–TO conjugate (500 nM) for target C or U in AP site-containing duplexes (500 nM): (**A**), AP–RNA duplex and (**B**) AP–DNA/RNA hybrid. Other solution conditions are the same as those given in Figure 2. Excitation, 504 nm. Temperature, 20 °C.

We performed the preliminarily experiments for detection of microRNAs based on the binding-induced light-up response of ClNaph–TO conjugate for AP–DNA/RNA hybrids. A 21-meric AP–DNA probe was designed for the selective detection of the let-7d sequence among let-7 family [29], as shown in Figure 5A. Hybridization between this probe and let-7d allows the construction of the AP–DNA/RNA hybrid containing an orphan C opposite an AP site. Meanwhile, the hybridization with other let-7 sequences leads to the formation of orphan U-containing hybrids with several mismatch base pairs (Table S1). We observed the significant light-up response of ClNapht-TO for the let-7d-containing hybrid (Figure 5A). This response is much larger than those for other let-7 sequences, which clearly shows that ClNaph–TO conjugate enables the selective detection probe for let-7d over other let-7 members. The binding affinity of the conjugate for the let-7d-containing hybrid (K_d = 23 nM) was found to be comparable to that for the model AP–DNA/RNA hybrid (cf. Figure S8). Our assay can be applied to the analysis of various microRNAs by using the AP site-containing DNA probe whose sequence is designed so as to be complementary to the target sequence. Figure 5B shows the fluorescence response of the conjugate in combination with the 22-meric AP–DNA probe for let-7i. Selective light-up detection for let-7i was achieved due to the selective binding of the conjugate to target C in the AP–DNA/RNA hybrid formed between the AP–DNA probe and let-7i. These results show the applicability of the ClNaph–TO conjugate for the selective detection of target microRNA sequences based on the hybridization for the construction of the AP–DNA/RNA hybrid as well as the binding-induced light-up response of the conjugate for the orphan C in the resulting hybrid.

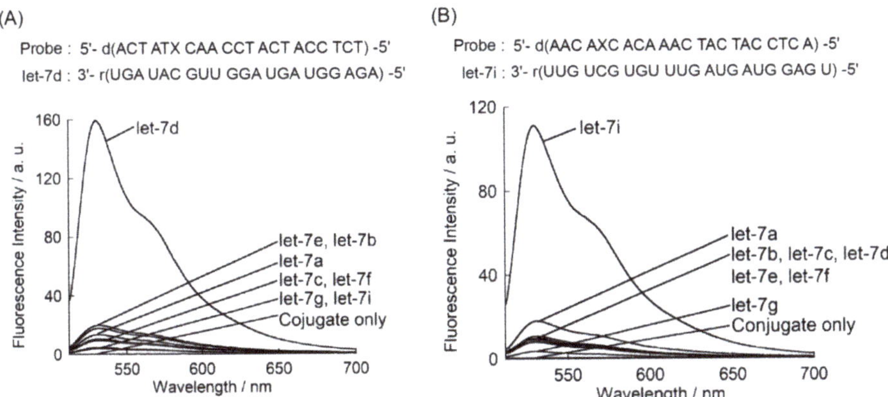

Figure 5. Selective detection of (**A**) let-7d and (**B**) let-7i sequences by ClNaph–TO conjugate in combination with AP–DNA probe (X = AP site (Spacer C3)). Sequences of other let-7 members are shown in Table S1. [ClNaph–TO], [target microRNA], [AP–DNA probe] = 500 nM. Other solution conditions for the fluorescence response of the conjugate are the same as those given in Figure 2. Excitation, 504 nm. Temperature: 20 °C.

4. Conclusions

In summary, we report that ClNaph served as a strong and highly selective binder for the orphan C opposite an AP site in DNA duplexes. In addition, we demonstrated the usefulness of ClNaph as the APL unit in the conjugate with a TO unit for the design of light-up probes for the detection of C-related mutations in DNA and microRNA sequences. These results obtained could provide insights needed to design this class of light-up conjugates suitable for the analysis of DNA and microRNA sequences. As shown in our previous works [19,20], the spacer length and structure have a large impact on the binding and fluorescence sensing abilities of APL–TO conjugates. We will examine these concerns for further improvement of the APL–TO conjugates in order to develop the practical assays.

Supplementary Materials: The following are available online at http://www.mdpi.com/2076-3417/10/12/4133/s1, Figure S1: ^{1}H NMR spectra of the conjugates; Figure S2: ESI-MS spectra of the conjugates; Figure S3: Fluorescence response of ClNaph to target dSpacer-containing AP–DNA duplexes. Inset: Titration curves for the binding of ClNaph to target nucleobases; Figure S4: Titration curves for the binding of ClNaph–TO conjugate to target nucleobases; Figure S5: Fluorescence response of ClNaph unit in the conjugate to target AP–DNA duplexes. Figure S6: Chemical structure of ClNaph–TO2 conjugate and its fluorescence response to target AP–DNA duplexes; Figure S7: Titration curves of the binding of ClNaph–TO conjugate to target C in the AP–DNA/RNA hybrid and AP–RNA duplex; Table S1: Sequences of let-7 members used in this study. Figure S8. Titration curves for the binding of ClNaph–TO conjugate to the let-7d/DNA probe hybrid.

Author Contributions: C.-x.W. and Y.S. conceived and designed the experiments. C.-x.W. and T.S. performed the experiments. C.-x.W. and Y.S. wrote the study. N.T. and S.N. supervised the project. All authors have read and agreed to the published version of the manuscript.

Funding: Y.S., N.T. and S.N. acknowledge the support from Scientific Research (S) (No. 22225003) from Japan Society for the Promotion of Science (JSPS).

Conflicts of Interest: The authors declare no conflict of interest.

References

1. Granzhan, A.; Kotera, N.; Teulade-Fichou, M.-P. Finding needles in a basestack: Recognition of mismatched base pairs in DNA by small molecules. *Chem. Soc. Rev.* **2014**, *43*, 3630–3665. [CrossRef] [PubMed]
2. Song, G.; Ren, J. Recognition and regulation of unique nucleic acid structures by small molecules. *Chem. Commun.* **2010**, *46*, 7283–7294. [CrossRef]
3. Sato, Y.; Nishizawa, S.; Yoshimoto, K.; Seino, T.; Ichihashi, T.; Morita, K.; Teramae, N. Influence of substituent modifications on the binding of 2-amino-1,8-naphthyridines to cytosine opposite an AP site in DNA duplexes: Thermodynamic characterization. *Nucleic Acids Res.* **2009**, *37*, 1411–1422. [CrossRef] [PubMed]
4. Qing, D.; Xu, C.-Y.; Sato, Y.; Yoshimoto, K.; Nishizawa, S.; Teramae, N. Enhancement of the binding ability of a ligand for nucleobase recognition by introducing a methyl group. *Anal. Sci.* **2006**, *22*, 201–203.
5. Wang, C.-X.; Sato, Y.; Kudo, M.; Nishizawa, S.; Teramae, N. Ratiometric fluorescent signaling of small molecules, environmentally sensitive dye conjugates for detecting single base mutations in DNA. *Chem. Eur. J.* **2012**, *18*, 9481–9484. [CrossRef] [PubMed]
6. Sato, Y.; Ichihashi, T.; Nishizawa, S.; Teramae, N. Strong and selective binding of amiloride to an abasic site in RNA duplexes: Thermodynamic characterization and microRNA detection. *Angew. Chem. Int. Ed.* **2012**, *51*, 6339–6372. [CrossRef]
7. Sato, Y.; Toriyabe, Y.; Nishizawa, S.; Teramae, N. 2,4-Diamino-6,7-dimethylpteridine as a fluorescent ligand for binding andsensing an orphan cytosine in RNA duplexes. *Chem. Commun.* **2013**, *49*, 9983–9985. [CrossRef]
8. Fakhari, M.A.; Rokita, S.E. A new solvatochromic fluorophore for exploring nonpolar environments created by biopolymers. *Chem. Commun.* **2011**, *47*, 4222–4224. [CrossRef] [PubMed]
9. Suda, H.; Kobori, A.; Zhang, J.; Hayashi, G.; Nakatani, K. N'-Bis(3-aminopropyl)-2,7-diamino-1,8-naphthyridine stabilized a single pyrimidine bulge in duplex DNA. *Bioorg. Med. Chem.* **2005**, *13*, 4507–4512. [CrossRef] [PubMed]
10. Verma, R.K.; Takei, F.; Nakatani, K. Synthesis and photophysical properties of fluorescence molecular probe for turn-on-type detection of cytosine bulge DNA. *Org. Let.* **2016**, *18*, 3170–3173. [CrossRef]
11. Arambula, J.F.; Ramisetty, S.R.; Baranger, A.M.; Zimmerman, S.C. A simple ligand that selectively targets CUG trinucleotide repeats and inhibits MBNL protein binding. *Proc. Natl. Acad. Sci. USA* **2009**, *106*, 16068–16073. [CrossRef] [PubMed]
12. Zeglis, B.M.; Barton, J.K. A mismatch-selective bifunctional rhodium-oregon green conjugate: A fluorescent probe for mismatched DNA. *J. Am. Chem. Soc.* **2006**, *128*, 5654–5655. [CrossRef] [PubMed]
13. Sato, T.; Sato, Y.; Iwai, K.; Kuge, S.; Nishizawa, S.; Teramae, N. Synthetic fluorescent probes capable of selective recognition of 3'-overhanging nucleotides for siRNA delivery imaging. *Chem. Commun.* **2015**, *51*, 1421–1424. [CrossRef]
14. Tanabe, T.; Sato, T.; Sato, Y.; Nishizawa, S. Design of a fluorogenic PNA probe capable of simultaneous recognition of 3'-overhang and double-stranded sequences of small interfering RNAs. *RSC Adv.* **2018**, *8*, 42095–42099. [CrossRef]
15. Sato, Y.; Kaneko, M.; Sato, T.; Nakata, S.; Takahashi, Y.; Nishizawa, S. Enhanced binding affinity of siRNA overhang-binding fluorescent probes by conjugation with cationic oligopeptides for improved analysis of the siRNA delivery process. *ChemBioChem* **2019**, *20*, 408–414. [CrossRef] [PubMed]
16. Xu, Z.; Morita, K.; Sato, Y.; Dai, Q.; Nishizawa, S.; Teramae, N. Label-free aptamer-based sensor using abasic site-containing DNA and a nucleobase-specific fluorescent ligand. *Chem. Commun.* **2009**, *42*, 6445–6447. [CrossRef] [PubMed]
17. Sato, Y.; Nishizawa, S.; Teramae, N. Label-free molecular beacon system based on DNAs containing abasic sites and fluorescent ligands that bind abasic sites. *Chem. Eur. J.* **2011**, *17*, 11650–11656. [CrossRef]
18. Sato, Y.; Zhang, Y.; Nishizawa, S.; Seino, T.; Nakamura, K.; Li, M.; Teramae, N. Competitive assay for theophylline based on an abasic site-containing DNA duplex aptamer and a fluorescent ligand. *Chem. Eur. J.* **2012**, *18*, 12719–12724. [CrossRef]

19. Sato, Y.; Kudo, M.; Toriyabe, Y.; Kuchitsu, S.; Wang, C.-X.; Nishizawa, S.; Teramae, N. Abasic site-binding ligands conjugated with cyanine dyes for "off-on" fluorescence sensing of orphan nucleobases in DNA duplexes and DNA-RNA hybrids. *Chem. Commun.* **2014**, *50*, 515–517. [CrossRef] [PubMed]

20. Sato, Y.; Saito, H.; Aoki, D.; Teramae, N.; Nishizawa, S. Lysine linkage in abasic site-binding ligand-thiazole orange conjugates for improved binding affinity to orphan nucleobases in DNA/RNA hybrids. *Chem. Commun.* **2016**, *52*, 14446–14449. [CrossRef]

21. Nygren, J.; Svanvik, N.; Kubista, M. The interactions between the fluorescent dye thiazole orange and DNA. *Biopolymers* **1998**, *46*, 39–51. [CrossRef]

22. Armitage, B.A. Cyanine dye-DNA interactions: Intercalation, groove binding, and aggregation. *Top. Curr. Chem.* **2005**, *253*, 55–76.

23. Dong, H.; Lei, J.; Ding, L.; Wen, Y.; Ju, H.; Zhang, X. MicroRNA: Function, detection, and bioanalysis. *Chem. Rev.* **2013**, *113*, 6207–6233. [CrossRef]

24. Sato, Y.; Zhang, Y.; Seino, T.; Sugimoto, T.; Nishizawa, S.; Teramae, N. Highly selective binding of naphthyridine with a trifluoromethyl group to cytosine opposite an abasic site in DNA duplexes. *Org. Biomol. Chem.* **2012**, *10*, 4003–4006. [CrossRef] [PubMed]

25. Puglisi, J.D.; Tinoco, I. Absorbance melting curves of RNA. *Methods Enzymol.* **1989**, *180*, 304–325.

26. Carreon, J.R.; Stewart, K.M.; Mahon, K.P., Jr.; Shin, S.; Kelly, S.O. Cyanine dye conjugates as probes for live cell imaging. *Bioorg. Med. Chem. Lett.* **2007**, *17*, 5182–5185. [CrossRef]

27. Carreon, J.R.; Mahon, K.P., Jr.; Kelly, S.O. Thiazole orange-peptide conjugates: Sensitivity of DNA binding to chemical structure. *Org. Lett.* **2004**, *6*, 517–519. [CrossRef]

28. Shaw, N.N.; Arya, D.P. Recognition of the unique structure of DNA:RNA hybrids. *Biochimie* **2008**, *90*, 1026–1039. [CrossRef]

29. Johnson, S.M.; Grosshans, H.; Shingara, J.; Byrom, M.; Jarvis, R.; Cheng, A.; Labourier, E.; Reinert, K.L.; Brown, D.; Slack, F.J. RAS is regulated by the let-7 microRNA family. *Cell* **2005**, *120*, 635–647. [CrossRef]

MDPI

St. Alban-Anlage 66

4052 Basel

Switzerland

Tel. +41 61 683 77 34

Fax +41 61 302 89 18

www.mdpi.com

Applied Sciences Editorial Office

E-mail: applsci@mdpi.com

www.mdpi.com/journal/applsci